MW00723299

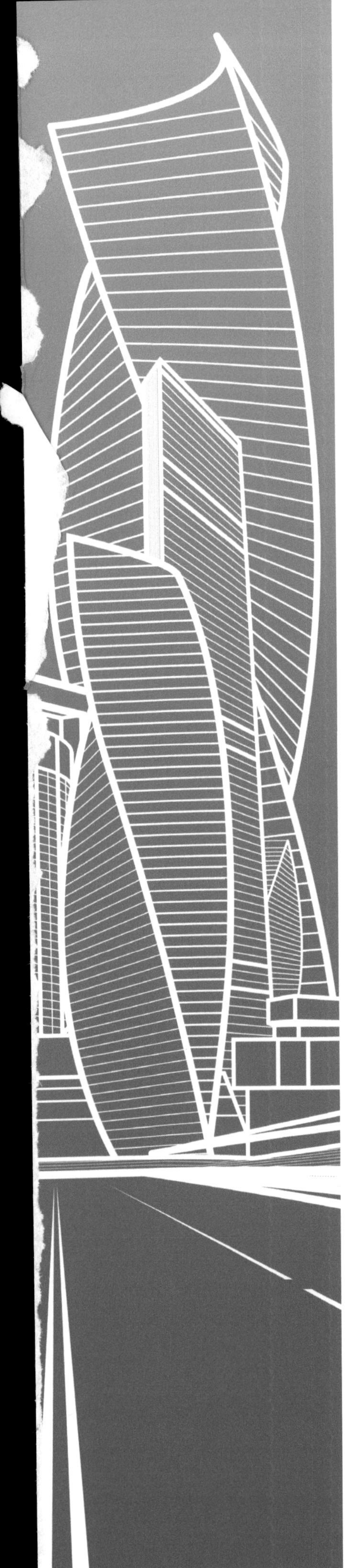

PE Civil

Exam Review Guide: Breadth

First Edition

by

School of PE

Muhammad Elgammal, PE, PMP

Daniel A. Howell, PhD, PE

Jeffrey S. MacKay, PE

Anthony Oyatayo, PE, MBA

Amr M. Sallam, PhD, PE

Nicholas G. Siegl, PE, ENV SP

George A. Stankiewicz, PE

Timothy R. Sturges, PE

with contributions by Alan Esser, PE, D.GE; Sukh Gurung, PhD, PE; Amir Mousa, MSCE, PE; and Francisco Perez, PE

School of PE™
A Division of EDUMIND

Dublin, Ohio

PE Civil Exam Review Guide: Breadth
First Edition

First edition, October 2019

ISBN 978-1-970105-00-1 (hardcover)

Library of Congress Control Number: 2019905508

Cover Art by Panimoni/Shutterstock.com
Cover Design by SPi Global

Printed in the USA

School of PE
An imprint of EduMind, Inc.
425 Metro Place N, Suite 450, Dublin, OH 43017
www.schoolofpe.com

Download the TotalAR app to access additional interactive features such as quizzes, videos, 3D animations, and flash cards.

Scan this QR code with a smartphone or tablet to download the TotalAR app.

Once you have downloaded the app, scan this TotalAR code to activate AR features in the book:

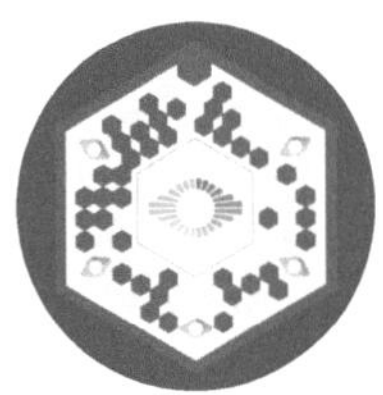

When the code is recognized, you will be prompted to enter your book's unique alphanumeric access key:

Unique Access Key:
LZZ8EMUT

Contents

About the Authors

Muhammad Elgammal, PE, PMP

Mr. Elgammal holds a BS in Civil Engineering and an MS in Critical Infrastructure. He is a registered professional engineer in the states of New Jersey and New York, and is also a certified project management professional. Elgammal has more than six years of experience in the industry, which includes material testing and inspection, construction operations, estimating, project control, and site design. In addition, he is also an adjunct professor at the New Jersey Institute of Technology. Elgammal has taught construction and material courses for School of PE since 2016.

Daniel A. Howell, PhD, PE

Dr. Howell holds BS, MS, and PhD degrees in Civil Engineering with a structural focus. He is a practicing structural engineer with experience in both public and private firms. With more than 16 years' design experience, he is also knowledgeable in structural dynamics, seismic analysis, bridge design, structural strengthening and retrofit, as well as new bridge design. Howell has been teaching for School of PE since 2011.

Jeffrey S. MacKay, PE

Mr. MacKay is a licensed civil engineer and partner at NTM Engineering, Inc. in Dillsburg, Pennsylvania. He received bachelor's and master's degrees from Drexel University and immediately began working in water resources engineering consulting. His background includes engineering design, environmental permits, policy development, and teaching. He has spent much of the past ten years developing policies, manuals, and training for the Pennsylvania Department of Transportation on subjects such as highway drainage systems and stormwater management, municipal separate storm sewer system (MS4) and national pollutant discharge elimination system (NPDES) permits, erosion and sediment pollution control, and hydrologic and hydraulic modeling. MacKay has been an instructor for School of PE since 2006.

Anthony O. Oyatayo, PE, MSSE, MBA

Mr. Oyatayo has a BS in Mechanical Engineering, an MS in Systems Engineering, and an MBA. He has ten years of work experience in different industries, including petroleum, theme park entertainment, and aerospace and defense. At School of PE, Oyatayo has proven teaching experience in multiple subjects.

Amr M. Sallam, PhD, PE

Dr. Sallam holds a BS in Civil Engineering with honors from Alexandria University. He received his MS in Civil Engineering (Geotechnical) from Alexandria University in 1998, and completed his PhD in Civil Engineering (Geotechnical) at the University of South Florida in 2004. With 22 years' experience, Sallam has served as project engineer, senior engineer, project manager, and principal in charge on numerous geotechnical and structural projects throughout Florida, the southeastern US, and internationally. His expertise includes geotechnical consulting for shallow and deep foundations for high-rise heavily loaded structures, bridges, deep and staged excavations, soil improvement, retaining and mechanically stabilized earth (MSE) walls, global stability, time domain reflectometry, stormwater and groundwater modeling, developing on karst geology, and developing on difficult soils. Sallam's expertise also includes numerical modeling using 2D and 3D finite and discrete element modeling. He has taught for School of PE since 2009.

Nicholas G. Siegl, PE, ENV SP

Mr. Siegl is a project manager with Gannett Fleming, Inc. and is a licensed professional engineer in Pennsylvania, Delaware, and Texas. He has worked in transportation engineering design and construction since 1980. His training experience includes the development and delivery of training courses for the Pennsylvania Department of Transportation (PennDOT), Penn State University, and School of PE. His area of expertise is highway design and construction engineering. He serves as a subject matter expert for the development of PennDOT's Intermediate Highway Design Skills course and the Right-of-Way for Project Managers course. Siegl has taught review courses for the PE Civil exam since 1999, and has served as a School of PE instructor since 2005.

George A. Stankiewicz, PE

Mr. Stankiewicz has two undergraduate degrees and a master's degree in management with more than 35 years' experience in the construction industry. His broad background in projects for federal, state, and local governments, universities, and major corporate clients has strengthened his skills as a leader. As an adjunct instructor for the Graduate School of Civil Engineering at New Jersey Institute of Technology, he has taught courses in construction management, systems in building design, and project scheduling. Stankiewicz has worked with School of PE since 2008 and has proven teaching experience in surveying, construction, materials, engineering economics, and project management.

Timothy R. Sturges, PE

Mr. Sturges holds a BE in Civil Engineering from Youngstown State University, an MS in Civil Engineering from the University of Akron, and an MS in Public Administration from Cleveland State University. He formerly held the position of committee research coordinator for the Transportation Research Board (TRB) standing committee on concrete pavement construction (AFH50) and is currently employed as a construction area engineer for the Ohio Department of Transportation, District 12. He has been an instructor for School of PE since 2017.

Acknowledgments

We would like to sincerely thank our reviewers, who provided valuable feedback for this book.

Book Reviewers:

Jahan R. Aghakasiri
Rodrigo Antunes, PhD, PE
Julienne Bautista
Andrew Caulum
Ozge Cavusoglu
Meridith Conser
Arturo Espinoza
Jessica Ganjian
Tesfai Giorgis
Peter Gorgas, PE
Mark Harris
John Hoobler
Jamie Killgore
Aritra Kundu
Christopher Magruder, PE
Krizia Maximilian
Nathan McVey
Brian Palmiter, Jr.
Donald Shaw III, PE
Jennifer Sommerfeld
John Strub
Gangadhar Tumu, PE
Jennifer White

Chapter Reviewers:

Alan Esser, PE, D.GE
Mark Fulkerson, PhD, PE
Steven F. Mangin, PE
Vijay Maramreddy, PE
Sidney H. May, PE
Amir Mousa, MSCE, PE
Halil Sezen, PhD, PE

Editorial Staff:

Senior Editorial Manager: Angela M. Pugh
Editors: Teresa Barensfeld, Amy Eagleeye
Production Editor: Katie McDonough
Editorial Assistant: Landen Stafford
Layout/Typesetting: SPi Global
Program Director: Chris Miller

Introduction

At School of PE, we know that an engineer isn't likely to be an expert in all aspects of engineering, so we employ multiple instructors to teach the material in their areas of expertise—this sets us apart from our competitors. We have received positive feedback from our students about the learning experience that we provide. In fact, we have prepared tens of thousands of students for the PE Civil exam since we started in 2004. We believe that learning from experts in their respective areas builds great engineers.

We've carried this approach over to our books, with subject matter experts authoring each chapter. We have asked the authors, who are also School of PE instructors, to include only the essential material that students should study to do well on the exam—nothing more, nothing less. That's why our book might be smaller in size than those of our competitors—we want our readers to get the essentials without spending valuable time studying unnecessary information.

About This Book

We've organized this book in a straightforward way that makes it easy to find what you're looking for. Generally, it is designed along the NCEES PE Civil (Breadth) exam specifications. In each chapter, you will find a table of contents followed by an exam guide, which outlines the topics that appear on that section of the exam, and the approximate number of questions on the test in that knowledge area. For example, chapter 1, "Project Planning," lists the four main topics covered on the test in that area, and the approximate number of project planning questions you may expect to see on the exam (4).

TIP

AR features are best viewed on tablets (iOS or Android) due to the larger screen size, although they can also be viewed on other mobile devices, such as smartphones.

You can navigate the book more easily with our tabbed pages within each chapter. This helps to differentiate one chapter from another so you can quickly find what you're looking for. Using time wisely is important when studying for the exam. Each chapter can function independently. For instance, you could begin by studying site development, put the book down, and then come back and skip to project management. Get to know this book and your other resources well. We also added additional space in the margins so that you can take notes as you go.

This book contains several useful tools to help you prepare for the PE Civil exam. For instance, our use of augmented reality (AR) features is designed to enhance your learning experience and memory recognition. You simply need to download the School of PE AR app–TotalAR (see p. iii). This will allow you to gain access to the book's enhanced content. When you see TotalAR (TAR) codes, like those in the box on the next page, throughout the book, look for the image inside the icon to determine the type of AR content it represents. For example, with the TotalAR app open on a smartphone or tablet, pointing to an AR icon with a magnifying glass will search the contents of the book and tell you the page number where the searched term appears. An AR icon with x = ? in the center leads to a quiz that can be taken on your device. In addition, if you see this symbol (◌) on an image in the book, scan the image with the TotalAR app to display AR content. We will continually add AR content to the book, so scan the TAR code with this symbol (◌) in the center to see an updated list of images in the book that are AR targets. TAR codes in a blue circle indicate content that is complimentary for a limited time, while TAR codes in an orange circle indicate content that is available for an additional fee.

When you scan your first TAR code in the book, you will be prompted to enter a unique access key, which you will find on page iii. The various types of AR content provided include quizzes, 3D content, flash cards, search, and an up-to-date errata list. The best way to learn the material is to work through the example problems and study their solutions—both those in the book and those available through the TotalAR app.

ICON	USE	ICON	USE
	AR List: Scan this TAR code to view a list of all AR content in the book. (The code also appears on p. xxvi, Brief Contents.)		**Meet the Author:** Scan this TAR code to watch a brief video introduction from the chapter authors. (This code also appears in each chapter.)
X	**Errata List:** Scan this TAR code to view a list of any errors in the book. (The code also appears on p. xxvi, Brief Contents.)	X=?	**Quiz:** Scan this TAR code to test your knowledge of the presented concepts. (This code also appears in each chapter.)
	Provide Feedback: Scan this TAR code to provide feedback or report an error. (This code also appears in each chapter.)		**Search:** Scan this TAR code to search a dynamic index to find content in the book. (This code also appears in each chapter.)
	Flash Cards: Scan this TAR code to review important terms. (This code also appears in each chapter.)		**Unit Conversion:** Scan this TAR code to access a unit conversion tool. (This code also appears in each chapter.)
NCEES	**NCEES Links:** Scan this TAR code to view important NCEES links. (The code also appears on p. xxv, following the NCEES exam syllabus.)		**Video Lectures:** Scan this TAR code to access video lectures from School of PE instructors. (This code also appears in each chapter, and is paid content.)
	Questions for Subject Matter Experts: Scan this TAR code to send School of PE subject matter experts your questions. (This code also appears in each chapter, and is a paid service.)		**Tutoring:** Scan this TAR code to sign up for one-on-one tutoring with a School of PE instructor. (This code also appears in each chapter, and is a paid service.)

● = complimentary content (for a limited time) ● = paid content

A Final Note

We have tried our best to make this book error free; it's been technically reviewed, edited, and tested. However, we are only human, and errors can happen. If you spot one, please report it to us so we can improve the content. Wherever you see the TAR code with in the center, simply target it with your smartphone or tablet to be taken to the error-reporting page. We will verify the correct information and add it to the errata page, which will be used to issue errata immediately. Errata can also be accessed from the book's website: https://publications.schoolofpe.com/pecivil/v1/webupdates.

We welcome your feedback and want to hear how well this book prepared you for the PE Civil (Breadth) exam. If you have any suggestions or comments, please send them to us at publications@edumind.com.

Best of luck!

The School of PE team

About the Exam

The NCEES PE Civil exam includes the following five disciplines: **Construction, Environmental, Geotechnical, Structural**, and **Transportation**. The exam itself is open book, meaning students can bring additional materials to be used for reference. The exam is eight hours long and contains 80 questions—40 in the four-hour morning session and 40 in the four-hour afternoon session. The morning session, also known as the breadth session, is administered for all examinees and includes questions in project planning, means and methods, soil mechanics, structural mechanics, hydraulics and hydrology, geometrics, materials, and site development. The afternoon session, known as the depth session, is divided into five sections, each corresponding to one of the five disciplines. During exam registration, examinees will select their discipline preference for the depth session, which only includes problems relevant to the chosen discipline. The breadth portion of the exam covers the material discussed in this book and follows the NCEES Principles and Practice of Engineering Examination CIVIL BREADTH Exam Specifications. These specifications include exam topics, subtopics, and the approximate number of questions per topic.

Exam-Taking Experience

Francisco Perez, PE

School of PE Editors

Taking a two-part, eight-hour exam of this kind may seem daunting. Everything you do, from the moment you leave your house to the time you enter the exam location, can affect your frame of mind on the day of the exam. Therefore, you will want to make this process as smooth and well planned as possible. The last thing you would want, for example, is to put in months of preparation for the exam, only to realize once you've arrived that you brought the wrong reference materials, didn't properly label the ones you did bring, or brought the wrong calculator (NCEES-approved calculator list: https://ncees.org/exams/calculator/).

A few helpful tips to consider:

- Double check that all of your references are organized and ready to go.
- Secure all references you are bringing with you in a wheeled crate for ease of transport.
- Fill your car's gas tank the night before the exam.
- Plan your route to the exam location.
- Allow enough travel time to the exam.
- Turn off the TV, put away your phone, and get plenty of sleep the night before.
- Bring $20 to $30 for food and parking.
- Leave mobile phones, smartwatches, fitness trackers, and other electronic devices in your vehicle. They are not allowed in the exam room.
- Have proper identification and your exam authorization ready and easily accessible.

Proctors will be assigned to each section of the room. They are very helpful and friendly, and are there to make sure that all examinees follow the rules at all times. Hence, knowing what to expect and preparing accordingly will help you avoid an uncomfortable situation.

Here are some items you are permitted to bring to the exam:

- Handheld, nonelectronic magnifying glass (no case)
- Identification
- NCEES-approved calculator
- Two straight edges
- Earplugs, to help tune out distracting noises
- Aspirin, ibuprofen, or any other personal medication you may need (in original bottles)
- A snack and/or nonalcoholic beverage
- A watch to keep track of time during the exam (no smartwatches)

Understandably, you may need to visit the restroom during the exam. Proctors will be quietly walking around and will address examinees who raise their hands to request a restroom break. The proctor will make sure that all of your exam materials are left closed and on your desk before you can leave. An optional scheduled break will be midway (40 questions) through the exam. If the exam authorization form does not state otherwise, examinees may leave the exam site during their break. However, NCEES cannot guarantee that there are food options nearby. Examinees may walk back to their parked vehicles for lunch.

Study Tips

For an exam of this magnitude, you will undoubtably want to implement effective study habits for the best results. A study schedule can help you stay on track. You will want to find a quiet room or office where you can study and organize the references that you plan to use during the exam.

A common misconception is that examinees should mainly study material that is in line with their strengths. In doing so, some examinees believe that these correctly answered questions will lead them to pass the exam. This approach, however, has often been reported as poorly effective. To yield a higher chance of success, examinees should expand their level of comfort beyond that of their strengths. Be aware of your weaknesses and become more comfortable with them. The fundamentals in each chapter are important; study them and become comfortable with each.

Completing as many practice problems as possible will increase your level of confidence going into the exam. However, it is not enough to work through problems and simply memorize their solutions. Make your approach to solving the problems as realistic to the exam as possible. Read each example thoroughly and ask yourself, "What subject fundamentals are applicable to this problem and what references do I have that contain relevant information?" Well-organized, tabbed, and diverse reference materials will be critical during the exam because you can quickly find the information you need.

When solving exam questions regarding load and resistance factor design (LRFD) and allowable stress design (ASD), the question will typically specify if it should be solved using LRFD or ASD. It's encouraged to become familiar with both approaches. Exam questions may require references to current codes and standards. The NCEES PE Civil exam is intended to be administered nationwide. Therefore, codes and standards referenced or required in the exam will not be specific to any state. The NCEES PE Civil exam adopts the current International Building Code (IBC), which is a good reference for examinees to bring along to the exam.

Exam Prep

The exam is open book, but you must be mindful of your resources and not be too adventurous. Only bring references that you are very familiar with to avoid wasting time searching for information. Also, you can have personal notes for reference, but they must be securely bound in order to maintain them neatly within your examination area. Loose notes and documents fastened with staples will not be allowed into the exam. A three-ring binder is a convenient method for securing personal notes. Examinees commonly place tabs in reference materials for quick access during the exam. However, if tabs appear to be loose and not permanently secured to books or notes, you may be asked to remove them or surrender the reference altogether. To avoid this, it is a good idea to adhere all tabs with a piece of transparent tape on either side.

As with any exam, taking too much time on any given problem will reduce the amount of time left for other problems. Therefore, it is a good idea to start by going through the exam completely and answering all questions that do not require lengthy calculations first. This will help ensure that you reserve most of your time for more complex or difficult problems thereafter.

HELPFUL HINTS FROM SCHOOL OF PE STUDENTS WHO HAVE TAKEN THE PE CIVIL EXAM

1. "Make sure your reference material is organized. Using a different color tab for each subject helped me."

2. "In my preparation, I extracted all of the formulas provided in the School of PE notes, and created my own formula sheet binder and divided all the sheets into the separate topics. This, combined with my pre-tabbed reference manual, were the main references I used during the exam (plus a few key textbooks). This formula sheet binder saved me a lot of time." (Please see Appendix A.9 for a list of the equations that appear in this book, organized by topic.)

3. "Tab your reference material so you know where to find things. I made a binder full of practice problems organized by topic, which really helped. Don't take reference material in that you didn't use to study."

4. "Do all of the easy problems first! I can't stress this enough. I went through with my notes and did over half of the morning problems within the first 1.5 hours of the exam because I had the exact material I needed. Starting the exam knowing you already have around 20/40 problems answered correctly and still have 2.5 hours is such a stress relief."

5. "Numbering problems easiest to hardest before starting the exam was the most helpful tip I received. It helped my mentality to get some questions answered in order to relax and think about the rest of the exam problems clearly. I also recommend acquiring all of the code books that are on the NCEES list so that you don't miss an easy lookup problem."

6. "Watch for units. Read the question very carefully and answer what is asked. Do the theory questions first and don't waste time on a single question."

7. "During the exam, when rating your questions as 1, 2, or 3 based on difficulty and time needed, consider also marking the discipline/topic and/or reference needed to answer. This helped me save time instead of switching binders constantly for the 1 and 2 questions, especially in the afternoon transportation section where I had over 12 reference manuals."

8. "When you are studying, pay attention to principles, not just practice problems. If you find yourself stuck on a problem, just reverse-engineer it with the available solutions to see which one works best."

NCEES Principles and Practice of Engineering Examination CIVIL BREADTH Exam Specifications

	Approximate Number of Questions
I. Project Planning	**4**
A. Quantity take-off methods	
B. Cost estimating	
C. Project schedules	
D. Activity identification and sequencing	
II. Means and Methods	**3**
A. Construction loads	
B. Construction methods	
C. Temporary structures and facilities	
III. Soil Mechanics	**6**
A. Lateral earth pressure	
B. Soil consolidation	
C. Effective and total stresses	
D. Bearing capacity	
E. Foundation settlement	
F. Slope stability	
IV. Structural Mechanics	**6**
A. Dead and live loads	
B. Trusses	
C. Bending (e.g., moments and stresses)	
D. Shear (e.g., forces and stresses)	
E. Axial (e.g., forces and stresses)	
F. Combined stresses	
G. Deflection	
H. Beams	
I. Columns	
J. Slabs	
K. Footings	
L. Retaining walls	
V. Hydraulics and Hydrology	**7**
A. Open-channel flow	
B. Stormwater collection and drainage (e.g., culvert, stormwater inlets, gutter flow, street flow, storm sewer pipes)	
C. Storm characteristics (e.g., storm frequency, rainfall measurement and distribution)	
D. Runoff analysis (e.g., Rational and SCS/NRCS methods, hydrographic application, runoff time of concentration)	
E. Detention/retention ponds	
F. Pressure conduit (e.g., single pipe, force mains, Hazen-Williams, Darcy-Weisbach, major and minor losses)	
G. Energy and/or continuity equation (e.g., Bernoulli)	

VI. Geometrics **3**

A. Basic circular curve elements (e.g., middle ordinate, length, chord, radius)
B. Basic vertical curve elements
C. Traffic volume (e.g., vehicle mix, flow, and speed)

VII. Materials **6**

A. Soil classification and boring log interpretation
B. Soil properties (e.g., strength, permeability, compressibility, phase relationships)
C. Concrete (e.g., nonreinforced, reinforced)
D. Structural steel
E. Material test methods and specification conformance
F. Compaction

VIII. Site Development **5**

A. Excavation and embankment (e.g., cut and fill)
B. Construction site layout and control
C. Temporary and permanent soil erosion and sediment control (e.g., construction erosion control and permits, sediment transport, channel/outlet protection)
D. Impact of construction on adjacent facilities
E. Safety (e.g., construction, roadside, work zone)

Source: NCEES website

Disclaimer: This NCEES syllabus is current as of August 2019. However, NCEES periodically updates its exam syllabuses. Any necessary changes may be incorporated in future editions of this text, which would be available for separate purchase. It is recommended that you purchase the edition of this text that corresponds to the most current version of the NCEES syllabus.

Brief Contents

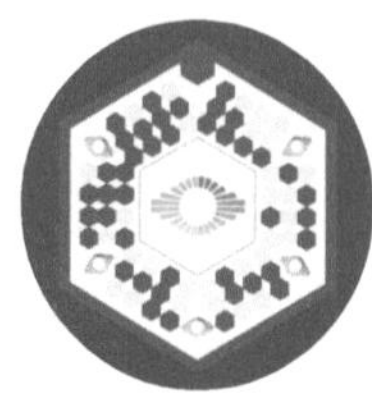

Project Planning

1

George A. Stankiewicz, PE

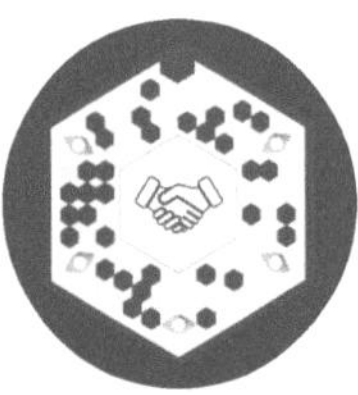

CONTENTS

Project Planning

EXAM GUIDE

I. Project Planning

A. Quantity take-off methods
B. Cost estimating
C. Project schedules
D. Activity identification and sequencing

Approximate Number of Questions on Exam: 4

NCEES Principles & Practice of Engineering Examination, Civil Breadth Exam Specifications

COMMONLY USED ABBREVIATIONS

AOA	activity on arrow
AON	activity on node
EF	early finish date for an activity
LF	late finish for an activity
OC	on center
PDM	precedence diagram method
SFCA	square feet of contact area
WBS	work breakdown structure

INTRODUCTION

The following axioms guide the understanding of the estimating process:

- **Estimating** is a complex process involving the collection of available and relevant information relating to the scope of a project, expected resource usage, and ongoing fluctuations in project costs.
- The estimating process involves evaluating abstract information through a process of visualizing the design for the intended construction project. This visualization translates the design into a monetary calculation of the project's final cost.
- At the beginning of a project, the estimate does not have a high degree of accuracy because little information is known. As the design progresses, information develops and accuracy increases.
- Estimating project costs involves considerable effort to gather and decipher large amounts of information. The construction of a building project includes the estimator's collection and review of all of the architectural and engineering plans, specifications, site data, the contractor's resource records (such as labor, material, and equipment), the owner's documents, and locally sourced cost information, among others.
- Building construction projects differ considerably from product manufacturing, in that construction company products are one of a kind. In conjunction with the dollar value of the hard cost, the valuation of project time is critical. Substantial effort in schedule timeline planning is required before a cost estimate is presented, to arrive at the owner's and contractor's mutual goals.

TIP

The most common blunder during quantity take-off estimating is to omit the zero position during the count.

1.1 QUANTITY TAKE-OFF METHODS

A **construction cost estimate** is a detailed process of interpreting construction plans and specifications to calculate costs for the material, labor, and equipment quantities and to compare them to similar constructed projects to establish an estimate for the proposed work. Uncertainty and risks are common due to the fact that material, labor, and equipment costs change over time and each building construction project is unique with often-unpredictable site conditions. The costs from previous projects are a key advantage used to minimize the risk and forecast future projected costs. Further, the estimator's quantity take-off for the material amount introduces risk into the accuracy of the count. To help with the analysis, sketch the work so as to better visualize the quantity take-off.

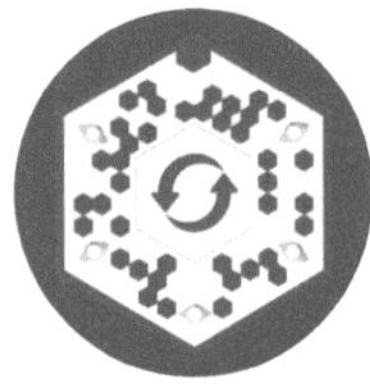

TIP

Remember that distractors are included in the exam questions to test your engineering judgment relevant to the subject matter.

Example 1.1: Counting the Zero Position

A capital improvement project requires the installation of a property line fence along the 250-ft northern boundary line. A decorative aluminum fence is constructed of posts spaced at 10-ft centers and an ornate picket infill panel. The labor hours required to install the posts and infill panels are 2 hr and 4 hr, respectively.

Bill of Materials:	
Aluminum Posts	\$645.35 each
Picket Infill Panel	\$1,985.50 each
Placed Concrete	\$498/yd^3
Ironworker	\$78/hr

The material cost for this scope of work is most nearly:

A. \$65,771.25
B. \$66,416.60
C. \$68,402.10
D. \$71,065.58

TIP

Note that although the labor hours are given, the question asks for the material costs only; labor hours are a distractor.

Solution

Using the bill of materials, extract the material cost only and multiply quantities by the given costs. The placed concrete cost is given per cubic yard (yd^3), but there are no volume quantities, which indicates a distractor.

Aluminum posts: $26 \times \$645.35 = \$16,779.10$

Picket infill panels: $25 \times \$1,985.50 = \$49,637.50$

Grand total: \$66,416.60

Remember to sketch the solution to help visualize the problem statement.

Answer: B

Example 1.2: Estimating Take-Off Quantities

A 450-ft length stormwater diversion channel is to be lined with concrete to prevent erosion. With 10% waste, the material unit costs of concrete and curing compound based on other recent projects in the area with similar volumes are \$98/yd^3 and \$40 per 5 gal, respectively. Project specifications require an application rate of curing compound at 1 gal per 300 ft^2.

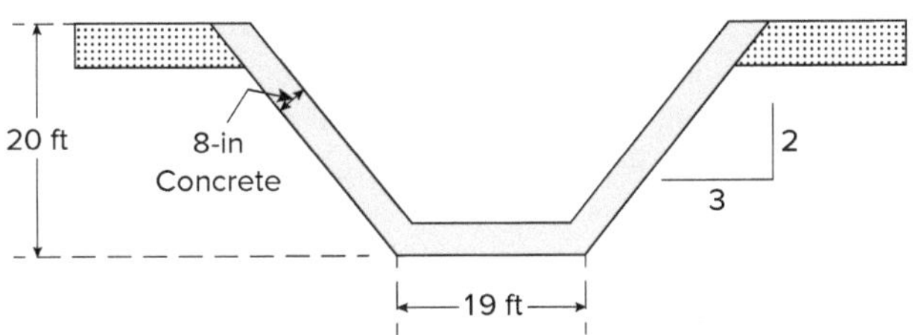

Example 1.2 *(continued)*

The total material cost for delivered concrete is most nearly:

A. \$85,000
B. \$90,160
C. \$99,176
D. \$109,074

Solution

$$\text{Horizontal length of side slope} = 20\text{ ft} \times \left(\frac{3}{2}\right) = 30\text{ ft}$$

$$\text{Slope length} = \sqrt{[(20\text{ ft})^2 + (30\text{ ft})^2]} = 36.06\text{ ft}$$

$$\text{Cross-sectional area of lining} = [(2 \times 36.06\text{ ft}) + 19\text{ ft}]\left(\frac{8\text{ in}}{12\text{ in/ft}}\right) = 60.74\text{ ft}^2$$

$$\text{Volume of lining} = \frac{(60.74\text{ ft}^2 \times 450\text{ ft})}{27\text{ ft}^3/\text{yd}^3} = 1{,}012\text{ yd}^3$$

$$\text{Delivered volume (add 10\% waste)} = 1{,}012\text{ yd}^3 \times 1.10 = 1{,}113\text{ yd}^3$$

$$\text{Material cost} = \$98/\text{yd}^3 \times 1{,}113\text{ yd}^3 = \$109{,}074$$

Answer: D

Example 1.3: Estimating Take-Off Quantities

Refer to the problem statement and diagram in Example 1.2 to solve this problem.

The total material cost for the concrete curing compound is most nearly:

A. \$1,120
B. \$1,160
C. \$1,200
D. \$1,240

Solution

Compute the surface area to which the curing compound will be applied. The curing compound is similar to a fluid, an applied paint that helps retard the evaporation rate of the water. The concrete mixture combines portland cement and water to bind the fine and coarse aggregates. The curing compound helps develop the strength of concrete by increasing the amount of time for the chemical bond between the cement and water to form.

$$\text{Surface area of canal} = [19\text{ ft} + (36.06\text{ ft} \times 2)] \times 450\text{ ft} = 41{,}004\text{ ft}^2$$

$$\text{Quantity of curing compound} = \frac{41{,}004\text{ ft}^2}{300\text{ ft}^2/\text{gal}} = 136.68\text{ gal}$$

Calculate 10% waste: $136.68\text{ gal} \times 1.10 = 150.35\text{ gal}$

Convert to purchase within 5-gal containers: $\frac{150.35\text{ gal}}{5\text{ gal}} = 30.07$ containers

Material cost = 31 containers × \$40/container = \$1,240

We cannot deviate from the manufacturer's specification and application rate. This question illustrates the importance of rounding up to meet the product specifications. The 7/100 of a 5-gallon container in the example is enough to support the manufacturer's position that the coverage rate was not met. Always round up in this situation.

Answer: D

1.1.1 Detailing Formwork

Concrete formwork is considered a mold that is used to form the concrete for the desired building structure. At the time of placement, the concrete is in a liquid state, which allows the mixture to flow freely to fit the mold. The concrete formwork must be strong enough to contain the liquid concrete and the various forces (wind and workers) needed to complete the work. The various members used in formwork construction are the plywood sheathing, studs, wales, wall ties, and braces. Example 1.4 calls out these members, but only asks for the number of 2-in × 4-in × 10-in studs to construct the shape.

Example 1.4: Detailing Formwork

The total number of nominal 2-in × 4-in × 10-ft, 0-in wood studs needed to construct the concrete formwork as shown in the figure is most nearly:

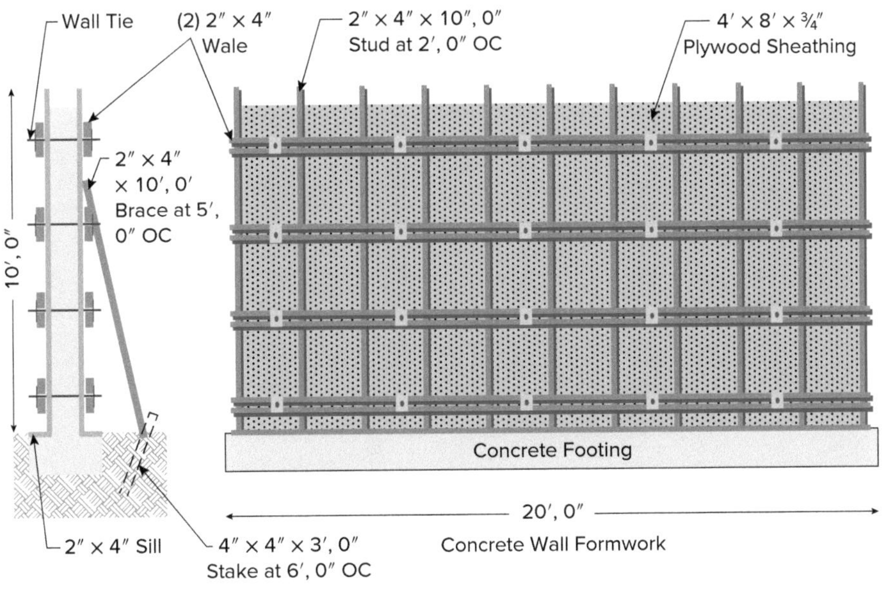

A. 33 pieces
B. 62 pieces
C. 63 pieces
D. 68 pieces

Solution

The drawing represents a concrete wall form and all of the structural elements. The section view shows the representation on both sides of the wall, which is constructed to form a four-sided box to contain the concrete placement. The nominal 2- × 4-in studs are used in all locations called out in the drawing. Separate all the elements and compute the quantity using the various lengths for each. Note that the section view thickness of the wall is not dimensioned, and as such, the ends are not entered into the calculations. Also, the spacing of the brace is shown as 5 ft, 0 in on center (OC), where the full length of one side of the wall must be laterally braced as shown. The dimensions require five braces along the 20-ft, 0-in length of wall to assure stability and safety of the formwork construction.

2- × 4-in stud length = 10 ft, 0 in

Vertical studs = 11 count × 10-ft length × 2 sides = 22 pieces

Horizontal wales = 4 wales × 2 at each location × 2 sides × 2 lengths = 32 pieces

Sill = 2 sides × 2 lengths per side = 4 pieces

Braces = 5 lengths along one side at 5 ft OC = 5 pieces

Example 1.4 *(continued)*

Total = 22 pieces + 32 pieces + 4 pieces + 5 pieces

Total = 63 pieces

Answer: C

Example 1.5: Determining Reinforcement Weight

The details for the placement of a 15-ft, 6-in (length) grade beam, 2 ft, 4 in (width) × 3 ft, 0 in (height), are shown in section A in the following figure. According to *ACI 318-14: Building Code Requirements for Structural Concrete*, all reinforcement steel exposed to earth requires a minimum 3-in cover. The total weight of reinforcing steel (lb) needed for the construction of the grade beam is most nearly:

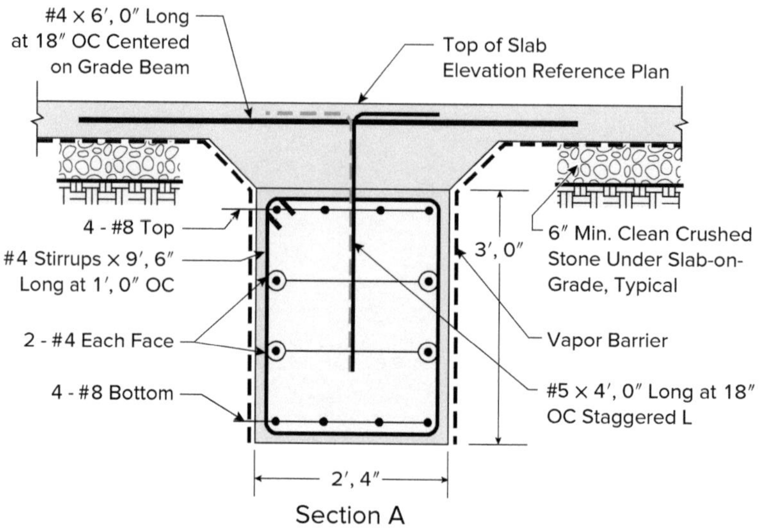

Section A

BAR SIZE	WEIGHT (LB/FT)	AREA (IN^2)
#3	0.376	0.11
#4	0.668	0.20
#5	1.043	0.31
#6	1.502	0.44
#7	2.044	0.60
#8	2.670	0.79
#9	3.400	1.00

A. 480 lb
B. 510 lb
C. 545 lb
D. 565 lb

Solution

Step 1: Examine section A and focus only on the information within the 2-ft, 4-in (width) × 3-ft, 0-in (height) grade beam. As shown, a grade beam or footing is part of the building's foundation system. The grade beam consists of a reinforced concrete beam typically spanning over unsuitable soils and resting on a bearing or supporting structure within the foundation system. Notice that ACI 318-14 is referenced. Follow the rule that all reinforcement steel exposed to earth requires a minimum 3-in cover. This requires the length of the reinforcement inserted into the 15-ft, 6-in wall to be adjusted to a 15-ft, 0-in dimension to accommodate the 3-in cover on either end of the beam.

TIP

Avoid computation traps due to the number of distractors found in the section view; for example, the #4 reinforcement bar that is 6-ft, 0-in is part of the slab pour and not the grade beam.

Example 1.5 *(continued)*

Step 2: Create a table of information to organize the location, quantity, length, weight, and extended total for the reinforcement weight. Make certain to count the zero position for both the stirrups and staggered L bar spacing. Multiply and find the extended total.

LOCATION	QUANTITY	LENGTH (FT)	WEIGHT (LB/FT)	EXTENDED TOTAL WEIGHT (LB)
#4 Stirrups at 1 ft, 0 in OC	16	9 ft, 6 in	0.668	101.54
#4 Longitudinal Bar, Each Face	4	15 ft, 0 in	0.668	40.08
#8 Top Bar	4	15 ft, 0 in	2.670	160.20
#8 Bottom Bar	4	15 ft, 0 in	2.670	160.20
#5 Staggered L Bar at 18 in OC	11	4 ft, 0 in	1.043	45.89
			Total	**507.91**

Answer: B

1.1.2 Computing Geometric Quantities

The following provides a time-saving method for computing various quantities of geometric shapes. Use the sketch together with the equations to compute the quantities.

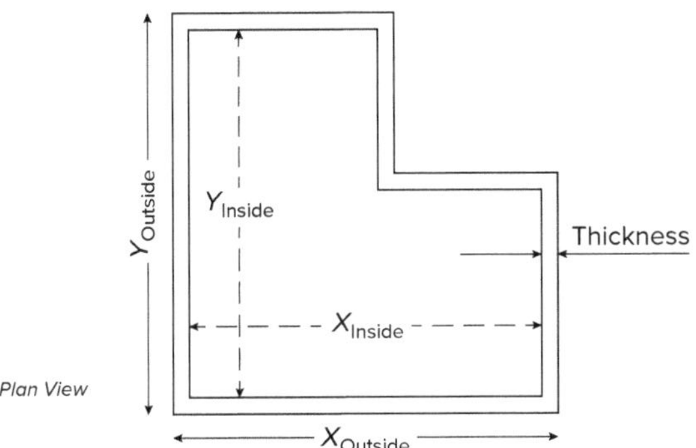

FIGURE 1.1 Geometric Shape Properties

Equations to calculate geometric properties of a **dimensional shape**:

$$X_{inside} = X_{outside} - 2_{thickness}$$ **Equation 1-1**

$$Y_{inside} = Y_{outside} - 2_{thickness}$$ **Equation 1-2**

$$\text{Outside perimeter} = 2(X_{outside} + Y_{outside})$$

$$\text{Inside perimeter} = 2(X_{inside} + Y_{inside})$$

To calculate the **cross-sectional area**, use the following:

$$\text{Cross-sectional area} = 2_{(thickness)}(X_{outside} + Y_{inside}) = 2_{(thickness)}(X_{inside} + Y_{outside})$$ **Equation 1-3**

To calculate the outside perimeter of a shape with recesses along a face, use the following:

$$\text{Outside perimeter} = 2\ (\text{length} + \text{width} + \text{recess})$$ **Equation 1-4**

$$\text{Inside perimeter} = \text{outside perimeter} - [4\ (2 \times \text{thickness})]$$ **Equation 1-5**

Use the mean perimeter to calculate the volume of the shape:

$$\text{Mean perimeter} = \text{outside perimeter} - \left[4\left(2 \times \left(\frac{\text{thickness}}{2}\right)\right)\right] \quad \textbf{Equation 1-6}$$

Example 1.6: Determining Outside Perimeter

The plan view for the foundation of a proposed building construction is shown below. The cast-in-place foundation wall is designed to be 10-in thick and 10-ft high.

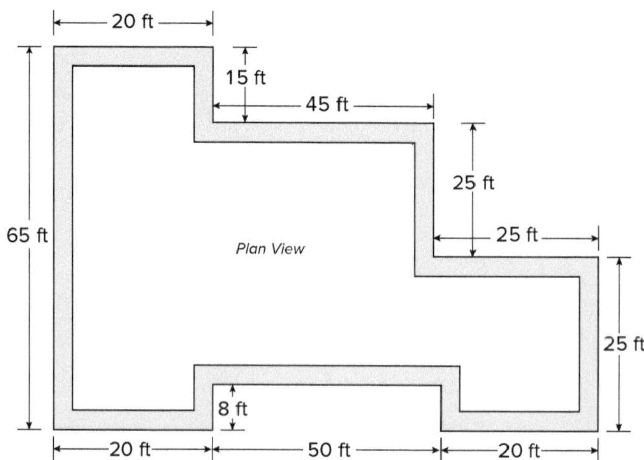

The outside perimeter of the wall is most nearly:

A. 277 ft
B. 301 ft
C. 326 ft
D. 347 ft

Solution

Use Equation 1-4, and apply it to the shape in the figure.

Outside perimeter = 2 × (length + width + recess along the face)

= 2 × (90 ft + 65 ft + 8 ft)

Outside perimeter = 326 ft

A recess is shown along the bottom wall in the figure above. The wall contains a straight line with a recess or indent, increasing the overall length of the line.

Answer: C

Example 1.7: Determining Required Volume of Concrete

Refer to the problem statement and diagram in Example 1.6 to solve this problem.

The theoretical volume of concrete (in cubic yards) required to construct the wall is most nearly:

A. 85 yd^3
B. 90 yd^3
C. 91 yd^3
D. 100 yd^3

Example 1.7 *(continued)*

Solution

Use Equations 1-4 and 1-6, and apply them to the shape in the figure.

$$\text{Outside perimeter} = 2 \times (90\text{ ft} + 65\text{ ft} + 8\text{ ft}) = 326\text{ ft}$$

$$\text{Mean perimeter} = 326\text{ ft} - (4)\left[2 \times \left(\frac{10\text{ in}}{12\text{ in/ft}} \div 2\right)\right] = 322.67\text{ ft}$$

$$\text{Volume of concrete} = 322.67\text{ ft} \times \left(10\text{ ft} \times \frac{10\text{ in}}{12\text{ in/ft}}\right) = \frac{2{,}685.5\text{ ft}^3}{27\text{ ft}^3/\text{yd}^3} = 99.58\text{ yd}^3$$

Answer: D

1.2 COST ESTIMATING

In this section, the geometric properties of various shapes will be evaluated to determine summary quantities and develop cost using historic productivity standards. Employers use measures of worker productivity to establish standards from which future work can be estimated and competitively bid. The cost estimator's focus is to use standard measures in recognized quantities (such as square foot, cubic yard, and square yard) and to develop project quantity summaries to establish cost proposals.

Often building contractors analyze project cost and make comparisons using a total building square foot (SF) measure multiplied by a square foot cost. This gross oversimplification for a complex structure provides a convenient way for developers to plan a construction project within a specific market. For example, the 50,000-square-foot cost for a commercial building in a metropolitan area can be quickly estimated as $10 million on the basis of a cost of $200/SF for the architectural/engineering, plumbing, mechanical, and electrical costs. Further, published costs establish the value of a property and provide guidelines for the valuation of an investment in a particular market.

Specialty trades also examine and quantify cost estimates using specific take-off quantities. One type of specific measure is **square feet of contact area (SFCA)**, which is used to measure the area where concrete contacts the forms. This is calculated by measuring the total linear feet of the concrete forms and multiplying it by the height of the forms. The SFCA measure is used by concrete contractors and civil engineers to determine the material and labor quantities to develop time and cost productivities. The time needed to construct formwork is often used in project schedules to evaluate the project's overall duration.

Example 1.8: Estimating Formwork—SFCA

Refer to the problem statement and diagram in Example 1.6 to solve this problem.

The required cast-in-place formwork square feet of contact area (SFCA) needed for the foundation wall construction is most nearly:

A. 3,260 SFCA
B. 4,520 SFCA
C. 6,450 SFCA
D. 8,450 SFCA

Example 1.8 *(continued)*

Solution

Use Equations 1-4 and 1-5, and apply them to the shape in the figure in Example 1.6.

Outside perimeter = 2 × (90 ft + 65 ft + 8 ft) = 326 ft

$$\text{Inside perimeter} = 326\text{ ft} + (4)\left(2 \times \left(\frac{-10\text{ in}}{12\text{ in/ft}}\right)\right)$$

Inside perimeter = 319.33 ft

Outside forms = 326 ft × 10 ft high = 3,260 SFCA

Inside forms = 319.33 ft × 10 ft high = 3,193.33 SFCA

Total formwork = 3,260 SFCA + 3,193.33 SFCA = 6,453 SFCA

(Note that the SFCA is defined elsewhere in this section.)

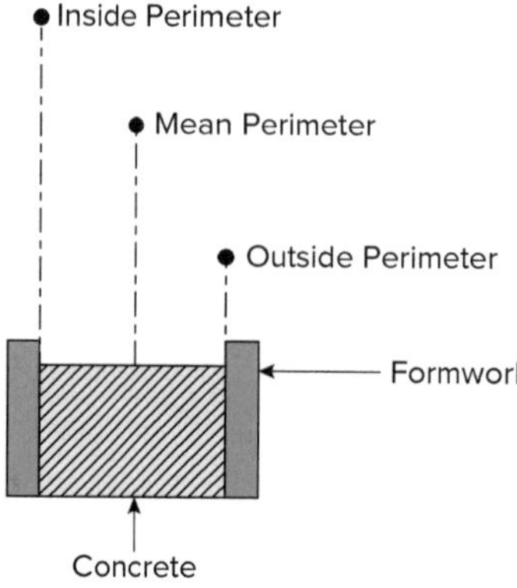

Answer: C

Example 1.9: Estimating Formwork—SFCA

A concrete crew will use available steel form panels measuring 2-ft, 6-in (width) × 4-ft, 0-in (height) to construct a 40-ft, 6-in (length) × 3-ft, 2-in (height) × 1-ft, 6-in (width) concrete knee wall. The SFCA for the formwork is most nearly:

A. 105 ft^2
B. 134 ft^2
C. 266 ft^2
D. 368 ft^2

Solution

Sketch and visualize the problem statement.

The SFCA consists of the surface of the formwork touched by the concrete. The formwork consists of a closed box shape to contain the concrete as it is placed. Therefore, apply the equation to calculate the contact area:

(40.5 ft + 1.5 ft + 40.5 ft + 1.5 ft) × 3.167 ft = 266.03 ft^2

Example 1.9 *(continued)*

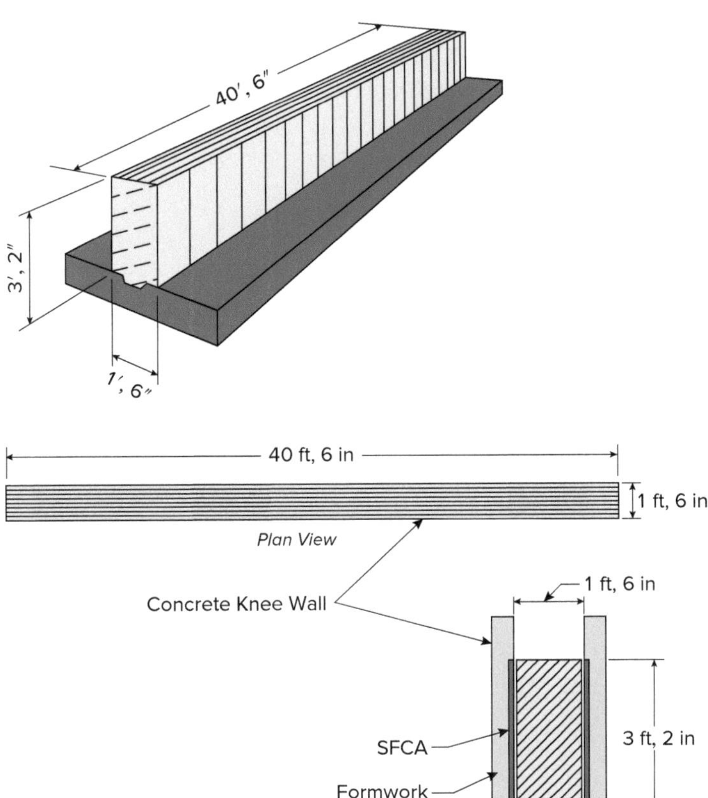

Answer: C

1.2.1 Take the "Outs" Out

The estimator is careful to ensure that the quantities are correctly measured. During the quantity take-off, scale measures of the material lengths, areas, and volumes are computed. Overall area quantities where the material is not part of the building's features are addressed for accuracy. For example, when computing the brick quantity for a masonry façade, an area of a door or window opening in the wall will be deducted from the total measure. In estimating terms, this reduction in materials is known as taking the "outs" out.

In all, estimating has inherent blunders to be aware of and avoid.

Reminders to avoid common estimating blunders:

- Remember to count the zero position when needed.
- Always take the "outs" out whenever there are quantities where material is not needed.
- When estimating, round up material quantities to ensure that the work will have enough materials to be completed.
- Include the given waste amount only when it is given. Do not introduce waste percent amounts not given in the exam problem statement.
- Sketch the work to make sure you are confident in the interpretation of the question and your answer.

Example 1.10: Taking the "Outs" Out

An electrical contractor is preparing to place a concrete duct bank encasement for the high-voltage feeders between manholes #11 and #12. The total distance is 323 ft. The duct bank is 2 ft × 3 ft and holds nine 4.5-in schedule 80 PVC conduits. The amount of concrete to be ordered is most nearly:

A. 60 yd^3
B. 63 yd^3
C. 72 yd^3
D. 75 yd^3

TIP

Remember to always take the "outs" out.

Solution

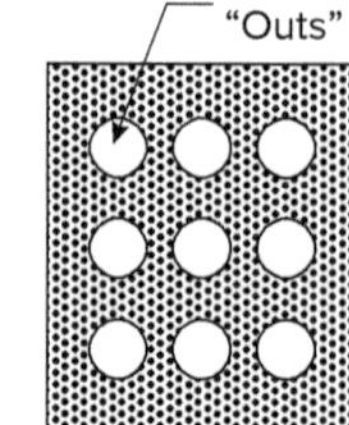

The schedule 80 PVC pipe is 4.5 in outside diameter (OD). The total volume of the duct bank encasement minus the conduit provides the solution.

$$\text{Total volume} = 2 \text{ ft} \times 3 \text{ ft} \times 323 \text{ ft} = \frac{1{,}938 \text{ ft}^3}{27 \text{ ft}^3/\text{yd}^3} = 71.77 \text{ yd}^3$$

$$\text{Total deduction} = 9 \times \left(\frac{\pi(4.5 \text{ in}/12 \text{ in/ft})^2}{4}\right) \times 323 \text{ ft} = \frac{321 \text{ ft}^3}{27 \text{ ft}^3/\text{yd}^3} = 11.89 \text{ yd}^3$$

$$\text{Total concrete} = 71.77 \text{ yd}^3 - 11.89 \text{ yd}^3 = 59.88 \text{ yd}^3 \approx 60 \text{ yd}^3$$

Sketch the solution. Note that the conduit volume reduces the total concrete volume of the material in the duct bank encasement.

Answer: A

1.2.2 Building Materials—Roof Surface

Roof pitch is determined by finding the amount of rise per foot run. It is represented by a triangular shaped drawing and expressed in inches, 4/12, 5/12, and so forth; the higher the number, the steeper the pitch or angle of incline.

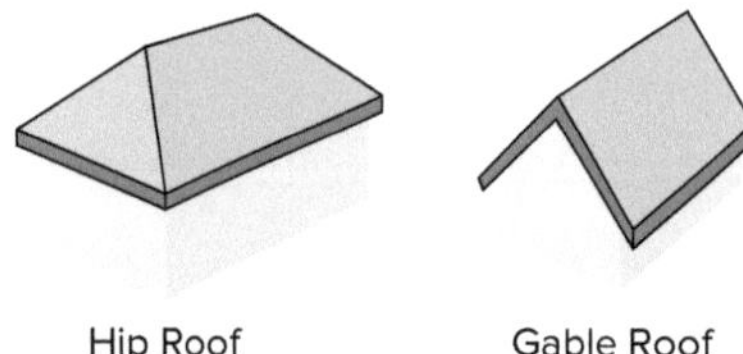

FIGURE 1.2 Hip and Gable Roofs

Use the axiom that the area of a hip roof is the same as the area of a gable roof.

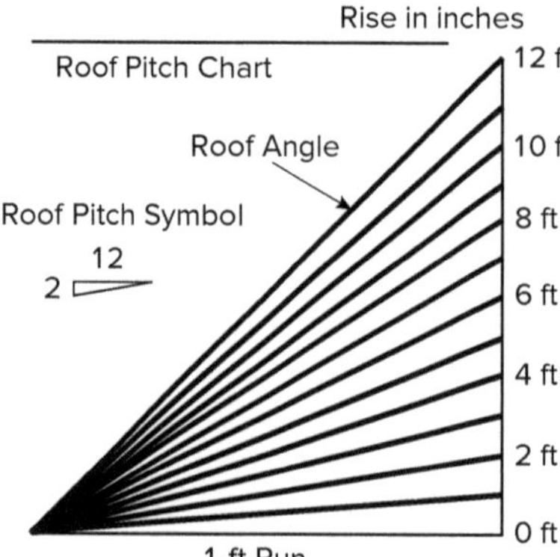

FIGURE 1.3 Roof Pitch Chart

Example 1.11: Building Materials—Roof Surface

The plan view for a roof on a wood frame house is shown. Using a waste factor of 15%, the quantity of plywood sheets (nominal dimensions 4 ft × 8 ft) required for sheathing the hip roof having a 6/12 pitch is most nearly:

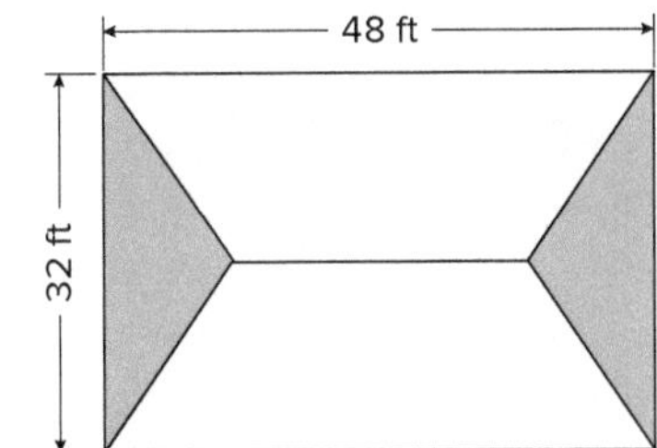

A. 48 sheets
B. 54 sheets
C. 62 sheets
D. 64 sheets

Solution

Based on the given dimensions, the center dimension to the roof ridge is 16 ft and the ridge height of the roof is computed by using 6/12 (rise over run). Sketch the problem statement and visualize the plan view as a three-dimensional object. Using the proportion of 6/12, multiply the centerline dimension and obtain the height of the ridge.

> **TIP**
>
> An important fact to remember is that the area of a hip roof is the same as the area of a gable roof.

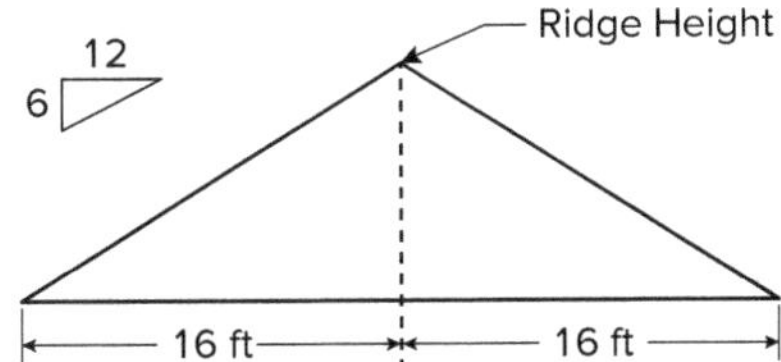

Centerline dimension × (rise over run) = ridge height

16 ft (6/12) = 8-ft ridge height

The length of the roof slope $= \sqrt{(16\text{ ft})^2 + (8\text{ ft})^2} = 17.89$ ft

Compute using:

(2 sides) (48 ft) (17.89 ft) = 1,717.44 ft^2

Plywood sheathing is available in 4 ft × 8 ft = 32 ft^2 per sheet

$$\frac{(1{,}717.44\text{ ft}^2 + 15\%\text{ waste})}{32\text{ ft}^2\text{ per sheet}} = 61.72 = 62\text{ sheets}$$

Answer: C

1.2.3 Estimating Brick Quantities

Estimating Brick Masonry

There are various methods to estimate material quantities on a project. The generally used estimating procedure is the wall-area method. It consists of multiplying the net wall area (gross areas minus areas of openings; take the "outs" out) by known quantities of measured brick material required per square foot. An important part of the analysis is determining the area of brick and the amount of mortar surrounding each unit area of wall that relies on both brick size and joint width.

The first step in estimating is to review the problem statement and determine which size of brick is provided for the surface area calculation. There are two dimensions to choose from:

Nominal dimensions. Nominal dimensions apply to brick and are the result of the specified dimension of the brick plus the thickness in inches of the mortar joint. Generally, these dimensions will be rounded to produce modules of 4 or 8 in The nominal dimension is typical of many construction materials. For example, dimensional lumber is called out as a nominal 2 in × 4 in; however, the actual dimension is 1½ in × 3½ in.

Specified dimensions and/or actual dimensions. Specified dimensions are the manufactured dimensions of the brick, *without* consideration for mortar. Actual dimensions are the measurements of the brick as it is manufactured.

Bond Pattern

For most brick sizes, one-half running bond is the basic pattern when laying a brick wall façade; that is, approximately half of the brick's length overlaps the brick below. This pattern is the most frequently used and is often referred to as the running bond or stretcher bond.

The definition of a masonry tooled joint is a joint for which the mortar between the bricks is shaped by using a concave or U-shaped jointing tool while the mortar is still in its plastic state. The purpose is to create a weathertight barrier preventing water infiltration.

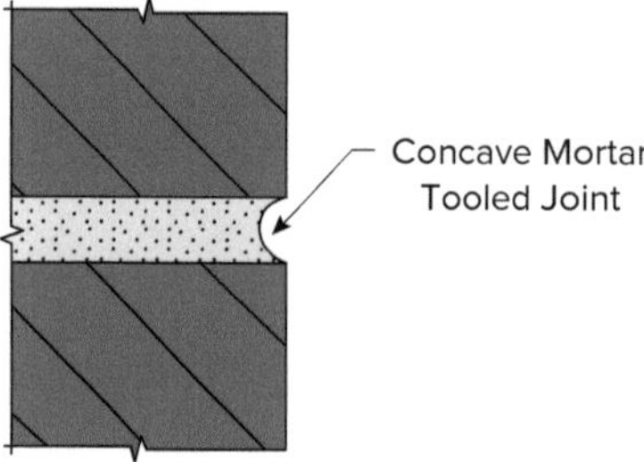

FIGURE 1.4 Concave Mortar Joint

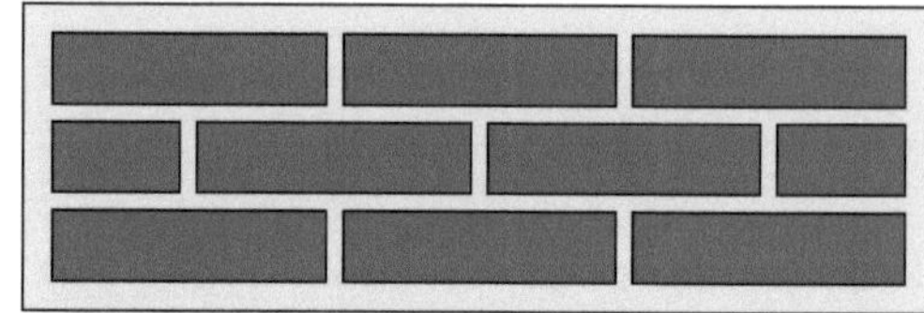

FIGURE 1.5 One-half Running Bond Masonry Wall Brick Pattern

Example 1.12: Estimating Brick Quantities

A brick veneer is planned for the front elevation of a structure 126-ft wide × 12-ft high with 26% of the exterior surface allotted for openings. The specified size of the selected brick is 2⅝ in (height) × 3⅝ in (width) × 7⅝ in (length), and the owner has selected a ⅜-in concave tooled mortar joint. The quantity of bricks needed (count) for a running bond pattern is most nearly:

A. 6,700 bricks
B. 7,500 bricks
C. 9,000 bricks
D. 9,500 bricks

Solution

Step 1: Compute the length of the brick-and-mortar joint to determine the area of each brick. Add the ⅜-in mortar joint to the top and one side only.

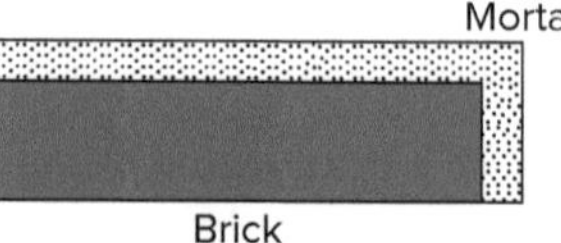

Example 1.12 *(continued)*

$7\frac{5}{8} + \frac{3}{8} = 8$ in (length)
$2\frac{5}{8} + \frac{3}{8} = 3$ in (height)

Step 2: Compute the area of the brick veneer.

Total area = 126 ft × 12 ft = 1,512 ft^2

$$\text{Total area of bricks} = 8\text{ in} \times 3\text{ in} = \frac{24\text{ in}^2}{144\text{ in}^2/\text{ft}^2} = 0.167\text{ ft}^2$$

$$\text{Number of bricks} = \frac{1{,}512\text{ ft}^2}{0.167\text{ ft}^2} = 9{,}054\text{ bricks}$$

Step 3: Take the "outs" out: reduce the amount by 26%.

9,054 bricks × 0.74 = 6,700 bricks

Answer: A

Example 1.13: Estimating Painting Quantities

A painting contractor uses a productivity standard table to calculate a cost proposal. The total cost ($) to paint a rectangular-shaped floorplan conference room that is 120 ft × 200 ft with a 2% reduction for wall openings and a $2,000 allowance for specialty trim work is most nearly:

PRODUCTIVITY STANDARD FOR SURFACES, 12-FT HIGH				
ACTIVITY	MATERIAL COVERAGE (FT^2/GAL)	MATERIAL COST ($ PER GAL)	LABOR PRODUCTIVITY (FT^2/HR)	BURDENED LABOR COST ($/HR)
Ceiling Surface	230	38.50	225 ft^2/hr	38.50
Wall Surfaces	265	45.60	245 ft^2/hr	42.50
Bookcases	125	66.80	125 ft^2/hr	72.85
Tables	100	75	100 ft^2/hr	86.44
Wall Openings	125	65	50 ft^2/hr	64.50

A. $9,500
B. $10,500
C. $11,500
D. $13,500

Solution

Compute the areas that are given in the problem statement for both the wall and ceiling.

Room perimeter = 120 ft + 200 ft + 120 ft + 200 ft = 640 ft

Wall area of all walls = 640 ft × 12 ft = 7,680 ft^2

2% for wall openings = 7,680 ft^2 × 0.02 = 153.6 ft^2

Wall area less 2% for openings = 7,680 ft^2 – 153.6 ft^2 = 7,526.4 ft^2

Ceiling surface area = 120 ft × 200 ft = 24,000 ft^2

TIP

Note that there are no bookcases or tables mentioned in the problem statement as these are considered distractors.

Example 1.13 *(continued)*

Use the computed surface areas to determine the required quantities.

PRODUCTIVITY STANDARD FOR SURFACES, 12-FT HIGH			
	CEILING	WALL SURFACES	WALL OPENINGS
Material Coverage (ft^2/gal)	230	265	125
Material Cost ($ per Gallon)	38.50	45.60	65.00
Material ($/$ft^2$)	0.1674	0.1721	0.52
Labor Productivity (ft^2/hr)	225 ft^2/hr	245 ft^2/hr	50 ft^2/hr
Burden Labor Cost ($/hr)	38.50	42.50	64.50
Labor ($/$ft^2$)	0.1711	0.1735	1.29
Total $/$ft^2$ (Labor + Materials)	0.3385	0.3456	1.81
Total ft^2	24,000	7,526.4	153.6
Total Cost	$8,124.00	$2,601.12	$278.02
Grand Total: $11,003.14			

Using the grand total plus the $2,000 allowance gives $13,003.14.

Answer: D

Example 1.14: Equipment Production

An earthwork excavation operation is under contract to transport a quantity of 3,000 yd^3 of borrow material. The excavation crew consists of an excavator with a production rate of 200 yd^3/day, a loader production of 250 yd^3/day, and three dump trucks moving a total amount of 150 yd^3/day. The crew formation required to minimize the duration of the operation will need to add which of the following?

A. 1 excavator
B. 1 loader
C. 2 loaders
D. 1 dump truck

Solution

Examine all of the work activities and establish the baseline provisions issued in the problem statement:

Examine using 1 excavator: Duration = 3,000 yd^3/200 yd^3/day = 15 days

Examine using 1 loader: Duration = 3,000 yd^3/250 yd^3/day = 12 days

Examine using 3 trucks: Duration = 3,000/150 yd^3/day = 20 days

The activity duration is governed by the lowest production rate of a total of 20 days.

By inspection, this is an unbalanced workflow crew where the loader is not working with full capacity; the production rate of this crew could be adjusted by increasing the number of trucks from 3 to 4 trucks.

The outcome would allow a balanced mix of resources, whereby the adjusted crew becomes 1 loader, 1 excavator, and 4 trucks.

Accordingly, the activity duration for the 4 trucks = 3,000 yd^3/200 yd^3/day = 15 days, which matches the production of the excavator.

Answer: D

Example 1.15: Estimating Labor Cost

A contractor will place 90 yd^3 of concrete for a housekeeping pad on the rooftop of a 36-ft tall building. Site conditions dictate that the safest and best method of placement is to use a crane and a 2-yd^3 bucket. To perform the task efficiently, five union laborers are needed: one at the concrete truck, three at the point of placement, and one on the portable internal vibrator. The wage rate for laborers is \$22/hr. (The union overtime rate is 1.5 times the wage rate after an 8-hr day.) The time needed for the operation is calculated by the following: setup is 15 min; cycle time consists of load = 3 min, plus swing, dump, and return = 6 min. Thus, the total cycle time = 9 min. The demobilize operation is 10 min. Supervision is done by the superintendent. Allow a 10% factor for inefficiencies during the cycle time. Crane rental cost is \$1,800 per 8-hr day. The total labor cost per cubic yard for the concrete crane placement of the 90 yd^3 of concrete is most nearly:

A. \$8.75
B. \$9.78
C. \$10.12
D. \$10.65

Solution

Identify (by underlining) relevant cost items and calculate summary quantities.

> **TIP**
>
> Although the overtime rate is given, it is not needed in the calculations. Also, because the question asks for the total labor costs, the crane rental cost is excluded.

Number of cycles: 90 yd^3/2 yd^3/bucket = 45 cycles

Total cycle time: 45 cycles × 9 min/cycle = 405 min

Inefficiency (labor delays, etc.):

10% of total cycle time = 405 min × 10% = 40.5 min ≈ 41 min

Setup and demobilize, subtotal = 15 min + 10 min = 25 min

Total operation time: 405 + 41 + 25 = 471 min or 7.85 hr ≈ 8 hr

Amount of time needed (adjusted to workday) = 8 hr

Total labor cost per 90 yd^3 = 5 laborers × 8 hrs × \$22/hr = \$880

$$\text{Cost per cubic yard} = \frac{\$880}{90\ \text{yd}^3} = \$9.78/\text{yd}^3$$

Answer: B

1.3 PROJECT SCHEDULES

The essential organizational requirements for the construction of a project schedule are the following:

- A **project network** depicts the sequence of linked activities necessary to complete a project. Each activity has a distinct and measurable scope of work with a specific, calculated duration and an exact start and finish time. An **activity** is a discrete work action within a project that will consume resources during its performance. The number of activities that are identified in a project plan are often dependent on the duration and usually limited to no more than 20 days in length for ease of analysis on large projects. An activity may also be referred to as a scope of work item.
- The activities of a project network diagram are represented by nodes with connected lines and arrows to show the interrelationship of activities. The graphical representation of a network diagram immediately communicates the logic of the project to the participants to use for continuous planning and progress monitoring.

- The purpose of **network scheduling** is to find the critical path in the linked activities that determines the total project duration. The critical path analysis through the network uses the activity's **float** to determine the flexibility in the start and finish times. Float is better described by the use of the word "flexibility" in the start and finish time of an activity. Float varies for each activity as the project progresses. Every update of the schedule calculates the flexibility in the start and finish times of an activity.
- The outcomes of project activities are evaluated based on the float results, which can be a negative duration (indicating the activity is behind schedule), zero (indicating no flexibility in the start and finish times), or a positive duration (indicating the activity has flexibility in the start and finish times).

1.3.1 Calculating Project Duration

Cost estimating and project scheduling are managed side by side during the preparation of a construction project cost proposal. The estimator's quantity take-off provides the needed calculations to obtain the duration for the overall project scope of work and to identify all project activities.

The estimator's material square foot measure for the brick needed for a masonry façade is evaluated against the mason's productivity in assembling a brick wall and used to determine the activity's duration. The project scheduler relies on the estimator's quantities to develop the duration and number of work crews to accomplish the project's tasks. The estimator relies on the scheduler's ability to develop the overall activity network and quantify the overall project duration so as to assign the general conditions and project management costs.

The company's accounting records track the specific work activities, which are then made into productivity tables. The tables guide the calculations for the project scheduler to determine the individual work activity durations. Knowing the total scope of work and the amount of time needed to complete each task, a network of activities can be linked to develop the master plan. After the owner and contractor mutually agree to the overall schedule, the plan can be implemented in the form of a binding agreement or contract. To protect the owner's interests, the construction contract will have provisions for late project delivery, which may include contractor penalties.

The importance of the accuracy of the project schedule activity duration calculations cannot be overstated. The legal consequences for the contractor not meeting the project's timeline are costly, as a project delay will impact the owner's ability to use the building. The duration of an activity is based on the scope of work and measured against task productivity and crew size. The general format of the activity duration equation is as stated:

$$\text{Duration}_{\text{activity}} = \frac{\text{Scope of Work}}{\text{Productivity} \times \text{Crew Size}} \qquad \textbf{Equation 1-7}$$

Variations on the equation depend on the activity's level of difficulty. The equation accommodates the productivity factor in the denominator by allowing variations. For example, the productivity in constructing a 5-ft-high wall is greater than in a 50-ft-high wall. The costs escalate due to the amount of equipment and labor needed for scaffolding, hoisting, safety provisions, and the like. If the activity has 30 days allocated for completion through the total network, the volume of work must be assessed, and the number of crews adjusted to fit the timeline. Worker productivity is affected by other factors, such as work crowding (having too many workers in a small space), learning curves (productivity increasing when the tasks are repetitive), and out-of-sequence work (workers unable to have a continuous workflow, being interrupted). These are a few of the considerations that the scheduler must incorporate to adjust for the quantities provided by the estimator.

The following example illustrates the impact that productivities have on the duration of a project.

Example 1.16: Calculating Project Duration

A concrete contractor's productivity records show that the construction duration for a cast-in-place foundation formwork is 1,000 ft^2/crew-day. The number of days required to complete a 200,000-ft^2 formwork for a cast-in-place concrete foundation using three crews is most nearly:

A. 55 days
B. 60 days
C. 70 days
D. 75 days

Solution

Apply the equation. The scope of work is found in the referenced productivity records for a cast-in-place foundation formwork, which is 1,000 ft^2/crew-day. By using the given area computation and crew size, the total duration of an activity can be found.

$$\text{Duration}_{\text{activity}} = \frac{\text{Scope of Work}}{\text{Productivity} \times \text{Crew Size}}$$

$$\text{Duration} = \frac{200{,}000 \text{ ft}^2 \text{ formwork}}{1{,}000 \text{ ft}^2/\text{crew-day} \times 3 \text{ crews}}$$

Duration = 67 days

Answer: C

Example 1.17: Scheduling Predecessor Table

NO.	ACTIVITY	PREDECESSOR	DURATION (HOURS)
1	Excavate Trench	—	8
2	Place Water Pipe	Excavate Trench	4
3	Test Pipe	Place Water Pipe	4
4	Backfill Trench	Test Pipe	8

The total duration (days) for a utility contractor working 8 hours per day to complete the installation of an underground water pipe as shown in the table is most nearly:

A. 1 day
B. 3 days
C. 16 days
D. 24 days

Solution

This example illustrates the step-by-step progression in the analysis of a project schedule given a predecessor table with activity, predecessor, and duration columns.

The first step in this iterative process of analysis is to identify the activity and determine the predecessor or the before-task impact on the remaining activities. The results are taken to the second step, which shows that the activities are all sequential and there are no concurrent or parallel tasks. As a result, the durations are added, namely, $8 + 4 + 4 + 8 = 24$ hr. Note that the question requests the solution in days, and the contractor works 8 hr/day (given); therefore, the conversion is made and the answer derived: 24 hours/8 hr/day = 3 days.

Answer: B

1.3.2 Units of Time and Equations

Typically, the activity duration is given and labeled in the schedule as calendar days or workdays. Durations can be derived and presented in days, weeks, months, or annual quarters, and, on urgent projects, in hours.

> **Calendar days** are consecutive days on the calendar, including holidays and weekends.
>
> **Workdays** are Monday through Friday, not including public holidays and weekends.
>
> An **ordinal** number refers to the numerical position of a day, for example, first, second, third, and so forth. In formal set theory, an ordinal number is one of the whole numbers.

The calculations for project scheduling use equations to find the activity duration. In project scheduling analysis, there are no logic statements such as if, and, or but.

Network analysis abbreviations and system equations are as follows:

ES = Earliest date that an activity can start

EF = Early finish date for an activity

LS = Late start for an activity

LF = Late finish for an activity

Equations:

$EF = ES + \text{Duration}$ **Equation 1-8**

$LS = LF - \text{Duration}$ **Equation 1-9**

TF = Total Float or flexibility in an activity start date

Total Float Equations:

$\text{Float} = LS - ES$ **Equation 1-10**

or

$\text{Float} = LF - EF$ **Equation 1-11**

Alternate: $TF = LF - ES - \text{Duration}$ **Equation 1-12**

1.3.3 Precedence Relationships

Finish-to-start: The most common precedence relationship. The predecessor activity must finish before the successor activity can start. (A finishes, and then B starts.)

Finish-to-finish: The predecessor activity must finish before the successor activity can finish. (A and B finish at the same time; the durations do not need to be the same.)

Start-to-start: The predecessor activity must start before the successor activity can start. (A and B start at the same time; the durations do not need to be the same.)

Start-to-finish: The predecessor activity must start before the dependent activity can finish, or the completion of the successor depends upon the initiation of the work of the predecessor. (A must start before B can be finished.)

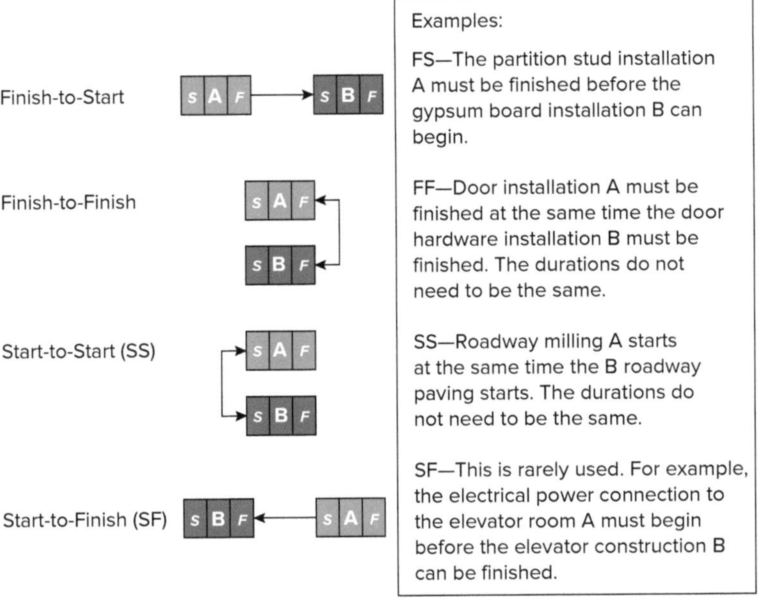

FIGURE 1.6 Precedence Relationships

1.3.4 Project Scheduling—Types of Methods

Also known as **activity-on-node** (AON), the precedence diagram method (PDM) can have any of the following four kinds of precedence:

- Finish-to-start (FS)
- Finish-to-finish (FF)
- Start-to-start (SS)
- Start-to-finish (SF)

There are two types of scheduling methods for project schedule analysis; their major distinctions are detailed as follows.

Method 1: PDM. Due to the four precedence relationships, the PDM is better referred to as computer-oriented scheduling because of its use in the type of scheduling that is similar to that done by most popular computer programs.

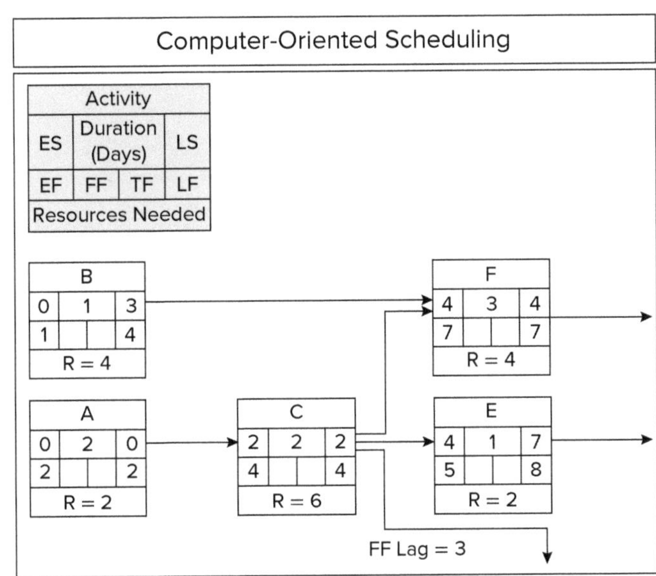

FIGURE 1.7 Precedence Diagram

Method 2: Arrow Diagram. This method of project scheduling is best remembered for its simplicity and ease of use for individuals. An extremely accurate project schedule can be developed by drawing circles and arrows on a page. It is referred to as the people-oriented approach.

- Also known as **activity-on-arrow** (AOA) or activity-on-branch
- May have dummy tasks
- Finish-to-start precedence only

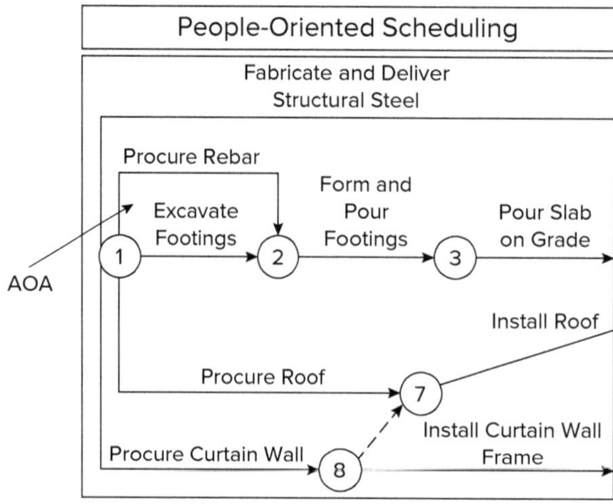

FIGURE 1.8 Activity on Arrow

1.3.5 Lead–Lag Relationships

A **lead relationship** is when the successor task begins before the predecessor task is complete. Very often, scheduling activities are adjusted with a lead influence due to many variables where successor work can start before the predecessor task is 100% complete. The overlap shown in the illustration below identifies the time savings.

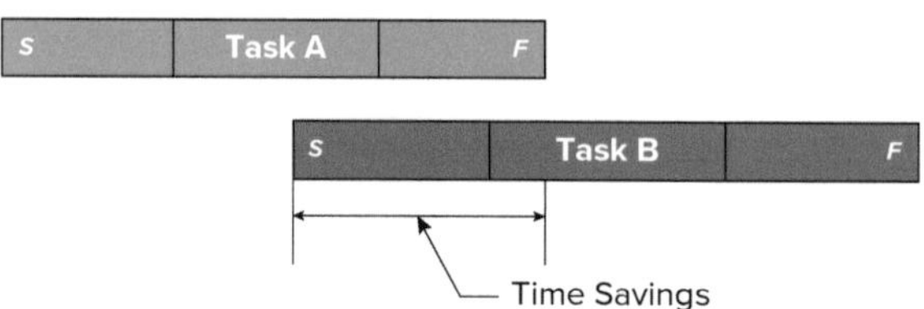

FIGURE 1.9 Lead Relationship

A **lag relationship** is when a successor task does not start immediately upon the completion of the predecessor task. Task A starts and finishes, and then a time lag is introduced where the successor task B starts. An example is when concrete is cast in place into formwork and all work must stop for a duration until the concrete strength develops, which is the result of age hardening.

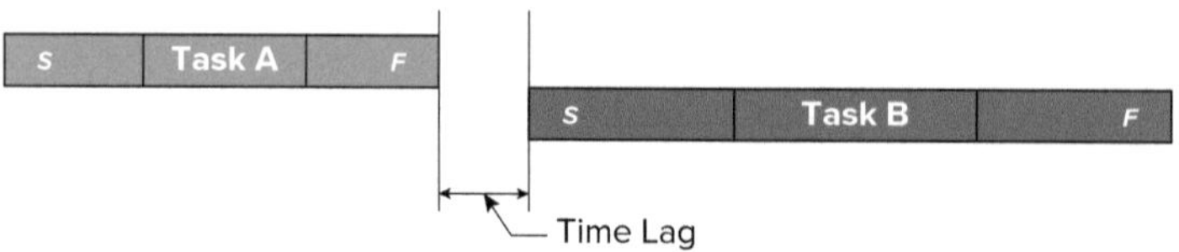

FIGURE 1.10 Lag Relationship

1.3.6 Arrow Diagramming Method

The key elements of an AOA diagram are the following:

- AOA (branch) with nodes as dependencies uses finish-to-start precedence only. AOA problems are solved essentially the same way as activity-on-node (AON, or PDM) problems—with forward and backward passes to determine the earliest and latest activity dates.
- When AOA was developed in the 1950s, a methodology was needed to resolve logic problems in the overall network. This logic conflict was resolved and led to the development of **dummy activities**, which maintain proper logic for various construction activities. If two activities have the same starting and ending events, a dummy activity shows a dependency on either of the two that otherwise cannot be shown by the activity network. The dummy activity provides only logic among the activities when it is used in the network.
- A dummy activity is treated as an activity, represented as a dotted line with an arrowhead indicating the direction of flow on the AOA network diagram. Any activity following the dummy cannot be started until the activity or activities preceding the dummy are completed. The dummy activity does not consume time or resources, but is a valid and calculable path in the network. The dummy activity is only used in the AOA network diagram.

1.3.7 Activity-on-Arrow—Graphic Diagrams

Graphical Precedence Definitions for AOA Network Diagrams

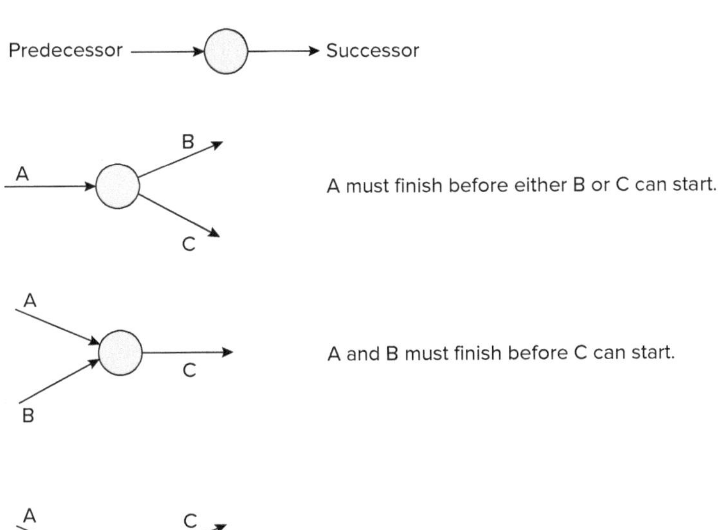

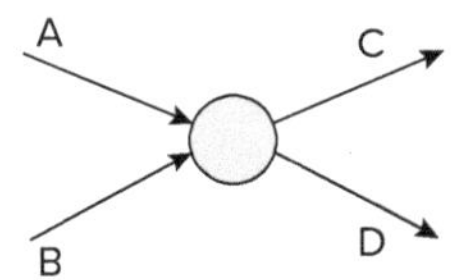

A and B must finish before either C or D can start.

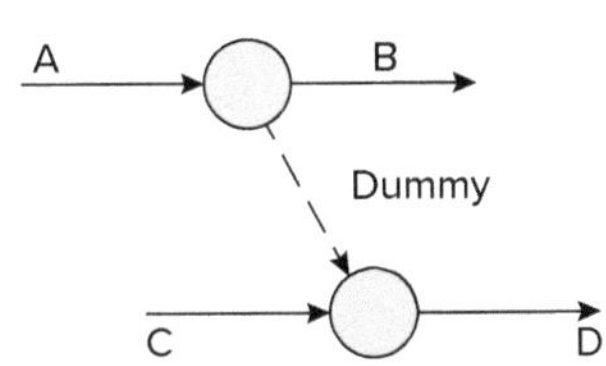

A must finish before B can start; both A and C must finish before D can start due to the dummy activity that is identified as a dashed line.

FIGURE 1.11 Graphical Diagram of AOA Network

> **TIP**
>
> Always focus on the arrowhead when reading the diagram. The arrowhead helps to resolve the number of activities leading to a node and to analyze the critical path through the network.

1.3.8 Activity-on-Arrow—Network Diagram

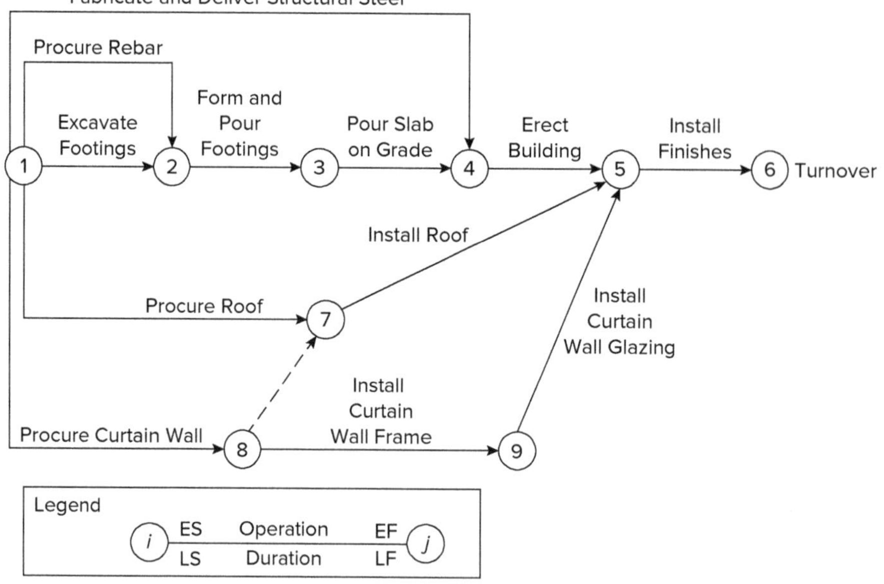

FIGURE 1.12 AOA Network Diagram

The AOA network provides the simplest of scheduling formats, in that all activities are a finish-to-start relationship only. The dummy activity, represented by the dashed line with arrowhead, identifies that there is a relationship between the activities, although there are no resources or time associated with the activity. The network illustrates the importance of procurement and material delivery in the schedule. The procurement component in the network tracks the delivery of the roof and curtain wall, while the dummy activity places a logic link indicating that the two activities must be completed before the installation of the roof can begin.

1.4 ACTIVITY IDENTIFICATION AND SEQUENCING

Example 1.18: Determining the Critical Path

Determine the number of paths and the duration of each, and identify the critical path.

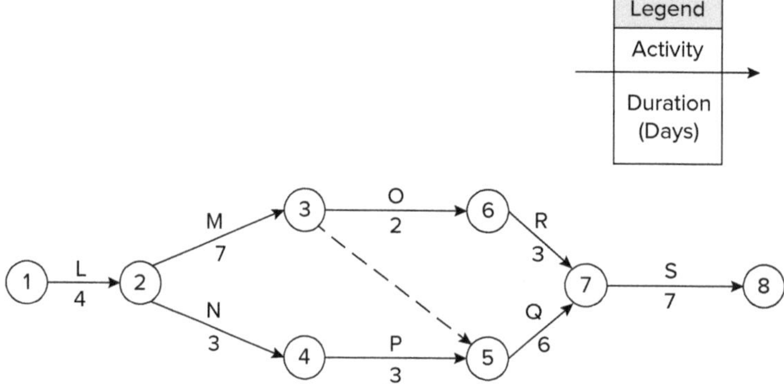

Solution

A critical path is a series of activities that determines the earliest possible project completion date or the longest path through the project network. The project will be completed in 24 days.

PATHS	DURATION	CRITICAL PATH
LMORS	4 + 7 + 2 + 3 + 7 = 23	
LNPQS	4 + 3 + 3 + 6 + 7 = 23	
LM-QS	**4 + 7 + 6 + 7 = 24**	**Critical path**

Answer: The critical path is LM-QS.

Example 1.19: Activity Identification and Sequencing

For the project network shown, the total number of paths through the network, number of critical paths, and project duration (days) are most nearly:

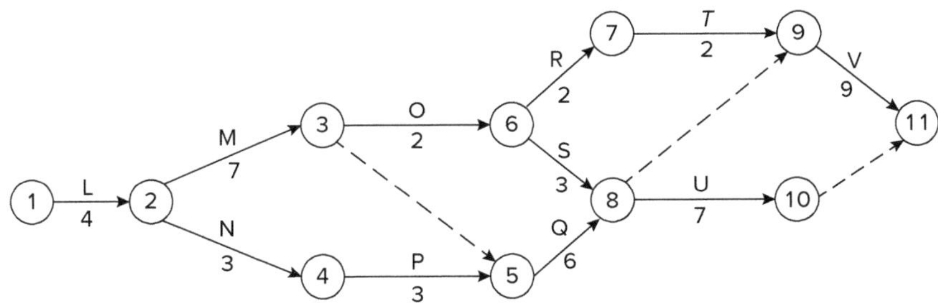

A. 7, 2, and 26
B. 6, 1, and 26
C. 5, 1, and 25
D. 5, 3, and 24

Activity duration (in days)

Solution

List the number of paths and calculate the duration for each.

L M O R T V = 4 + 7 + 2 + 2 + 2 + 9 = 26 days, critical

L M O S U = 4 + 7 + 2 + 3 + 7 = 23 days

L M O S V = 4 + 7 + 2 + 3 + 9 = 25 days

L M Q V = 4 + 7 + 6 + 9 = 26 days, critical

L M Q U = 4 + 7 + 6 + 7 = 24 days

L N P Q U = 4 + 3 + 3 + 6 + 7 = 23 days

L N P Q V = 4 + 3 + 3 + 6 + 9 = 25 days

> **TIP**
>
> The network analysis reveals that there are two critical paths. All paths can be critical, as there is no fixed rule that limits the number of critical paths.

Answer: A

1.4.1 Work Breakdown Structure

A **work breakdown structure** (WBS) is a representation of a workflow organization used to coordinate all of the work necessary to complete a project. Constructed to allow for clear, logical groupings—either by activity or project deliverable—a WBS is arranged in a top-down hierarchy. A WBS typically represents the work identified in the approved project scope of work, and serves as the foundation for schedule development and cost estimating.

The goals of developing a WBS and WBS dictionary are (1) the WBS allows the project team to dynamically and rationally plan out the project to completion, evaluating all process scenarios; (2) the WBS is a repository for the information about work effort that needs to be accomplished within a project; and (3) the WBS structures activities into coherent logical groupings that will lead to the project objectives, starting from initial concept and moving throughout the project life cycle. The WBS and WBS dictionary are not the activity schedule, but rather the concepts for its development. A sample of a WBS is found in the following illustration.

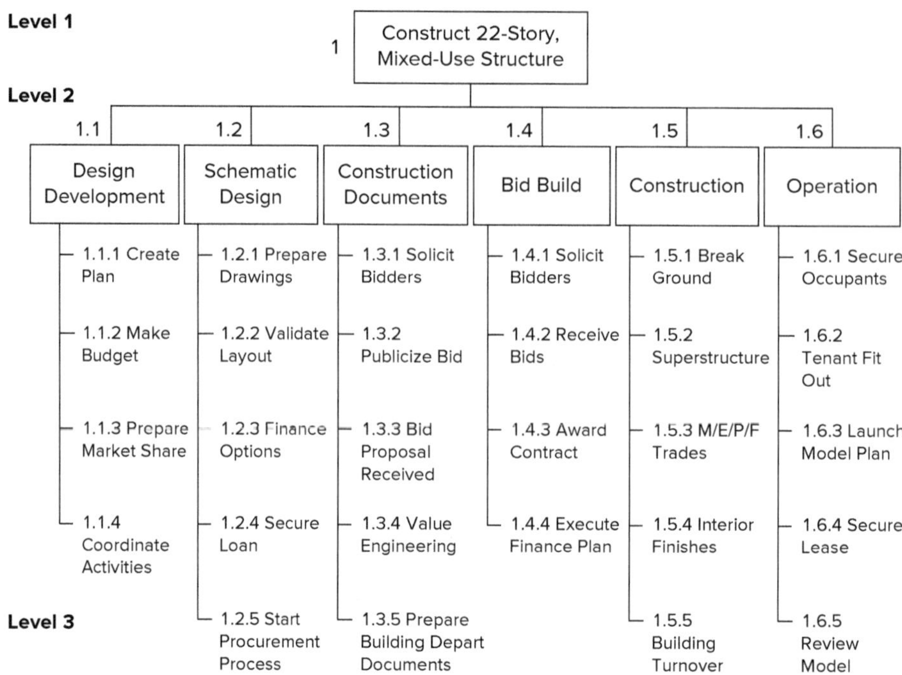

FIGURE 1.13 WBS Hierarchy Example

Example 1.20: Activity Relationships—Precedence Tables

The following activity relationship tables provide the detail for the project network as shown in the diagram. The activity relationship table that provides the best fit to the project network diagram is:

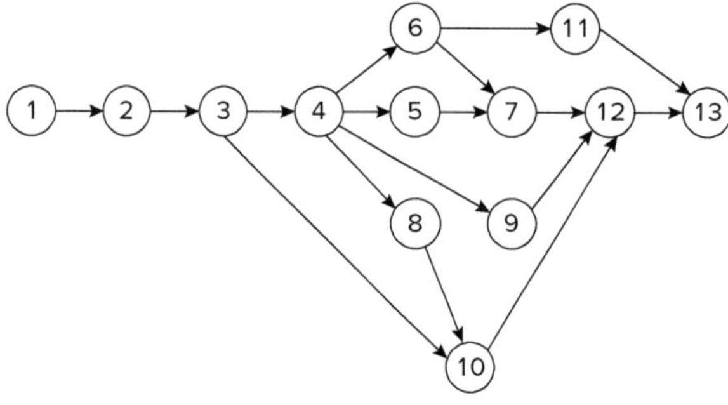

A. 1
B. 2
C. 3
D. 4

Example 1.20 *(continued)*

Project Network Diagram

1:

Number	Activity	Successor
1	Start	2
2	B	3,4
3	C	4,10
4	D	5,6,8,9
5	E	7,8
6	F	7,11
7	G	12
8	H	10
9	I	11,12
10	J	12
11	K	13
12	L	13
13	Finish	

2:

Number	Activity	Predecessor
1	Start	
2	B	1
3	C	2
4	D	3
5	E	4
6	F	4
7	G	5,6
8	H	4
9	I	4
10	J	3,8
11	K	6
12	L	7,9,10
13	Finish	11,12

3:

Number	Activity	Successor
1	Start	2
2	B	3
3	C	4,10
4	D	5,6,8,910
5	E	7
6	F	7,11
7	G	12
8	H	10
9	I	12
10	J	12,13
11	K	13
12	L	13
13	Finish	

4:

Number	Activity	Predecessor
1	Start	
2	B	1
3	C	2
4	D	3
5	E	3,4
6	F	4
7	G	5,6
8	H	4
9	I	4,5,6
10	J	3,8
11	K	6
12	L	7,9,10
13	Finish	11,12

Solution

A project network is a flowchart depicting the sequence in which a project's events are to be completed by showing the events and their dependencies. The question arranges the network diagram and requests selection for the best fit to the given tables. Inspection of the tables identifies either a successor or predecessor relationship. A predecessor is an activity that must be completed (or be partially completed) before a specified activity can begin.

Dependencies are created in project scheduling as the predecessor and successor fields. The predecessor is the task completed prior to the current task, and the successor is the task completed after the current task. Examination reveals that Table 2 is arranged to match the network diagram.

Answer: B

1.4.2 Project Scheduling

Project scheduling: Sequential process to allocate resources to execute all activities in the project

Considerations for evaluating project networks, step by step:

1. Define activities or tasks according to the project objectives in a WBS.
2. Identify an activity as an individual unit of work with a fixed start and a fixed finish.
3. Identify precedence relationships or dependencies (FS, FF, SS, SF).
4. Estimate the time required to complete each task.
5. Draw an AOA diagram inserting dummy activities if required.
6. Apply a critical path analysis to calculate earliest and latest starting times, earliest and latest completion times, float (slack) times, and critical path.
7. Develop a Gantt chart, or a graphical view of the project to quickly analyze progress.
8. Continuously monitor and revise the time estimates along the project duration to keep the project on target.

Total float or total flexibility: The total float is the time between an activity's early finish time and late finish time as determined by the calculations for the network diagram. The equation for total float is either the late start minus the early start (LS – ES), or the late finish minus the early finish (LF – EF).

During the network analysis, (1) all of the linked paths are enumerated; (2) the duration is calculated; and (3) the longest path or the earliest time of project completion is derived to determine the critical path.

The critical path is the path with all linked activities that from FS flow, defined through the network and the path's activities, have no float or flexibility in their start or finish time, or the total float is calculated as zero.

A delay in an activity on the critical path with a total float equaling zero will delay the total project, day for day. If an activity in a network has a negative total float (or is behind schedule), then the entire project is delayed by that amount. An activity in a network when the total float is a positive duration indicates that the activity has flexibility in the start and finish time. Remember that total float has an effect on the total project duration.

TIP

As a general rule, activities on the critical path will have a free float of zero, which matches the activities' total float of zero.

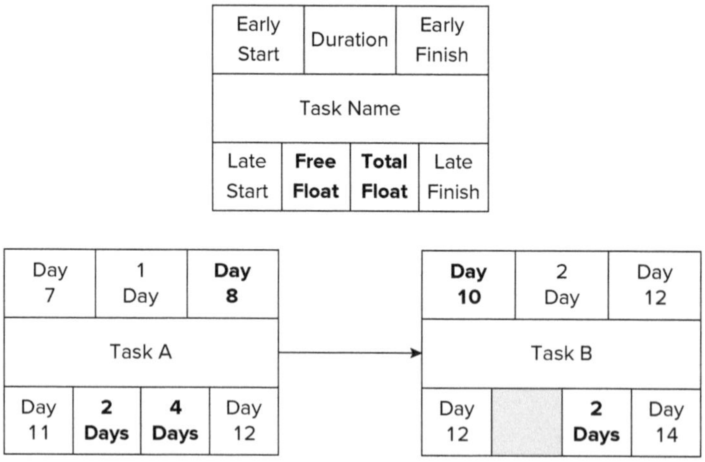

Free float is an evaluation performed on a network activity that shows the amount of time a single activity can be delayed without delaying the early start of its successor activity. For example, activity A and its successor B are evaluated, and activity A's early finish is subtracted from the successor activity B's early start time. The results are either a zero free float, indicating that activity A must start on the planned time, or a positive number, indicating that activity A can vary its start without impacting the start time of activity B's early start time. When reviewing free float, activity B's early start time is not affected by activity A's early start time.

Early Start	Duration		Early Finish
	Task Name		
Late Start	**Free Float**	**Total Float**	Late Finish

Day 7	1 Day		**Day 8**
	Task A		
Day 11	**2 Days**	**4 Days**	Day 12

→

Day 10	2 Day		Day 12
	Task B		
Day 12		**2 Days**	Day 14

FIGURE 1.14 Example of Total Float and Free Float

Calculating free float example:

$$\text{Free float}_{\text{Task A}} = \text{ES}_{\text{successor}} - \text{EF}_{\text{Task A}} \quad \textbf{Equation 1-13}$$

$$\text{Free float}_{\text{Task A}} = \text{day } 10 - \text{day } 8$$

$$\text{Free float}_{\text{Task A}} = 2 \text{ days}$$

In the example, task A has the ability to start on day 7 or day 8 without affecting the early start of task B; task B will start on day 10 with or without the delay of starting task A. Note that in this example, there is no successor to task B, and therefore there is no free float calculation.

Calculating total float example:

Use the equations: late start minus early start (LS – ES), or late finish minus early finish (LF – EF)

Task A

Total float = LS – ES

Total float = day 11 – day 7

Total float = 4 days

Task B

Total float = LS – ES

Total float = day 12 – day 10

Total float = 2 days

Critical Path

- A critical path is the series of activities that determines the earliest possible date of project completion or the longest path through the project network.
- A critical path is defined as those activities with zero float or slack in a project network.
- **Gantt chart:** Developed by Henry Gantt in the 1910s, a Gantt chart represents a visual horizontal bar chart that shows the activities in a project schedule. Gantt charts depict individual activity descriptions, durations, and start and finish dates of a project. A Gantt chart shows a graphical depiction that enables one to quickly evaluate the project's progress.

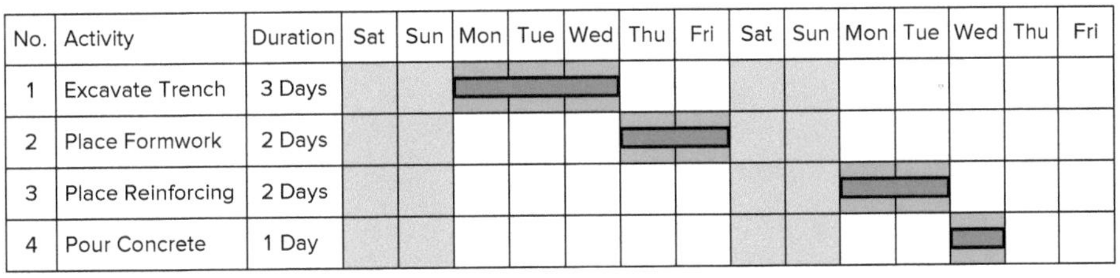

No.	Activity	Duration	Sat	Sun	Mon	Tue	Wed	Thu	Fri	Sat	Sun	Mon	Tue	Wed	Thu	Fri
1	Excavate Trench	3 Days														
2	Place Formwork	2 Days														
3	Place Reinforcing	2 Days														
4	Pour Concrete	1 Day														

FIGURE 1.15 **Example Gantt chart.** The orange bar in the blue progress bars represents critical path activities. If an activity on the critical path is delayed, then the total project duration is impacted and delayed day by day.

Critical Path Method—Essentials

- The critical path method is used to predict the duration of a project by analyzing sequences of activities to determine which one has the least amount of schedule flexibility (float).
- Early dates are calculated by a forward pass using a specified start date.

> **TIP**
>
> There are many sign conventions used to display information in schedule analysis. Always interpret the project schedule using the legend provided with the diagram. A sample of a legend is shown at right with all of the relevant information needed for analysis.

- Late dates are calculated by a backward pass using a specified completion date (usually the early finish date).
- Start date and calculated early finish and late finish dates are generated for each activity.

<table>
<tr><td>Early Start</td><td>Duration</td><td>Early Finish</td></tr>
<tr><td colspan="3">Task Name</td></tr>
<tr><td>Late Start</td><td>Float</td><td>Late Finish</td></tr>
</table>

FIGURE 1.16 Legend Sample

1.4.3 Calculating Forward Pass

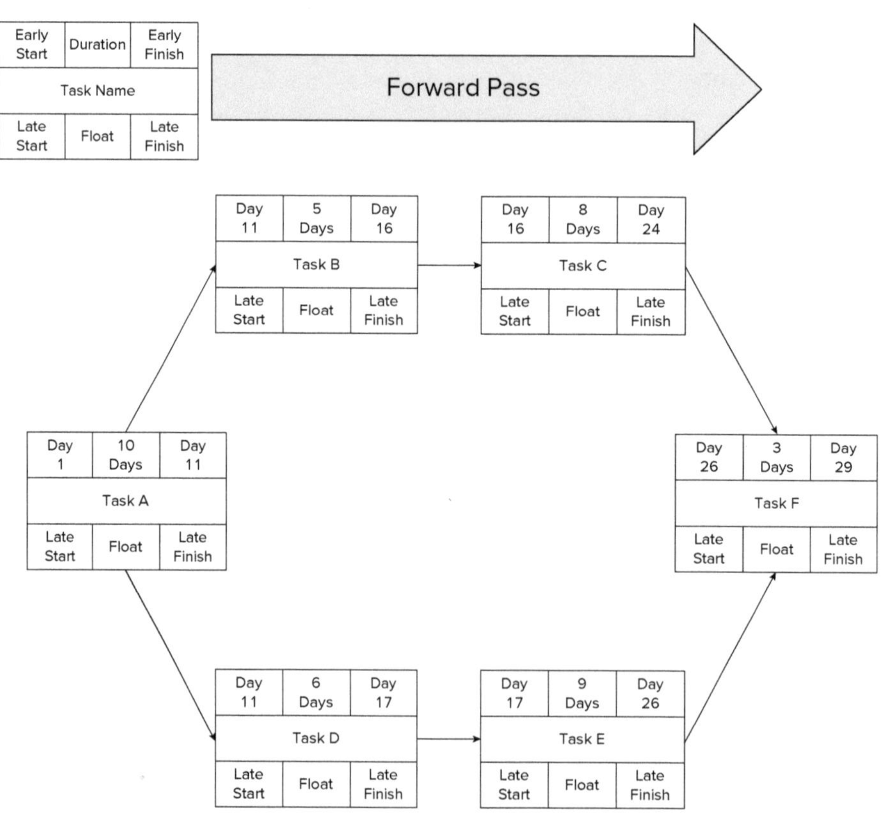

FIGURE 1.17 Calculating Forward Pass

The forward pass moves from left to right, calculating when tasks can end using their early start date and the expected duration. Use the equation EF = ES + duration. The objective is to calculate the longest path through the network. When a choice needs to be made, such as in task C and task E, always use the larger value for the successor task, task F.

The critical path is A–D–E–F, and the total duration of the project is 29 days.

1.4.4 Calculating Backward Pass

The early finish for the forward pass is also the late finish for the project. In the backward pass, we move from right to left using the late finish and the duration to determine the late start. Use the equation LS = LF – duration. When a choice needs to be made, such as in task B and task D, always use the smaller value for the following task, task A. The purpose of the backward pass is to find the late start and prepare for the total float calculations.

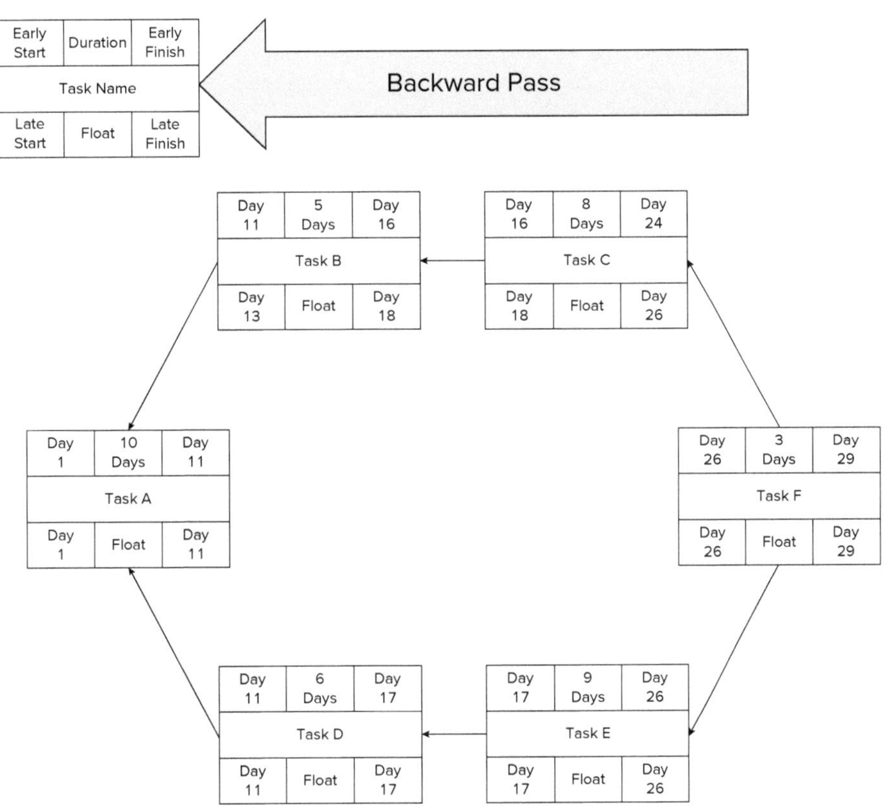

FIGURE 1.18 Calculating Backward Pass

1.4.5 Calculating Total Float

Float can be calculated by finding the difference between the early start and the late start. Activities on the critical path are known to have zero float, by definition, and the tasks A-D-E-F have zero float. In this case, only tasks B and C have positive float, two days in both cases. However, if task B is two days late in starting, task C loses its float. Float is a property of a network fragment. When calculating the forward and backward pass, there was a structured pattern to follow (that is, moving from the left or right). During the calculations for total float, the equation steps do not need to follow a pattern.

Use the most common format of the equation:

Float = LS – ES

or

Float = LF – EF

Early Start	Duration	Early Finish
	Task Name	
Late Start	Float	Late Finish

Calculating Float

Day 11	5 Days	Day 16
	Task B	
Day 13	2 Days	Day 18

Day 16	8 Days	Day 24
	Task C	
Day 18	2 Days	Day 26

Day 1	10 Days	Day 11
	Task A	
Day 1	0 Days	Day 11

Day 26	3 Days	Day 29
	Task F	
Day 26	0 Days	Day 29

Day 11	6 Days	Day 17
	Task D	
Day 11	0 Days	Day 17

Day 17	9 Days	Day 26
	Task E	
Day 17	0 Days	Day 26

FIGURE 1.19 Calculating Total Float

Example 1.21: Resource Sequencing

An activity and relationship table is provided for the network diagram.

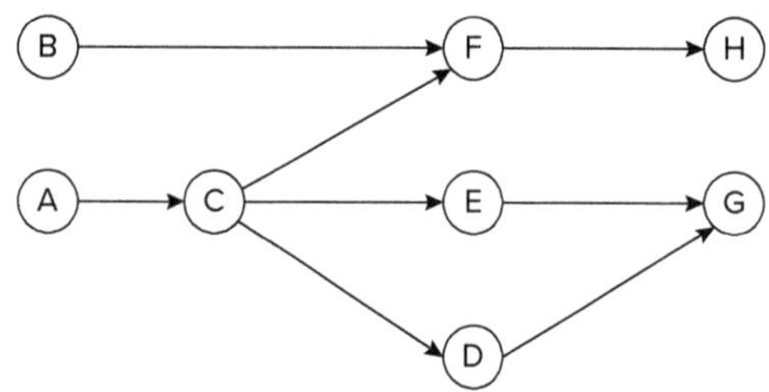

ACTIVITY	SUCCESSOR	DURATION DAYS	WORKERS PER DAY
A	C	2	2
B	F	1	4
C	F, E, D	3	6
D	G	2	4
E	G	1	4
F	H	3	4
G	Finish	1	2
H	Finish	2	4

The workday when the maximum number of workers is present is:

A. 3
B. 6
C. 7, 8
D. 12

Solution

Determine the number of paths and the critical path duration.

Example 1.21 *(continued)*

BFH $= 1 + 3 + 2 = 6$

ACFH $= \mathbf{2 + 3 + 3 + 2 = 10}$ **Critical Path**

ACEG $= 2 + 3 + 1 + 1 = 7$

ACDG $= 2 + 3 + 2 + 1 = 8$

Extend the activity table to determine the number of workers needed per day. Worker usage can be accumulated and presented in the table format below. Identify the critical path to build confidence in the solution. The columns are totaled to show that on day 6, the maximum number of workers is needed, which is 12.

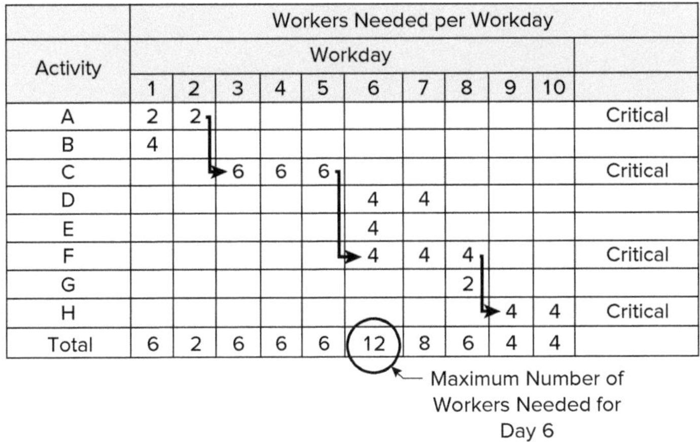

Activity	Workers Needed per Workday										
	Workday										
	1	2	3	4	5	6	7	8	9	10	
A	2	2									Critical
B	4										
C			6	6	6						Critical
D						4	4				
E						4					
F						4	4	4			Critical
G								2			
H									4	4	Critical
Total	6	2	6	6	6	12	8	6	4	4	

Maximum Number of Workers Needed for Day 6

Answer: B

1.4.6 Resource Leveling

Resource leveling is a project management tool used to examine a project for an unbalanced use of resources over time, and for resolving over-allocations of planned resources or unintended resource conflicts.

When originally performing project planning activities, the project manager will schedule certain activities to start simultaneously to fit the overall project schedule. However, availability of resources may not be known during the initial planning phase. An imbalance may occur regarding the availability of materials, workers, or equipment. The activities will need to be rescheduled to manage the newly found project constraint. Limitations placed on the allocation of one or more resources are defined as **constraints**. An example of a constraint can be as broad as time, scope of work, or money.

TIP

Resources are often denoted as the labor, material, and equipment needed for a construction project. Resources also include the nontangible items, such as money, budget, risk, teamwork, timelines, and the like.

Project planning resource leveling is the process of resolving activity scheduling conflicts through a reorganization of the project schedule activities. Resource leveling can also be used to balance the resources over the short- or long-term constraints.

The project manager's implementation of resource leveling requires delaying activities until resources are available without impacting the overall project schedule. However, an important rule to follow when leveling resources is to avoid delaying tasks that are on the critical path; otherwise, a delay in the project finish date could be a result. The overall objective of resource leveling is to analyze the activities' free float or total float and use their schedule flexibility in order to avoid delaying the project. When implemented, resource leveling can take the project's scope of work and balance it against the available resources to fit them within the given duration.

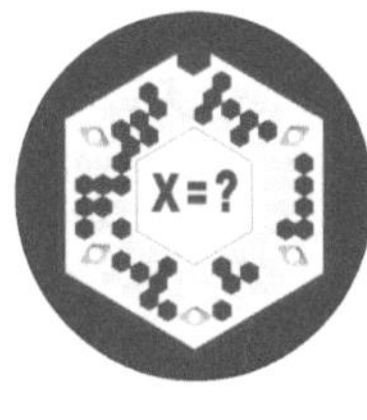

Examples 1.22–1.23: Resource Leveling

The project network with activities A, B, and C, along with their durations, are shown. Activity A has three days of total float, and activity C has two days of total float. Activity A requires two workers, B requires four workers, and C requires two workers. Apply the project network to the following:

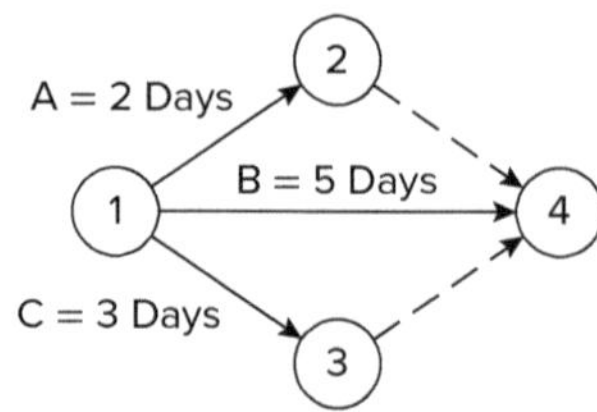

Example 1.22

If all of the activities start on day 1, the number of workers needed on day 4 is most nearly:

A. 2
B. 4
C. 6
D. 8

Example 1.23

If activity C is delayed 2 days, its total float, the number of workers needed on day 4 is most nearly:

A. 2
B. 4
C. 6
D. 8

Solution

Calculate the number of paths, the duration of each, and the total float for each. Since the schedule uses activities that have total float to adjust for the resource usage, the delay of activity C by two days does not impact the schedule duration and allows for six workers every day for the total duration; thus, it is level. The initial schedule has an imbalance in the daily use of workers; thus, it is not level.

PATH	DURATION	CRITICAL	TOTAL FLOAT
A	2		$5 - 2 = 3$
B	5	Critical	$5 - 5 = 0$
C	3		$5 - 3 = 2$

Construct histograms to evaluate the individual questions.

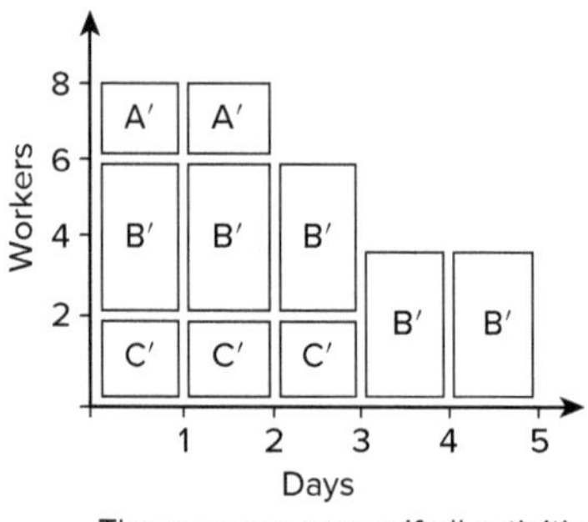

The resource usage if all activities start on day 1

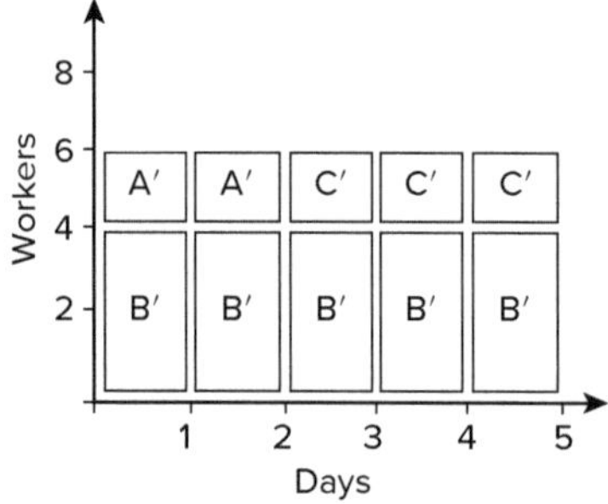

The resource usage if activity C is delayed 2 days (its total float)

Answer: B (1.22), C (1.23)

REFERENCE

1. ACI (American Concrete Institute) Committee 318. *ACI 318-14: Building Code Requirements for Structural Concrete*. Farmington Hills, MI: ACI, 2014.

Means and Methods

2

George A. Stankiewicz, PE

CONTENTS

EXAM GUIDE

II. Means and Methods

A. Construction loads
B. Construction methods
C. Temporary structures and facilities

Approximate Number of Questions on Exam: 3

NCEES Principles & Practice of Engineering Examination, Civil Breadth Exam Specifications

COMMONLY USED ABBREVIATIONS

ABC	aggregate base course
A	area
i	hydraulic gradient (H/L)
k	permeability
MGD	million gallons per day
OSHA	Occupational Safety and Health Administration
psf	pounds per square foot
Q	flow quantity

INTRODUCTION

Some common considerations when exploring the means and methods of construction for a particular project are highlighted in this chapter. The leading consideration for a building construction site is life safety.

The means of construction comprise the equipment that is used to accomplish a task. The methods of construction are the techniques or procedures used to accomplish the construction of a structure. The phrase means and methods of construction is commonly used to help define the lines of responsibility for the construction contractor. The means and methods of a construction project, the contractor's responsibility, lead to the contractor's overall sequencing, performance techniques, and procedures of construction. In addition, temporary strategies required to construct the project must guarantee life-safety precautions. There is no reliance on the owner or designers associated with the project.

During construction, the contractor's shoring and formwork cost can exceed the cost of the concrete and reinforcing steel combined. The design of safe and economical shoring is essential to a successful project where the contractor's means and methods will prevail. The contractor also depends on the falsework used for the temporary works associated with the casting of concrete structures. The purpose of concrete falsework is to support the structure true to line and grade until it has gained sufficient strength to be self-supporting.

The contractor's responsibility to the owner is that the construction project meets the governing codes such as the International Building Code (IBC) and all of the included references to the construction standards, such as those of the American Concrete Institute (ACI), American Institute of Steel Construction (AISC), American Society of Civil Engineers (ASCE), ASTM International, American National Standards Institute (ANSI), and so on. These construction standards uniformly set the guidelines for quality assurance for the owner.

2.1 CONSTRUCTION LOADS

Gravity and lateral loads are primary considerations in any building design because they define the nature and magnitude of the external forces that a building must permanently resist during its life cycle. Construction loads are defined by the loads imposed on a partially completed or temporary structure during, and as a result of, the construction process. Construction loads are the materials, personnel, and equipment imposed on the temporary or permanent structure during the construction process.

Construction loads impose life-safety hazards until the structure is complete and ready for occupancy. A principal construction material used in a variety of applications in all construction projects is concrete, an age-hardening product. At the time of placement, cast-in-place concrete is liquid and has zero strength. As time progresses, concrete gains strength due to the hydraulic bond between the portland cement and fine/coarse aggregates, which is activated chemically by water. The compressive strength of concrete is confirmed and measured 28 days after placement to meet the requirements of the building code for evaluating its strength. During the age-hardening process, precautions are needed to ensure that the design strength is maintained throughout the structure and its connecting elements.

Similarly, rigid frame structures (such as steel or wood construction) need all of the structural elements placed and connected to allow gravity and lateral load paths to be fixed and achieve static equilibrium. The frame structure, in the configuration of beams and columns, cannot resolve the lateral loads until all of the diaphragm elements and shear walls are connected. Until then, the structure has the highest life-safety risks for construction workers.

2.1.1 Mechanical Properties of Materials

The importance of construction loads, the mechanical properties of materials, and the ways in which load analysis impacts the assembly of a structure are the focus of this section.

All properties of materials have a unique stress-strain relationship where each can be described using the terms found below. An exclusive stress-strain curve is developed for each material as it is subjected to deformation (strain = stretch) at distinct levels of compressive or axial loading (stress = force).

Brittle materials break without any plastic deformation or stretch, and fracture readily without absorbing any energy. Examples include glass and most ceramics.

Ductile materials deform before they fracture. They are often referred to as pliable, supple, springy, and soft. Examples of ductile materials are aluminum, tin, copper, mild steel, and lead. Ductile materials can be stretched without breaking and drawn into thin wire lines.

The **proportional elastic limit** is the maximum stress that can be applied without causing permanent deformation. Once the force is released, the material's strain is relieved, and the material returns to its original form.

The **modulus of elasticity** is the ratio of the applied stress to strain exhibited within the proportional limit of a material, also known as Young's modulus. Young's modulus is evaluated in either tension or compression.

Poisson's ratio is the mathematical analysis of the measure of the Poisson effect: if a material is compressed in one direction, it expands in the other two directions; on the other hand, it contracts if the material is stretched. It is the ratio of the amount of lateral strain of a material to the longitudinal strain when an axial force is applied within the proportional limit and is used in equations associated with strength of materials. Poisson's ratio of lateral strain to the axial strain for most common metals is approximately 0.0 to 0.5. The ratio is often described as being similar to when a rubberband is being stretched. When the longitudinal force is applied, the rubberband becomes thinner. When the force is removed, the rubberband returns to its original shape.

Strain is the response to an applied load that allows material to be displaced from its original dimension. The increase in strain predicts the amount of stress or force on a material. **Stress** is a measure of force per unit area acting on a material in three observable actions. **Tensile stress** is observed on a material when force is added, and the material stretches or increases in length. When a force is acting to shorten or decrease its length, it is known as **compressive stress. Shear stress** is a force that causes sections or layers to move in opposite directions. The maximum point in a linear relationship where the stress is proportional to the strain is known as the **proportional limit**.

Toughness, often related to ductility, is a material's ability to absorb impact or shock load in the plastic range without deforming or fracturing. Toughness is the material's ability to bend and its resistance to fracture when stresses are applied. The characteristic of the material is that it will deform, bend, and stretch rather than break. Toughness relies on a combination of strength and ductility.

The **ultimate strength** of a material is the maximum force that will cause complete failure or the breaking strength of the material. The **yield point** is the point on the material's stress-strain relationship where there is an increase in stretch without an increase in force. This is the point where the material is permanently deformed. Yield strength is a point on the stress-strain relationship where the maximum force can be applied without permanent deformation.

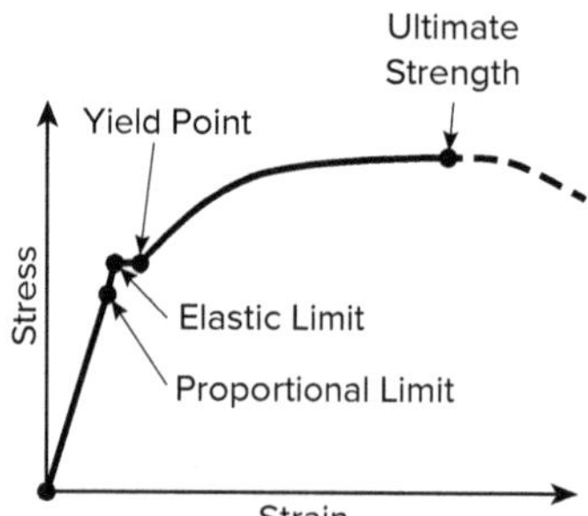

FIGURE 2.1 Stress-Strain Relationship for Steel

The evaluation of the stress-strain relationship for steel is found in the example of a wire gauge experiment. Using a thin-gauge, straight, steel wire, the proportional/elastic limit of the stress-strain relationship is observed when the wire is bent between two fingers with just enough force for the wire to rebound to its original straight shape when the force is removed. The yield point is exhibited when a greater force is applied and the wire will not rebound, but has become permanently deformed. Next, the ultimate strength is demonstrated by placing the wire in an apparatus and applying an axial load of approximately 5,000 lb, thereby pulling the wire apart. Steel's modulus of elasticity is 29 million pounds per square inch, with an ultimate strength of 58,000 psi. The strength is found when the wire is measured using the cross-sectional area.

The next action is to bend the wire back and forth until the wire breaks. This observation exhibits the phenomenon known as fatigue. The cyclic bending action broke the wire, which was caused by the microcracks in the wire moving further away, causing the cracks to grow and separate. All materials have microcracks, and in engineering, cyclical loading on materials is accounted for during the design of systems.

Example 2.1: Strength of Materials

The breaking strength of a material is also known as its:

A. ultimate strength.
B. yield point.
C. proportional limit.
D. elastic limit.

Solution

The solution to this question is best understood by a review of the stress-strain curves for the mechanical properties of materials, as shown in Figure 2.1. As the stress or force increases, the strain or stretch of the material increases. The figure represents steel, but all materials have their own developed stress-strain curves. The ratio of stress to strain within the elastic range is the modulus of elasticity, or Young's modulus, *E*. This modulus is consistently about 29,000,000 psi for all structural steels. Generally, the ultimate strength varies dependent on the steel's chemical composition. For example, the ultimate or breaking point for ASTM A36 steel is in the range of 50,000 to 80,000 psi.

Answer: A

Example 2.2: Simple Beam Analysis

The net result on the amount of bending moment and shear force when compared between a uniformly distributed load on a simple beam and an equivalent concentrated load on a simple beam is most nearly:

A. 0.25
B. 1
C. 2
D. 2.5

Solution

This example is presented as a non-quantitative question as it describes the potential effect of the net forces but requires a numerical response. The example can be solved by applying the structural beam equation for a simple beam having a uniform load and a concentrated load. Typically, during construction, loads are moved along a structure and the movement develops differing load and moment reactions. Determine the shear and moment, and develop the potential of three cases where the net effect can be applied to find an answer.

Compute the reaction force (R_{Left} and R_{Right}) for a simple beam in three derived cases (see case figures below). Calculate using the equations where w = weight per length, P = load, l = length, and a and b are equal to the length from the left and right reaction force, respectively. R_{Left} and R_{Right} describe the location of the reaction force positioned on the ends of the beam.

Case 1: $\frac{wl}{2}$; Example: $\frac{800\text{ lbf/ft} \times 4\text{ ft}}{2} = 1{,}600\text{ lbf}$

Case 2: $\frac{P}{2}$; Example: $\frac{3{,}200\text{ lbf}}{2} = 1{,}600\text{ lbf}$

Case 3: $R_{\text{Left}} = \frac{pa}{l}$; Example: $\frac{3{,}200\text{ lbf} \times 3.5\text{ ft}}{4} = 2{,}800\text{ lbf}$

$R_{\text{Right}} = \frac{Pb}{l}$; Example: $\frac{3{,}200\text{ lbf} \times 0.5\text{ ft}}{4} = 400\text{ lbf}$

The moment reaction is calculated and adjusted for units using inch-lbf for each case by calculation:

Case 1: $\frac{wl^2}{8}$; Example: $\frac{800\text{ lbf/ft}}{12\text{ in/ft}} \times \frac{48\text{ in}^2}{8} = 19{,}200\text{ in-lbf}$

Example 2.2 *(continued)*

Case 2: $\frac{Pl}{4}$; Example: $\frac{3{,}200 \text{ lbf} \times 48 \text{ in}}{4} = 38{,}400 \text{ in-lbf}$

Case 3: $\frac{Pab}{l}$; Example: $\frac{3{,}200 \text{ lbf} \times 6 \text{ in} \times 42 \text{ in}}{48 \text{ in}} = 16{,}800 \text{ in-lbf}$

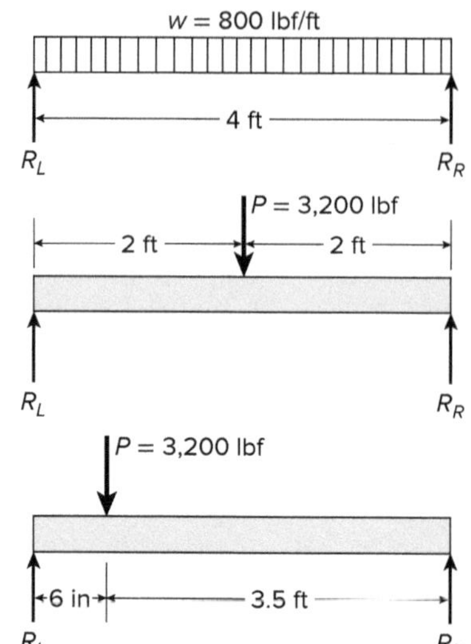

Simple Beam—Uniform Load
Case 1: The left and right reaction forces are equal.

Simple Beam—Concentrated Load
Case 2: The left and right reaction forces remain equal.

Simple Beam–Concentrated Load on Any Point
Case 3: The reaction force nearly doubles when the load is closer to the maximum point of shear.

LOADING CONDITION	CASE 1	CASE 2	CASE 3
Reaction or Shear at R_{Left} (lbf)	1,600	1,600	2,800
Reaction or Shear at R_{Right} (lbf)	1,600	1,600	400
Maximum Bending Moment (in-lbf)	19,200	38,400	16,800

In conclusion, three case evaluations of the force locations along the beam demonstrate that both the shear and moment can be doubled when compared to a uniform load and concentrated load on the beam.

Answer: C

2.1.2 Construction Loads—Concrete Placement

According to the ACI [1], form dead loads are the actual weight of the forms, plus the weight of fresh (that is, wet) concrete. Form live loads are the weight of workers, equipment, and material storage. The minimum live load is 50 pounds per square foot (psf). Wet concrete is like water—it exerts a lateral pressure, which increases with the depth of the form casting a triangular load distribution, similar to any granular material. The lateral load on vertical formwork is used to compute the strength of the formwork and has no bearing on the characterization of load paths traveling through the structural system of a structure.

Example 2.3: Concrete Placement

The placement of wet concrete cast in place during construction is considered a:

A. live load.
B. dead load.
C. construction load.
D. lateral load.

Answer: B

2.1.3 Crane Lifting

When computing force vectors, the equations describe an applied quantity with a magnitude, direction, and angle of application. Force is a vector quantity that is measured by the amount of pressure, tension, weight lifted, and strain, as well as the direction in which the force is being applied.

Example 2.4: Crane Lifting

The selected rigging technique is to use wire-rope slings that are attached to a spreader beam to lift a 40,000 lbf heating, ventilation, and air-conditioning (HVAC) unit to the rooftop dunnage supports. The configuration of the hoisting equipment for the crane pick is shown in the figure below. The total rated tension (lbf) capacity in sling B is most nearly:

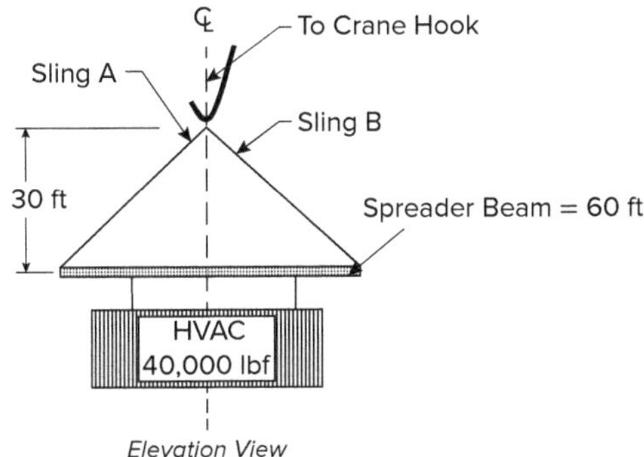

A. 14,142 lbf
B. 20,000 lbf
C. 28,280 lbf
D. 40,000 lbf

Solution

The crane has the capacity to lift the gravity weight of the 40,000-lb weight. The center of mass of an object is found at the center of gravity of an object being lifted. It is a theoretical point where the entire mass of an object is concentrated.

Using the centerline as shown and recognizing it to be the center of gravity ensures that forces can be reconciled through the rigging configuration. The mechanical principle behind the spreader beam is that it is identified to be infinitely stiff and is not affected by the forces being transferred through it. Therefore, the left and right reaction forces can be computed using simple beam analysis on the spreader beam. Once the vertical force vectors are found, the tension force in sling B can be computed using the angle of application. The geometry of the shape identifies the sling angle as a horizontal-to-vertical 1:1 relationship (30-ft wide by 30-ft high) or a 45° angle to the horizon.

A free-body diagram will help with the analysis. Sketch a free-body diagram of the force vectors represented in the problem statement and solve for the tension in sling B.

Free-Body Diagram

The calculations to find the resultant force using the vertical vector are as follows:

$$R_{\text{right}} = \frac{40{,}000 \text{ lbf}}{2} = 20{,}000 \text{ lb}$$

$$\text{Tension}_{\text{B}} = \frac{20{,}000 \text{ lbf}}{\sin 45^{\circ}}$$

$$\text{Tension}_{\text{B}} = 28{,}284.27 \text{ lbf}$$

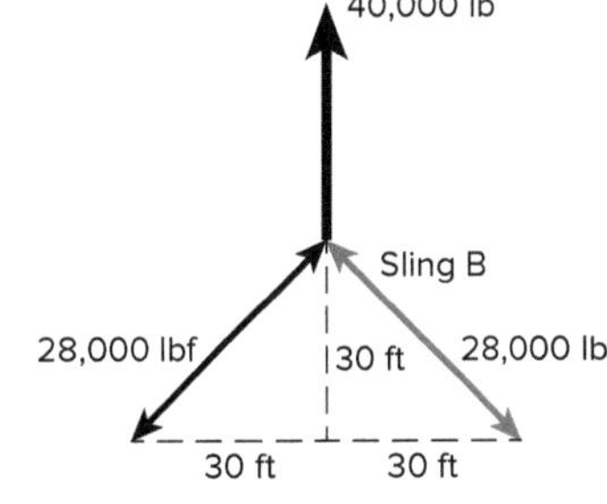

Answer: C

TIP

Notice the coincidence between the centerline of the symmetrical loading and the center of gravity. The center of gravity is the point at which a force may be applied to cause linear energy in lifting an object without any angular distortion.

2.1.4 Rigid Body in Equilibrium

In order for a rigid body to stay in equilibrium, the net force and the net moment on the object must both be zero. Usually, the analysis is performed with the motion of objects in two dimensions (*xy*), in which case the conditions for static equilibrium are:

$$\Sigma F_x = 0, \quad \Sigma F_y = 0 \qquad \textbf{Equation 2-1}$$

In physics, a material body can be considered to consist of a very large number of particles. Physics tends to analyze the particles of the body. However, engineering mechanical analysis defines a rigid body as one that does not deform; that is, the distance between the individual particles making up the rigid body remains unchanged under the action of external forces. The body is a whole and an indivisible singular unit. Thus, a rigid body under the action of a force tends to rotate about an axis as a singular unit. In order for a body to be at rest, mathematically both the resultant force is zero and the forces acting on a body do not let it rotate.

Conditions for a Rigid Body in Equilibrium

In comparison to the forces on a distinct particle, the forces on a rigid body are usually not concurrent and may cause the body to rotate due to the moments created by the forces.

Due to these physical factors, an indeterminate solution is the conclusion in resolving the force vectors.

For a rigid body to remain in equilibrium, the net force as well as the net moment about any reference point O must be equal to zero. Therefore, the solution can be determined using the equations of static equilibrium, namely:

$$\Sigma F = 0 \text{ and } \Sigma M_O = 0 \qquad \textbf{Equation 2-2}$$

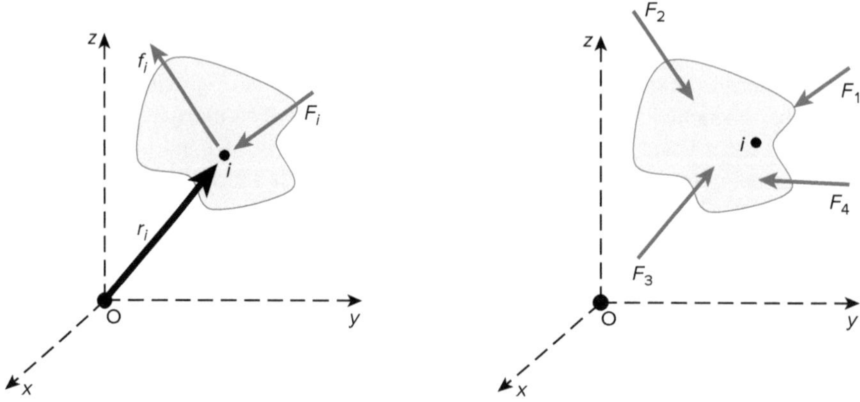

FIGURE 2.2 Idealization of Indeterminate (left) and Determinate Forces (right)

2.2 CONSTRUCTION METHODS

In preparation for construction, or earthwork, the first action taken by a contractor on the project site is to clear the right of way. Next is the control survey, which, when completed, locates the legal limits of construction. Then, the earthwork begins with the removal and disposal of all vegetation and debris. Often, the removal and disposal of structures, underground structures, contaminated soil, pavements, abandoned utilities, and the like are performed during the initial stages for preparation of the construction project, according to the contractor's means and methods.

2.2.1 Construction Methods—Earthwork Equipment

The methods used in earthwork construction require the contractor to optimize both the swelling and compaction behavior of fill material. Compaction is a soil densification process achieved by the application of mechanical work energy, which results in improving the structural engineering properties of soils.

It is essential to control the soil's compaction properties to achieve the most efficient and cost-effective earthwork method, which is always influenced by the dry density and water content of the fill materials. The ideal condition during placement is to match the fill material in-place unit weight as closely as possible to the maximum dry unit weight. However, this condition rarely exists, and the compaction effort varies in time and cost due to the properties of the soil swell/shrinkage factors.

The swell/shrinkage factors of soils are the geotechnical properties exhibited by individual soils, and are discussed in chapter 3. This section focuses mainly on the placement and management of soils and earthwork. Soils can be categorized based on (1) particle size classifications (either fine or coarse grained); (2) gradation (well or poor); and (3) plasticity index, which is the primary variable in evaluating swell/shrinkage factors.

The soil's water content is a significant factor influencing the swell/shrinkage of both fine- and coarse-grained soils. The contractor's earthwork equipment becomes the main factor in meeting the specification requirement for structural fill. As a vibrating roller compactor applies energy to loosely placed soil, the air voids and moisture in the soil compact to a maximum level for optimum soil/moisture content. The following earth compaction graph illustrates the maximum achievable density for the compacting effort. The *y*-axis shows the percent dry density (PCF), and the *x*-axis shows the percent moisture. The natural water content of soil is critical to the compaction; soil that is either too wet or too dry evades the compaction effort and its effect on the swell/shrinkage factor.

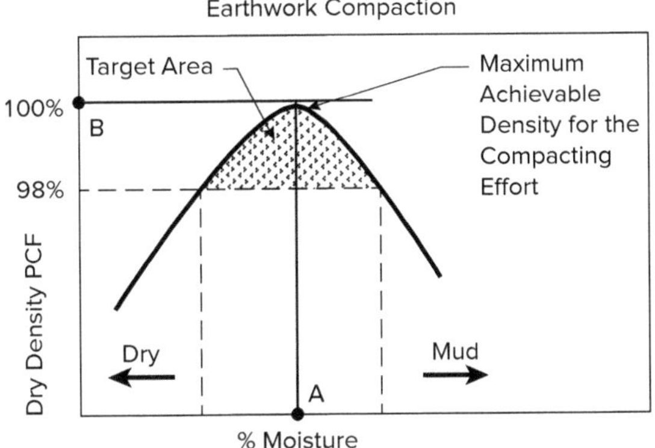

FIGURE 2.3 **Soil Density and Moisture Content Relationship for Optimal Compaction**

Maximum density is found at the intersection of points A (optimum moisture content) and B (optimum dry density).

Moisture in the soil is essential for proper compaction. Water serves as a lubricant within soil, allowing the particles to slide together. Too little moisture (dry soil) results in inadequate compaction—the particles cannot move past each other to achieve compaction and densification. Too much moisture (mud) leaves water-filled voids and weakens the load-bearing capacity.

During the process of machine compaction, most soils require a range of moisture content for proper densification to fit the geotechnical material properties and the optimum unit weight. The drier the soil, the more resistant it is to compaction. In a water-saturated state (also known as mud), the voids between particles are partially filled with water and lack cohesion.

Compaction is achieved by inputting energy to expel the air and water in the soil's voids. The reduction of the voids creates the following changes in the material:

- Increase in unit weight
- Decrease in compressibility
- Decrease in permeability

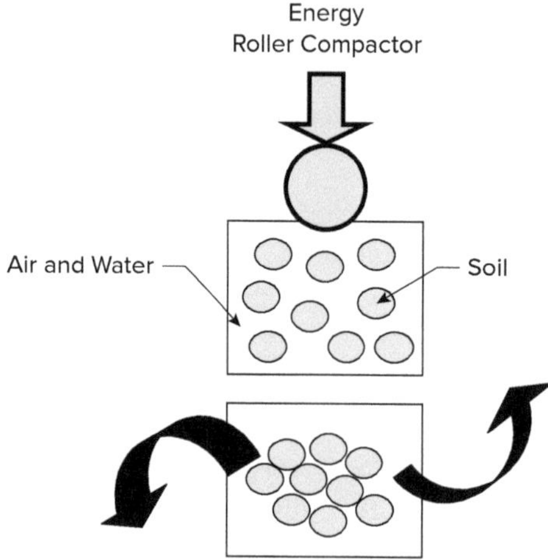

FIGURE 2.4 Effects of Compaction

Example 2.5: Earthwork Equipment

Which of the following pieces of construction equipment can be used most efficiently to compact fine-grained clay soils for a large roadway project?

A. Self-propelled sheepsfoot roller
B. Smooth drum vibratory roller
C. Pneumatic-tire roller
D. Midsize tamping-foot (padfoot) roller

The purpose of compaction is to obtain the optimum density of the soil's properties, as close to zero air voids as possible. There are several kinds of special-purpose construction equipment. The following list identifies the most common and the purpose for each.

Solution

Sheepsfoot rollers can be described as smooth drum rollers with feet or extensions protruding from the drum. There are many varieties of self-propelled and towed versions. The sheepsfoot roller is most effective for the compaction of plastic soils such as clay or silt, compacting from the bottom of each lift toward the top: as the drum turns, the foot comes out, and the result is a loose layer on top. High contact pressures make the feet penetrate through the loose material, compacting it with the foot tip. As the soil material compacts, the feet visibly rest on the surface of the soil.

TIP

It is important to note that the exam's nonquantitative questions often rely on a best fit for the answer. It can be argued that all of the answers appear to be appropriate, but only one is the best fit that addresses all of the nuances in the problem statement.

Pneumatic-tire rollers use rubber tires to provide kneading action on the soil, generally compacting from the top of the lift downward, akin to sealing the surface of the soil. The rubber tire contact area and the ground contact pressure cause a kneading action similar to fingers pressing into bread dough. Considered light-to-medium duty and available as self-propelled machines, they are used primarily for the compaction of granular base material, as well as hot mix asphalt.

Example 2.5 *(continued)*

Vibratory rollers are smooth drum rollers typically used for granular and mixed soil materials. The vibratory roller works on a principle of soil particle reorganization resulting from the energetic vibration forces generated by the drum hitting the solid surface. The mechanical properties of the vibratory compactor are that they generate three forces: (1) static pressure, using the gravity weight of the compactor; (2) impact force as the drum is lifted by the vibrating mass; and (3) vibration of the drum against the solid surface. The soil particles vibrate and rearrange themselves; the result is that the voids between particles become smaller and the soil's density increases. The best application for the vibratory compactor is the compaction of granular and mixed soils, from large rocks to fine sand.

Tamping-foot, or padfoot, rollers combine the vibratory roller and the sheepsfoot roller in one machine. Tamping-foot rollers are similar to sheepsfoot rollers, but have tapered feet, or pads, that penetrate soil, compacting from the bottom to the top for uniform density. However, unlike the sheepsfoot roller, the tapered pads can walk out of the lift without fluffing the top soil. Therefore, the top of the lift is also being compacted. In addition, the combined forces of gravity and the vibratory impact work in a top-down way to compact the soil. Due to the tapered foot shape and the vibration, tamping-foot rollers achieve a kneading effect. Tamping-foot rollers can compact soils having as much as 50% cohesive content (such as silts and clays) and are effective on all soils except clean sand.

Vibratory plate compactors are small engine-powered walk-behind machines (similar in size and appearance to a lawn mower) that use powerful vibratory action to compact soil, sand, or gravel. The small-engine eccentric-drive mechanism propels the compactor, which has a bottom-mounted steel plate with an area ranging from 1.5-to-3 ft^2. Plate compactors can differ in weight, with ranges weighing from 150 pounds to 1,700 pounds. The unit is used on soil or gravel prior to pouring concrete or laying asphalt surfaces.

Answer: A

Example 2.6: Earthwork Equipment

Which of the following equipment should be used within 3 ft of a 250-ft long basement masonry wall when placing aggregate base course (ABC) fill material to achieve 95% compaction?

A. Sheepsfoot roller
B. Vibrating compacting roller
C. Pneumatic-tire roller
D. Vibratory plate compactor

Solution

Analysis of this question requires examining each piece of equipment and its purpose. The sheepsfoot roller (A) is most effective for compacting plastic soils like clay or silt. The vibrating compacting roller (B) is typically used for granular and mixed soil materials, and it is very heavy and too large to be used next to a building with a newly

Example 2.6 (continued)

completed masonry wall. The pneumatic-tire roller (C) has rubber tires to provide a kneading action of soil or subgrade, and pneumatic-tire rollers generally compact from the top of the lift downward and are too large to be close to the newly completed masonry wall. The best fit to the answer is the vibratory plate compactor (D) because the compaction placement of ABC requires little compaction as the stone is relatively compacted when placed and spread. The ABC particles vary in size from ¾ in down to dust. Typically, a small hand-operated tamper, such as a vibratory plate compactor, can be used successfully.

Answer: D

> **TIP**
>
> All static equilibrium calculations use the center of gravity.

2.2.2 Center of Gravity

For a static, rigid body, the physical force of gravity acts on all particles of the body, whereby the accumulated measured weight of the body acts at a single central point. This unitary and measurable point is called the center of gravity. Analyzing the force diagram for a rigid body, a single force of gravity can be placed at the center of gravity.

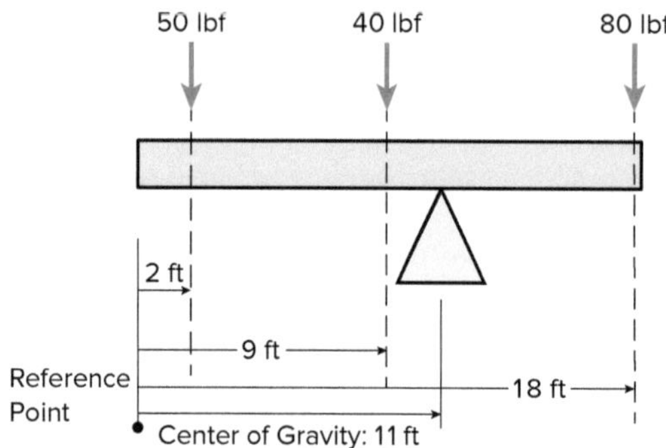

FIGURE 2.5 Calculating the Center of Gravity

The distances from the center of the main object and the two additional weights from the reference point:

- Center of beam = 9 ft away from reference point
- Left weight = 2 ft away from reference point
- Right weight = 18 ft away from reference point

Moment is equal to force times perpendicular distance $M = F \perp D$

- Weight at center = 9 ft × 40 lb = 360 ft-lb
- Left weight = 2 ft × 50 lb = 100 ft-lb
- Right weight = 18 ft × 80 lb = 1,440 ft-lb

Sum the moments: 360 ft-lb + 100 ft-lb + 1,440 ft-lb = 1,900 ft-lb

Sum the weights: 40 lb + 50 lb + 80 lb = 170 lb

Find the center of gravity: $\dfrac{1{,}900\text{ ft-lb}}{170\text{ lb}} = 11.18\text{ ft}$

The center of gravity is 11.18 ft from the reference point.

2.2.3 Crane Stability

The allowable bearing pressures are either given or can be obtained from a variety of published sources to determine, without tests, the jobsite conditions (for example, IBC or OSHA). Once the soil-bearing pressure is determined, outrigger stability can be calculated. Also, the crane manufacturer's content in the crane operator's manual cannot be deviated from and must be strictly followed.

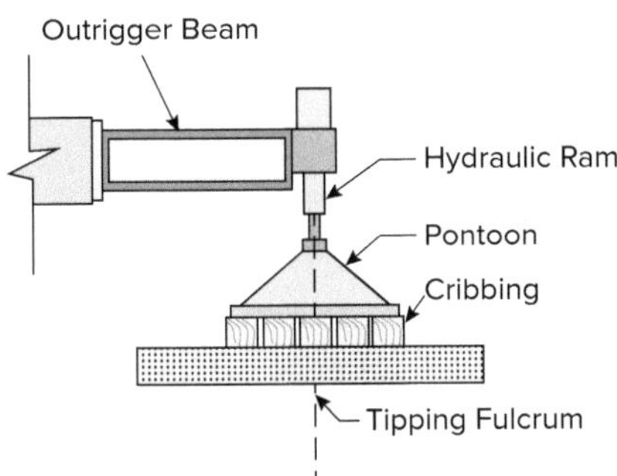

FIGURE 2.6 Crane Outrigger Placement

Example 2.7: Crane Stability

The jobsite placement of a hydraulic crane is based on the data provided in the manufacturer's specifications. The crane outriggers are equipped with steel pontoons with a contact area of 307 in^2. The main jack reaction is 63,500 lb, which exerts 213 psi ground-bearing pressure. The crane is set up on compacted gravel soil with bearing pressures given as 10 tons/ft^2. The minimum contact area (ft^2) of cribbing needed under each outrigger while lifting is most nearly:

A. 2.13 ft^2
B. 2.67 ft^2
C. 3.07 ft^2
D. 3.26 ft^2

Solution

Determine bearing force: 213 psi × 144 in^2/ft^2 = 30,672 lb/ft^2

Pressure exerted below outrigger pad contact area: $\dfrac{307\text{ in}^2}{144\text{ in}^2/\text{ft}^2} = 2.13\text{ ft}^2$

Allowable soil bearing pressure given: 10 tons/ft^2 or 20,000 lb/ft^2

Determine the required contact area ratio: $\dfrac{30{,}672\text{ lb/ft}^2}{20{,}000\text{ lb/ft}^2} = 1.53$

Increase the size of the bearing contact area using timber pads with a minimum size of: 2.13 ft^2 × 1.53 = 3.26 ft^2

Some data points in this problem are distractors that are not relevant to the solution steps: 63,500 lb, since the required ground-bearing pressure is 213 psi, and the pontoon area of 307 in^2, which is used for the bearing area only.

Answer: D

Example 2.8: Tipping Fulcrum

A truck crane that has 143 ft of boom length with a 125-ft radius is lifting a load, fully extending the outriggers for the jobsite conditions. The tipping load (in 1,000 pounds-force, or kips) based on the crane data provided is most nearly:

A. 10 kips
B. 15 kips
C. 18 kips
D. 22 kips

Given parameters:

- Weight of crane = 220,000 lb
- Weight of boom = 24,000 lb
- Distance from the tipping fulcrum to the boom center of gravity radius = 52 ft
- Distance from crane center of gravity to tipping fulcrum (centerline of outrigger) = 17 ft
- Distance from lifting load to tipping fulcrum = 114.5 ft

Solution

Develop a free-body diagram. Place the given parameters and mark the known values to visualize the problem statement. The loads are provided in a format indicating that they are all perpendicular to the horizontal grade.

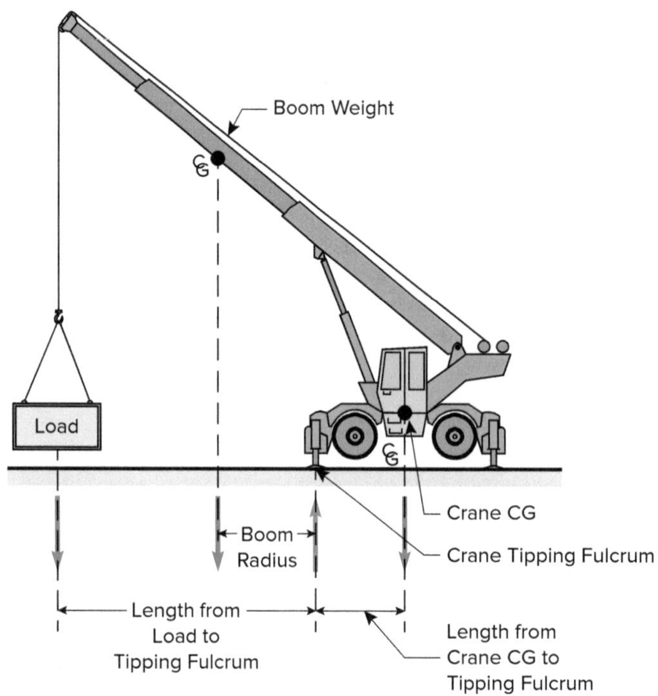

Free Body Diagram—
arrows show the resolutions of the forces
that achieve equilibrium

Identify the components shown in the figure:

W_{crane} = weight of crane = 220,000 lb

W_{boom} = weight of boom = 24,000 lb

Bcg = boom center of gravity = 52 ft

CG_{tf} = 17 ft, from crane center of gravity to tipping fulcrum (centerline of outrigger)

L_{dis} = 114.5 ft, distance from load to tipping fulcrum

Means and Methods

School of PE

Example 2.8 *(continued)*

Determine the stability relationship for the given diagram. The equal sign in the stability equation below (Equation 2-3) is considered the fulcrum point, where the balance of the weights is achieved on both sides.

$$\text{Stability} = (\text{lifting load} \times L_{dis}) + (W_{boom} \times Bcg) = (W_{crane} \times CG_{tf})$$ **Equation 2-3**

Rewrite the equation to find the lifting load (convert the weights to kips):

$$\text{Lifting load} = \frac{(220 \text{ kips} \times 17 \text{ ft}) - (24 \text{ kips} \times 52 \text{ ft})}{114.5 \text{ ft}} = 21.7 \text{ kips}$$

Note that the problem statement asks for the "tipping load" or the load that will make the crane tip.

Illustrations of the plan view and elevation of the crane outrigger configuration follow.

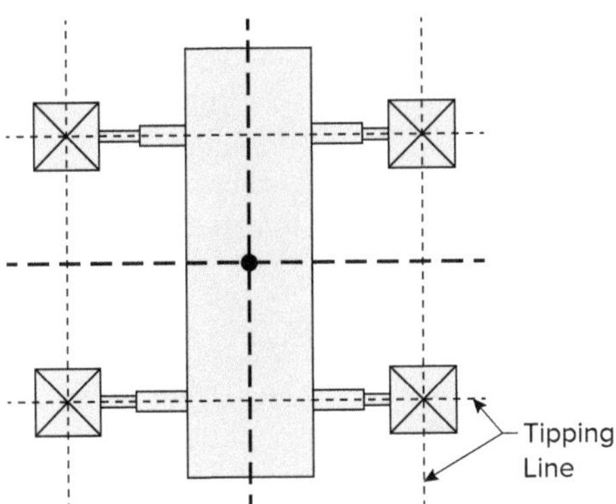

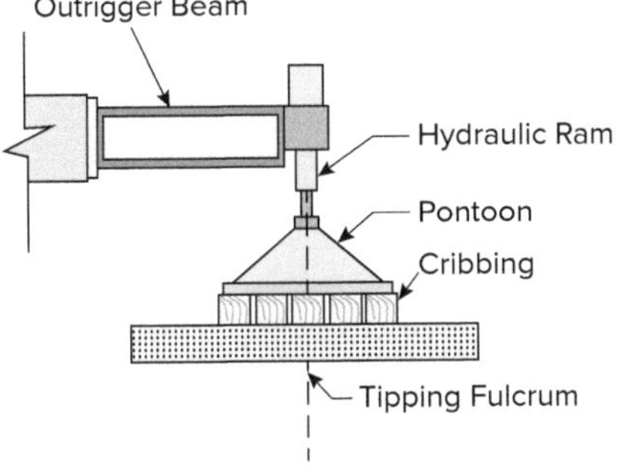

Answer: D

2.3 TEMPORARY STRUCTURES AND FACILITIES

2.3.1 Static Equilibrium

Static equilibrium occurs when an object is at rest. Static refers to the object being stationary, and equilibrium refers to the object either having no net forces acting upon it or having all of its net forces balanced—equal and opposite. The state of an object relative to equilibrium is determined by the net forces acting upon it. This is established in the general format of the equations for static equilibrium:

$$\Sigma F = 0 \text{ and } \Sigma M_O = 0$$ **Equation 2-2** (p. 42)

Newton's third law of motion, the law of action and reaction, states that for every force acting on a body, the body exerts a force having equal magnitude in the opposite direction along the same line of action as the original force.

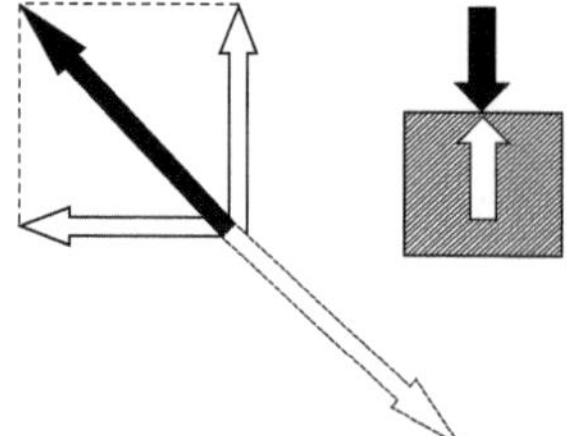

FIGURE 2.7 Equal and Opposite Forces

Vector quantities possess both magnitude and direction. A force has both magnitude and direction. Force is a vector quantity. USCS units are pounds (lb) and SI units are newtons (N). Forces can cause motion or act to keep an object at rest. A resultant force is the single force that represents the vector sum of two or more forces.

2.3.2 Falsework

The standard definition for falsework is any temporary structure used to support a permanent structure while it is not self-supporting. Falsework also includes temporary support structures for a variety of construction elements. A typical building construction site contains many examples of falsework when building the permanent structure: (1) formwork is used to mold concrete to form a desired shape; (2) scaffolding is available for workers to access the structure being constructed; and (3) earthen shoring, which is a temporary structural reinforcement, is used when workers are in a trench excavation for utility installations.

Temporary structures are utilized in various construction operations, such as the following:

1. Concrete formwork construction
2. Scaffolding
3. Shoring of formwork
4. Cofferdam construction
5. Building underpinning
6. Diaphragm/slurry walls for reinforced excavations
7. Earth-retaining structures
8. Construction groundwater dewatering

Temporary structures are critical elements of a construction plan. A temporary structure in construction affects the life safety of the workers as well as the general public.

Architectural or engineering plans and specifications for the permanent structure are submitted for building code compliance review, for the welfare of the general public. Falsework plans are developed by the constructor and follow the building code guidelines and OSHA standards to assure the safe construction of the permanent structure.

Example 2.9: Falsework: Cofferdams

A square, double-wall, sheet-pile cofferdam with dimensions (out-to-out) of 50 ft on each side was erected to allow for the construction of a bridge abutment at station 273+52.30. The top-of-pile elevation is 377.5 ft and the bottom-of-footing elevation is 346.50 ft. The 6-ft space between the double-wall sheet piles was filled with soil with a permeability of 0.035 ft/day established in situ. The river bottom is impermeable and the mean high tide water level is 366.50 ft. The volume of water to be pumped per day (in million gallons per day, MGD), using a factor of safety of two, is most nearly:

A. 5.3 MGD
B. 0.053 MGD
C. 0.0053 MGD
D. 0.00053 MGD

Example 2.9 *(continued)*

Solution

Sketch the problem statements and locate the bottom-of-footing elevation (346.50 ft) and high tide (366.50 ft). These dimensions establish the amount of surface area to determine the amount of seepage.

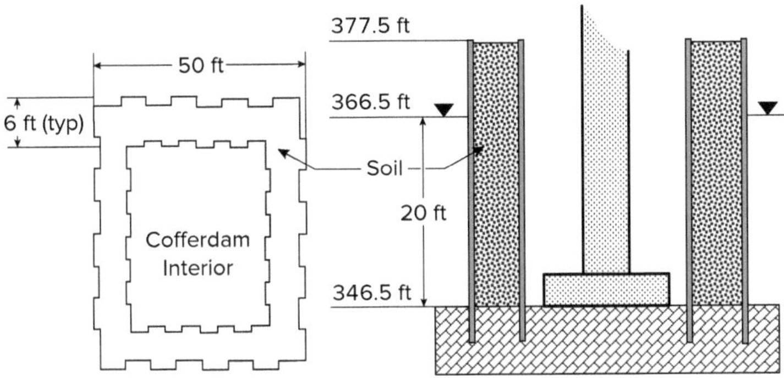

The movement of groundwater through an aquifer is given by Darcy's law, where:

$Q = kiA$ **Equation 2-4**

Q = flow quantity

k = coefficient of permeability, given as 0.035 ft/day

i = hydraulic gradient $\left(i = \frac{\Delta h}{l}\right)$

A = cross-sectional area perpendicular to the flow direction

Compute the seepage surface area, which is the inside surface of the cofferdam, or 38 ft × 38 ft:

$A = (50\text{ ft} - 6\text{ ft} - 6\text{ ft}) \times 4\text{ sides} \times 20\text{ ft}$

$A = 3{,}040\text{ ft}^2$

Height of river water = mean high water level − abutment bottom of footing = 366.5 ft − 346.5 ft

H = height of river water = 20 ft

L = length of soil between steel sheathing = 6 ft

$$\text{Hydraulic gradient} = \frac{H}{L} = \frac{20\text{ ft}}{6\text{ ft}}$$

Hydraulic gradient = 3.33

Assemble the equation and solve by including the factor of safety of two:

$Q = kiA$

$Q = 0.035\text{ ft/day} \times 3.33 \times 3{,}040\text{ ft}^2$

Example 2.9 *(continued)*

$Q = 354.66 \text{ ft}^3/\text{day}$

Convert ft^3/day to gallon/day (1 ft^3 = 7.48 gal)

$Q = 354.66 \text{ ft}^3/\text{day} \times 7.48 \text{ gal/ft}^3$

$Q = 2{,}652.90 \text{ gal/day}$

$$\begin{aligned}\text{Answer} &= Q \times (\text{factor of safety of two}) \\ &= 2{,}653 \text{ gal/day} \times 2 \\ &= 5{,}306 \text{ gal/day}\end{aligned}$$

$\text{Answer} = 0.0053 \text{ MGD}$

Answer: C

2.3.3 Load Paths

The evaluation of the load path in the illustrations given in Example 2.10 require that the analysis of the gravity and lateral forces be resolved to a state known as static equilibrium. The lateral wind force is placing a load on the rigid frame structure where the forces travel through the frame and are resolved at the base of the foundation.

One technique is to view the structure from a perspective of failure, given that the wind will have a tendency to lift the building and rotate it in a clockwise direction.

Example 2.10: Load Paths

Which one of the figures below most nearly depicts the load path for lateral resistance upon the static load case? (Arrows depict the direction of force only.)

A. Load Path Diagram

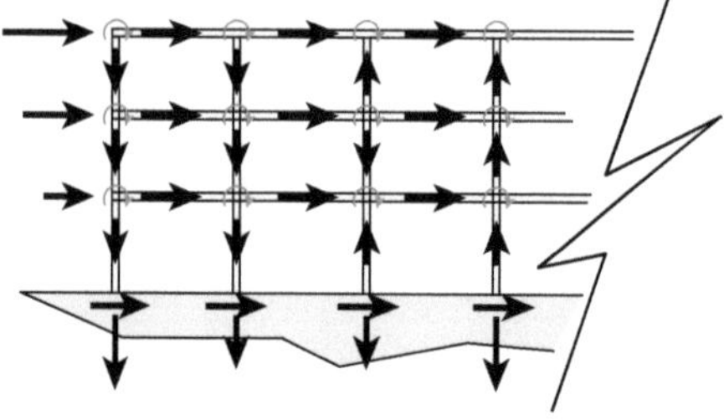

B. Load Path Diagram

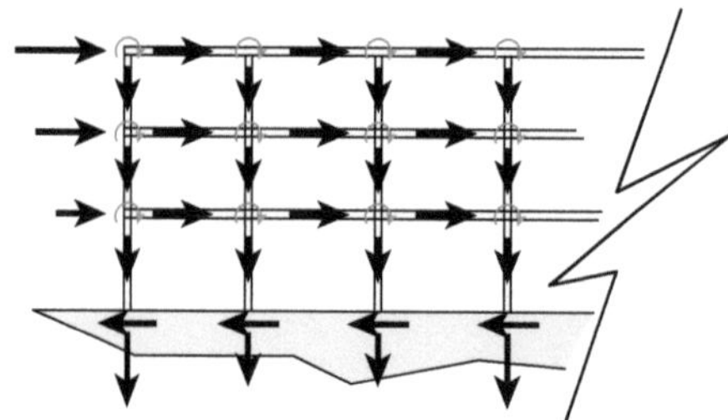

Example 2.10 *(continued)*

C. Load Path Diagram

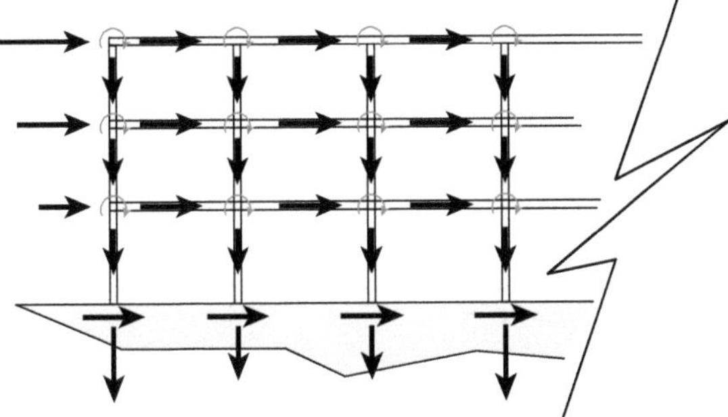

D. Load Path Diagram

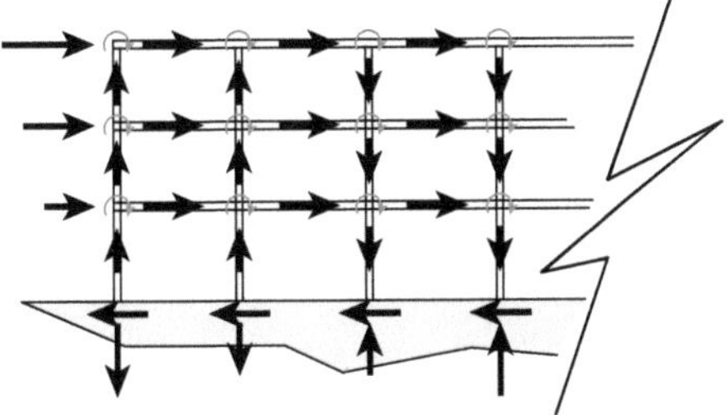

A. A
B. B
C. C
D. D

Solution

The location of the forces where static equilibrium is resolved and in equal and opposite values is found in Figure D, which properly resolves the forces and allows the gravity and lateral forces imposed on the structure to achieve static equilibrium.

The right side shows a downward gravity force reconciled by the upward force of the earth; the left side shows an upward force on the frame and a downward force holding the frame to the earthen footing. The left-facing lateral forces at the base of the columns are equal and opposite to the lateral wind force acting on the structure.

Answer: D

Example 2.11: Tributary Area

An extract from the structural drawings for a first-floor rigid frame structure is given. The floor system must support its own weight of 60 lb/ft^2 and a live load of 125 lb/ft^2. The unfactored load for column B3 is most nearly:

Example 2.11 *(continued)*

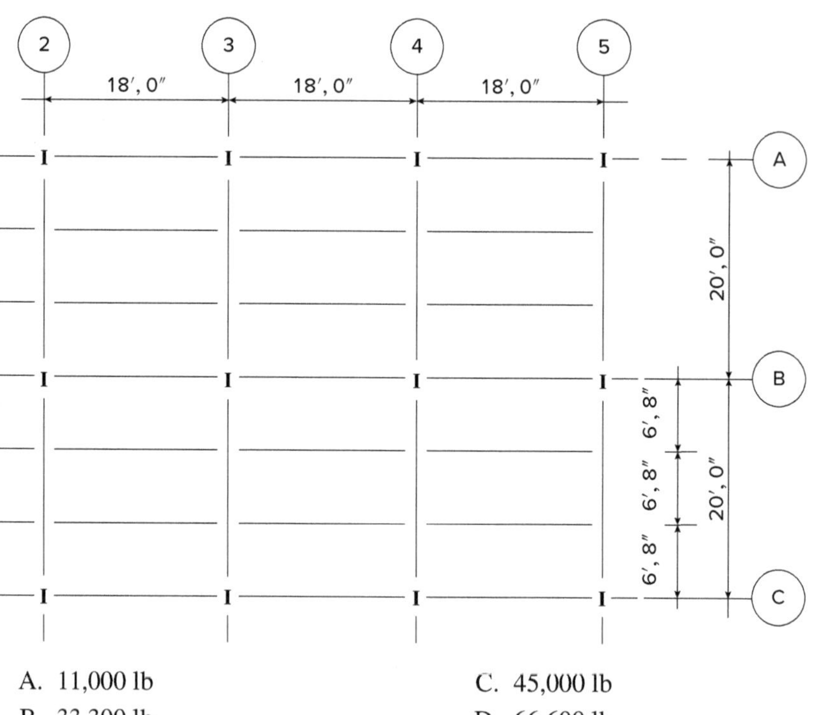

A. 11,000 lb

B. 33,300 lb

C. 45,000 lb

D. 66,600 lb

Solution

Determine the equivalent surface area (or the tributary area) surrounding column B3. The structural bay spacing is 18 ft, 0 in × 20 ft, 0 in. The combined live and dead load is 185 lb and is considered the unfactored load. The tributary area is defined as the area that contributes to the load on a member supporting that area.

$(18 \text{ ft} - 0 \text{ in}) \times (20 \text{ ft} - 0 \text{ in}) \times (185 \text{ lb/ft}^2) = 66{,}600 \text{ lb}$

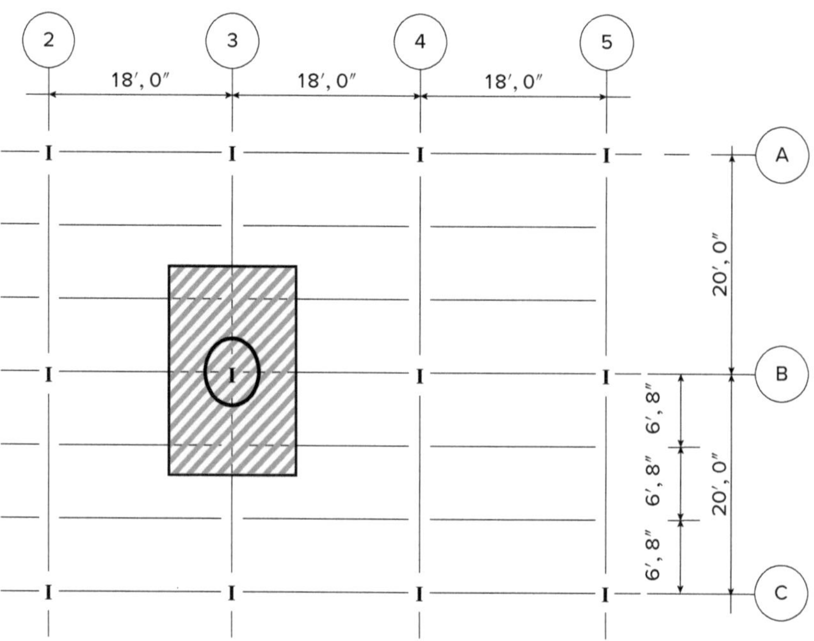

Answer: D

Example 2.12: Temporary Bracing

A normal-weight concrete deadman anchor restrained from sliding with dimensions 3 ft (width) × 3 ft (length) × 3 ft (height) is used to support a formwork diagonal brace that has a compressive load of 1,500 lbf. The brace attaches to a center-pin clevis, which is 8 in above the block and is located 12 ft away from the face of the formwork and attached to a wale 16 ft above the center pin. The factor of safety from overturning is most nearly:

A. 1.2
B. 1.5
C. 2.2
D. 2.9

Solution

Sketch the problem statement.

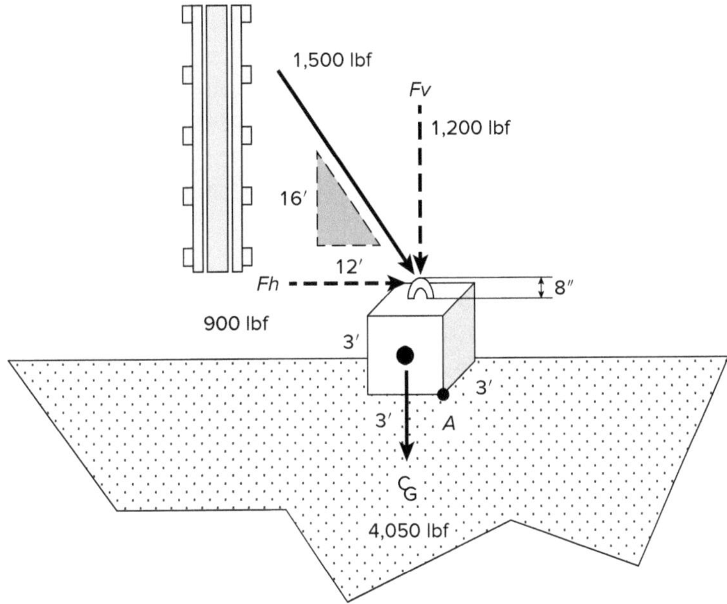

Determine the weight of the deadman anchor using normal-weight concrete, which is a standard measure of 150 lb/ft^3.

$$\text{Weight} = 150\ \text{lb/ft}^3 \times 3\ \text{ft} \times 3\ \text{ft} \times 3\ \text{ft}$$
$$\text{Weight} = 4{,}050\ \text{lbf}$$

Use the diagonal load to determine the force found in both the horizontal and vertical components. Notice that the brace geometry forms a 3-4-5 triangle, simplifying calculations. The NCEES uses the 3-4-5 triangle, as it is a special right triangle with numbers that are easy to remember; no calculation is necessary to verify that it is a right triangle. Multiples of 3-4-5, such as 6-8-10 or 12-16-20, help in the assessment of the problem and make the solution quicker to find.

$$Fh = 1{,}500\cos(53^\circ) = 900\ \text{lbf}$$
$$Fv = 1{,}500\sin(53^\circ) = 1{,}200\ \text{lbf}$$

Determine the factor of safety. The numerator develops the moments that resist or hold down the deadman anchor while the denominator separates the force holding down the anchor due to the force trying to overturn the anchor.

Example 2.12 *(continued)*

Summing the moments about point A on the deadman anchor, the factor of safety against overturning is calculated as follows:

$$\text{Factor of safety} = \frac{\text{resisting moments}}{\text{overturning moments}} \qquad \textbf{Equation 2-5}$$

$$= \frac{4{,}050\ \text{lbf}\,(1.5\ \text{ft}) + 1{,}200\ \text{lbf}\,(1.5\ \text{ft})}{900\ \text{lbf}\,(3.67\ \text{ft})}$$

$$\text{Factor of safety} = \frac{7{,}875\ \text{ft-lbf}}{3{,}303\ \text{ft-lbf}}$$

$$\text{Factor of safety} = 2.384$$

Answer: C

TIP

Every part of the problem statement must be evaluated to determine its importance in developing a solution.

Note that the problem statement—"deadman anchor restrained from sliding"—verifies that the equations for static equilibrium can be used to solve this problem, as the sum of the forces in the x and y axes is equal to zero and allows it to be a determinate solution. Without this phrase, the force would not be concurrent and could cause movement of the anchor.

Example 2.13: Concrete Wall Forms

The design for a concrete wall formwork consists of vertical strong backs (studs) spaced 24 in apart that are supported by horizontal wales spaced every 18 in. Eight strands of number 9 wire formwork ties (wire tie tensile breaking capacity = 700 lb per strand) are used to resist the hydrostatic pressure of the 4,000-psi concrete exerting 480 psf on the 1¾-in plyform sheathing. The formwork ties are anchored every fourth stud. The factor of safety at each location of the form wall tie is most nearly:

A. 1.0
B. 1.3
C. 2.3
D. 2.5

Solution

Determine the area held by the tie wire (contributory or tributary area). The wall dimensions are not provided; however, the description of the contributory spacing indicates a rectangular area that is 18 in tall × 72 in wide (the wire is anchored at every fourth stud holding three spans).

$$\text{Area} = \frac{(72\ \text{in})(18\ \text{in})}{144\ \text{in}^2/\text{ft}^2}$$

$$= 9\ \text{ft}^2$$

$$\text{Force} = (480\ \text{psf})(9\ \text{ft}^2)$$

$$= 4{,}320\ \text{lb}$$

Each computed area of the formwork is known as its contributory, which is held by eight strands of wire ties per each corner or contributory area. The factor of safety is computed:

$$\text{Factor of safety} = \frac{(8\ \text{strands})(700\ \text{lb/strand})}{4{,}320\ \text{lb}}$$

$$\text{Factor of safety} = \frac{5{,}600\ \text{lb}}{4{,}320\ \text{lb}}$$

$$\text{Factor of safety} = 1.296$$

Sketch the problem statement to visualize the contributory area:

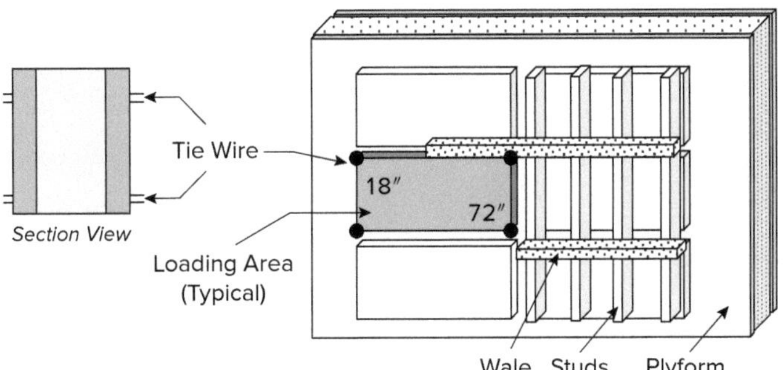

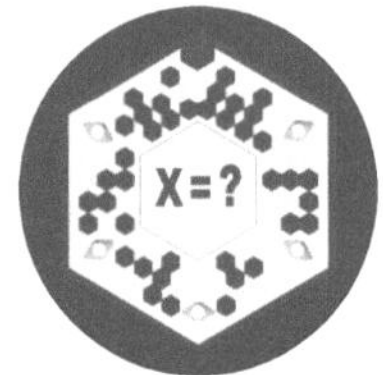

Answer: B

REFERENCE

1 ACI (American Concrete Institute) Committee 347. *ACI 347R-14: Guide to Formwork for Concrete.* Farmington Hills, MI: ACI, 2014.

Soil Mechanics

Amr M. Sallam, PhD, PE

Alan Esser, PE, D.GE

Sukh Gurung, PhD, PE

3

CONTENTS

CONTENTS (*continued*)

EXAM GUIDE

III. Soil Mechanics

A. Lateral earth pressure
B. Soil consolidation
C. Effective and total stresses
D. Bearing capacity
E. Foundation settlement
F. Slope stability

Approximate Number of Questions on Exam: 6

NCEES Principles & Practice of Engineering Examination, Civil Breadth Exam Specifications

COMMONLY USED ABBREVIATIONS

A	area of applied pressure
C_c	compression index
C_r	recompression index
C_v	coefficient of consolidation (length^2/time)
c	cohesion
EFD	equivalent fluid density
e_o	initial void ratio
F_O	driving forces (from the active earth pressure, surcharge, and hydrostatic pressure)
F_R	resisting forces (from friction on the wall base and passive earth pressure)
F_S	friction force
H	thickness of the soil layer
H_d	length of the shortest drainage path
NC	normally consolidated
OC	overconsolidated
R_a	active earth pressure
$R_{a,h}$	horizontal component of the active earth pressure
$R_{a,v}$	vertical component of the active earth pressure
R_p	passive resistance
S_c	primary consolidation settlement
T_v	time factor
u	pore water pressure
W	vertical weight

COMMONLY USED SYMBOLS

ϕ	angle of shearing resistance of soil
β	angle of backfill slope with a horizontal line
θ	angle of the back of the wall with a horizontal line
σ'_o	in situ effective overburden pressure at the midpoint of the clay layer
σ'_f	final effective pressure ($\sigma'_f + \Delta\sigma$)
σ_v	total vertical stress
σ'_v	effective vertical stress
γ_w	unit weight of water

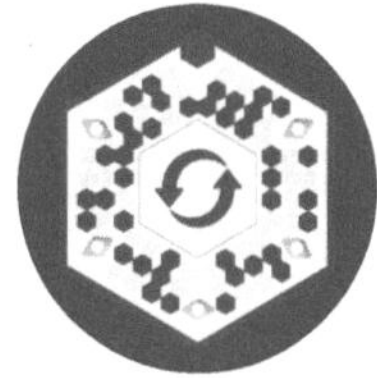

3.1 LATERAL EARTH PRESSURE

In general, unsupported vertical cuts in most soils are unstable, especially for permanent structures. Retaining walls or earth-retaining structures are needed to support vertical cuts in soils. The force exerted by the supported soil on the retaining wall is called the soil's lateral earth pressure. Retaining walls should be designed to withstand the thrust resulting from the lateral earth pressure with specific safety factors to protect against multiple failure modes including sliding, overturning, global instability, and bearing capacity failure.

Lateral earth pressure at any depth in the soil mass is dependent on the vertical effective stress at that depth multiplied by the lateral earth pressure coefficient K, which is dependent on the soil shear strength and the movement of the wall relative to the soil.

$$K = \frac{\sigma'_h}{\sigma'_v}$$ **Equation 3-1**

K = lateral earth pressure coefficient

σ'_h = lateral earth pressure at any depth in the soil mass

σ'_v = vertical effective stress at any point in the soil mass

Three cases of lateral earth pressure are encountered:

1. **At-rest earth pressure:** In some design cases, movement is restrained and the wall does not move (Fig. 3.1a). The earth pressure in this case is greater than the active case. Designers do not work with at-rest earth pressure unless the field conditions mandate using it. Examples of cases where at-rest pressure is used include basement walls braced at the top and the bottom, and the sides of box culverts. The at-rest earth pressure coefficient, calculated from Equation 3-2, is used to determine the at-rest earth pressure and resultant forces.
2. **Active earth pressure:** Where outward movement of the wall is allowed, the earth pressure is reduced to the active state. Failure is assumed to occur along a planar surface, as shown in Figure 3.1b. An active earth pressure coefficient, calculated from Equations 3-7 and 3-8, is used to determine the active earth pressure and resultant forces.
3. **Passive earth pressure:** This pressure is the resistance of soils in the passive zone where the wall will be pushed against the soil. Soil passive resistance should be looked at as a reserve. With the increase of active forces on walls, enough passive resistance will be developed from the reserve to stabilize the wall along with friction resistance at the base of the wall. Failure is assumed to occur along a planar surface as shown in Figure 3.1c. A passive earth pressure coefficient, calculated from Equations 3-12 and 3-13, is used to determine the passive earth pressure and resultant forces.

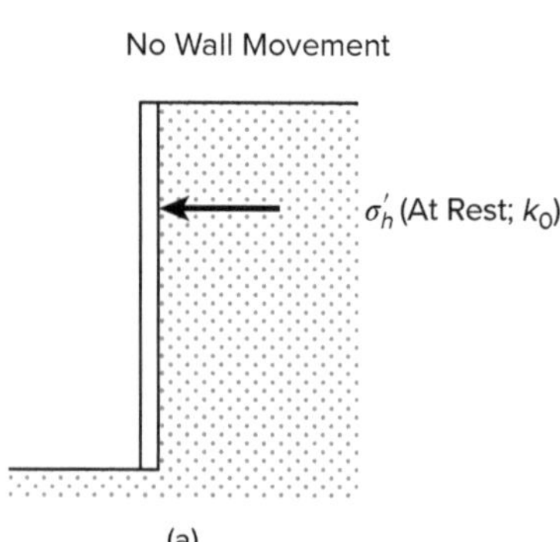

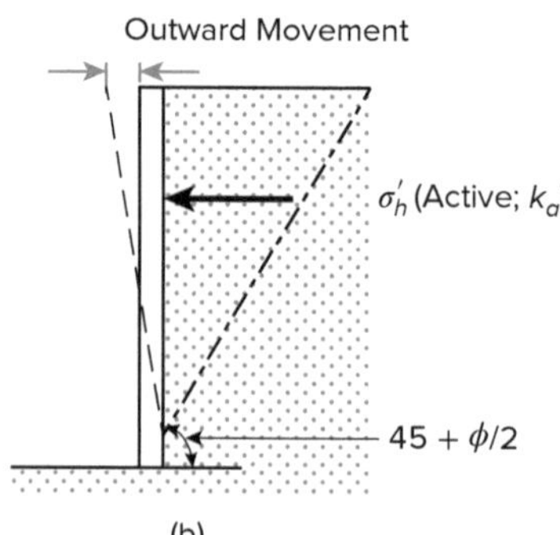

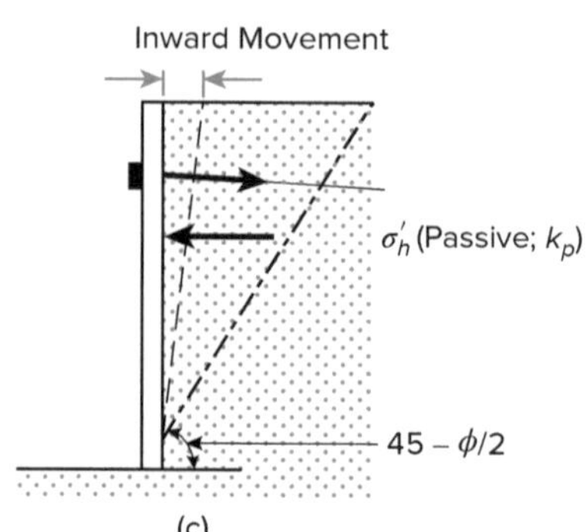

FIGURE 3.1 Types of Earth Pressure on Retaining Walls

There are two main theories used to calculate the earth pressure on walls: the **Coulomb** and **Rankine** theories. Based on the target accuracy of the calculations, the designer selects the theory to apply. A wall in the most generic form is shown in Figure 3.2, in which the following angles are shown:

ϕ = the angle of shearing resistance of the supported soil
β = the angle of backfill slope with a horizontal line
δ = the angle of friction between the back of the wall and the supported soil
θ = the angle of the back of the wall with a horizontal line

Coulomb theory takes into consideration all of the above angles when calculating the earth pressure coefficient; therefore, it provides a more accurate value. Rankine theory neglects the friction between the soil and the wall, assumes the back of the wall is vertical ($\theta = 90°$), and takes into account the back-slope angle (β). Most designers use Rankine theory in earth pressure calculations.

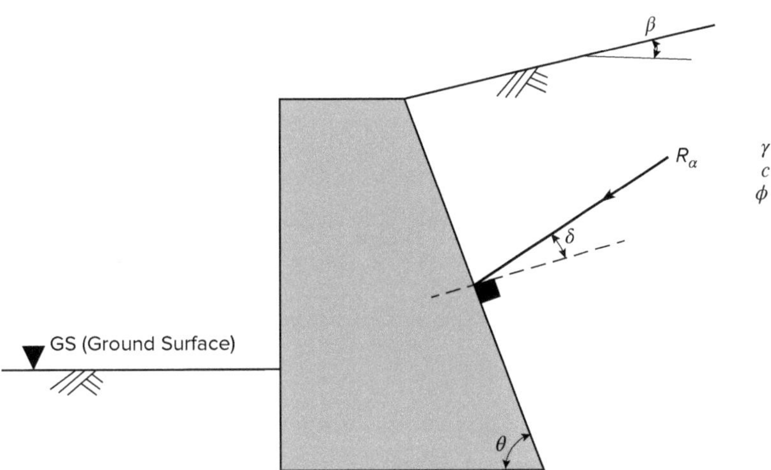

FIGURE 3.2 Sketch of a Generic Retaining Wall

3.1.1 At-Rest Earth Pressure

Under conditions of zero horizontal displacement of the wall, the at-rest earth pressure coefficient can be calculated using the following equation:

$K_O \approx 1 - \sin\phi'$[granular soils and normally consolidated (NC) clays] **Equation 3-2**

3.1.2 Active Earth Pressure—Rankine Theory

The following equation is used to determine the active earth pressure (p_a) for all soils.

$$p_a = K_a \gamma H - 2c\sqrt{K_a}$$ **Equation 3-3**

p_a = active earth pressure

K_a = active earth pressure coefficient

γ = soil unit weight

H = height of retaining wall

c = soil cohesion

For saturated clay soils, $\phi = 0$, $K_a = 1$:

$$p_a = \gamma H - 2c$$ **Equation 3-4**

For granular soils, $c = 0$:

$$p_a = K_a \gamma H$$ **Equation 3-5**

The active earth pressure distribution behind a retaining wall is triangular, as shown in Figure 3.3. If there is a surcharge or live load, q_s, on top of the backfill behind the wall, a uniform rectangular pressure is added (Fig. 3.3). The active resultant is the area of the active earth pressure diagrams, and it is expressed in force per linear foot of the wall.

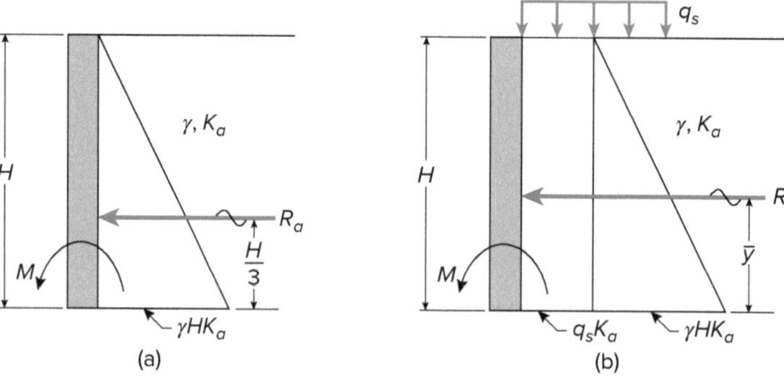

FIGURE 3.3 Active Earth Pressure with and without Surcharge (No Cohesion)

If the soil has cohesion, the active pressure consists of a negative rectangular pressure due to the cohesion and a positive triangular pressure due to the soil. The negative pressure due to cohesion causes separation to develop to a critical depth (depth of a tension crack) where the horizontal pressure is zero (Fig. 3.4). The depth of the tension crack can be estimated by:

$$Z_{cr} = \frac{2c}{\gamma\sqrt{k_a}}$$ **Equation 3-6**

Z_{cr} = critical depth to zero active earth pressure

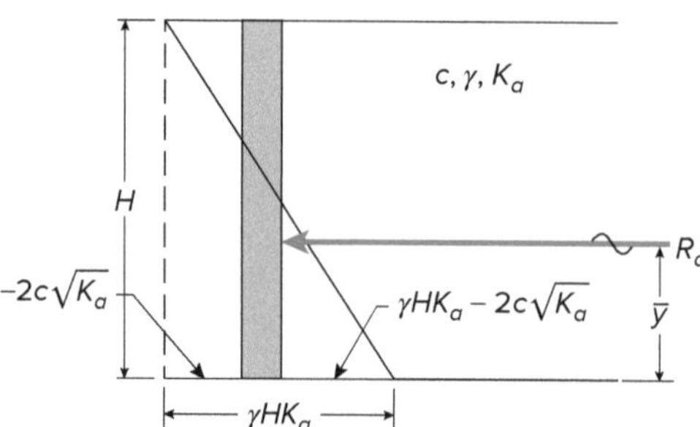

FIGURE 3.4 Active Earth Pressure with Cohesion

If there is water behind the retaining wall, then the hydrostatic water pressure triangle must be added. Also, the unit weight of the soil below the water needs to be the effective or submerged unit weight (Fig. 3.5).

Soil Mechanics

School of PE

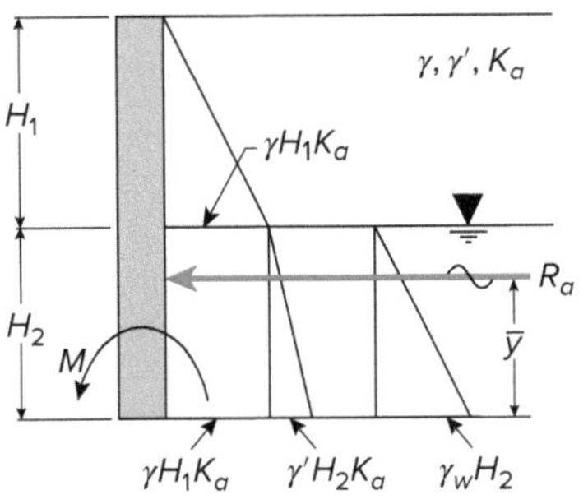

FIGURE 3.5 Active Earth Pressure with Water Behind the Wall (No Cohesion)

For level backfill ($\beta = 0$), the active earth pressure coefficient is calculated from the following equation:

$$K_a = \frac{1}{K_p} = \frac{1 - \sin\phi}{1 + \sin\phi} = \tan^2\left(45° - \frac{\phi}{2}\right)$$ **Equation 3-7**

K_a = active earth pressure coefficient

For a sloping backfill ($\beta \neq 0$), the Rankine active earth pressure coefficient is calculated from the following equation:

$$K_a = \cos\beta\left(\frac{\cos\beta - \sqrt{\cos^2\beta - \cos^2\phi}}{\cos\beta + \sqrt{\cos^2\beta - \cos^2\phi}}\right)$$ **Equation 3-8**

3.1.3 Passive Earth Pressure—Rankine Theory

For a level backfill ($\beta = 0$), the following equation is used to determine the passive earth pressure (p_p) for all soils:

$$p_p = K_p\gamma H + 2c\sqrt{K_p}$$ **Equation 3-9**

p_p = passive earth pressure

For saturated clay soils, $\phi = 0$, $K_p = 1$: $p_p = \gamma H + 2c$ **Equation 3-10**

For granular soils, $c = 0$: $p_p = K_p\gamma H$ **Equation 3-11**

If the soil has cohesion, the passive earth pressure behind a retaining wall consists of a rectangular pressure due to the cohesion added to the triangular soil pressure as shown in Figure 3.6. The **passive resistance force** is the resultant area of the passive earth pressure diagrams and is expressed in force per linear foot of the wall. For level backfill ($\beta = 0$), the passive earth pressure coefficient is calculated from the following equation:

$$K_p = \frac{1}{K_a} = \frac{1 + \sin\phi}{1 - \sin\phi} = \tan^2\left(45° + \frac{\phi}{2}\right)$$ **Equation 3-12**

For a sloping backfill ($\beta \neq 0$), the Rankine passive earth pressure coefficient is calculated from the following equation:

$$K_p = \cos\beta\left(\frac{\cos\beta + \sqrt{\cos^2\beta - \cos^2\phi}}{\cos\beta - \sqrt{\cos^2\beta - \cos^2\phi}}\right)$$ **Equation 3-13**

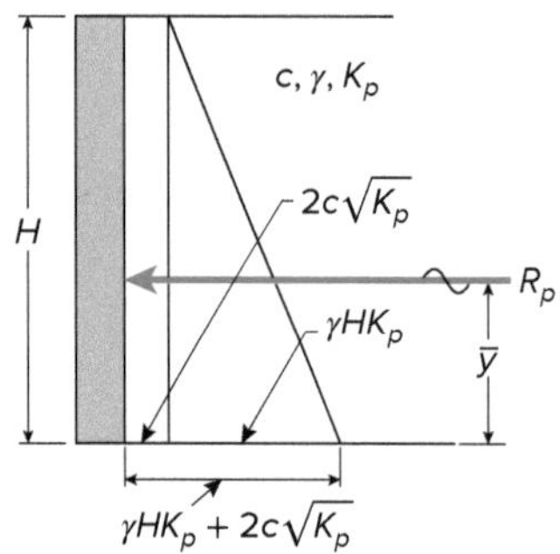

FIGURE 3.6 Passive Earth Pressure with Cohesion

3.1.4 Earth Pressure Coefficients—Coulomb Theory

Coulomb theory takes into consideration the friction between the backfill soil and the wall, the slope of the back of the wall, and the sloping backfill.

Coulomb theory uses the following expressions for the active and passive earth pressure coefficients:

$$K_a = \frac{\sin^2(\theta + \phi)}{\sin^2\theta \sin(\theta - \delta)\left(1 + \sqrt{\dfrac{\sin(\phi + \delta)\sin(\phi - \beta)}{\sin(\theta - \delta)\sin(\theta + \beta)}}\right)^2} \quad \textbf{Equation 3-14}$$

$$K_p = \frac{\sin^2(\theta - \phi)}{\sin^2\theta \sin(\theta + \delta)\left(1 - \sqrt{\dfrac{\sin(\phi + \delta)\sin(\phi + \beta)}{\sin(\theta + \delta)\sin(\theta + \beta)}}\right)^2} \quad \textbf{Equation 3-15}$$

θ = back of the wall angle with the horizontal (see Fig. 3.2)

δ = angle of friction between the wall and the soil backfill with the horizontal (see Fig. 3.2)

For sloping backfills, the vertical and horizontal components may be calculated as:

$$(R_a)_v = R_a \sin(90 - \theta + \delta) \quad \textbf{Equation 3-16}$$

$$(R_a)_h = R_a \cos(90 - \theta + \delta) \quad \textbf{Equation 3-17}$$

Example 3.1: Earth Pressure Coefficients—Rankine Theory

A 10-ft-high gravity retaining wall with flat backfill (β = 0) retains a clean sand for which γ = 120 lb/ft^3 and ϕ = 32°. Using Rankine's earth pressure theory, calculate the total active earth pressure, the active resultant, and the overturning moment about the toe.

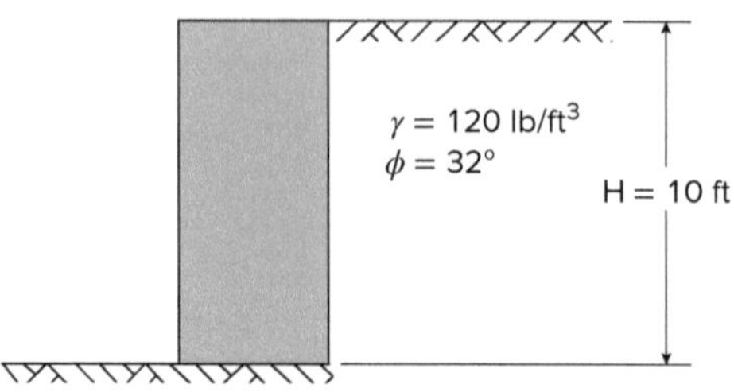

Example 3.1 *(continued)*

Solution

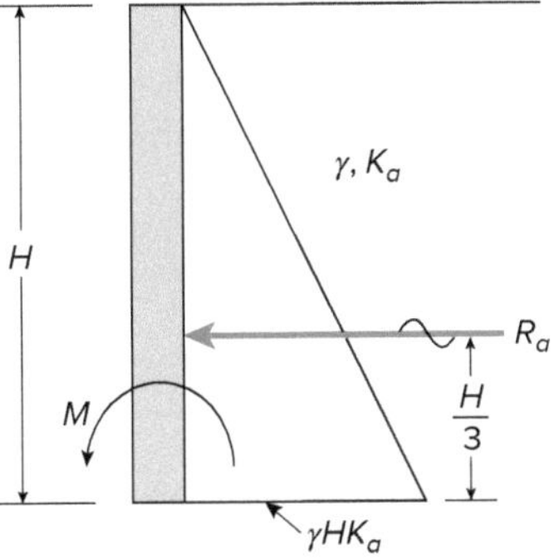

Calculate the active earth pressure coefficient:

$$K_a = \tan^2\left(45° - \frac{\phi}{2}\right) = \tan^2\left(45° - \frac{32°}{2}\right) = 0.307$$

Calculate the active earth pressure and resultant force:

$$p_a = \gamma H\, K_a = (120\ \text{lb/ft}^3)(10\ \text{ft})(0.307) = 368\ \text{lb/ft}^2 \text{ per foot of the wall}$$

$$R_a = \frac{1}{2} p_a\, H = \frac{1}{2}(368\ \text{lb/ft}^2)(10\ \text{ft}) = 1{,}842\ \text{lb/ft}$$

Calculate the overturning moment:

$$M = R_a\left(\frac{H}{3}\right) = 1{,}842\ \text{lb/ft}\left(\frac{10\ \text{ft}}{3}\right) = 6{,}140\ \frac{\text{ft-lb}}{\text{ft}}$$

3.1.5 Rigid Retaining Walls

Retaining walls are designed by assigning initial dimensions based on experience, and then using trial-and-error adjustments of the assumed dimensions to satisfy the stability requirements. The retaining walls must provide sufficient safety factors to protect against sliding, overturning, bearing failure, and global failure, and should be economical. Rigid retaining walls depend on the weight of the wall stem and base and the weight of the retained backfill over the toe and heel of the wall to develop friction-resisting forces and stability moments to counteract active earth pressures and overturning moments.

Gravity retaining walls are constructed with plain concrete or stone masonry and depend primarily on their weight for stability. **Cantilever retaining walls** are used to retain higher unsupported soils and are generally made of reinforced concrete that consists of a thin stem, a base slab, and an optional key. They depend primarily on the weight of retained soil above the base slab for stability. **Sheet pile walls** are also used to support soils; however, they depend on passive earth pressure for stability.

Retaining walls should be designed to provide the required safety factors for both overturning and sliding modes of failure. The safety factors are calculated as follows:

Overturning about toe: $$FS_{OT} = \frac{\Sigma M_R}{\Sigma M_0}$$ **Equation 3-18**

M_R = moments that resist overturning (from soil backfill, wall weight, and passive earth pressure)

M_O = moments that contribute to overturning (from active earth pressure, surcharges, and hydrostatic pressure)

Sliding: $$FS_{SL} = \frac{\Sigma F_R}{\Sigma F_0}$$ **Equation 3-19**

F_R = resisting forces (from friction on the wall base and passive earth pressure)

F_0 = driving forces (from the active earth pressure, surcharge, and hydrostatic pressure)

The factor of safety (FS) against overturning (moments taken about the toe) is typically required to be at least 1.5 if passive resistance is neglected, and greater than 2.0 if passive resistance or cohesion is included.

FS against overturning can be increased by increasing the width of the wall base.

The FS against sliding is typically required to be at least 1.5 and at least 2.0 if passive resistance or cohesion is included. It can be increased by increasing the width of the wall base or by adding a key to the base of the wall.

Passive resistance and cohesion may not be reliable and are often neglected in stability analysis. The weight of the soil over the toe of the wall is generally very small and often neglected when computing the total weight.

Figure 3.7 presents the typical driving and stability forces needed for wall stability analysis. The following forces are identified:

For sliding check:

- Driving forces: active earth pressure (R_a), which is the area of the earth pressure diagram
- Resisting forces: friction force (F_s) resulting from the vertical weights (W), friction between the wall base and foundation soil (δ), adhesion between the wall base and the foundation soil (c_A), and passive resistance (R_p)

For overturning check:

- Overturning moments: the moment from the active earth pressure (R_a applied at $H/3$) taken about the toe of the wall
- Stability moments: moments from the total weight (W) of the wall, the soil backfill, and the moment from the passive resistance (R_p applied at $D/3$) taken about the toe of the wall

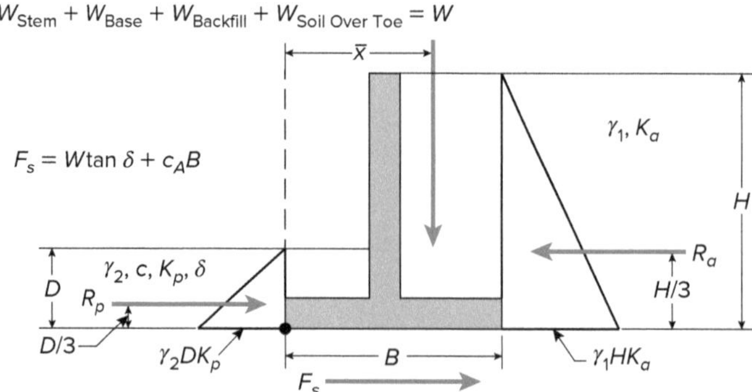

FIGURE 3.7 Typical Forces on Cantilever Retaining Walls (Horizontal Backfill)

Figure 3.8 shows the forces acting on a wall with the backfill sloping at angle β from the horizontal. When Rankine theory is used, the resultant of the active force (R_a) is assumed to act parallel to the surface of the backfill. The resultant force is resolved into horizontal ($R_{a,h}$) and vertical ($R_{a,v}$) components. The horizontal ($R_{a,h}$) component is the driving force for sliding and overturning. The vertical ($R_{a,v}$) component is added to the vertical forces in calculating the friction resistance. Its moment, taken about the toe of the wall, adds a stabilizing moment.

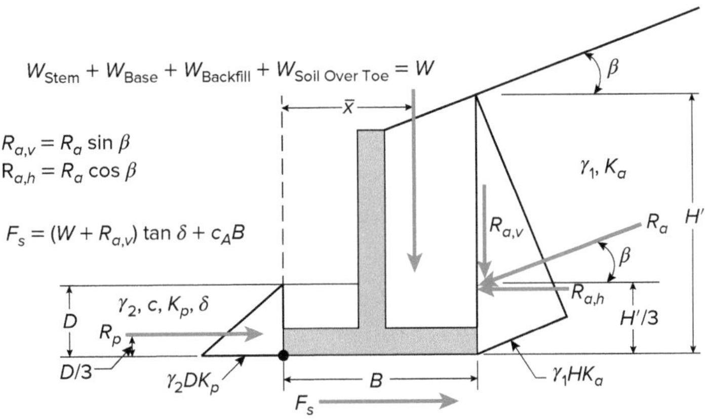

FIGURE 3.8 Typical Forces on Cantilever Retaining Walls (Sloping Backfill)

Example 3.2: Rigid Retaining Walls

Use Rankine's earth pressure theory to check the stability (FS against sliding and overturning) of the retaining wall shown in the figure below. The unit weight of concrete is 150 lb/ft^3. The water table is well below the bottom of the footing and does not affect the bearing capacity. Neglect passive pressure and the weight of the backfill over the toe.

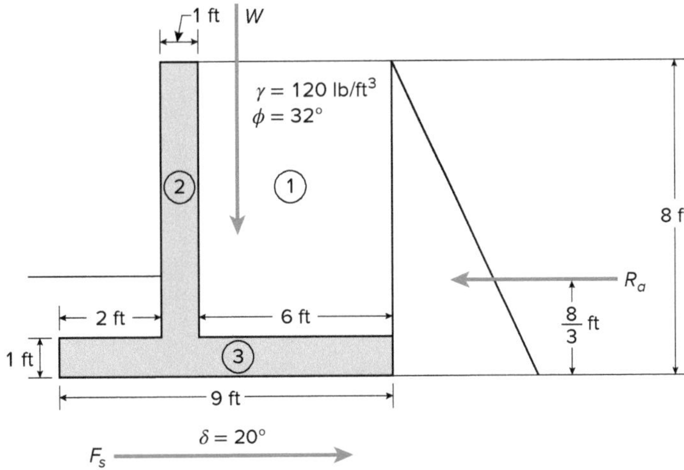

Solution

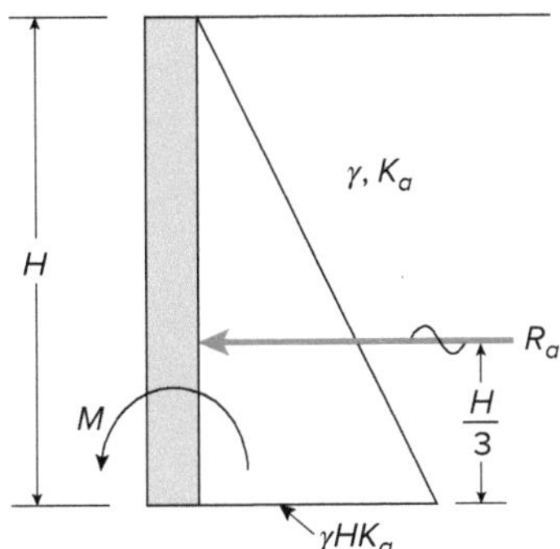

Determine the resultant active earth force, R_a:

$$K_a = \tan^2\left(45° - \frac{30°}{2}\right) = 0.333$$

$$R_a = \frac{\gamma H^2 K_a}{2} = \frac{(120 \text{ lb/ft}^3)(8 \text{ ft})^2(0.333)}{2} = 1{,}279 \text{ lb/ft}$$

Example 3.2 *(continued)*

Determine the sum of the vertical forces and the moment about the toe (complete the table below):

	W (LB/FT)	ARM (FT)	M_R (FT-LB/FT)
1	$6 \times 7 \times 120 = 5{,}040$	6	$5{,}040 \times 6 = 30{,}240$
2	$1 \times 7 \times 150 = 1{,}050$	2.5	$1{,}050 \times 2.5 = 2{,}625$
3	$9 \times 1 \times 150 = 1{,}350$	4.5	$1{,}350 \times 4.5 = 6{,}075$
	$\Sigma V = 7{,}440$ lb/ft		$\Sigma M_R = 38{,}940$ ft-lb/ft

For factor of safety against sliding:

$$F_s = \Sigma V \tan \delta = (7{,}440 \text{ lb/ft}) \tan 20° = 2{,}708 \text{ lb/ft}$$

$$(\text{FS})_{\text{slide}} = \frac{\Sigma F_R}{\Sigma F_O} = \frac{F_s}{R_a} = \frac{2{,}708 \text{ lb/ft}}{1{,}279 \text{ lb/ft}} = 2.12$$

For factor of safety against overturning:

$$(\text{FS})_{OT} = \frac{\Sigma M_R}{\Sigma M_O} = \frac{\Sigma M_R}{R_a \times \frac{H}{3}} = \frac{38{,}940 \text{ ft-lb/ft}}{(1{,}279 \text{ lb/ft})\left(\frac{8 \text{ ft}}{3}\right)} = 11.4$$

3.1.6 Equivalent Fluid Density

It is convenient to express the lateral pressures acting on retaining walls in terms of an equivalent fluid density (EFD, γ_{eq}). This simplifies calculations and relieves the structural engineer from having to estimate the weight of the backfill and retained soils in order to interpret the lateral earth pressure for design.

3.1.6.1 Drained Backfill

Free-draining granular backfill typically weighs less than the retained soil. The equivalent fluid density should be calculated based on the weight of the granular backfill in the entire Rankine zone (45° + ϕ/2) behind the wall. If the entire Rankine zone is not backfilled with granular fill, the weight used to calculate the equivalent fluid density should be increased in proportion to the size and shape of the backfill zone. The equivalent fluid density is given by:

$$\gamma_{\text{eq}} = k\gamma_t \qquad \textbf{Equation 3-20}$$

γ_t = the total (moist) unit weight of the backfill or retained soil, whichever controls

$K = k_o$ for at-rest pressure, k_a for active pressure, or k_p for passive pressure

3.1.6.2 Saturated Backfill

When groundwater rises behind a retaining wall, the backfill remains saturated. The effect of the groundwater must be added. For saturated backfill, the equivalent fluid density is given by:

$$\gamma_{\text{eq}} = k\gamma_{\text{sat}} + (1 - k)\gamma_{\text{w}} \qquad \textbf{Equation 3-21}$$

γ_{sat} = the saturated unit weight of the soil

Example 3.3: Saturated Backfill—Equivalent Fluid Density

Soil used to backfill a cantilever retaining wall has a friction angle of 30° and unit weights γ_t = 115 lb/ft^3 and γ_{sat} = 118 lb/ft^3. Determine the Rankine active equivalent fluid density (EFD) for drained and submerged conditions.

Example 3.3 *(continued)*

Solution

The Rankine active earth pressure coefficient (Equation 3-7) is:

$$k_a = \tan^2\left(45° - \frac{\phi}{2}\right) = \tan^2\left(45° - \frac{30°}{2}\right) = 0.333$$

EFD for the drained condition (Equation 3-20):

$$\gamma_{eq} = k\gamma_t = 0.333 \times 115\ \text{lb/ft}^3 = 38\ \text{lb/ft}^3$$

EFD for the saturated condition:

$$\gamma_{eq} = k\gamma_{sat} + (1 - k)\gamma_w = 0.333 \times 118\ \text{lb/ft}^3 + (1 - 0.333) \times 62.4 = 81\ \text{lb/ft}^3$$

In recommendations, these values would be rounded to 40 lb/ft^3 and 80 lb/ft^3.

3.2 SOIL CONSOLIDATION

Settlement of fine-grained soils occurs in three stages:

1. Immediate or elastic settlement occurs rapidly. Analysis is based on the theory of elasticity.
2. Primary consolidation settlement occurs due to the expulsion of water from soil pores.
3. Secondary compression (creep) occurs as soil particles readjust, reorient, and crush. Figure 3.9 presents the stages of settlement for fine-grained soils.

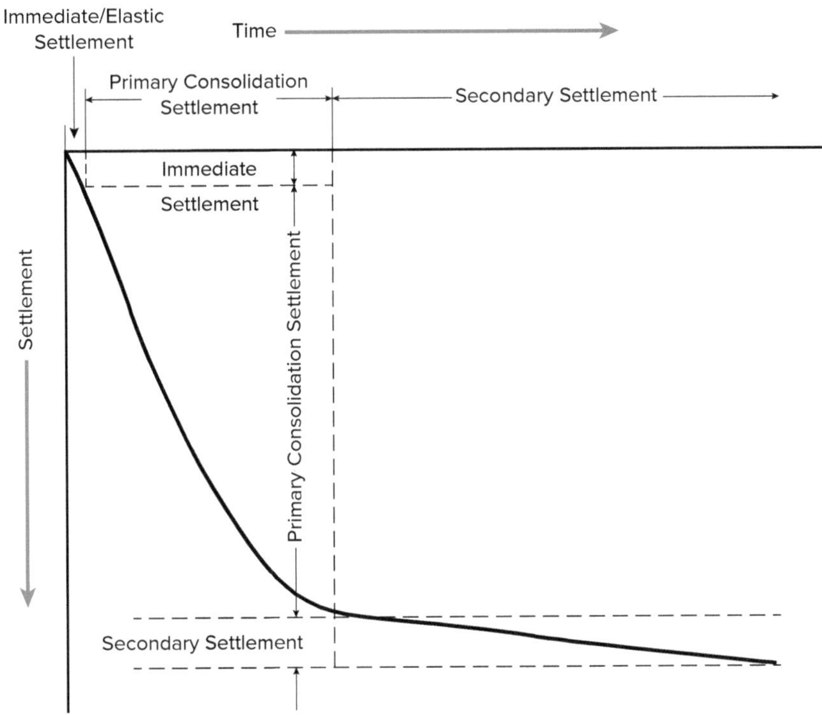

FIGURE 3.9 Three Phases of Settlement in Fine-Grained Soils

Consolidation is the decrease in volume of a fully saturated fine-grained soil with low permeability under applied pressure. **Primary consolidation** is due to dissipation of excess pore pressure (Δ_u) with time. The extended time required for primary consolidation in fine-grained soils (silts and clays) is due to their low permeability. Coarse-grained soils (sands and gravels) undergo consolidation settlement, but at a much faster rate due to their higher permeability.

Secondary compression occurs after the excess pore pressure (Δ_u) is dissipated and the soil particles undergo readjustment, reorientation, and sometimes crushing.

The process of consolidation is often confused with the process of compaction. Compaction increases the density of an unsaturated soil by mechanically reducing the volume of air voids. Consolidation is a time-related process of increasing the density of a saturated low-permeability soil under pressure by squeezing water from soil voids.

When pressure increases on a fully saturated low-permeability fine-grained soil, initially at time $t = 0$, the water in the pores takes all of the pressure and the soil particles are not affected by it. Over time, the water carrying the excess pressure will be squeezed from the pores, and the pressure will be transferred to the soil particles. The reduction in void space as the water is squeezed from the soil results in consolidation settlement. This process will continue until the excess water pressure is discharged and the excess pressure is transferred to the soil particles. At that time ($t = \infty$), which could be months or years, the primary consolidation process is complete.

3.2.1 Stress History of Clay Soils

Fine-grained soils have memory of stress changes. The maximum pressure a fine-grained soil has been subjected to in recent history can be estimated from laboratory tests. This is a unique property of fine-grained soils.

When fine-grained soils are encountered during field exploration, undisturbed (Shelby tube) samples should be collected for consolidation testing (oedometer test). The in-place effective vertical stress (σ_0') should be recorded. This is the current vertical stress on the soil (also called overburden pressure). The undisturbed sample is then extruded from the Shelby tube and is trimmed and put into the loading machine. It is then saturated and loaded in multiple increments of pressure (Fig. 3.10). The vertical compression of the clay is recorded for each increment until primary compression is complete or for 24 hours.

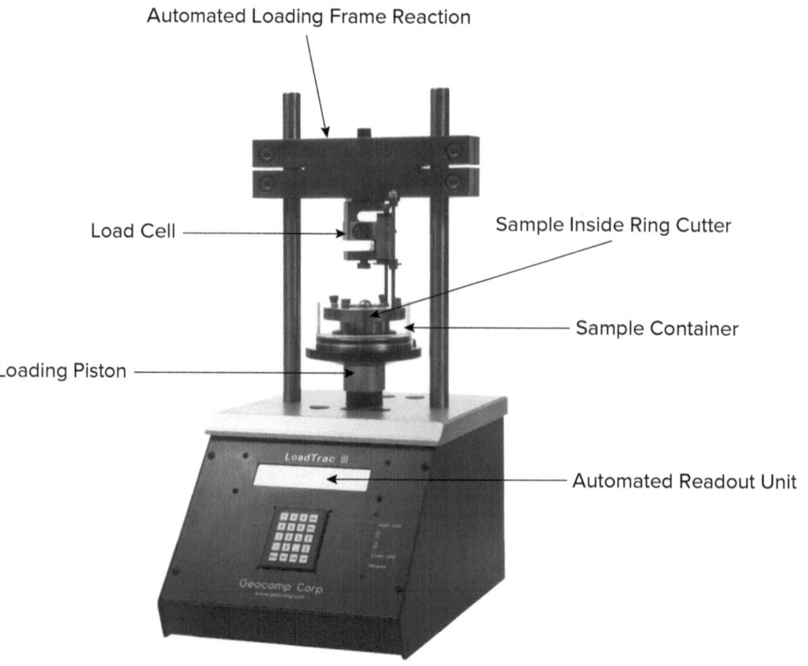

FIGURE 3.10 Fully Automated Consolidation Test Equipment with Computerized Loading Frame and Data Collection System

Photo courtesy of GeoComp Corporation.

The **final compression** readings are used to estimate the primary consolidation settlement. A typical plot of void ratio or strain versus effective vertical pressure (log scale) for fine-grained soils is shown in Figure 3.11. Each point on this plot is the final reading after the end of primary consolidation or after 24 hours of applying a pressure increment. A plot of vertical compression readings versus time for a pressure increment is used to estimate the time rate of settlement.

Figure 3.11 is a plot of void ratio versus log effective stress for a clay soil. The curve from point A to B is known as the recompression curve. The relatively flat slope for this part of the plot represents recompression of the sample that has undergone swell as a result of the stress relief experienced when it was removed from the ground and/or the effect of overconsolidation of the in-field clays. A significant change in the slope occurs at point B. The vertical effective stress at this point is the preconsolidation pressure. This is the highest stress to which the soil has been consolidated in the recent past. The curve from point B to C is called the virgin compression curve. The plot from point C to D is called the swelling, or rebound, curve, which represents the swell or rebound that the soil undergoes as a result of unloading. The slope of this curve is similar to the slope of the curve from point A to B.

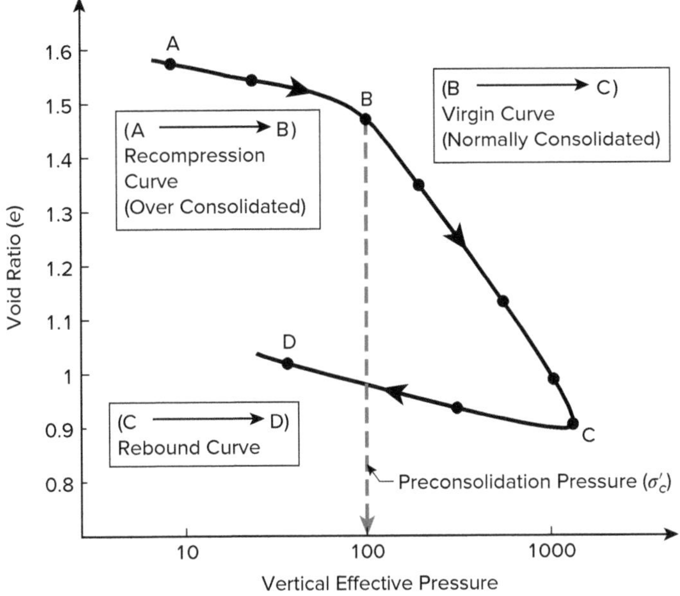

FIGURE 3.11 Typical One-Dimensional Consolidation Test

Based on the value of the preconsolidation pressure (σ'_c) compared with the current overburden pressure, clays are classified into two categories:

- Normally consolidated (NC) clay: The present effective overburden pressure is the maximum pressure the soil has been subjected to in the recent past ($\sigma'_o \approx \sigma'_c$).
- Overconsolidated (OC) clay: The present effective overburden pressure is less than what the soil has seen in the recent past. The past maximum effective overburden pressure is called the preconsolidation pressure ($\sigma'_o < \sigma'_c$).

σ'_o = present effective overburden pressure (in situ stress)

σ'_c = preconsolidation pressure

The overconsolidation ratio (OCR), which represents the level of preconsolidation of overconsolidated clays, is defined as follows:

$$OCR = \frac{\sigma'_c}{\sigma'_o}$$ **Equation 3-22**

3.2.2 Consolidation of Normally Consolidated (NC) Clay

For NC clay, the primary consolidation settlement is calculated using the compression index, C_c (see Fig. 3.12):

$$S_c = \Sigma\left(\frac{C_c}{1+e_o}\right)H\log\left(\frac{\sigma'_f}{\sigma'_o}\right)$$ **Equation 3-23**

S_c = primary consolidation settlement

σ'_o = in situ effective overburden pressure at the midpoint of the clay layer

σ'_f = final effective overburden pressure ($\sigma'_o + \Delta\sigma$)

$\Delta\sigma$ = the stress increase at the midpoint of the clay layer

H = thickness of the soil layer (If the compressible stratum is thick, it should be split into sublayers to improve the accuracy of the settlement calculations.)

C_c = compression index (slope of the virgin compression curve)

e_o = initial void ratio

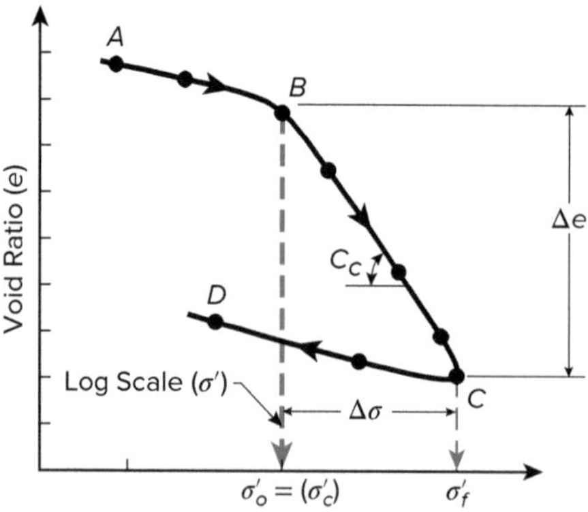

FIGURE 3.12 One-Dimensional Consolidation Test of NC Clay

$$C_C = \frac{\Delta e}{\log \frac{\sigma'_f}{\sigma'_o}}$$ **Equation 3-24**

Example 3.4: Consolidation of NC Clay

A square mat foundation will be constructed at the ground surface of the soil profile shown in the figure below. The foundation will increase the vertical stress at the midpoint of the NC clay layer by 450 lb/ft^2. Determine the primary consolidation settlement of the NC layer.

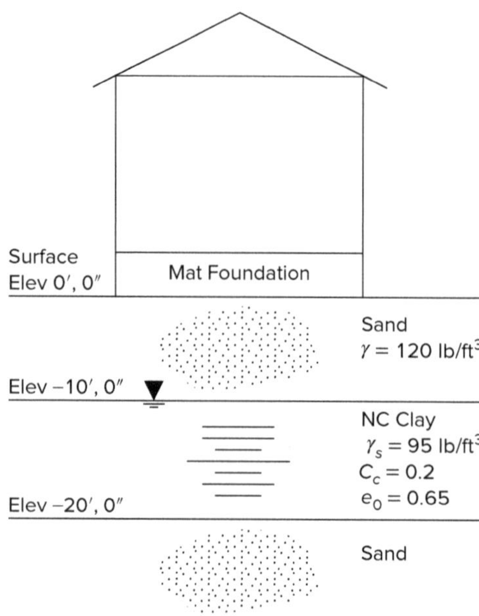

Soil Mechanics

School of PE

Example 3.4 *(continued)*

Solution

Determine the initial and final effective overburden pressures at the midpoint of the NC layer.

$$\sigma_o' = \left(120\,\text{lb/ft}^3\right)(10\text{ ft}) + \left(95\,\text{lb/ft}^3 - 62.4\,\text{lb/ft}^3\right)(5\text{ ft}) = 1{,}363\text{ lb/ft}^2$$

$$\sigma_f' = 1{,}363\text{ lb/ft}^2 + 450\text{ lb/ft}^2 = 1{,}813\text{ lb/ft}^2$$

Determine the primary consolidation settlement of the normally consolidated clay layer.

$$S_c = \left(\frac{C_c}{1+e_o}\right)H\log\left(\frac{\sigma_f'}{\sigma_o'}\right) = \left(\frac{0.2}{1+0.65}\right)(10\text{ ft})\log\left(\frac{1{,}813\text{ lb/ft}^2}{1{,}363\text{ lb/ft}^2}\right)$$

$S_c = 0.15$ ft or 1.8 in

3.2.3 Consolidation of Overconsolidated (OC) Clay

Two cases are shown for overconsolidated clays in Figure 3.13 based on the values of the in situ effective stress (σ_o'), the preconsolidation pressure (σ_c'), and the final pressure (σ_f').

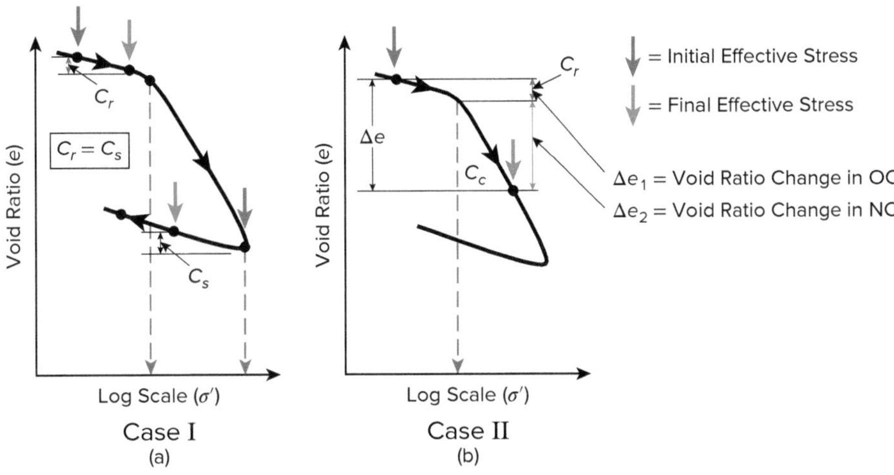

FIGURE 3.13 One-Dimensional Consolidation Tests of OC Clays (Case I and Case II)

3.2.3.1 Case I: $\sigma_f' \leq \sigma_c'$

For overconsolidated (OC) clays with a final effective overburden pressure that is less than the preconsolidation pressure, primary consolidation settlement is calculated using Equation 3-25:

$$S_c = \sum\left(\frac{C_r}{1+e_o}\right)H\log\left(\frac{\sigma_f'}{\sigma_o'}\right) \qquad \textbf{Equation 3-25}$$

S_c = primary consolidation settlement

σ_o' = in situ effective overburden pressure at the midpoint of the clay layer

σ_f' = final effective overburden pressure ($\sigma_o' + \Delta\sigma$)

H = thickness of the soil layer (If the compressible stratum is thick, it should be split into sublayers to improve the accuracy of the settlement calculations.)

C_r = recompression index (slope of the recompression line)

e_o = initial void ratio

3.2.3.2 Case II: $\sigma'_f > \sigma'_c$

For OC clays with a final effective overburden pressure that is higher than the preconsolidation pressure, primary consolidation settlement is calculated with a two-stage solution, using both the recompression index, C_r, and the compression index, C_c:

$$S_c = \Sigma\left[\left(\frac{C_r}{1+e_o}\right)H\log\left(\frac{\sigma'_c}{\sigma'_o}\right) + \left(\frac{C_c}{1+e_o}\right)H\log\left(\frac{\sigma'_f}{\sigma'_c}\right)\right]$$ **Equation 3-26**

$\Delta e_1 \Rightarrow$ $\Delta e_2 \Rightarrow$

S_c = primary consolidation settlement

σ'_o = initial effective overburden pressure at the midpoint of the clay layer

σ'_f = final effective overburden pressure ($\sigma'_o + \Delta\sigma$)

H = thickness of the soil layer (If the compressible stratum is thick, it should be split into sublayers to improve the accuracy of the settlement calculations.)

C_r = recompression index (slope of the recompression curve)

C_c = compression index (slope of the virgin compression curve)

e_o = initial void ratio

3.2.4 Empirical Relationships of C_c and C_r

If time and budget do not allow consolidation tests to be performed to estimate the consolidation parameters needed for calculations, the compression and recompression indices can be estimated from the following empirical relationships:

TIP

It should be noted that the margin of error in using these equations could be in the order of ± 40%.

For all clays: $C_c = 1.15(e_o - 0.35)$ and C_r = 5% to 10% of C_c

For undisturbed clay of low to moderate sensitivity: $C_c = 0.009(LL - 10)$

e_o = initial void ratio

LL = liquid limit (%)

3.2.5 Rate of Consolidation

In addition to estimating the primary consolidation settlement, it is important for the engineer to consider the settlement of clays with time. The goal is to be able to answer the question of how long it will take for a percentage of the consolidation settlement to occur or how much settlement will occur after a certain time. Consolidation could continue for months or even years depending on the thickness of the consolidating layer and drainage conditions.

The average degree of consolidation, U, is defined as the percent of target settlement relative to the full primary consolidation settlement.

$$U = \frac{S_t}{S_c}\%$$ **Equation 3-27**

S_t = settlement of the layer at time t

S_c = total primary consolidation settlement

The average degree of consolidation of a saturated clay layer is a function of the nondimensional time factor, T_v, which is an important factor in estimating the time for a certain degree of consolidation as follows:

$$T_v = \frac{C_v t}{H_d^2}$$ **Equation 3-28**

T_v = time factor (directly correlated to the degree of consolidation U, from Table 3.1)

c_v = coefficient of consolidation (length/time^2)

t = time for the target degree of consolidation

H_d = length of the shortest drainage path (H_d for single (one-way) drainage; $\frac{H_d}{2}$ for double (two-way) drainage—see Fig. 3.14)

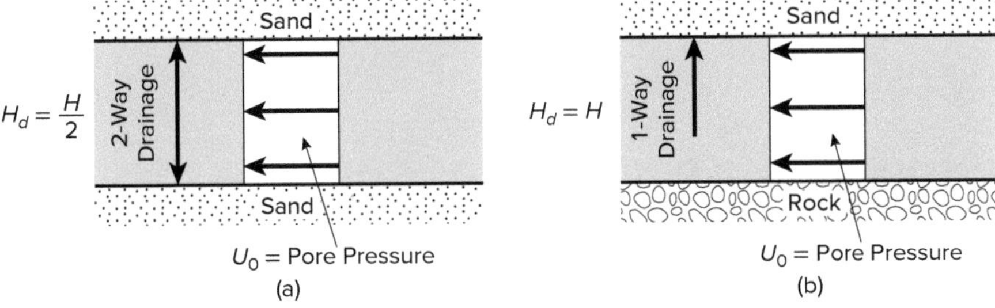

FIGURE 3.14 Two-Way and One-Way Drainage Paths

TABLE 3.1 Average Degree of Consolidation, U, Versus Time Factor, T_v, for Uniform Initial Increase in Pore Water Pressure

U %	T_V
0	0.000
10	0.008
20	0.031
30	0.071
40	0.126
50	0.197
60	0.287
70	0.403
80	0.567
90	0.848
93.1	1.000
95.0	1.163
98.0	1.500
99.4	2.000
100.0	Infinity

Source: *Soils and Foundations Reference Manual (FHWA-NHI-06-088)*, 2006 [1].

Example 3.5: Rate of Consolidation

A 15-ft-thick clay is bounded by sand at the top and bottom. The clay has a coefficient of consolidation of 0.3 ft^2/day. Determine the time when 50% and 90% of the total settlement will occur.

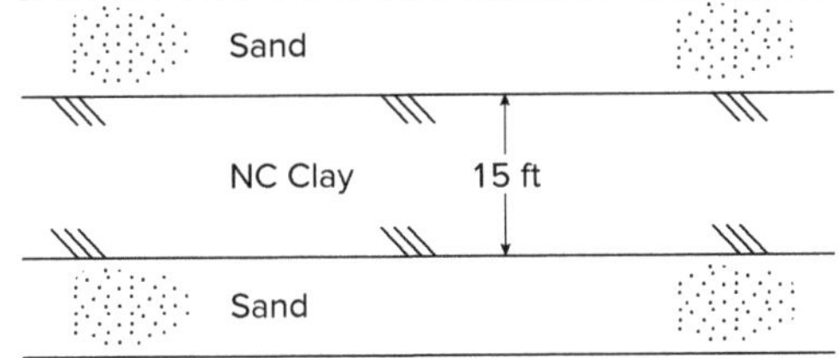

Solution

Double drainage $\rightarrow H_d = \frac{15\text{ ft}}{2} = 7.5\text{ ft}$

Example 3.5 (continued)

From Table 3.1:

For $U = 50\% \rightarrow T_v = 0.197$

For $U = 90\% \rightarrow T_v = 0.848$

Calculate the time for 50% and 90% of consolidation to occur:

$$t_{50} = \frac{T_v H_d^2}{C_v} = \frac{0.197 \times 7.5}{0.3} = 37 \text{ days}$$

$$t_{90} = \frac{T_v H_d^2}{C_v} = \frac{0.848 \times 7.5}{0.3} = 159 \text{ days}$$

3.3 EFFECTIVE AND TOTAL VERTICAL STRESSES

There are two types of vertical stresses in a soil mass: (1) stresses due to the soil's own weight and (2) stresses developed from applying loads or stresses from foundations, retaining walls, embankments, tanks, fills, and so forth. This section explains the concept of stresses in a soil mass due to its own weight.

3.3.1 Total Vertical Stress

Total vertical stress, σ_v, is the summation of stresses generated from the weight of the soil above a certain level in the ground (z):

$$\sigma_v = \Sigma \gamma_i Z_i$$ **Equation 3-29**

σ_v = total vertical stress

γ_i = total unit weight of the soil layer(s) (The unit weight could be dry, moist, or saturated.)

z_i = thickness of soil layer(s)

For example, the total vertical stress at point A for the soil profile shown in Figure 3.15 is calculated as:

$$\sigma_v = \gamma_1 Z_1 + \gamma_2 Z_2$$ **Equation 3-30**

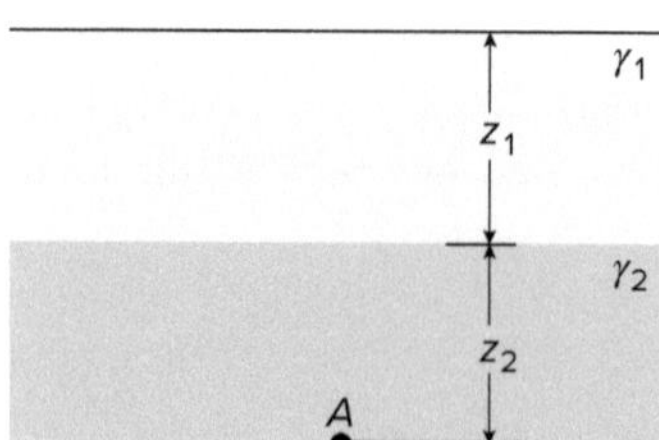

FIGURE 3.15 Total Vertical Stress

3.3.2 Pore Water Pressure

Pore water pressure is the result of the buoyant force, u, exerted by water in the soil mass. In most cases, the pore water pressure is hydrostatic. Seepage cases or water under artesian pressure need to be carefully evaluated.

3.3.2.1 Hydrostatic Pore Water Pressure

In hydrostatic conditions (Fig. 3.16a), the pressure head is equal to the distance between the point of interest and the groundwater surface (phreatic surface). The pore water pressure is calculated as follows:

$$u = \gamma_w Z_w \qquad \textbf{Equation 3-31}$$

u = pore water pressure

γ_w = unit weight of water, 62.4 lb/ft^3

z_w = depth from the point of interest to the groundwater surface

3.3.2.2 Seepage or Artesian Pore Water Pressure

In case of upward/downward seepage or artesian conditions (Fig. 3.16b), the pore water pressure is calculated as follows:

$$u = \gamma_w h_p \qquad \textbf{Equation 3-32}$$

u = pore water pressure

γ_w = unit weight of water

h_p = pressure head at the point of interest

TIP

The pressure head is the distance between the point of interest and the free water surface (in a piezometer or pore pressure transducer).

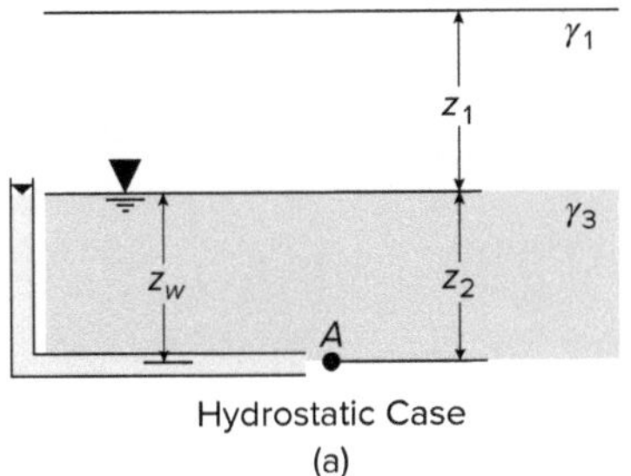

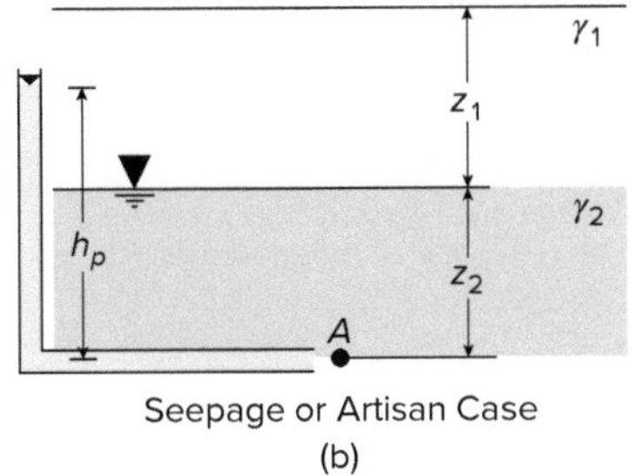

FIGURE 3.16 Examples of Pressure Head for Hydrostatic (left) and Seepage (right) Cases

Example 3.6: Pore Water Pressure

Calculate the pore pressure at point A for both cases shown below.

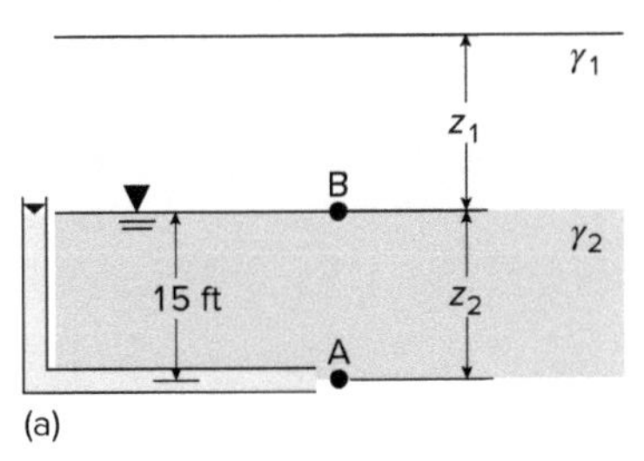

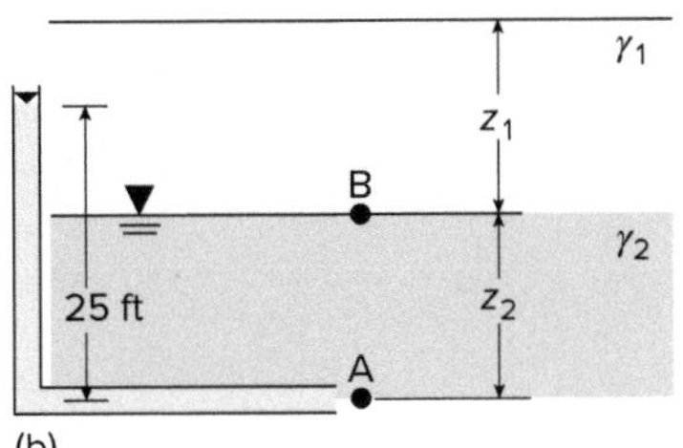

Solution

(a) Hydrostatic condition: $u = \gamma_w z_w = 62.4 \text{ lb/ft}^3 \times 15 \text{ ft} = 936 \text{ lb/ft}^2$

(b) Seepage condition: $u = \gamma_w h_p = 62.4 \text{ lb/ft}^3 \times 25 \text{ ft} = 1{,}560 \text{ lb/ft}^2$

Note that $u = 0$ at point B in both cases.

3.3.3 Effective Vertical Stress

The **effective vertical stress** is the portion of the total stress that will be supported through the soil's particle contact. The vertical effective stress, σ'_v, in a soil element at a depth, z, is the total vertical stress, σ_v, imposed by the total weight above minus the pore water pressure, u. It is simply the total stress minus the buoyancy effect.

$$\sigma'_v = \sigma_v - u$$ **Equation 3-33**

σ'_v = effective vertical stress

σ_v = total vertical stress

u = pore water pressure

Example 3.7: Vertical Stresses

Assuming hydrostatic conditions, determine the total and effective vertical stresses at points A and B for the soil profile shown:

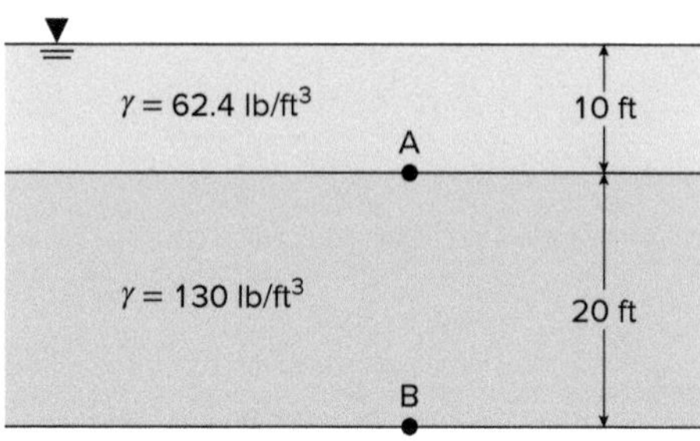

Solution

The condition shown is in a waterway (such as a lake, river, or sea) because the water surface is above the ground surface.

Total vertical stress:

$$\text{Point A: } \sigma_v = \left(62.4 \text{ lb/ft}^3\right)(10 \text{ ft}) = 624 \text{ lb/ft}^2$$
$$\text{Point B: } \sigma_v = \left(624 \text{ lb/ft}^2\right) + \left(130 \text{ lb/ft}^3\right)(20 \text{ ft}) = 3{,}224 \text{ lb/ft}^2$$

Effective vertical stress:

$$\text{Point A: } \sigma'_v = 624 \text{ lb/ft}^2 - \left(62.4 \text{ lb/ft}^3\right)(10 \text{ ft}) = 0 \text{ lb/ft}^2$$
$$\text{Point B: } \sigma'_v = 3{,}224 \text{ lb/ft}^2 - \left(62.4 \text{ lb/ft}^3\right)(30 \text{ ft}) = 1{,}352 \text{ lb/ft}^2$$

In hydrostatic conditions, the vertical effective stress may also be calculated in one step by utilizing the unit weight (dry or moist) above the water table and utilizing the submerged (effective) unit weight for the soil below the water table. The submerged (effective) unit weight can be estimated as:

$$\gamma_{\text{submerged}} = \gamma' = (\gamma_{\text{saturated}} - \gamma_{\text{water}})$$ **Equation 3-34**

In Example 3.7:

Point B: $\sigma'_v = 20 \text{ ft} \times (130 \text{ lb/ft}^3 - 62.4 \text{ lb/ft}^3) = 1{,}352 \text{ lb/ft}^2$

3.4 BEARING CAPACITY

Foundations may be classified as shallow, intermediate, and deep. **Shallow foundations**, such as spread footings, strip footings, combined footings, and mat/raft foundations, are generally supported on shallow soils within a few feet of the ground surface. They rely on shallow soils to provide enough bearing capacity and acceptable total and differential settlements. **Deep foundations** extend deep into the ground to transfer the structural load to the deep competent soils, bypassing the shallow weak soils. Deep foundations are superior to shallow foundations in terms of load-carrying capacity; however, they are more expensive and require longer time to install, test, and approve. **Intermediate foundations** are shallow foundations supported on improved soils. Soil improvement can increase the bearing capacity and reduce the potential settlement of weak shallow soils. In many projects, soil improvement may provide a technically sound and economically viable foundation solution. This section will concentrate on shallow foundations on natural unimproved soils.

Generally, for shallow foundations, the foundation depth (D_f) is less than its width (B). For deep foundations, D_f is substantially larger than the foundation width B ($D_f/B > 10$).

Figure 3.17 presents the main components of a typical shallow foundation where: B is the width, L is the length, and D_f is the bearing depth measured to the bottom of the footing. P_{net} is the column load from the structure, W_s is the soil fill weight above the footing, and W_c is the weight of the concrete footing.

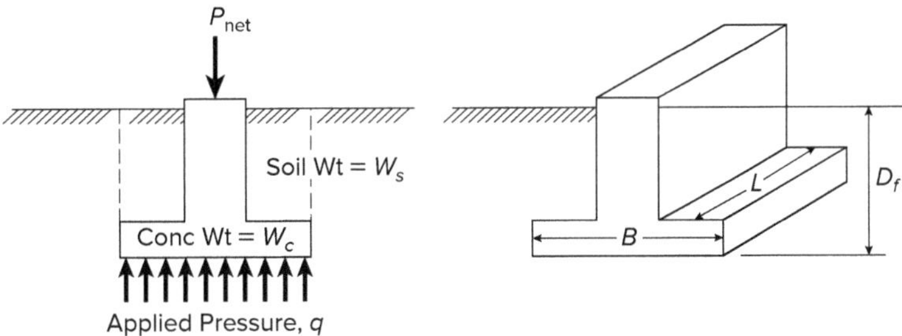

FIGURE 3.17 Shallow Spread Footing Dimensions

Table 3.2 provides examples of appropriate foundation choices for various soil conditions (not exhaustive). Also consider that driven piles may not be appropriate if they must penetrate soil layers with large obstructions (boulders, fill with construction debris, and so forth) that may cause deflection or pile tip damage, or prevent the pile from reaching design embedment depth.

TABLE 3.2 Foundation Types Correlated with Soil Conditions

SOIL CONDITIONS	APPROPRIATE FOUNDATION TYPE AND LOCATION	DESIGN COMMENTS
(1) El.0 Ground Surface; Compact Sand, (Deposit to Great Depth)	Installation Below Frost Depth or Where Erosion Might Occur	Spread footings are most appropriate for conventional foundation needs. A deep foundation such as piles could be required if uplift or other unusual forces (for example, seismic, effect of flood) could act.
(2) El.0; Firm Clay or Firm Silt and Clay (to Great Depth)	Installation Depth Below Frost Depth, or Below Zone Where Shrinkage and Expansion Due to Change in Water Content Could Occur	Spread footings are most appropriate for conventional foundation needs.

SOIL CONDITIONS	APPROPRIATE FOUNDATION TYPE AND LOCATION	DESIGN COMMENTS
(3) 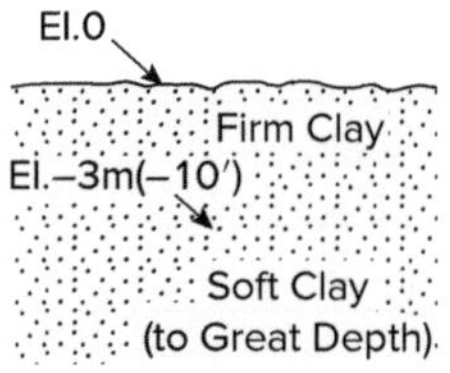 El.0 Firm Clay El.–3m(–10′) Soft Clay (to Great Depth)	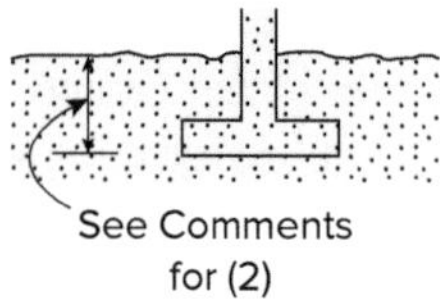 See Comments for (2)	Spread footings would be appropriate for a low-to-medium range of loads, if not installed too close to a soft clay layer. If heavy loads are to be carried, deep foundations might be required.
(4) 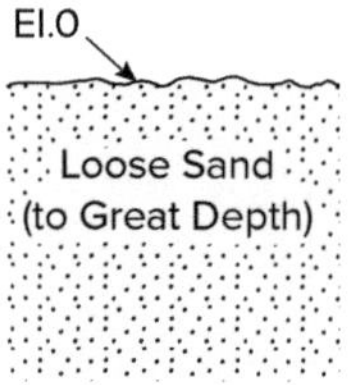 El.0 Loose Sand (to Great Depth)	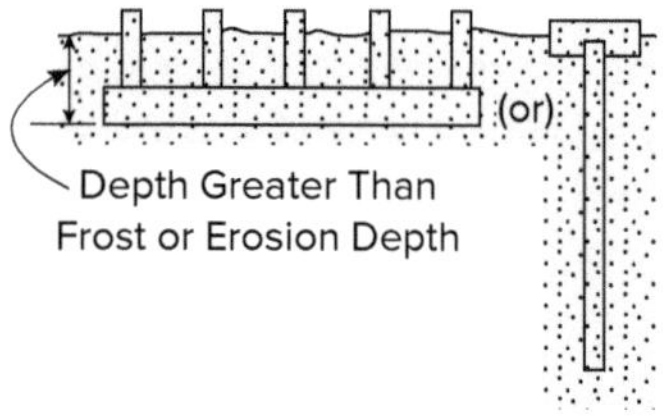 (or) Depth Greater Than Frost or Erosion Depth	Spread footing may settle excessively or require very low bearing pressures. Consider a mat foundation or ground improvement (for example, vibroflotation). Driven piles could be used and would densify the sand.
(5) 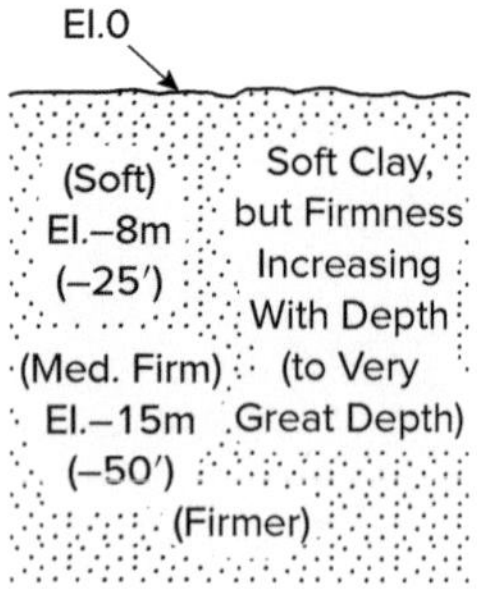 El.0 (Soft) El.–8m (–25′) (Med. Firm) El.–15m (–50′) (Firmer) Soft Clay, but Firmness Increasing With Depth (to Very Great Depth)	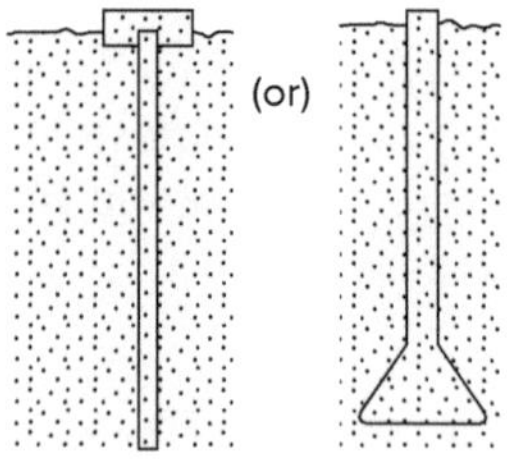 (or)	Spread footing is probably not appropriate. Friction piles or piers would be satisfactory if some settlement could be tolerated. Long piles would reduce settlement problems. Also consider mat or floating foundation.
(6) 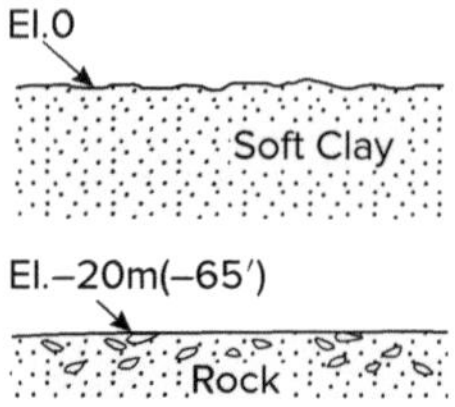 El.0 Soft Clay El.–20m(–65′) Rock	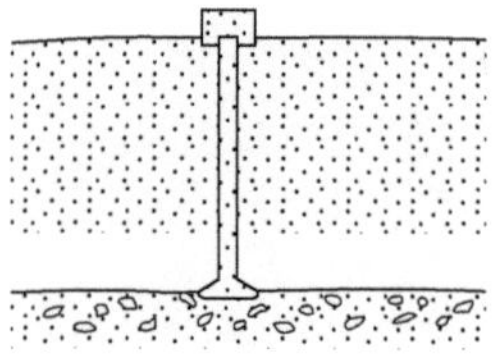	Deep foundation—piles, piers, caissons—bearing on or in rock.
(7) 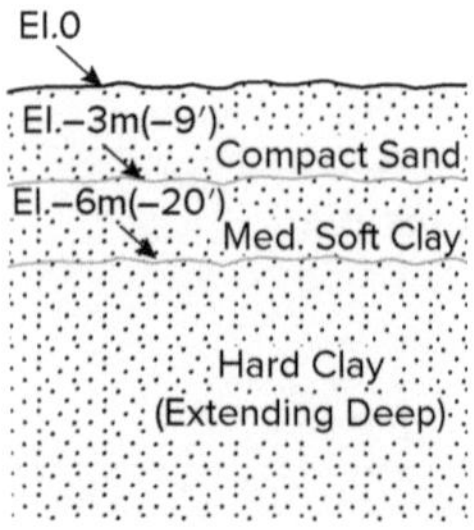 El.0 El.–3m(–9′) Compact Sand El.–6m(–20′) Med. Soft Clay Hard Clay (Extending Deep)	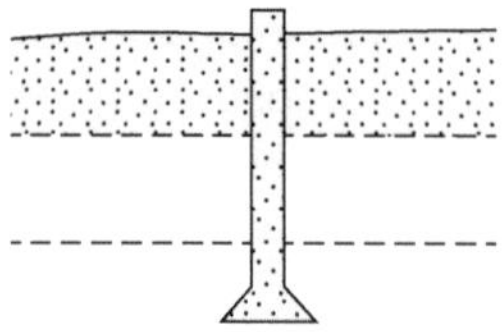	Spread footings in upper sand layer would probably experience large settlement because of the underlying soft clay layer. Consider drilled piers with a bell formed in a hard clay layer, or other pile foundation into a hard clay layer.
(8) 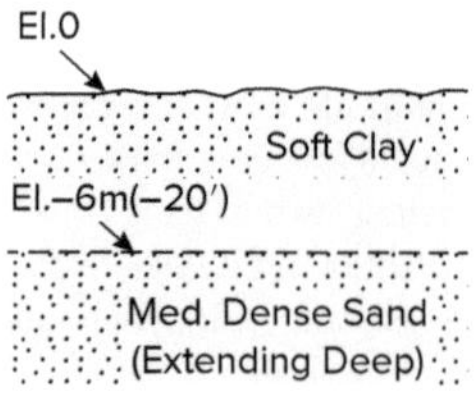 El.0 Soft Clay El.–6m(–20′) Med. Dense Sand (Extending Deep)	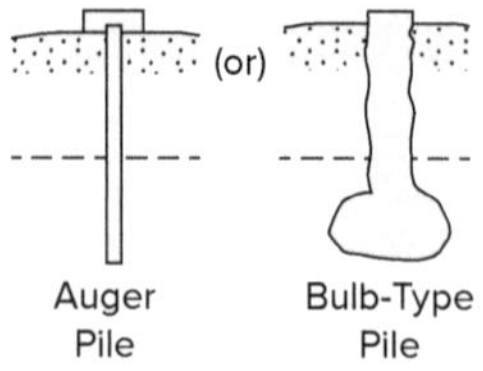 (or) Auger Pile Bulb-Type Pile	Deep foundation—cast-in-place piles. Augercast piles or bulb piles into a sand layer appear most appropriate.
(9) 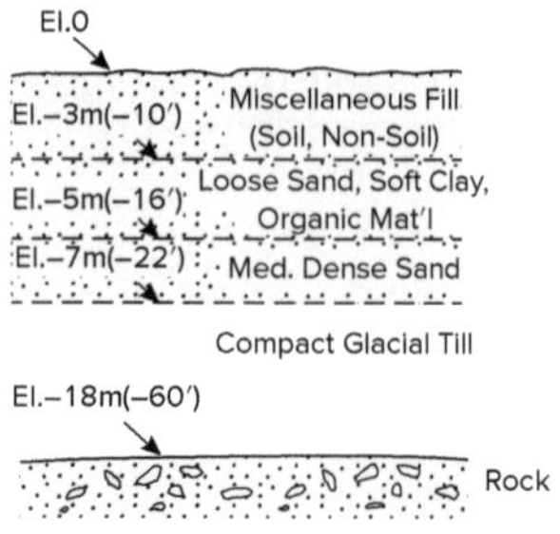 El.0 Miscellaneous Fill (Soil, Non-Soil) El.–3m(–10′) Loose Sand, Soft Clay, Organic Mat'l El.–5m(–16′) Med. Dense Sand El.–7m(–22′) Compact Glacial Till El.–18m(–60′) Rock	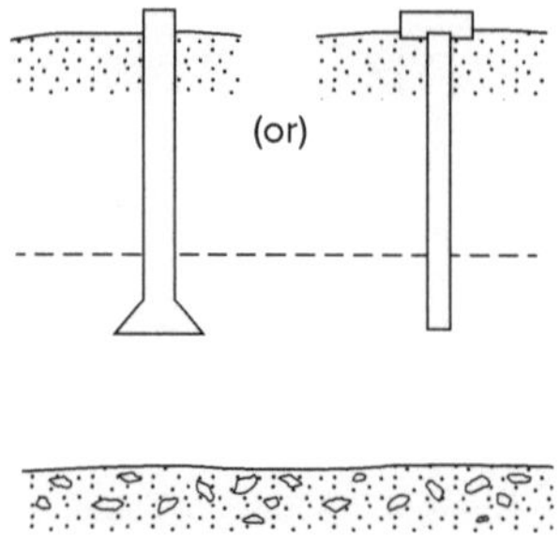 (or)	Deep foundations will extend into medium dense sand or, preferably, into compact glacial till. There is a strong possibility for drilled piers with bell constructed in till. Also, consider cast-in-place and driven concrete piles, timber piles, and pipe piles.

SOIL CONDITIONS	APPROPRIATE FOUNDATION TYPE AND LOCATION	DESIGN COMMENTS
(10) El.0 El.–2.5m(–8′) Miscellaneous Fill (Poor) Med. Dense Sand El.–12m(–40′) Med. Firm Clay El.–30m(–100′) Rock	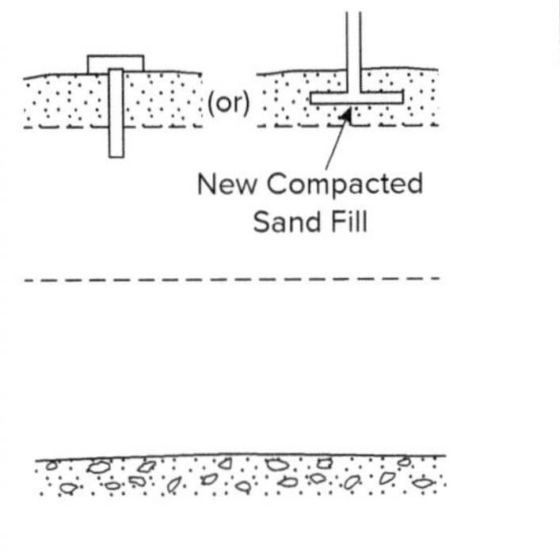 (or) New Compacted Sand Fill	Deep foundations penetrating through the fill are appropriate. With piles or piers, consider stopping in the upper zone of the sand layer so as to limit compression of the clay layer. Also consider replacing poor fill with compacted structural fill, and then using spread footings in the new fill.
(11) El.0 Soft Clay El.–12m(–40′) Med. Dense to Dense Sand El.–20m(–65′) Soft Clay (to Rock) El.–45m(–150′) Rock	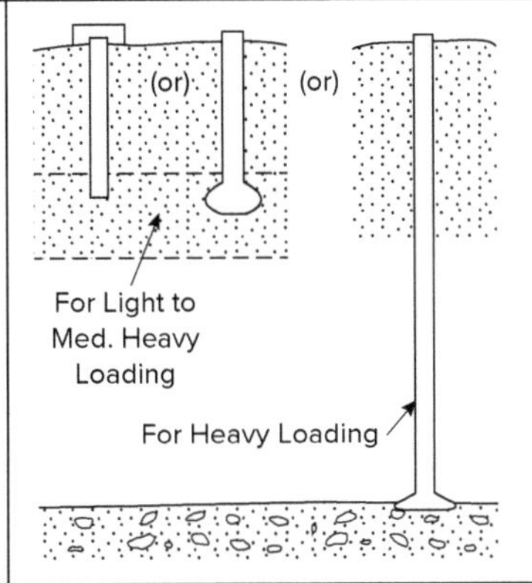 (or) (or) For Light to Med. Heavy Loading For Heavy Loading	If foundation loads are not too heavy, consider using piles of piers bearing in the upper zone of the sand layer and check for settlement. If foundation loads are heavy, consider driven piles (steel) or caissons to rock.
(12) El.0 Miscellaneous Soil and Non-Soil Fill El.–2.5m(–8′) Loose Sand and Soft Clay El.–5m(–16′) Rock	(or) Basement Sub-basement	For cases where rock is close to the ground surface, piers or piles may be utilized. If a basement is planned, consider extending the basement to the rock material.

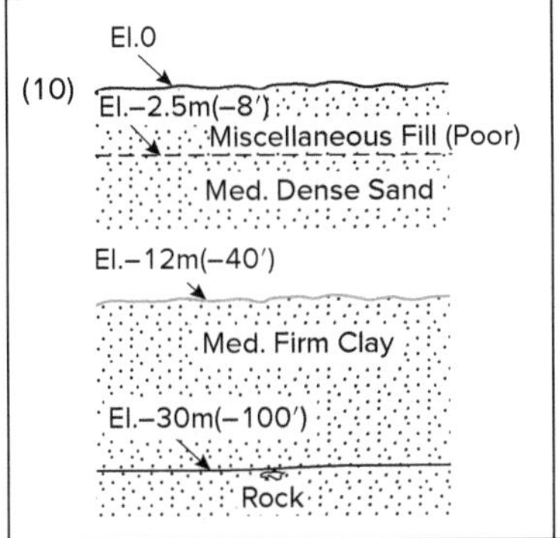

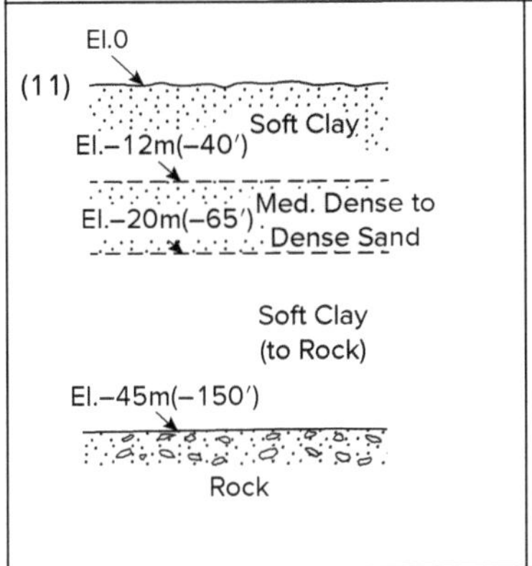

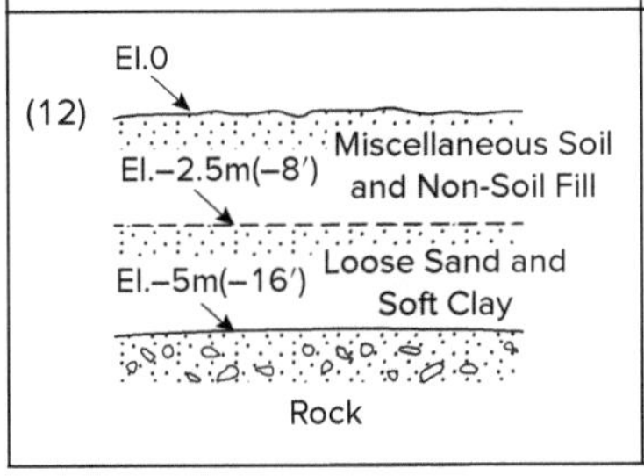

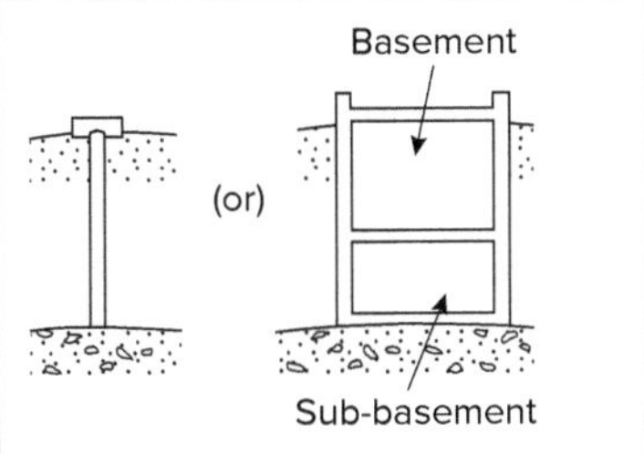

3.4.1 Structural Loads (Demand)

Most buildings consist of floors and roof slabs that carry their own weight and the dead and live loads. The loads are then transferred from the slabs to the edge beams, then to the columns, and then through the columns to the ground. Therefore, for foundation design purposes, a building can be represented by columns of different sizes carrying different loads. The gross bearing pressure is the total applied pressure of the foundation acting on the soil at the base of the foundation, including the column or net load (P_{net}), the weight of the footing (W_c), and the overlying soil (W_s). The gross bearing pressure is compared with the ultimate bearing capacity of the soil.

$$Q_g = \frac{P_g}{A}$$ **Equation 3-35**

$$P_g = P_{net} + W_c + W_s$$

Q_g = gross bearing pressure or contact pressure on the soil (the total demand)

P_g = gross vertical load

A = area of applied pressure (footing)

> **TIP**
>
> It is a common practice for foundation designers to use P_{net} and neglect the effect of the soil and concrete weight above the foundation bearing level.

The net bearing pressure is compared with the net bearing capacity of the soil. The net bearing pressure is the applied pressure, neglecting the weight of the footing and overlying soil.

$$Q_{\text{net}} = \frac{P_{\text{net}}}{A}$$ **Equation 3-36**

Q_{net} = net bearing pressure or net contact pressure (net demand)

P_{net} = net vertical load (column load)

A = area of applied pressure (footing)

3.4.2 Soil Bearing Capacity (Supply)

The structural demand must be satisfied by the soil bearing capacity with an appropriate factor of safety. Working with the gross and net loads requires comparing to the allowable and net allowable bearing capacities, respectively. This leads to the following design equations:

$$(q_{\text{net}})_{\text{all}} = \left(\frac{q_{\text{net}}}{FS}\right)_{\text{capacity}} \geq (Q_{\text{net}})_{\text{applied}}$$ **Equation 3-37**

and

$$q_{\text{all}} = \left(\frac{q_{\text{ult}}}{FS}\right)_{\text{capacity}} \geq (Q_{\text{g}})_{\text{applied}}$$ **Equation 3-38**

q_{ult} = ultimate bearing capacity of the foundation soil

q_{net} = net bearing capacity of the foundation soil

Bearing capacity failure is a catastrophic event; therefore, appropriate safety factors need to be utilized. The following sections explain how to estimate the bearing capacity of soils.

3.4.3 Terzaghi's Bearing Capacity Theory

The **ultimate bearing capacity** is theoretically the vertical bearing pressure at which a general shear failure will occur in the foundation soil. Terzaghi's general bearing capacity equation for an infinitely long (strip) foundation is given as:

$$q_{\text{ult}} = cN_c + \gamma D_f N_q + 0.5\gamma B N_\gamma$$ **Equation 3-39**

q_{ult} = ultimate bearing capacity

c = cohesion

D_f = depth of footing from the ground surface to the bottom of the footing

γ = unit weight of the soil

B = width of the footing

N_c, N_q, N_γ = bearing capacity factors based on ϕ (Table 3.3)

The net bearing capacity is corrected for the overburden pressure of the soil, giving the additional bearing capacity beyond the pressure applied by the soil before the foundation existed. Note that when the groundwater table is above the base of the footing, substitute the vertical effective overburden stress, σ'_D, at depth D_f for γD_f in Equation 3-39. The net bearing pressure is given by:

$$q_{\text{net}} = q_{\text{ult}} - \gamma D_f$$ **Equation 3-40**

The **net allowable bearing capacity** is the maximum net bearing pressure the soil can safely support with a reasonable factor of safety (typically 2 to 3 for foundations).

$$(q_{\text{net}})_{\text{all}} = \frac{q_{\text{net}}}{FS}$$ **Equation 3-41**

The bearing capacity of the soil and the bearing pressure from the structural load can be thought of in terms of supply and demand. The allowable bearing capacity is the available supply. This must be greater than or equal to the applied bearing pressure, which is the demand placed on the soil.

The N_c term in the general bearing capacity equation is the contribution from the cohesion of the soil. If the footing bears on a granular material where cohesion is equal to zero, then this term will equal zero and thus can be eliminated from the equation.

The N_q term is the contribution from the overburden pressure (or surcharge) of the soils from the ground surface to depth, D_f, equal to the depth of the footing. If the groundwater table is below the depth of the footing, then $\sigma'_D = \gamma D_f$.

The N_γ term is the contribution from the unit weight of the soils below the base of the footing and the footing size. Thus, γ in the N_γ term of the equation is the unit weight of the soil below the foundation. The effect of the groundwater table on the soil bearing capacity will be discussed later in this section.

Since Terzaghi, various researchers have recommended alternative bearing capacity factors and have applied various shape, depth, and inclination correction factors to each term of the equation. The modified form of the equation with the correction factors applied is referred to as the general bearing capacity equation. The bearing capacity factors and shape correction factors introduced in Tables 3.3 and 3.5 are attributable to Meyerhof, Vesic, and others. The general bearing capacity equation has the form:

$$q_{\text{ult}} = cN_c S_c + \gamma D_f N_q S_q + 0.5\gamma B N_\gamma S_\gamma$$ **Equation 3-42**

The first character in the subscript of the correction factors indicates the term the factor is applied to—c, q, or γ—and the second character indicates that the factor is a correction for depth, shape, or inclination from the vertical line of the applied load. Shape and inclination factors should not be used together.

TABLE 3.3 Bearing Capacity Factors

ϕ	N_C	N_Q	N_γ	ϕ	N_C	N_Q	N_γ
0	5.14	1.0	0.0	23	18.1	8.7	8.2
1	5.4	1.1	0.1	24	19.3	9.6	9.4
2	5.6	1.2	0.2	25	20.7	10.7	10.9
3	5.9	1.3	0.2	26	22.3	11.9	12.5
4	6.2	1.4	0.3	27	23.9	13.2	14.5
5	6.5	1.6	0.5	28	25.8	14.7	16.7
6	6.8	1.7	0.6	29	27.9	16.4	19.3
7	7.2	1.9	0.7	30	30.1	18.4	22.4
8	7.5	2.1	0.9	31	32.7	20.6	26.0
9	7.9	2.3	1.0	32	35.5	23.2	30.2
10	8.4	2.5	1.2	33	38.6	26.1	35.2
11	8.8	2.7	1.4	34	42.2	29.4	41.1
12	9.3	3.0	1.7	35	46.1	33.3	48.0
13	9.8	3.3	2.0	36	50.6	37.8	56.3
14	10.4	3.6	2.3	37	55.6	42.9	66.2
15	11.0	3.9	2.7	38	61.4	48.9	78.0
16	11.6	4.3	3.1	39	67.9	56.0	92.3
17	12.3	4.8	3.5	40	75.3	64.2	109.4
18	13.1	5.3	4.1	41	83.9	73.9	130.2
19	13.9	5.8	4.7	42	93.7	85.4	155.6
20	14.8	6.4	5.4	43	105.1	99.0	186.5
21	15.8	7.1	6.2	44	118.4	115.3	224.6
22	16.9	7.8	7.1	45	133.9	134.9	271.8

From *AASHTO LRFD Bridge Design Specifications* (*5th ed.*), 2010, by the American Association of State Highway and Transportation Officials, Washington, DC. Used by permission.

The general form of the bearing capacity equation (Equation 3-38) is only applicable for continuous footings. For other geometries, shape factors must be applied (Table 3.4, Equation 3-42).

TABLE 3.4 Shape Correction Factors

FACTOR	FRICTION ANGLE	COHESION TERM (S_c)	UNIT WEIGHT TERM (S_γ)
Shape Factors, S_c, S_γ, S_q	$\phi = 0$	$1 + \left(\frac{B_f}{5L_f}\right)$	1.0
	$\phi > 0$	$1 + \left(\frac{B_f}{L_f}\right)\left(\frac{N_q}{N_c}\right)$	$1 - 0.4\left(\frac{B_f}{L_f}\right)$

From *AASHTO LRFD Bridge Design Specifications* (5th ed.), 2010, by the American Association of State Highway and Transportation Officials, Washington, DC. Used by permission.

If the friction angle is not provided, but the SPT N value is, use Table 3.5 to estimate the angle of shearing resistance/friction angle, φ, for cohesionless/granular/coarse-grained soils from the corrected SPT N-value, $N1_{60}$. The friction angle can also be measured by laboratory tests such as triaxial and direct shear tests.

TABLE 3.5 Estimation of Friction Angle of Cohesionless Soils From Standard Penetration Tests

DESCRIPTION	VERY LOOSE	LOOSE	MEDIUM	DENSE	VERY DENSE
Corrected SPT $N1_{60}$	0	4	10	30	50
Approximate ϕ, degrees*	25–30	27–32	30–35	35–40	38–43
Approximate moist unit weight, (γ) pcf*	70–100	90–115	110–130	120–140	130–150

*Use larger values for granular material with 5% or less fine sand and silt.
Note: Correlations may be unreliable in gravelly soils due to sampling difficulties with split-spoon sampler.
Source: *Soils and Foundations Reference Manual (FHWA-NHI-06-088)*, 2006 [1].

Example 3.8: Bearing Capacity

Determine the ultimate and the net allowable bearing capacities for the continuous footing shown. Assume a factor of safety of 2.5.

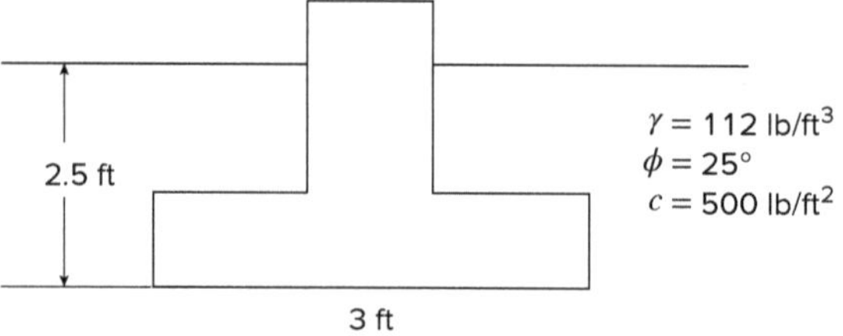

Solution

From Table 3.3: $N_c = 20.7$, $N_q = 10.7$, $N_\gamma = 10.9$

For continuous footings $B_f/L_f \approx 0$. From Table 3.5: $S_c = S_\gamma = 1.0$

Solve for the ultimate bearing capacity, q_{ult}:

$$q_{ult} = cN_cS_c + \gamma D_f N_q + 0.5\gamma B N_\gamma S_\gamma$$

$$q_{ult} = (500 \text{ lb/ft}^2 \times 20.7 \times 1) + (112 \text{ lb/ft}^3 \times 2.5 \text{ ft} \times 10.7) + (0.5 \times 112 \text{ lb/ft}^3 \times 3 \text{ ft} \times 10.9 \times 1) = 13{,}529 \text{ lb/ft}^2$$

Solve for the net bearing capacity, q_{net}:

$$q_{net} = q_{ult} - \gamma D_f$$

$$q_{net} = 13{,}529 \text{ lb/ft}^2 - 2.5 \text{ ft} \times 112 \text{ lb/ft}^3 = 13{,}249 \text{ lb/ft}^2$$

Example 3.8 (continued)

Solve for the net allowable bearing capacity, $(q_{\text{net}})_{\text{all}}$:

$$(q_{\text{net}})_{\text{all}} = \frac{q_{\text{net}}}{FS}$$

$$(q_{\text{net}})_{\text{all}} = \frac{13{,}249\ \text{lb/ft}^2}{2.5\ \text{ft}} = 5{,}300\ \text{lb/ft}^2$$

3.4.3.1 Bearing Capacity in Clay

For bearing capacity in cohesive soils, where $\phi = 0$, the bearing capacity factors $N_q = 1$ and $N_\gamma = 0$. Therefore, the ultimate and net bearing capacity equations for clay soils will simplify as follows:

$$q_{\text{ult}} = cN_cS_c + \gamma D_f$$ **Equation 3-43**

$$q_{\text{net}} = cN_cS_c$$ **Equation 3-44**

The undrained shear strength (s_u) of clay is equal to the cohesion (c), which is one-half of the unconfined compressive strength (q_{unc}), as shown:

$$S_u = c = \frac{q_{\text{unc}}}{2}$$ **Equation 3-45**

Example 3.9: Bearing Capacity in Clay

A 6-ft square footing is founded in a clay soil 3 ft below the ground surface. The soil parameters are $\gamma = 125\ \text{lb/ft}^3$, $c = 1{,}200\ \text{lb/ft}^2$, and $\phi = 0$. Determine the net allowable bearing capacity. Assume a factor of safety of 3.

Solution

From Table 3.3: $N_c = 5.14$

From Table 3.5: for $\phi = 0$, $S_c = 1 + \frac{B_f}{5L_f} = 1 + \frac{6\ \text{ft}}{5 \times 6\ \text{ft}} = 1.2$

$$q_{\text{net}} = cN_cS_c$$

$$q_{\text{net}} = 1{,}200\ \text{lb/ft}^3 \times 5.14 \times 1.2 = 7{,}402\ \text{lb/ft}^2$$

$$(q_{\text{net}})_{\text{all}} = \frac{q_{\text{net}}}{FS}$$

$$(q_{\text{net}})_{\text{all}} = \frac{7{,}402\ \text{lb/ft}^2}{3} = 2{,}467\ \text{lb/ft}^2$$

3.4.3.2 Bearing Capacity in Sand

For bearing capacity in cohesionless soils, where $c = 0$, the cohesion term of the bearing capacity equation will be zero. The ultimate and net bearing capacity equations for cohesionless soils will simplify as follows:

$$q_{\text{ult}} = \gamma D_f + 0.5\gamma B N_\gamma S_\gamma$$ **Equation 3-46**

$$q_{\text{net}} = \gamma D_f(N_q - 1) + 0.5\gamma B N_\gamma S_\gamma$$ **Equation 3-47**

Example 3.10: Bearing Capacity in Sand

A 3-ft × 3-ft square footing bears at the surface of a sand deposit for which $\phi = 32°$ and $\gamma = 125$ lb/ft^3. The water table is very deep. Determine the gross allowable force (lb) that this footing can carry. Assume a factor of safety of 2.

Solution

$$c = 0, \phi = 32°, N_\gamma = 30.2, S_\gamma = 1 - 0.4\frac{B}{L} = 1 - 0.4 \times \frac{3}{3} = 0.60$$

Calculate the ultimate bearing capacity in sand:

$$q_{\text{ult}} = \gamma D_f N_q + 0.5\gamma B N_\gamma S_\gamma$$

$$q_{\text{ult}} = 0 + 0.5 \times 125 \times 3 \times 30.2 \times 0.6 = 3{,}398 \text{ lb/ft}^2$$

Calculate the allowable bearing pressure:

$$q_{\text{all}} = \frac{q_{\text{ult}}}{FS}$$

$$q_{\text{all}} = \frac{3{,}398 \text{ lb/ft}^2}{2} = 1{,}699 \text{ lb/ft}^2$$

Calculate the allowable gross vertical force:

$$(P_g)_{\text{all}} = q_{\text{all}} A = (1{,}699 \text{ lb/ft}^2)(3)^2 = 15{,}289 \text{ lb}$$

3.4.4 Effect of Groundwater Table on Bearing Capacity

Case 1: The water table is located at or above the base of the footing ($D_f \geq D_1 \geq 0$).

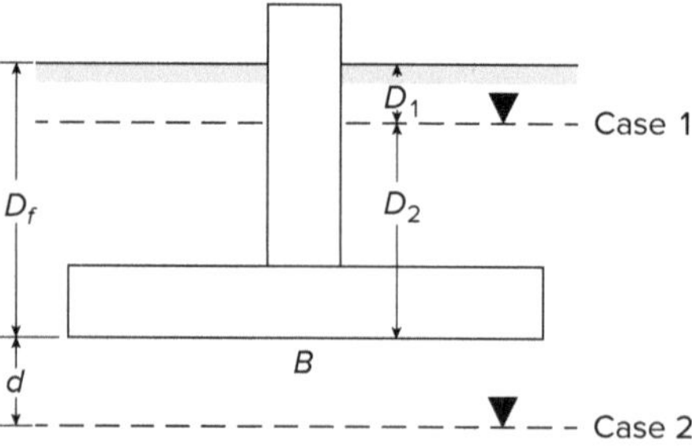

FIGURE 3.18 Effect of Water Table on the Ultimate Bearing Capacity of Soils

The factor γD_f in the N_q term is the effective vertical stress σ'_D at the base of the footing:

$$\sigma'_D = \gamma D_1 + \gamma' D_2 \qquad \textbf{Equation 3-48}$$

The value of γ in the N_γ term is the submerged (effective) unit weight:

$$\gamma' = \gamma_{\text{sat}} - \gamma_w \qquad \textbf{Equation 3-49}$$

Case 2: The water table is located between the base of the footing and a depth of B below the base of the footing ($B > d > 0$).

The value of γ in the N_γ term is taken as a weighted-average unit weight between the dry and the submerged unit weights:

$$\bar{\gamma} = \frac{d}{B}\gamma + \left(1 - \frac{d}{B}\right)\gamma' \qquad \textbf{Equation 3-50}$$

Case 3: The water table is located below the base of the footing at a depth greater than or equal to B ($d \geq B$).

When the water table is located below the base of the footing by a depth greater than or equal to B, the water table effect is negligible.

Example 3.11: Effect of Groundwater on Bearing Capacity

After temporary dewatering, a square footing measuring 5 ft × 5 ft will be constructed to bear on sand at a depth of 3.5 ft below the ground surface. Prior to construction, groundwater was at a depth of 2 ft and is expected to return to that depth when dewatering is terminated. Soil properties are $\gamma = 116\ \text{lb/ft}^3$, $\gamma_{sat} = 120\ \text{lb/ft}^3$, and $\phi' = 34°$. Determine the net allowable soil bearing pressure for the footing. Use a factor of safety of 2.

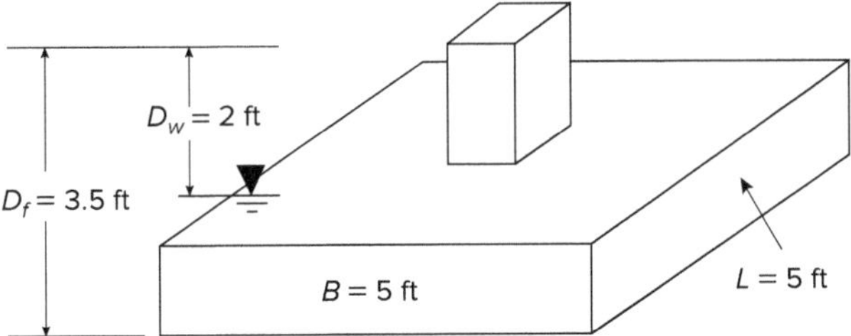

Solution

$$N_q = 29.4, N_\gamma = 41.1, S_\gamma = 1 - 0.4\frac{B}{L} = 1 - 0.4 \times \frac{5\ \text{ft}}{5\ \text{ft}} = 0.6$$

Net bearing capacity equation with shape factor:

$$q_{net} = \gamma D_f(N_q - 1) + 0.5\gamma B N_\gamma S_\gamma$$

Calculate net bearing capacity with the groundwater depth at 2 ft (case 1):

$$\sigma'_D = \gamma D_1 + \gamma' D_2 = 116\ \text{lb/ft}^3 \times 2 + (120\ \text{lb/ft}^3 - 62.4) \times 1.5 = 318.4\ \text{lb/ft}^2$$

$$\gamma' = \gamma_{sat} - \gamma_w = 120\ \text{lb/ft}^3 - 62.4 = 57.6\ \text{lb/ft}^2$$

$$q_{net} = \sigma'_D(N_q - 1) + 0.5\gamma' B N_\gamma S\gamma$$

$$q_{net} = 318.4\ \text{lb/ft}^2 \times (29.4 - 1) + 0.5 \times 57.6\ \text{lb/ft}^2 \times 41.1 \times 0.6 = 9{,}043 + 3{,}551$$

$$= 12{,}594\ \text{lb/ft}^2$$

Calculate net allowable bearing capacity:

$$(q_{net})_{all} = \frac{q_{net}}{\text{FS}}$$

$$(q_{net})_{all} = \frac{12{,}594\ \text{lb/ft}^2}{2} = 6{,}297\ \text{lb/ft}^2$$

3.4.5 Eccentric Loads on Rectangular Shallow Spread Footings

Eccentric loading is created when footing carries a moment M in addition to a vertical load P. The moments might result from wind loads on tall buildings, horizontal loads on retaining or basement walls, or eccentrically applied building loads. The combined moment and vertical loads affect both the demand exerted from the structural loads and the supply-bearing capacity of the foundation soil.

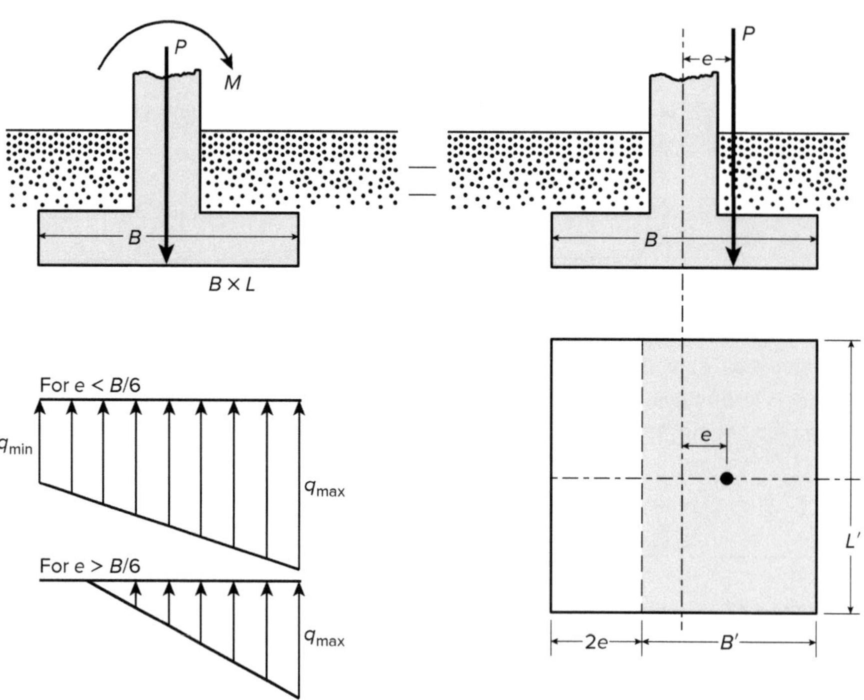

FIGURE 3.19 Eccentrically Loaded Foundations

Source: Das, Braja M. *Fundamentals of Geotechnical Engineering* (2nd ed.). Thomson, 2005. Used with permission.

In Figure 3.19, the eccentricity is shown in the B direction; however, it can occur in the L direction, or both. The equations given in this section should be used for eccentricity in the B or L direction.

3.4.5.1 Effect of Eccentricity on the Demand—Structural Loads

Eccentricity, $e = \dfrac{M}{P}$ **Equation 3-51**

For eccentricity inside the middle third of the footing ($e < B/6$):

$$q_{\min} = \frac{P}{BL}\left(1 - \frac{6e}{B}\right),\ q_{\max} = \frac{P}{BL}\left(1 + \frac{6e}{B}\right)$$ **Equations 3-52 and 3-53**

For eccentricity outside the middle third of the footing ($e > B/6$):

$$q_{\min} < 0,\ q_{\max} = \frac{4P}{3L(B - 2e)}$$ **Equation 3-54**

3.4.5.2 Effect of Eccentricity on the Supply—Soil Bearing Capacity

For eccentric footing, an approximate bearing capacity may be estimated using the effective footing width:

$B' = B - 2e_B$ or $L' = L - 2e_L$ **Equation 3-55**

e_L = eccentricity along the footing length

e_B = eccentricity along the footing width

To calculate the ultimate bearing capacity, the value of B' is used in the general bearing capacity equation when the eccentricity is in the B direction. L' is used when the high

Soil Mechanics

School of PE

eccentricity is in the L direction such that L' becomes smaller than B. The value of B' or L' should be used, when applicable, to determine the shape factors as defined in Table 3.5.

The ultimate, net, and allowable bearing capacity equations become:

$$q_{\text{ult}} = cN_cS_c + \gamma D_f N_q + 0.5\gamma B' N_\gamma S_\gamma$$ **Equation 3-42** (p. 85)

$$q_{\text{net}} = q_{\text{ult}} - \gamma D_f$$ **Equation 3-40** (p. 84)

$$(q_{\text{net}})_{\text{all}} = \frac{q_{\text{net}}}{\text{FS}}$$ **Equation 3-41** (p. 84)

Example 3.12: Safety Factor

A 3 ft × 6 ft rectangular footing has a gross load, P, of 40,000 lb and a moment, M, of 25,000 ft-lb. The foundation is located with its base 3 ft below the ground surface and is bearing on sand with an effective internal friction angle of 30° and a moist unit weight of 118 lb/ft^3. Assume the groundwater is very deep. Determine the safety factor against bearing capacity failure.

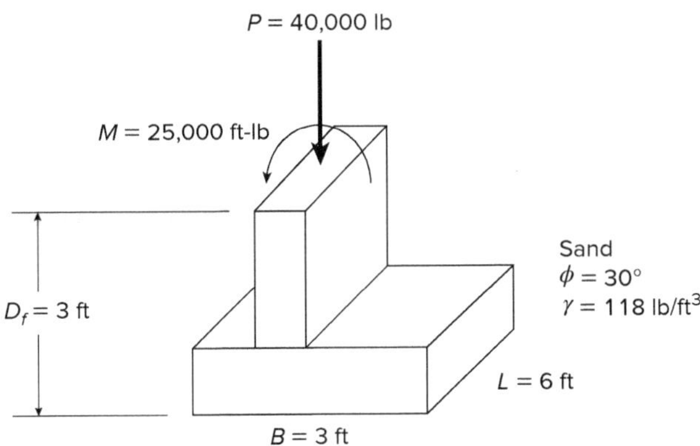

Solution

Determine eccentricity:

$$e_B = \frac{M}{P}$$

$$e_B = \frac{25{,}000 \text{ lb}}{40{,}000 \text{ ft-lb}} = 0.625 \text{ ft}$$

$$\frac{B}{6} = \frac{3 \text{ ft}}{6} = 0.5 \text{ ft} \qquad e_B > \frac{B}{6}$$

Determine effective footing width:

$$B' = 3 - 2 \times 0.625 = 1.75 \text{ ft}$$

Select bearing capacity factors and shape factors:

$$N_q = 18.4, N_\gamma = 22.4, S_\gamma = 1 - 0.4 \times \frac{1.75}{6} = 0.88$$

Example 3.12 (continued)

Calculate the ultimate bearing capacity:

$q_{ult} = \gamma D_f N_q + 0.5\gamma B' N_\gamma S_\gamma$

$q_{ult} = 118 \times 3 \times 18.4 + 0.5 \times 118 \times 1.75 \times 22.4 \times 0.88 = 6{,}513 + 2{,}035 = 8{,}548 \text{ lb/ft}^2$

Determine the maximum applied pressure, Q_{max}:

$$Q_{max} = \frac{4P}{3L(B-2e)} = \frac{4(40{,}000 \text{ lb})}{3(6 \text{ ft})(3 \text{ ft} - 2(0.625 \text{ ft}))} = 5{,}079 \text{ lb/ft}^2$$

Determine the safety factor:

$$\text{safety factor} = \frac{q_{ult}}{Q_{max}} = \frac{8{,}548 \text{ lb/ft}^2}{5{,}079 \text{ lb/ft}^2} = 1.68$$

3.5 FOUNDATION SETTLEMENT

All soils will experience settlement after load is applied. There are three types of soil settlement:

(1) Immediate/elastic settlement: occurs in all soil types during and right after construction; the dominant settlement in cohesionless, granular, coarse-grained soils

(2) Primary consolidation settlement: occurs mainly in clays due to the slow drainage of pore water with excess pressure due to loading

(3) Secondary/creep settlement: happens mainly due to particle rearrangement and crushing

3.5.1 Modified Hough Method for Estimating Settlement in Non-Cohesive Soils

Per the FHWA *Soils and Foundations Reference Manual* [1], the following steps are used in the modified Hough method to estimate settlement:

1. Determine the bearing capacity index (C') by using Figure 3.20 with the N-value and the visual and manual classification of the soil.
2. Compute settlement utilizing Equation 3-56. For thick layers, the settlement estimate may be improved by dividing the layer into sublayers and summing the settlements for all sublayers:

$$\Delta H = \sum \frac{H}{C'} \log_{10} \frac{p_o + \Delta p}{p_o} \quad \textbf{Equation 3-56}$$

ΔH = summation of settlements for all sublayers

H = thickness of a layer or sublayer

C' = bearing capacity index (Fig. 3.20)

p_o = existing effective overburden pressure at the center of the layer or sublayer

Δp = increase of pressure at the center of the layer or sublayer due to the external load

$p_o + \Delta p$ = final pressure applied to the layer or sublayer

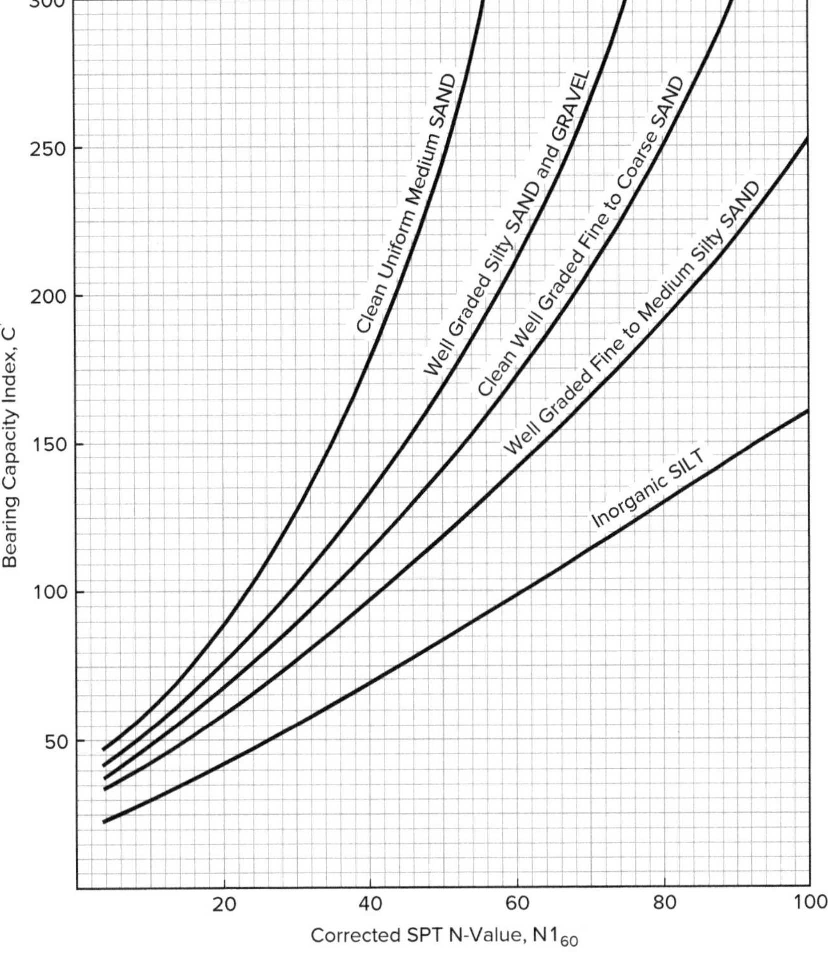

Note: The Inorganic SILT curve should generally not be applied to soils that exhibit plasticity because *N*-values in such soils are unreliable.

FIGURE 3.20 Bearing Capacity Index (*C*′) Values Used in Modified Hough Method for Computing Settlement in Non-Cohesive Soil

From *AASHTO LRFD Bridge Design Specifications* (*5th ed.*), 2010, by the American Association of State Highway and Transportation Officials, Washington, DC. Used by permission.

Example 3.13: Modified Hough Method to Estimate Soil Settlement

For the wide embankment shown below, utilize the modified Hough method to estimate the elastic settlement at the center of the embankment.

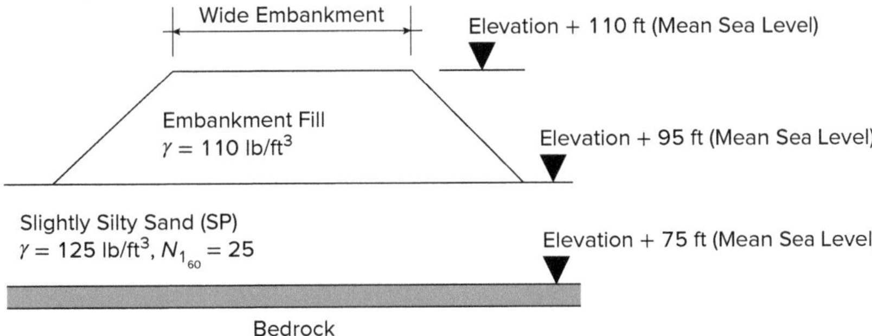

Solution

Estimate the current initial overburden pressure due to the soil's weight at the midpoint of the slightly silty sand layer (before the embankment construction):

$$p_o = \frac{20}{2} \times 125 \text{ lb/ft}^3 = 1{,}250 \text{ lb/ft}^2$$

Example 3.13 *(continued)*

Because the embankment is wide, no stress reduction will be calculated at the center of the slightly silty sand layer:

$\Delta p = 15 \times 110 = 1{,}650\ \text{lb/ft}^2$

Refer to Figure 3.20. For $N1_{60} = 25$ and the silty sand curve, the corresponding value of $C' \approx 68$.

Calculate the elastic settlement as follows:

$$\Delta H = H\left(\frac{1}{C'}\right)\log_{10}\frac{p_o + \Delta p}{p_o}$$

$$\Delta H = 20 \times \frac{1}{68}\log_{10}\left(\frac{1{,}250\ \text{lb/ft}^2 + 1{,}650\ \text{lb/ft}^2}{1{,}250}\right) = 0.107\ \text{feet} = 1.3\ \text{inches}$$

3.5.2 Primary Consolidation Settlement

Primary consolidation settlement occurs mainly in saturated, fine-grained, low-permeability soils such as clays. Initially, when excess pressure is applied to a clay layer in the field, porewater pressure increases, forcing it to drain out of the clay. However, because of the low-permeability nature of the clay, it might take months or years for the porewater, with excess pressure, to drain out of the clay. During this process, the excess pressure transfers to the clay skeleton causing a continuous reduction in volume over time, known as primary consolidation. The time needed for the primary consolidation to occur depends on the clay properties, such as permeability and consolidation parameters, as well as the stress level from loading. (See Section 3.2 on soil consolidation for more detailed information.)

3.5.3 Vertical Stress Increase in a Soil Mass Due to Loading

When pressure is applied at the surface of the soil or beneath a foundation at the foundation level, excess vertical, lateral, and shear stresses within the soil mass, influenced by the added pressure, result. The stress increase from the added pressure is higher in the soil near the applied pressure, especially directly under the loaded area. The stress increase typically dissipates with depth and distance away from the loaded area. Figure 3.21 shows a typical vertical stress distribution with depth under an embankment.

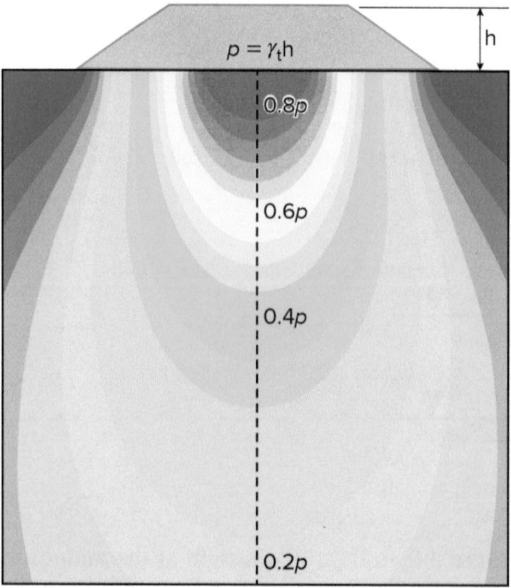

FIGURE 3.21 Schematic of Vertical Stress Distribution Under Embankment Loading

Source: *Soils and Foundations Reference Manual (FHWA-NHI-06-088),* 2006 [1].

The stress increase is needed to estimate different types of settlements. Most available solutions are based on the theory of elasticity and the Boussinesq (1885) solution, developed for point loads. The solutions depend on the shape of the footing or load application medium.

3.5.3.1 Point Load

The increase in vertical stress ($\Delta\sigma$) on a soil element located at depth z and radial distance r from the axis of the point load P are given by:

$$\Delta\sigma = \left(\frac{3P}{2\pi z^2}\right)\left(\frac{1}{1+(r/z)^2}\right)^{2.5}$$ **Equation 3-57**

For $r = 0$: $$\Delta\sigma = \left(\frac{3P}{2\pi z^2}\right)$$ **Equation 3-58**

3.5.3.2 Uniformly Loaded Infinitely Long/Continuous/Strip and Square Footings

A footing may be considered infinitely long when the length is more than ten times the width, which is also known as strip footing or wall footing. Figure 3.22 provides contour lines of the increase in vertical stresses under square and infinitely long footings. To utilize Figure 3.22, the dimensions of the field problem (foundation size and distance from the foundation to the point at which stresses are estimated) need to be scaled down to match the ½ B × ½ B grid of the figure. Once the point of interest is located, the value shown on the contour line represents the influence factor of the stress increase at this point relative to the applied pressure at the foundation level.

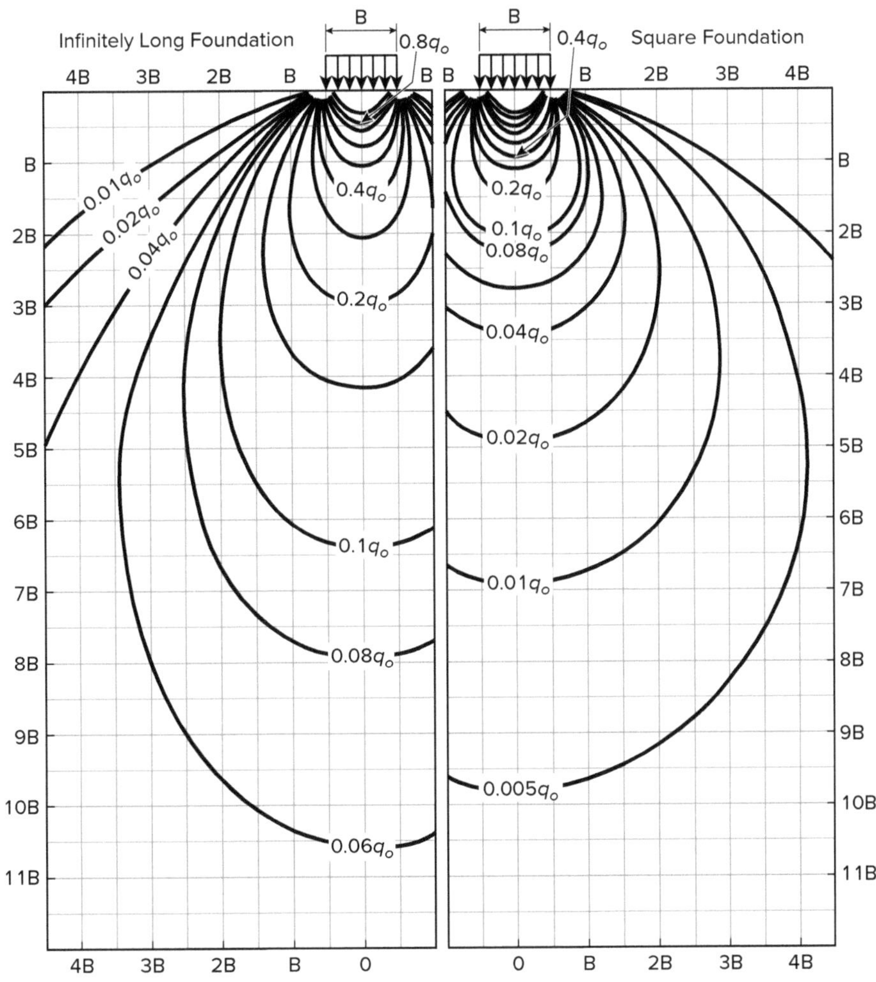

FIGURE 3.22 Vertical Stress Contours (Isobars) Based on Boussinesq's Theory for Continuous and Square Footings

Source: *Soils and Foundations Reference Manual (FHWA-NHI-06-088)*, 2006 [1].

3.5.3.3 Approximate (2:1) Stress Distribution

For preliminary analysis, a quick estimate of the stress increase beneath a loaded area can be made using a 2:1 distribution. In this approximate method, the original loaded area is assumed to have grown, with depth, from all sides at a rate of 1H:2V. Therefore, at a depth Z, a footing that is originally $B \times L$, will grow to be $(B + Z) \times (L + Z)$. The stress increase at that depth, Z, will be estimated by dividing the original load by the enlarged footing. The method is approximate because it assumes the stress increase at a specific depth is constant at all points relative to the load application area (center, corner, and outside of the footing).

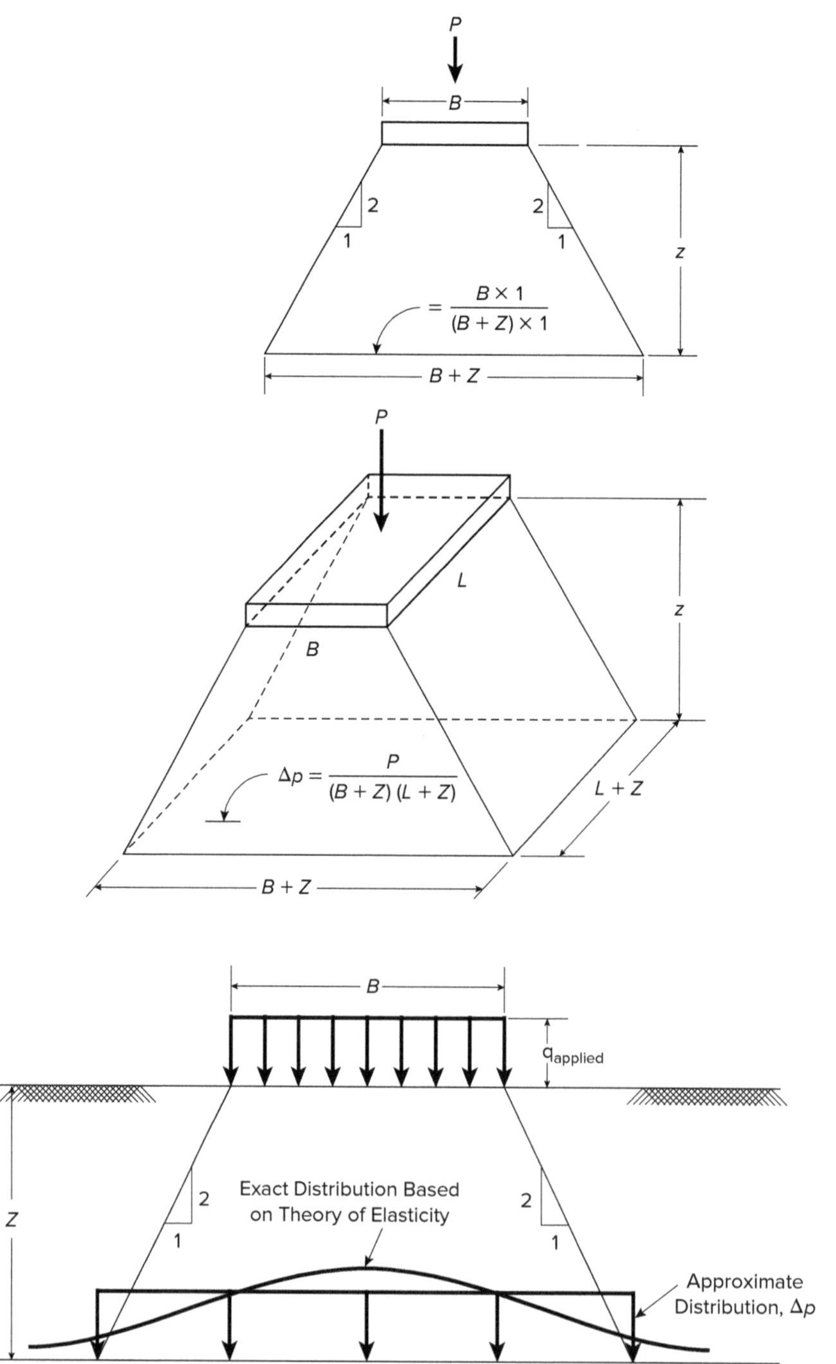

FIGURE 3.23 Distribution of Vertical Stress by the 2:1 Method (after Perloff and Baron, 1976)

Source: *Soils and Foundations Reference Manual (FHWA-NHI-06-088)*, 2006 [1].

For a rectangular area: $$\overline{\Delta\sigma} = \frac{qBL}{(B+z)(L+z)} = \frac{P}{(B+z)(L+z)}$$ **Equation 3-59**

For a circular area: $$\overline{\Delta\sigma} = \frac{qr^2}{\left(r+\frac{z}{2}\right)^2} = \frac{P}{\pi\left(r+\frac{z}{2}\right)^2}$$ **Equation 3-60**

For an infinitely long foundation wall: $$\overline{\Delta\sigma} = \frac{P(\text{line load unit})}{(B+Z)(1)}$$ **Equation 3-61**

3.5.3.4 Uniformly Loaded Rectangular Areas (Stress Under the Corner)

For uniformly loaded rectangular areas, the increase in vertical stress can be expressed in terms of an influence factor.

$$\Delta\sigma = qI$$ **Equation 3-62**

q = applied bearing pressure (P/A)

I = influence factor from charts or curves

Calculate: $m = B/z$ $n = L/z$

B = footing width

L = footing length

z = vertical depth below the base of the uniformly loaded area

Use m and n to get the influence factor I from Figure 3.24, and then calculate:

$\Delta\sigma = qI$ (under the corner of the footing)

In order to estimate the stress increase under the center of the footing:

- Divide the footing into four symmetrical equal areas such that the center of the footing becomes the corner of each of the quadrants.
- Calculate the influence factor under the corner for one of the quadrants. Note that m and n in this case are one-half of the overall footing dimensions.
- The vertical stress increase may be estimated as:

  $\Delta\sigma = 4qI$

Example 3.14: Uniformly Loaded Rectangular Areas

A flexible rectangular area measures 10 ft × 20 ft. It supports an applied pressure of 3,000 lb/ft^2. Determine the vertical stress increase due to the applied load at a depth of 20 ft below the corner of the rectangular area (Fig. 3.24).

Example 3.14 (continued)

Solution

$$m = \frac{B}{Z} = \frac{10\ \text{ft}}{20\ \text{ft}} = 0.5 \qquad n = \frac{L}{Z} = \frac{20\ \text{ft}}{20\ \text{ft}} = 1.0 \qquad \rightarrow \qquad I = 0.12$$

$$\Delta\sigma = Iq = 0.12 \times 3{,}000\ \text{lb/ft}^2 = 360\ \text{lb/ft}^2$$

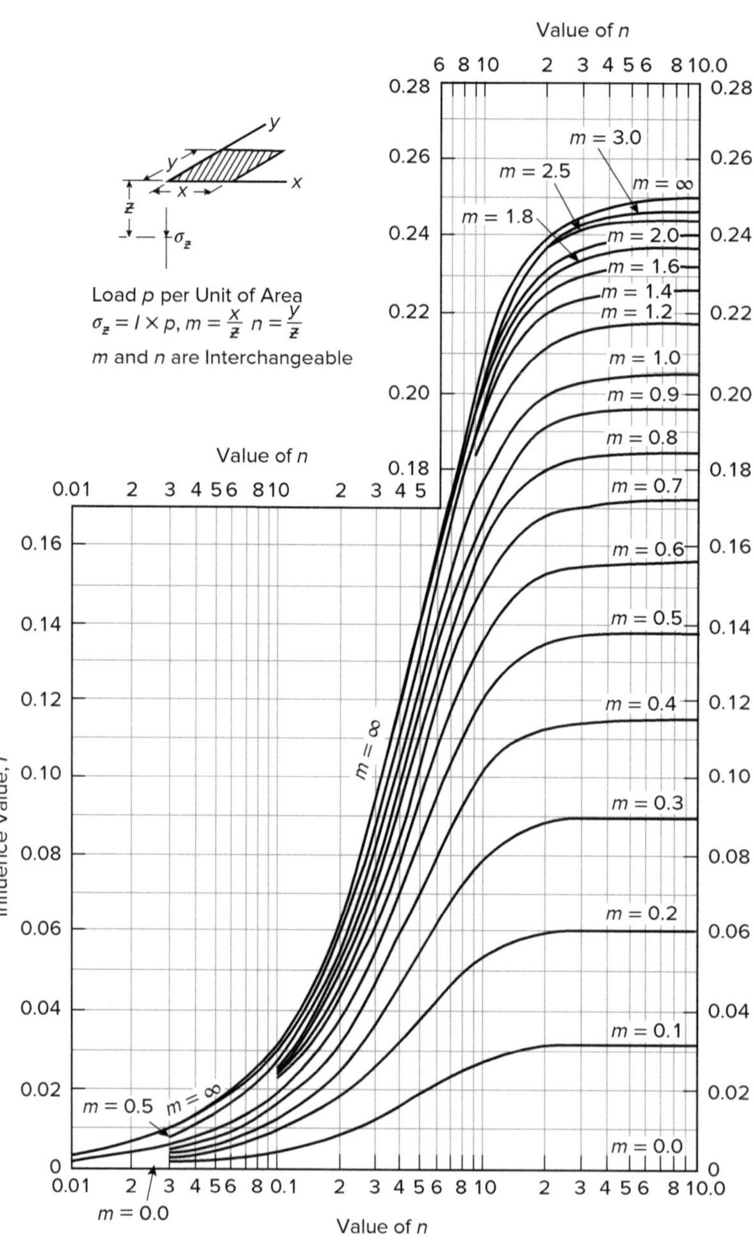

FIGURE 3.24 **Influence Factors for Finding Vertical Stress Under the Corner of a Rectangular Loaded Area**

Source: *Soil Mechanics Design Manual 7.01*, Naval Facilities Engineering Command, 1986.

Example 3.15: Uniformly Loaded Rectangular Areas

A square footing measures 50 ft × 50 ft. It supports a pressure of 6,000 lb/ft^2. Determine the increase in vertical stress due to the applied load at a depth of 25 ft below the center of the rectangular area using the Boussinesq stress contours (Fig. 3.22).

Solution

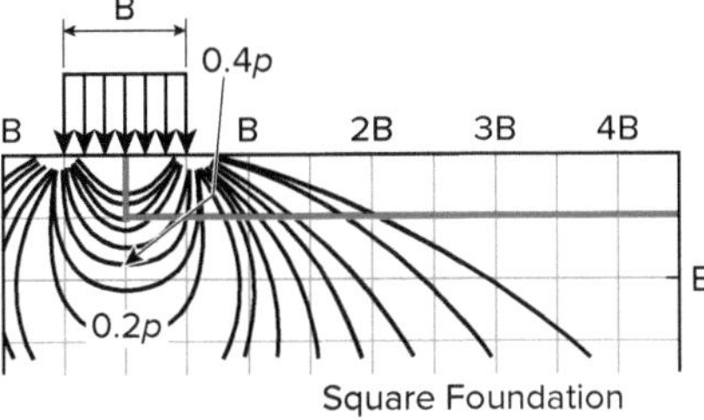

Source: *Soils and Foundations Reference Manual (FHWA-NHI-06-088)*, 2006 [1].

$x = 0$ ft

$z = 25$ ft

$B = 50$ ft

Horizontal distance from center of footing: $\frac{x}{B} = \frac{0 \text{ ft}}{50 \text{ ft}} \Rightarrow x = 0B$

Vertical distance from base of footing: $\frac{Z}{B} = \frac{25 \text{ ft}}{50 \text{ ft}} \Rightarrow Z = 0.5B$

Using Figure 3.24, $I = 0.7$

$\Delta\sigma = (0.7)(6{,}000 \text{ lb/ft}^2) = 4{,}200 \text{ lb/ft}^2$

3.6 SLOPE STABILITY

Slope stability is the potential for a slope in soil to withstand movement under given loading conditions. Stability is determined by comparing the shear stress mobilized along any potential failure surface to the available shear strength of the soil along that failure surface.

$$FS_{\text{slope stability}} = \frac{\text{available shear strength}}{\text{mobilized shear stresses}}$$ **Equation 3-63**

Increased shear stresses or decreased shear strength will decrease the safety factor. Increases in shear stress can be caused by external loading, lateral pressure, and transient forces. Decreases in shear strength may be due to pore water pressure, weathering effects, and organic materials.

A factor of safety as low as 1.25 may be used for highway embankment slopes. For certain extreme events such as seismic events and sudden drawdown, lower factors of safety may be acceptable. The factor of safety should be increased to a minimum of 1.30 for slopes whose failure would cause significant damage to major roadways such as regional routes and interstates, and 1.50 for slopes beneath bridge abutments; retaining structures; and slopes where buildings are on or near the slope.

3.6.1 Effect of Water on Slope Stability

Soft, saturated foundation soils or groundwater generally play an important role in geotechnical failures, and may be significant factors in slope failures.

In cohesionless soils, water has a negligible effect on the angle of internal friction (φ). However, the effective stress of saturated cohesionless soils is decreased below the water table, resulting in lower overall shear resistance (τ′). The increase in water content of clays decreases their cohesion, hence decreasing their shear strength (Fig. 3.25).

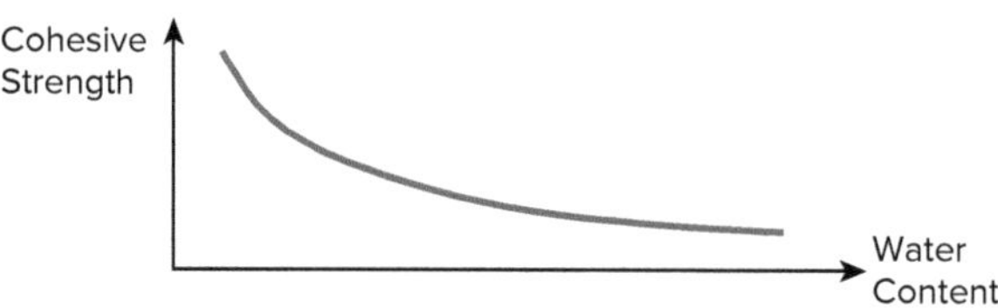

FIGURE 3.25 Effect of Water Content on Cohesive Strength of Clay

Source: *Soils and Foundations Reference Manual (FHWA-NHI-06-088)*, 2006 [1].

3.6.2 Infinite Slopes in Dry Cohesionless Soil

See Figures 3.26 and 3.27.

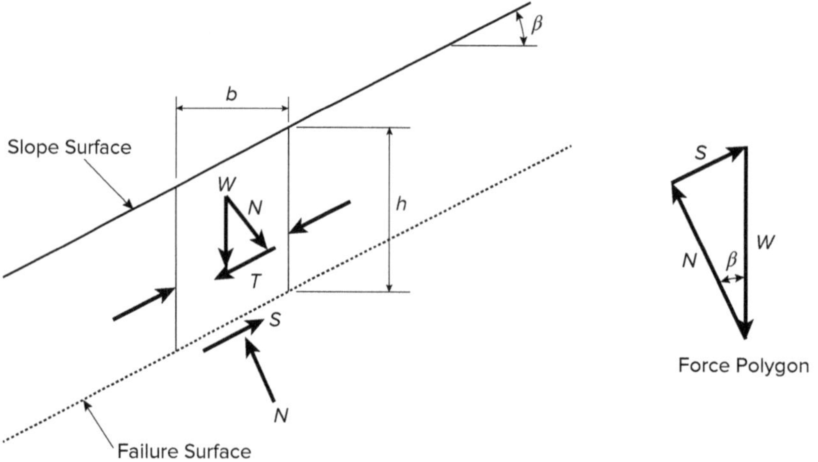

FIGURE 3.26 Infinite Slope Failure in Dry Sand

Source: *Soils and Foundations Reference Manual (FHWA-NHI-06-088)*, 2006 [1].

$$\text{FS} = \frac{S}{T} = \frac{N \tan \phi}{W \sin \beta} = \frac{(W \cos \beta) \tan \phi}{W \sin \beta} = \frac{\tan \phi}{\tan \beta}$$ **Equation 3-64**

FS = factor of safety

ϕ = internal friction angle

β = angle of slope from horizontal

3.6.3 Infinite Slopes in c-ϕ Soil with and Without Parallel Seepage

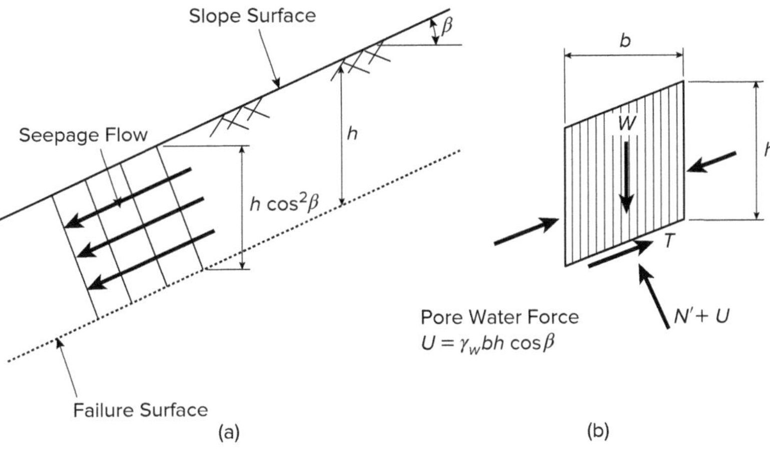

FIGURE 3.27 Infinite Slope Failure in a c-φ Soil with Parallel Seepage

Source: *Soils and Foundations Reference Manual (FHWA-NHI-06-088)*, 2006 [1].

$$FS = \frac{c' + h(\gamma_{sat} - \gamma_w)(\cos^2\beta)\tan\phi'}{\gamma_{sat}\, h \sin\beta \cos\beta}$$ **Equation 3-65**

FS = factor of safety

ϕ' = effective internal friction angle

c' = effective cohesion

β = angle of slope from horizontal

h = vertical height of the failure surface to the slope surface

γ_{sat} = saturated unit weight

γ_w = water unit weight

For $c' = 0$, the above expression reduces to:

$$FS = \frac{\gamma'}{\gamma_{sat}}\frac{\tan\phi'}{\tan\beta}$$ **Equation 3-66**

3.6.4 Taylor's Slope Stability Chart for Clays

The slope stability chart shown in Figure 3.28 is the Taylor's chart modified by Janbu. This chart applies to slopes of homogeneous saturated clay ($\phi = 0$). The chart is based on the following assumptions:

- No water is outside of the slope, or the slope is fully submerged.
- No surcharge or external loads are on the slope, and there are no tension cracks.
- Shear strength is from cohesion only and remains constant with depth.
- The slope failure is rotational and occurs along a circular arc.

The factor of safety is given by:

$$FS = \frac{N_o c}{\gamma_t H}$$ **Equation 3-67**

N_o = stability number (from Taylor's chart)

γ_t = total unit weight of soil (Use γ' for the submerge condition.)

c = cohesion (undrained shear strength) of the soil

H = height of the slope

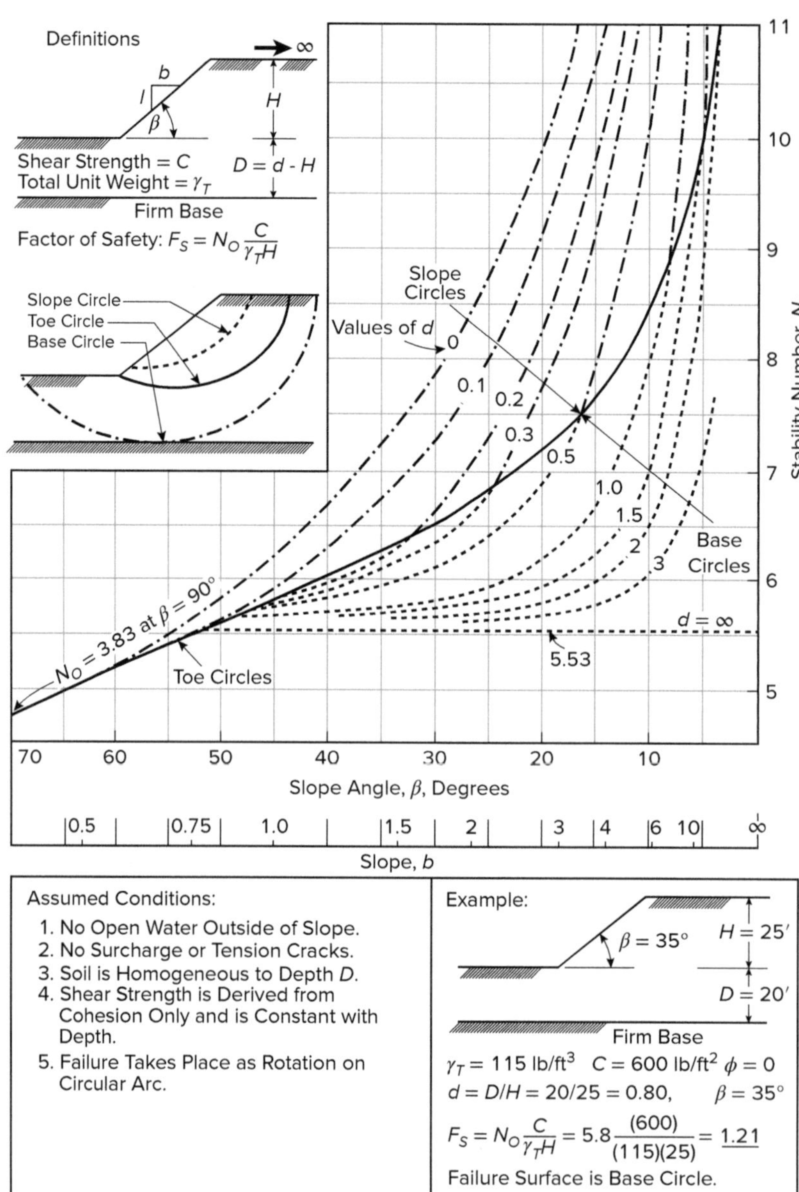

FIGURE 3.28 **Taylor's Chart Modified by Janbu for Saturated Clay Slopes**

Source: *Soil Mechanics Design Manual 7.01*, Naval Facilities Engineering Command, 1986.

The stability number is obtained from the chart for a depth factor d and slope angle β, relative to the horizontal. D and H are defined in Figure 3.28 for base, toe, and slope failures.

$$d = \frac{D}{H}$$ **Equation 3-68**

For slope angles greater than $\beta = 53°$, the failure surface will pass through the toe. For steeper slopes, extend the solid line in the chart linearly to $N_o = 3.83$ for $\beta = 90°$.

The minimum acceptable factor of safety is generally 1.3 to 1.5.

Example 3.16: Taylor's Slope Stability Chart

A temporary cut is proposed as shown. The cut will be made in clay soil with an undrained shear strength, $c = 625$ lb/ft^2, and $\gamma = 125$ lb/ft^3. If the factor of safety must be at least 1.5, what is the maximum slope angle for the proposed cut? Use Taylor's chart.

Example 3.16 (continued)

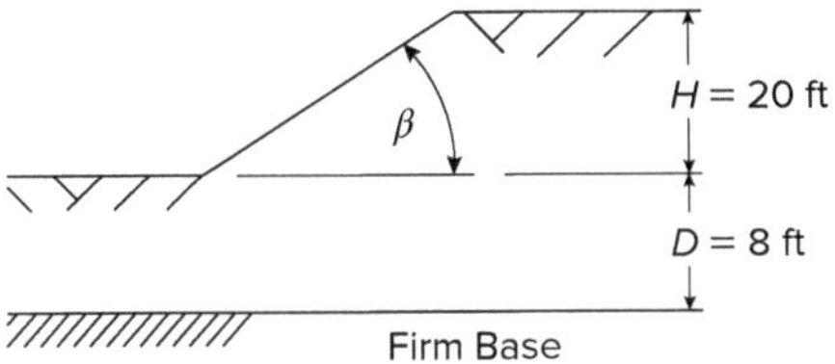

Solution

$$FS = \frac{N_o c}{\gamma H} \Rightarrow N_o = \frac{(FS)\gamma H}{c} = \frac{1.5(125\ \text{lb/ft}^3)(20\ \text{ft})}{625\ \text{lb/ft}^2} = 6.0$$

$$d = \frac{D}{H} = \frac{8\ \text{ft}}{20\ \text{ft}} = 0.4$$

From Taylor's slope stability chart, for $N_o = 6$ and $d = 0.4$:

The slope angle is $\beta \approx 35°$.

3.6.5 Method of Slices

In slope stability analysis, where the normal stress and shear strength vary along the failure surface, it is helpful to divide the failure mass into a series of vertical slices and to write equations based on the forces acting on the individual slices. For methods that consider moment equilibrium only, the plane of failure is typically assumed to be a circular surface, as shown in Figure 3.29. The factor of safety against global stability may be defined as:

$$\text{FS} = \frac{\text{Sum of Resisting Forces} \times \text{Moment Arm (R)}}{\text{Sum of Driving Forces} \times \text{Moment Arm (R)}} \qquad \textbf{Equation 3-69}$$

Since both the driving and the resisting forces have the same moment arm, Equation 3-69 may be simplified as:

$$\text{FS} = \frac{\text{Sum of Resisting Forces}}{\text{Sum of Driving Forces}} \qquad \textbf{Equation 3-70}$$

The process is repeated for multiple points of rotation and multiple radii of the failure surface.

TIP

The minimum safety factor for all trials is the safety factor for the slope.

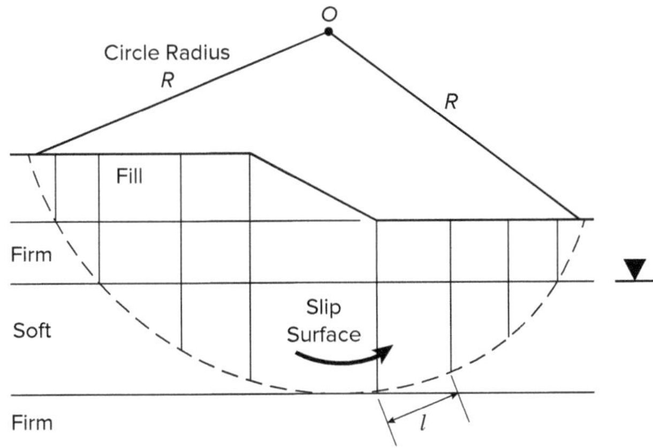

FIGURE 3.29 Geometry of Ordinary Method of Slices

Source: *Soils and Foundations Reference Manual (FHWA-NHI-06-088)*, 2006 [1].

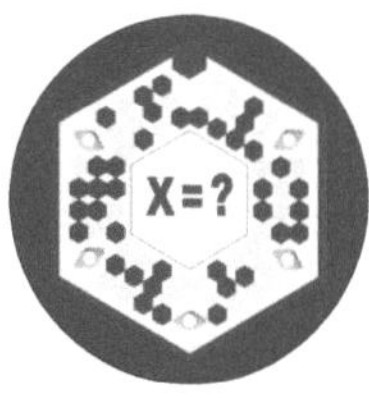

Forces on the individual slices include (1) weight of the soil mass (*W*), (2) pore pressures (*u*) if any, and (3) forces acting on the sides and bottom of the slice. The forces are resolved into components normal and parallel to the bottom of the slice.

In the ordinary method of slices, the following assumptions are made to make the calculations tractable: (1) side forces on each slice are equal and opposite, and are neglected, and (2) the bottom of the slice is straight.

For effective stress analysis, the factor of safety is given by:

$$F = \frac{\Sigma[c'l + (W\cos\alpha - ul)\tan\phi]}{\Sigma W \sin\alpha}$$ **Equation 3-71**

c' = effective cohesion

ϕ' = friction angle of soil

W = slice weight

u = average pore pressure acting on the bottom of the slice

l = length of the bottom of the slice

α = angle of the bottom of the slice measured from the horizontal

For total stress analysis, the factor of safety is given by:

$$F = \frac{\Sigma[c_T l + W\cos\alpha \tan\phi_T]}{\Sigma W \sin\alpha}$$ **Equation 3-72**

c_T = total stress cohesion

ϕ_T = friction angle of the soil

(other parameters are as previously defined)

REFERENCE

1. Samtani, N. C.; Nowatzki, E. A. 2006. *Soils and Foundations Reference Manual (FHWA-NHI-06-088).* Washington, DC: US Dept. of Transportation, Federal Highway Administration.

Structural Mechanics

Daniel A. Howell, PhD, PE

Amir Mousa, MSCE, PE

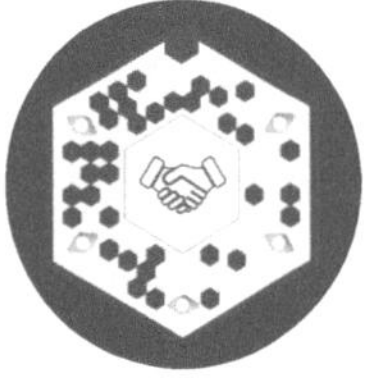

4

CONTENTS

CONTENTS (*continued*)

CONTENTS (*continued*)

EXAM GUIDE

IV. Structural Mechanics

A. Dead and live loads
B. Trusses
C. Bending (e.g., moments and stresses)
D. Shear (e.g., forces and stresses)
E. Axial (e.g., forces and stresses)
F. Combined stresses
G. Deflection
H. Beams
I. Columns
J. Slabs
K. Footings
L. Retaining walls

Approximate Number of Questions on Exam: 6

NCEES Principles & Practice of Engineering Examination, Civil Breadth Exam Specifications

COMMONLY USED ABBREVIATIONS

ASD	allowable strength design
d	effective depth
D	dead load
I	moment of inertia
K	ratio of lateral pressure to vertical pressure
L	live load
L_r	roof live load
LRFD	load and resistance factor design
M	externally applied moment on cross section
M_1	absolute value of the smaller moment at the top or bottom of a column
M_2	absolute value of the larger moment at the top or bottom of a column
r	radius of gyration
S	section modulus
P	applied load
Y	distance from the neutral axis

COMMONLY USED SYMBOLS

ϕ friction angle of the soil
f'_c concrete compressive strength at 28 days

4.1 DESIGN METHODOLOGY

There are two design methodologies that are currently in use for the PE Civil exam, based on the national design code for structural steel and reinforced concrete. The older of the two is referred to as allowable strength design (ASD) or allowable stress design. It is allowed to be used in the structural steel design portion of the exam. The other design methodology is known as load and resistance factor design (LRFD), which is utilized not only in reinforced concrete design, but also in structural steel design.

Thus, for structural steel design problems on the exam, the examinee has the option of choosing which method to use to solve the problem. The given information will be worded in ASD with the associated answer key, and on an additional page the given information will be worded in LRFD with the associated answer key. Although the answers may be slightly different mathematically for each method, the value in the Scantron sheet will be the same (that is, answer B) for ASD and for LRFD.

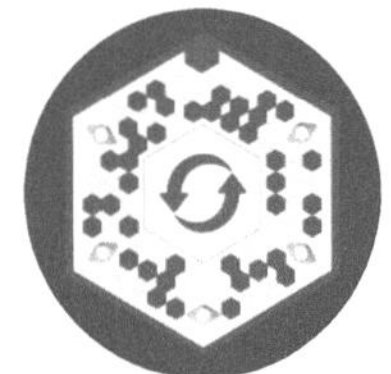

4.1.1 Allowable Strength Design (ASD)

ASD limits the calculated stress from applied loads to a fraction of the allowable stress of the material. The general equation for stress in ASD is:

$$f_{\text{Calculated}} \leq F_{\text{Allowable}}$$ **Equation 4-1**

The nomenclature shown is important; the lowercase value on the left-hand side is based on actual forces on the structure, whereas the uppercase value on the right-hand side is the limit of the stress based on the material. All stresses for ASD are assumed to occur within the linear-elastic portion of the material's stress-strain curve. This implies that when force is applied to a structure, it will deflect or deform a given amount. When the force is removed, the structure will revert to its previously undeformed shape with no permanent deformation.

4.1.1.1 Forces in Allowable Strength Design

In general, forces are summed based on the applied loads present with no additional factors for ASD. These loads are referred to as service loads. Once the total service loads are summed, the stress on the cross section is calculated. The one exception to this rule relates to overturning forces for stability analysis of a structure. Refer to Section 4.2.2.1.1 on overturning forces for this specific case.

The downside to ASD is in the uncertainty of the loading on the left-hand side of Equation 4-1. It is difficult to determine if the stress on an element is due to self-weight, wind, seismic load, live load, snow, rain, and so forth. There is only one total value on the left-hand side, and no weight is given to forces that may have higher or lower probabilities of exceedance. For example, truck loads on interstate highways vary frequently, which is the reason that weigh stations are present: to ensure overweight trucks do not damage bridges that may have a weight restriction.

4.1.1.2 Allowable Stress Values for Allowable Strength Design

The possible uncertainty of the loading on the left-hand side of Equation 4-1 is balanced to a degree with the allowable stress on the right-hand side. This is due to a safety factor, Ω, being applied to the ultimate capacity of the material. For example, structural steel for residential construction may have a yield stress of 36 ksi (248 MPa). However, the $F_{\text{Allowable}}$ in Equation 4-1 would be reduced by a safety factor, depending on the stress type (flexural, axial, shear, and so forth). Thus, if the loading were to exceed the calculated values, the safety factor, Ω, would account for the overstress. The safety factor used depends on the type of stress being investigated; that is, there will be different allowable stress values for bending, flexural, or axial stresses.

4.1.2 Load and Resistance Factor Design (LRFD)

Load and resistance factor design (LRFD) is the newer of the two methodologies, and it permits the designer to consider possible uncertainties of the applied loads. LRFD is based on a statistical distribution of both load and resistance, both of which are assumed to follow a normal distribution, as shown in Figure 4.1. In LRFD, forces and resulting stresses in materials are limited based on specific failure types (such as yielding, rupture, or buckling). For steel, the material yields when the nominal capacity of a member is reached, after which permanent displacements are present.

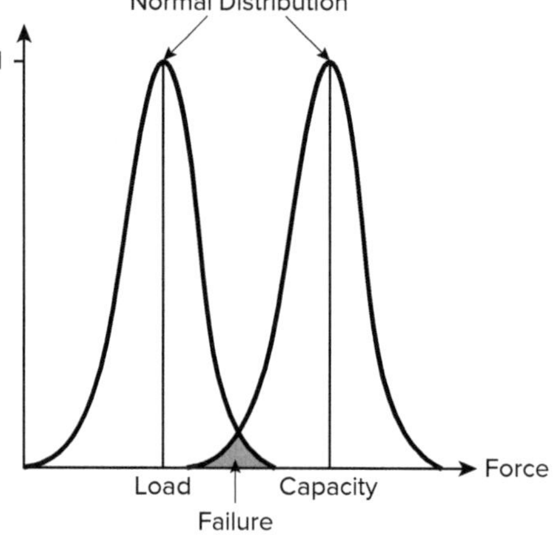

FIGURE 4.1 Load and Resistance Distribution

4.1.2.1 Forces in Load and Resistance Factor Design

LRFD differs from ASD on the loading side as well as the capacity side. The individual service loads (wind, dead, live, rain, snow, seismic, and so forth), defined as Q, are multiplied by load factors, γ. The applicable load types are defined in Table 4.1. There are seven load combinations, as shown in Equations 4-2 through 4-8, which are utilized in ASCE 7-10, "Minimum Design Loads and Associated Criteria for Buildings and Other Structures" (henceforth referred to as ASCE 7) [1]; the International Building Code [2]; and the American Concrete Institute (ACI) "Building Code Requirements for Structural Concrete" design codes (hereafter referred to as ACI 318-14) [3].

TABLE 4.1 Load Factor Definitions

LOAD TYPE	DEFINITION
D	Dead Load
L	Live Load
L_r	Roof Live Load
S	Snow Load
R	Rain Load
W	Wind Load
E	Seismic Load

> **TIP**
>
> The subscript is also a shortcut in the exam, as it indicates that the provided action includes all applicable load factors. Thus, when the subscript *u* is shown in the given information for a problem, it indicates that less work is required in LRFD, mainly that the user does not have to determine the applicable load factors for the specific problem.

$1.4D$ **Equation 4-2**

$1.2D + 1.6L + 0.5(\text{Max } (L_r, S, R))$ **Equation 4-3**

$1.2D + 1.6(\text{Max}(L_r, S, R)) + \text{Max}(1.0L, 0.5W)$ **Equation 4-4**

$1.2D + 1.0W + 1.0L + 0.5(\text{Max } L_r, S, R)$ **Equation 4-5**

$1.2D + 1.0E + 1.0L + 0.2S$ **Equation 4-6**

$0.9D + 1.0W$ **Equation 4-7**

$0.9D + 1.0E$ **Equation 4-8**

Applying Equation 4-3 for a typical interior floor in a building, the last term would not apply. Moment design would result in Equation 4-9.

$M_u = 1.2 \times M_D + 1.6M_L$ **Equation 4-9**

In Equation 4-9, the subscript *u* on the left-hand side indicates the ultimate (that is, factored) design action—for moment in this case.

4.1.2.2 Resistance Values in Load and Resistance Factor Design

The resistance side of LRFD is similar to ASD in that it results in a reduction on the nominal capacity of a material. The difference is that instead of dividing the nominal capacity by a safety factor, Ω (typically greater than 1.0), in LRFD the nominal capacity is multiplied by a strength reduction factor, ϕ (typically less than 1.0). In each case, this results in a reduction on the capacity side. The strength reduction factor, ϕ, varies depending on the stress being investigated. Thus, it will have different values depending on comparing flexural, shear, axial, and bearing stresses.

4.2 LOADING PER ASCE 7 2010/ACI 318-14

4.2.1 Load Definitions

Figure 4.2 shows three categories of loads based on where they are likely to appear on the PE Civil exam. This section is geared toward the morning breadth part of the exam. The first tier represents dead and live gravity loads that would be typical for numerical solutions in the exam.

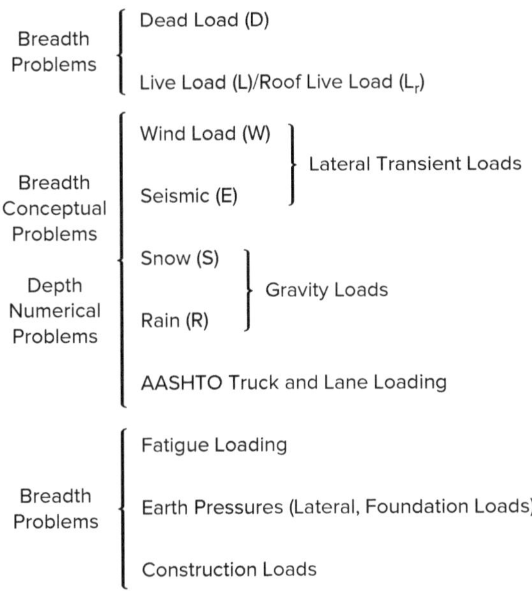

FIGURE 4.2 General Load Types

4.2.1.1 Dead Loads

Dead load encompasses the self-weight of an element and all permanent equipment attached to that element over the intended service life of the structure. Changes in building use may tend to blur the definition of a dead load because the engineer may not be able to foresee load location changes over the service life of a structure.

4.2.1.2 Live Loads

Live loads vary in magnitude and location over time. The critical aspect of live loads is to provide the larger live load for intended occupancy or use if a floor of a building may be a dual or multiuse location. Minimum live loads based on occupancy or use are shown in ASCE 7 2010. For example, if an office building has both lobbies and offices on the same floor, they represent 100 lb/ft^2 and 50 lb/ft^2 live loads, respectively. For this example, the larger of the two (100 lb/ft^2) would be used for the design.

4.2.1.3 Other Vertical Loads

Roof-level loads provide an additional force that designers are required to include in their projects. The roof level load includes rain, snow, and roof live load (for possible maintenance activity) forces. In general, code provisions require the roof slab to be designed for the largest of the three loads shown. Actual required forces are given in ASCE 7 for each type of loading.

4.2.1.4 Lateral Loads

Wind and seismic design provide for transient loads that may not always be present, but are always included in the design of a structure. The wind loading load factors were modified in the ASCE 7 2010 edition. Prior to that publication, the base wind loads based on geographic location were obtained and then multiplied by the appropriate wind load factor. The base wind loads were then changed to include the wind load factors; thus, the wind load factors are smaller than those in older versions of ASCE 7.

4.2.1.5 Construction Loads

Construction loads can often result in forces that exceed the intended forces for a structure or, in rare instances, result in a stress that would not be considered under the final as-built condition. An example of this relates to stresses in steel beams for bridge applications. As shown in Figure 4.3, the exterior beam of a bridge is shown with a cast-in-place concrete deck supporting a reinforced concrete barrier. Once the deck has been cast, the beam will resist mainly vertical dead and live loads. However, during casting of the deck and prior to placement of the reinforced concrete barrier, the steel beam must support the wet weight of the concrete. Based on the overhang to the far left, this results in torsional stresses in the beam that may not have been accounted for in the original design and that have the potential to lead to failure.

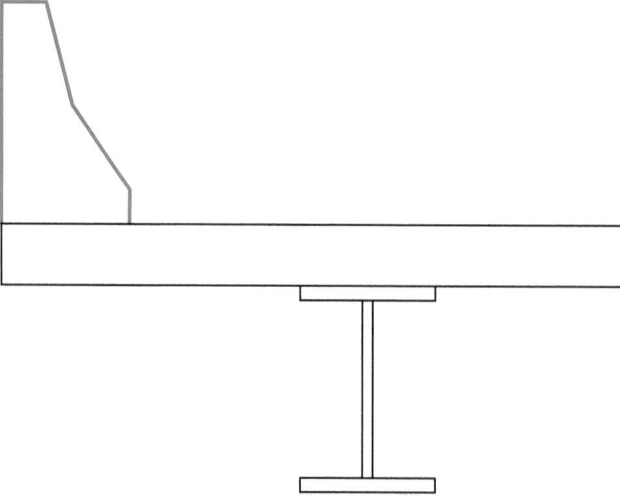

FIGURE 4.3 Steel Beam Construction Loads

4.2.2 Load Factors

Load factors are typically greater than or equal to 1.0. For Equation 4-3, note the variability of the factors for dead and live loads. The larger factor for a live load is due to its inherent variability. The previous example of weigh stations on interstate roads (Sec. 4.1.1.1) illustrates the chance for exceedance for a live load. The dead load includes a 20% increase, which would account for loads that may not be easily calculated. For example, in building design, the weight of movable partition loads or cubical sections is given a blanket loading that would cover typical self-weight values. However, if heavier than expected wall sections are used, this may result in an overload condition. Thus, the dead load factor is increased from 1.0 to 1.2.

4.2.2.1 Load Combinations for Maximum Effect

4.2.2.1.1 Overturning Due to Lateral Forces

Equations 4-7 and 4-8 indicate a dead load factor of 0.9. This condition is used to check overturning for typical building loads for overall stability. In both equations, the second load is a lateral force—wind or seismic. Why is the dead load reduced in this scenario? Figure 4.4 shows a typical building subjected to lateral wind loading, which will be checked for overturning and uplift of footing A.

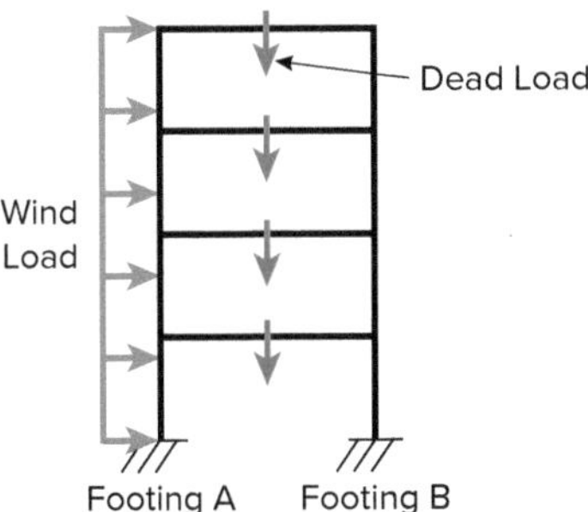

FIGURE 4.4 Lateral Force on Building for Overall Stability

The driving force that causes overturning is the wind load in Figure 4.4. The resisting force will come from the dead load of each floor load. The reduction in dead load is a conservative approach that will account for any dead loads that were overestimated in the design of the beams, columns, and foundations. For example, the dead load of a drop ceiling, HVAC ducts, electrical conduits, and ceiling lights would not be calculated on an individual basis for each component. An estimate of dead load for such equipment would be used. If the actual weight of equipment was on the order of 8 lb/ft^2, then the designer would use a slightly higher value of 12 lb/ft^2. This results in a slightly larger element when designing the beams, columns, and foundation elements, but it is a conservative design. For overturning, it would be aggressive to assume 12 lb/ft^2 is resisting the wind load when in reality the dead load is closer to 8 lb/ft^2. This is accounted for in the design codes with a dead load factor of 0.9 when it is used as resistance to uplift.

4.2.2.1.2 Gravity Loads

The two load combinations shown in Equations 4-2 and 4-3 indicate a dead load factor of 1.4 or 1.2, respectively. In most instances, the second equation that includes a 1.6 live load component will govern, unless the structure being considered has a significantly larger ratio of dead load to live load, such as an agricultural facility or possibly an offsite storage location where human occupation is low within the structure.

4.3 DETERMINATE STRUCTURAL ANALYSIS

A determinate structural analysis implies that there are enough equations of equilibrium to solve for the unknown boundary conditions for a given beam, truss, or frame structure. The degree of indeterminacy is calculated from the total number of unknowns minus three, which represent the equations of equilibrium.

4.3.1 Boundary Conditions for Two-Dimensional Beams and Trusses

There are three boundary conditions for a two-dimensional analysis, as indicated in Figure 4.5.

The roller support, represented by a circle, provides one restraint normal to the supporting surface. More importantly, the other degrees of freedom that are permitted include translation left to right, as well as rotation. The second boundary condition is denoted as a pin or hinge, which provides restraint in the horizontal and vertical directions.

FIGURE 4.5 **Boundary Conditions for Two-Dimensional Analysis**

The permitted degree of freedom for a pin support is rotation. The final boundary condition is a fixed support, represented as a series of slashed lines. The fixed support provides restraint from the horizontal and vertical directions, as well as rotation.

4.3.2 Resultant Forces

Support reactions are found by summing vertical forces or the moment about a point. However, in some instances, point loads must be determined from a uniform or triangular distributed loading. For the case of a uniformly distributed load, the resultant can be calculated as the weight per foot times the distance of the loading.

Example 4.1: Resultant Forces

The loading shown is −3.7 kips/ft over a distance of 13.5 ft. What is the resultant of the force?

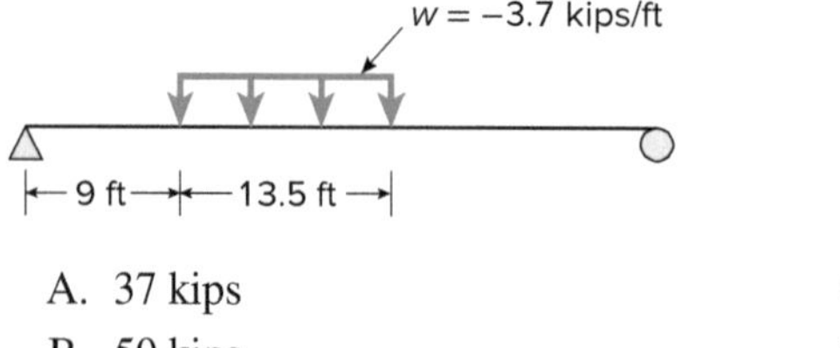

A. 37 kips
B. 50 kips
C. 24 kips
D. 96 kips

Solution

The resultant of this force acts at the center of that 13.5-ft distance or 6.75 ft from the start of the loading with a value of −3.7 kips/ft × 13.5 ft = 50 kips.

$$X = 9\text{ ft} + \frac{13.5\text{ ft}}{2} = 15.75\text{ ft}$$

Resultant Force:

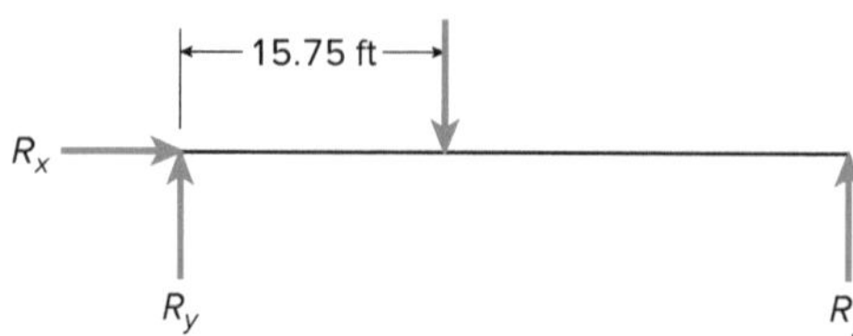

Answer: B

For a triangular loading, the same approach is used; however, the resultant location will act at two-thirds of the triangular distance from the low end or one-third of the triangular distance from the high end of the loading.

TIP

A common error in summing moments with triangular loading is to find the resultant correctly, but mix the two-thirds and one-third distances, which will vary depending on what end the moment is taken from on the beam.

Example 4.2: Resultant Forces

The slope of the triangular loading is shown as -1.2 kips/ft^2. Where does the resultant force act from the right-hand side?

A. 7 ft

B. 20 ft

C. 5 ft

D. 12 ft

Solution

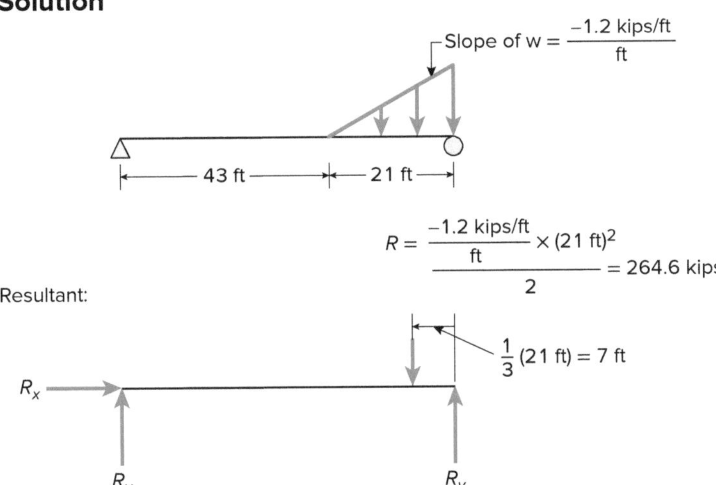

Summing the moments about the right-hand side, the moment arm for the triangular loading will be 7.0 ft from the right-hand side. If the triangular load is not located at the terminal end of the beam, then an additional distance would be added to the start of the loading.

Answer: A

The last type of loading combines the first two mentioned (uniform and triangular)—a trapezoidal loading. There is an equation to determine the resultant loading in this case, as shown in Equation 4-10.

$$\text{Resultant}_{\text{trapezoid}} = \frac{1}{2}(F_1 + F_2)h \qquad \textbf{Equation 4-10}$$

F_1 is the load at one end of the trapezoidal loading, F_2 is the load at the opposite end of the trapezoidal loading, and h is the horizontal distance between F_1 and F_2. The location of the resultant force for the trapezoidal load is possible; however, an easier solution is to decompose the trapezoidal load into equivalent rectangular and triangular loadings. This will result in two forces versus one, but the solution is easier to follow.

4.3.3 Equations of Equilibrium for Two-Dimensional Beams and Trusses

There are three equations of equilibrium, shown in Equations 4-11 through 4-13.

$$\Sigma F_v = 0 \text{ (Summation of forces in the vertical direction} = 0) \qquad \textbf{Equation 4-11}$$

$\Sigma F_h = 0$ (Summation of forces in the horizontal direction = 0) **Equation 4-12**

$\Sigma M_{\text{Point } x} = 0$ (Summation of moments about a point $x = 0$) **Equation 4-13**

Note that the point x in Equation 4-13 is arbitrary along the length of a beam or truss structure. In general, the extreme ends of the beam are chosen as a point to sum moments about, whereas in truss applications, a joint is chosen as the point to sum moments about.

Example 4.3: Finding Reactions

Find reactions at D and E.

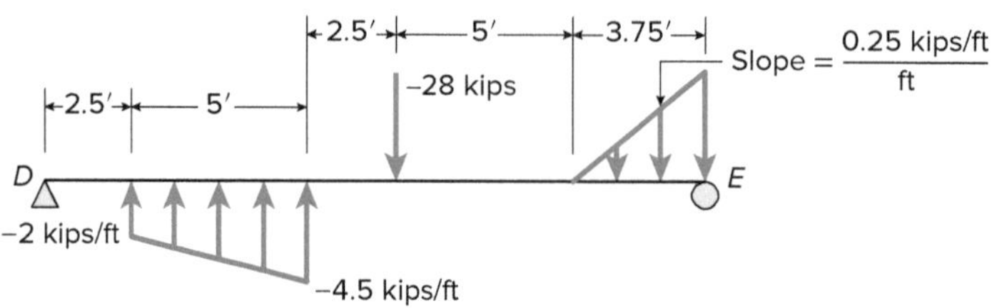

Solution

Resultant forces:

Trapezoidal loads decompose into:

1. Uniform load of −2.0 kips/ft Resultant = (−2.0 kips/ft)(5.0 ft) = 10 kips
2. Triangular load of −2.5 kips/ft Resultant $= \dfrac{(-2.5 \text{ kips/ft})(5.0 \text{ ft})}{2} = 6.25$ kips

Second triangular load:

$$\text{Resultant} = 0.25\,\frac{\text{kip}}{\text{ft}}/\text{ft}\,\frac{(3.75 \text{ ft})^2}{2} = 1.76 \text{ kips}$$

Free body with resultant forces:

$$\curvearrowleft + \Sigma M_E = 0 = (1.76 \text{ kips})(1.25 \text{ ft}) + (28 \text{ kips})(1.25 \text{ ft} + 7.5 \text{ ft}) - (6.25 \text{ kips})\left(1.25 \text{ ft} + 7.5 \text{ ft} + 4 \text{ ft} + \frac{4 \text{ in}}{12}\right)$$
$$-(10 \text{ kips})(1.25 \text{ ft} + 7.5 \text{ ft} + 4 \text{ ft} + 4/12 + 10/12) - (D_y)(18.75 \text{ ft}) = 0$$

$$D_y = 1.40 \text{ kips}\downarrow$$

$$+\uparrow \Sigma F_y = 0 \Rightarrow -1.40 \text{ kips} + 10 \text{ kips} + 6.25 \text{ kips} - 28 \text{ kips} - 1.76 \text{ kips} + E_y = 0$$

$$E_y = 14.91 \text{ kips}\uparrow$$

$D_x = 0$ by inspection

Structural Mechanics

School of PE

4.3.4 Shear and Moment Diagrams

A convenient transition from beam reactions to shear and moment diagrams lies in free-body analysis of a beam section. For the beam shown in Example 4.4, if the shear diagram was to be constructed, one method would be to provide a free body of the beam on each side of an applied load. On the free body, the unknown of interest (shear or moment) would be determined utilizing the equations of equilibrium. For simplicity, Example 4.4 shows the shear as the unknown on each free body.

Example 4.4: Shear and Moment Diagrams

Determine the shear diagram using free-body diagrams from the loadings shown.

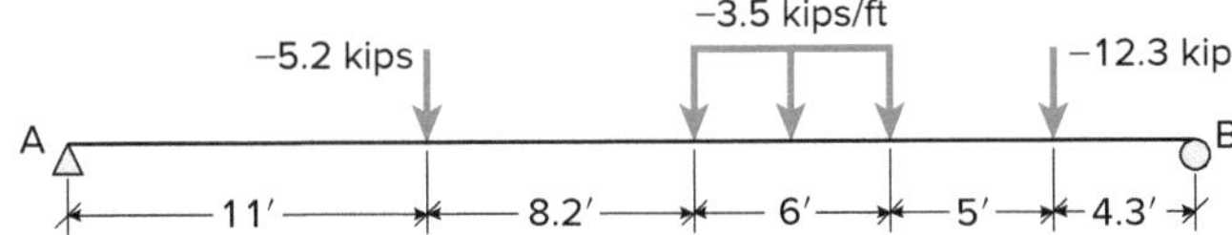

Solution

Free Body 1: (from Left-Hand Side to 10.9 ft from Support *A*)

Free Body 2: (from 11.1 ft to 19.1 ft from Support *A*)

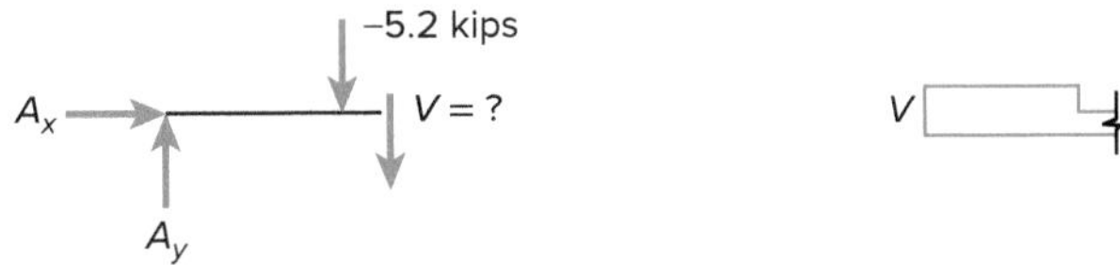

Free Body 3: (from 19.2 ft to 25.1 ft from Support *A*)

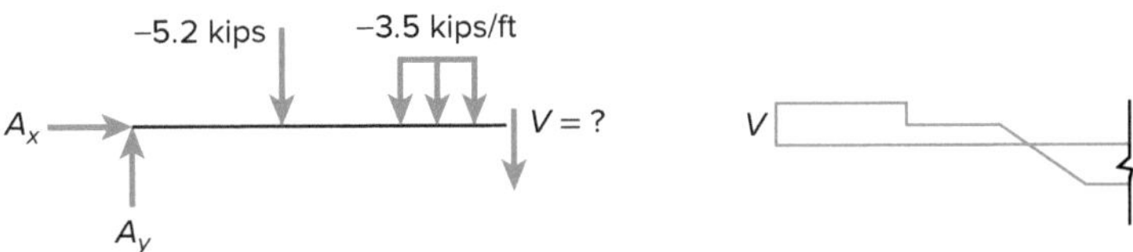

Free Body 4: (from 25.2 ft to 30.1 ft from Support *A*)

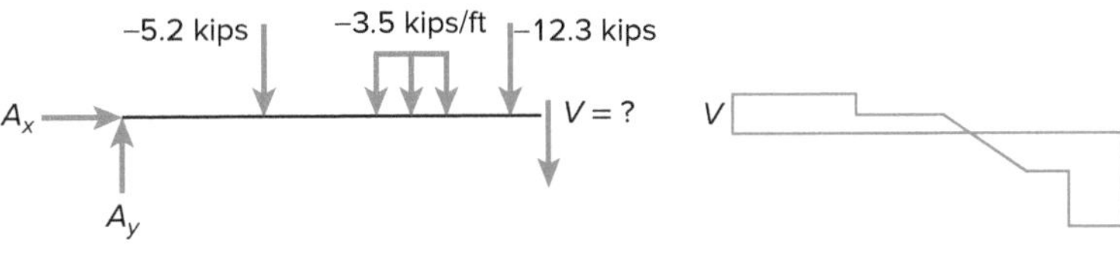

The shear diagram can be constructed from the shear values indicated from the four free-body diagrams. If the unknown in the free body was identified as a moment instead of a shear force, the moment diagram could be constructed in a similar manner. The downside of the free-body approach is the time commitment required for the solution. In Example 4.4, the loading shown includes several loads and thus four free-body diagrams need to be drawn and solved. It could be done, but the time required would likely be restrictive in the exam setting.

Another method is to create the shear and moment diagram simply from the applied loads shown. There are a few rules that will aid in this approach, identified in Table 4.2. The rules shown are described in more detail after Figure 4.6. Conversion from applied load to shear is an integral relationship, as is conversion from shear to moment.

TABLE 4.2 Rules for Shear and Moment Diagram Creation

SHEAR DIAGRAMS:
1. The shear diagram changes by the magnitude of the load. 2. Shear is constant along unloaded portions of a beam. 3. For a point load, the shear function is constant. For a uniform load, the shear function is linear. 4. The slope of a shear diagram at a point is equal to the magnitude of the load at that point.
MOMENT DIAGRAMS:
1. For constant shear, the moment function is linear. 2. For linear shear, the moment is a second-degree curve. 3. The change in the moment is equal to the area under the shear diagram. 4. Maximum (or minimum) moment occurs at points of zero shear. Identify those locations in a member length to select the worst (largest) moment. 5. An applied moment at a given location does not affect the shear diagram, only the moment diagram. 6. The slope of a moment diagram at a point is equal to the magnitude of shear at that point.

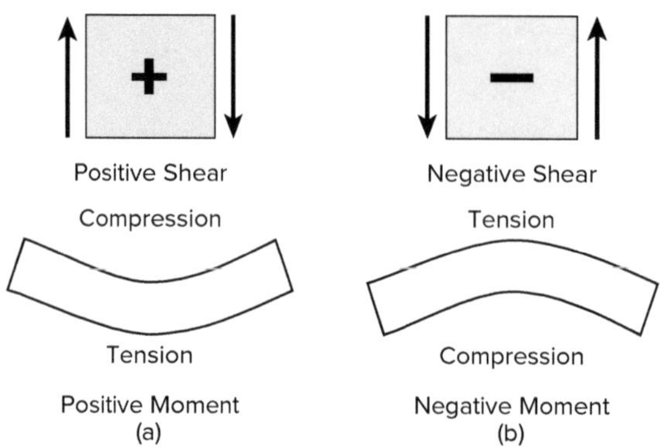

FIGURE 4.6 Sign Convention for Shear and Moment Diagrams

Rules 1 and 2 for shear diagrams indicate a relationship with respect to the slope of the curve for unloaded portions of the beam, in which the slope of the line is zero. This matches the relationship shown in Example 4.4. Rule 3 indicates the integral relationship from applied load to shear. For example, if a point load of −10 kips was applied to a beam, the shear diagram would be constant before and after the applied load, but would result in a vertical jump in the shear diagram. Transitioning to the moment diagram, the integral of a constant is a linear curve, that is, $-10x$. Thus, the moment diagram would have a linear curve with a slope equal to the value of the shear diagram. The same would hold for an applied linear load. For example, if a uniformly distributed load of −4.5 kips/ft was applied to a simply supported beam, the shear diagram would be a linear distribution with a slope equal to the value of the applied load, −4.5 kips/ft in this case. The final rule for shear diagrams indicates the integral relationship previously noted.

For the moment diagram, rules 1 and 2 define the integral relationship that follows from the previous discussion between applied load and shear. For rule 3, the numerical value of the moment can be calculated from the area under the shear diagram. Example 4.5 illustrates the process for a simply supported beam with a uniformly distributed load.

Example 4.5: Maximum Moment in a Beam Due to Shear Diagram

(1) Find the maximum moment in the beam.

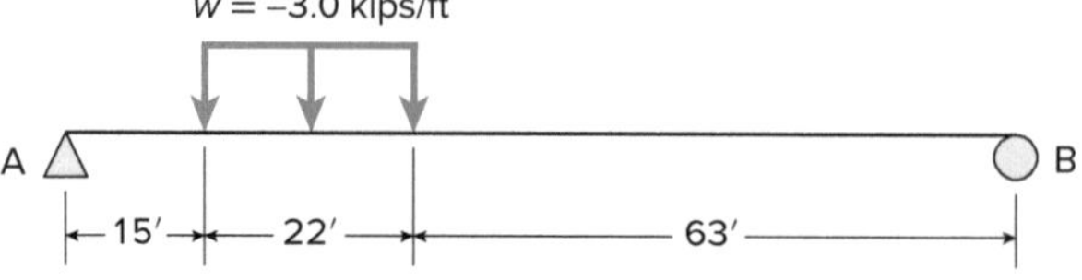

Example 4.5 *(continued)*

Solution (1)

$$\curvearrowleft + \Sigma M_A = 0 \Rightarrow -\left(3.0\,\frac{\text{kips}}{\text{ft}}\right)(22\text{ ft})\left(15\text{ ft} + \frac{22}{2}\right) + B_y(100\text{ ft}) = 0$$

$$B_y = 17.16\text{ kips}\uparrow$$

$$+\uparrow\Sigma F_y = 0 \Rightarrow A_y - \left(3.0\,\frac{\text{kips}}{\text{ft}}\right)(22\text{ ft}) + 17.16\text{ kips} = 0$$

$$\Rightarrow A_y = 48.84\text{ kips}\uparrow$$

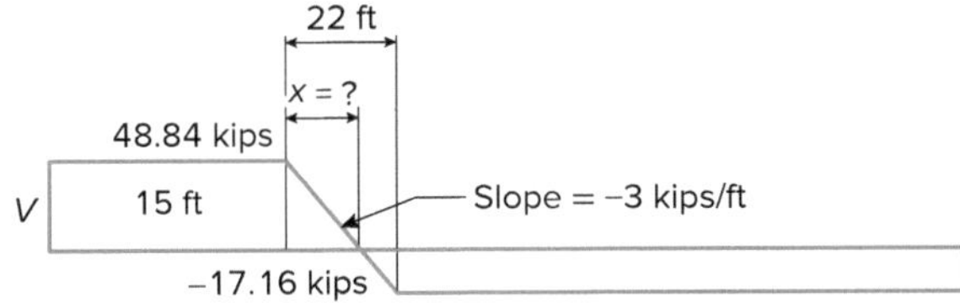

(2) What is the distance x where the shear changes from positive to negative? Use similar triangles.

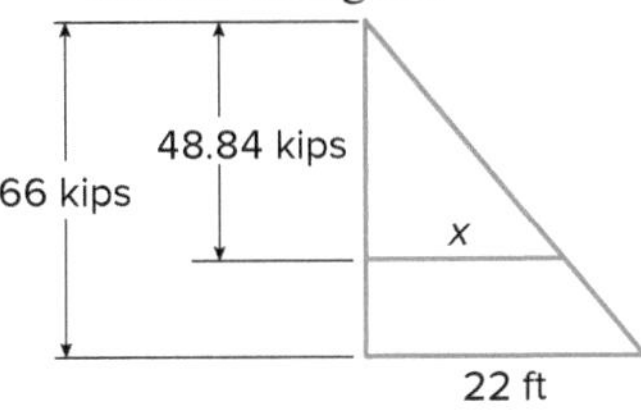

Solution (2)

$$\frac{22\text{ ft}}{66\text{ kips}} = \frac{x}{48.84\text{ kips}}$$

$$\Rightarrow x = \frac{22\text{ ft}(48.84\text{ kips})}{66\text{ kips}} = 16.28\text{ ft}$$

$$\text{Area under shear diagram} = (48.84\text{ kips})(15\text{ ft}) + \frac{1}{2}(16.28\text{ ft})(48.84\text{ kips})$$

$$= 1{,}130\text{ kips/ft}$$

The location of a maximum or minimum moment corresponds to a point where the shear diagram crosses the horizontal axis. In Example 4.5, there was only one crossing, resulting in a maximum positive moment. In problems where multiple crossings occur, there may be a local minimum positive or negative moment.

Rule 5 indicates that if a concentrated moment is situated along the length of a beam, it is neglected for the shear diagram, but must be accounted for in the moment diagram.

Exam questions regarding the general shape of a shear or moment diagram, from a given loading, require additional discussion. Example 4.6 indicates a simply supported beam with pin-roller boundary conditions subjected to an arbitrary loading, $w(x)$. The loading increases from left to right, but no numerical values are indicated.

Example 4.6: Shape of Shear Moment Diagram

What is the resulting shape of the shear and moment diagram?

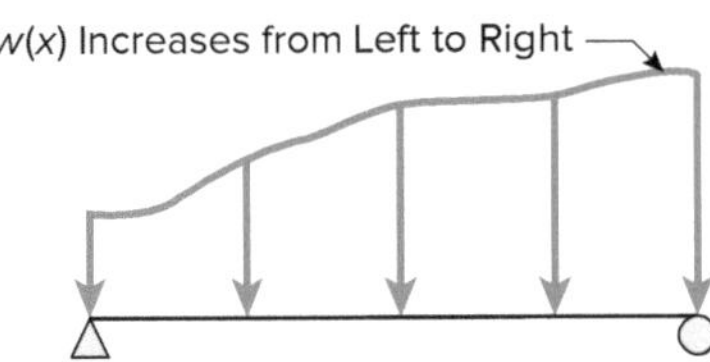

Example 4.6 (continued)

Step 1: Is the starting point at, above, or below zero?

Step 2: Is it a positive or negative slope, and is it moving from left to right?

Step 3: Concave up or concave down? Curve 1 or curve 2?

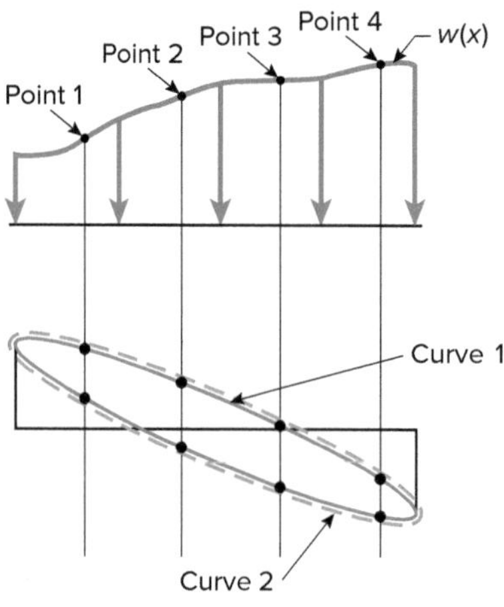

Solution

The smaller the applied load, the flatter the tangent point on the shear diagram. The top curve is correct.

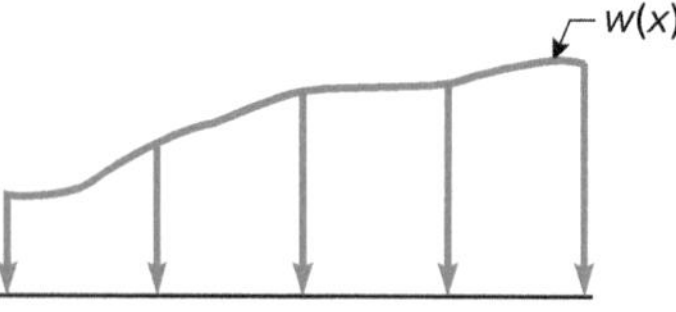

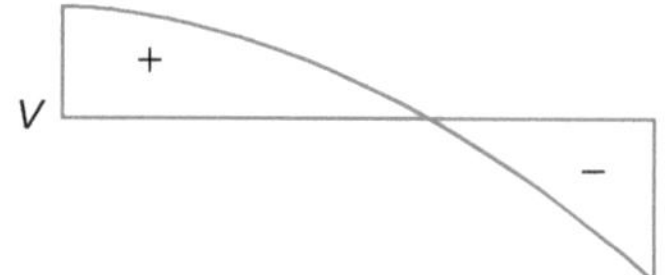

Moment diagram:

Step 1: Starting point = 0 due to boundary conditions (pin support)

Step 2: Positive or negative slope moving left to right? The positive slope is initially due to the shear diagram being above the x axis.

Step 3: Concave up or concave down? Tangent points?

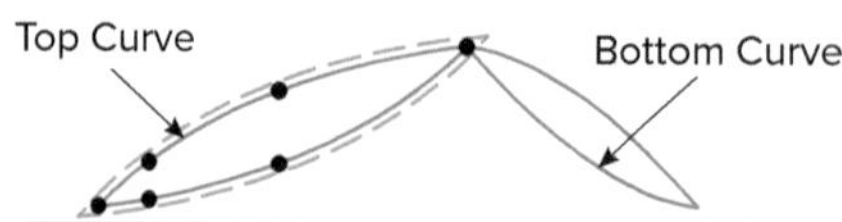

The top curve corresponds to values from the shear diagram.

There are three questions that give an indication of what the shape of the shear and the moment diagram will look like. The shear diagram will be the starting point for Example 4.6. The first question relates to the starting point on the left-hand side. Will the location be above the horizontal axis, zero, or below the horizontal axis? Based on the pinned support on the left-hand side, there will be an upward reaction; thus, the diagram should not start at zero. Assume a value above zero as the starting point. The second question relates to the slope of the curve, moving from left to right. Will the slope be positive or negative for this example? For this solution, the applied load is downward, which would indicate that the slope of the shear diagram will be negative. The last question relates to a concave up or concave down curve. These two options are shown in the Example 4.6 solution. To determine if the top or bottom curve is the correct solution, identify a few points on the loading curve. These are labeled as points 1–4 in the solution. The goal is to relate the applied load to the inclination or slope of a tangent point on the shear diagram at the same location. In essence, the smaller the value of the applied load, the flatter the tangent point will be on the shear diagram. As the value of the applied load increases, the tangent point on the curve will increase, going from a relatively flat tangent to a near-vertical tangent point as the value of the applied force increases. For Example 4.6, the top curve would be the correct solution.

For the moment diagram, the same three questions are posed from the shear diagram creation. On the left-hand side for the moment diagram, will the curve start above the horizontal axis, at zero, or below the horizontal axis? The curve will start and end at zero at the left- and right-hand sides because of the boundary conditions of a pin and/or roller support, both of which provide no restraint for rotation. The second question is: Moving from left to right, is the slope of the curve initially positive or negative? For this answer, refer back to the shear diagram, which is located above the horizontal axis on the left-hand side. This would indicate that the slope of the moment diagram is positive on the left-hand side. The last question—is the curve concave up or concave down?—is illustrated in the Example 4.6 solution. The top curve is the correct answer. Instead of comparing values on the shear diagram as done in the previous diagram, investigate the value of the shear diagram from left to right. The shear diagram starts out with a positive value, and then gets smaller and smaller until it crosses the horizontal axis into the negative portion of the curve. The tangent points on the moment diagram would then start with a larger incline and get progressively flatter as the value of the shear diagram decreases. At the crossing point in the shear diagram, the moment diagram would have a horizontal tangent. Beyond the tangent point, the moment diagram would have a tangent point that is relatively flat, but with a negative slope, because the shear diagram is negative but has a small value initially. To the right of this location, the value of the shear diagram continues to grow larger and more negative, resulting in a tangent point that is more inclined and closer to the vertical position.

Note that superposition of loading cases is possible for similar boundary conditions. In other words, for a simply supported beam with a uniformly distributed load and a concentrated moment, the deflection on the elastic curve of both load cases can be determined independently and simply added together afterward.

TIP

Appendix A.4 shows maximum deflections, elastic curve, shear, and moment diagrams for various loading cases for simply supported, cantilever, and fixed beam cases. These figures offer a wealth of information that can be used in exam problems.

The general shape of the shear and moment diagram for various loading types for a simply supported beam is shown in Figure 4.7. Note that for a concentrated moment along the length of a beam, a vertical change in the moment diagram results. If constructing such a diagram in the exam, the sign convention of the vertical jump becomes critical to the correct solution. A simple solution would be to pick either a vertical jump upward or a vertical jump downward. Continue the remainder of the diagram to determine if it does indeed close to zero on the right-hand side, if the boundary conditions of a pin or roller are present. If the diagram does not close to zero, then the direction of the vertical jump was initially incorrect. Alternatively, one may attempt to determine if the vertical jump is above or below by investigating the sign of the applied moment (clockwise or counterclockwise). However, this may not be clear at first glance, and in general may not result in the correct solution.

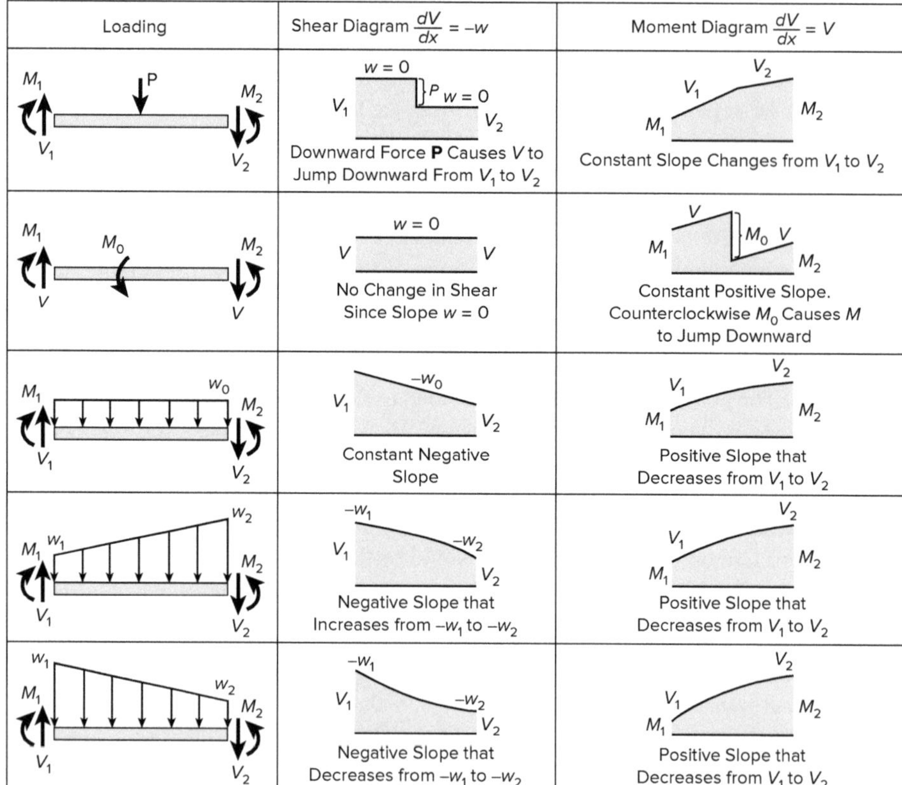

FIGURE 4.7 Shape of Shear and Moment Diagrams for Various Loading Types

Source: Hibbeler, Russell C. *Mechanics of Materials (3rd ed.)*, 1997. Reprinted by permission of Pearson Education, Inc., New York.

4.3.5 Influence Lines for Continuous Beams

Shear and moment diagrams provide information along the entire length of a beam. An influence line only provides information at one specific point in a span. It is useful for determining where loading should be placed for maximum effect. The image in Example 4.7 shows a nine-span continuous structure that is representative of a beam line in a typical building, or possibly a bridge structure. Determining the maximum reaction at the fourth support results in too many possible span-loading combinations to be feasibly included on the exam. For a uniformly distributed load, would loading all spans result in the largest reaction? How about loading the even spans only? Or possibly the odd spans only? Or possibly loading the first span, and then every other span afterward? The possible number of loading combinations makes comparison difficult. Influence lines remove the ambiguity of what spans or portions of a span to load for reaction, shear, and moment. The key concept for influence lines is the deflected shape of the continuous beam being investigated. A diving board, as illustrated in Figure 4.8,

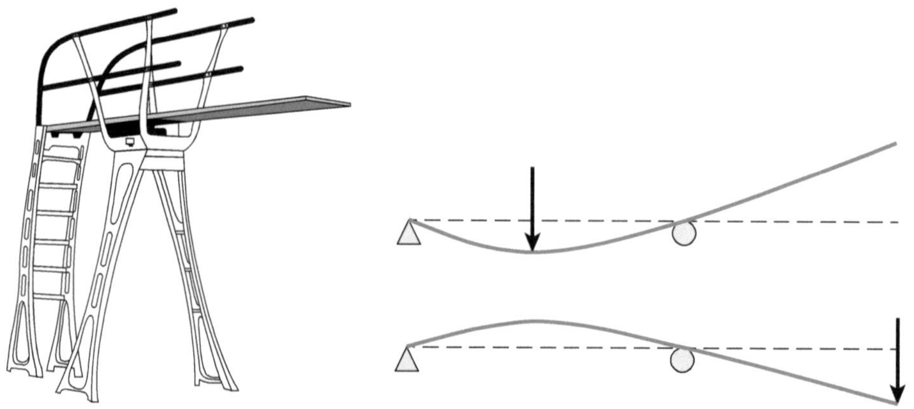

FIGURE 4.8 Diving Board Example for Influence Diagram

provides a good visual reference to indicate which direction one span will deflect, based on the location of the load, and more importantly, what effect that has on adjacent spans. The general rule of thumb is that downward deflection in one span results in upward deflection in adjacent spans. The corollary also holds true.

Example 4.7: Multispan Continuous Beam for Loading Purposes

What spans should be loaded with a uniformly distributed load for the maximum reaction at R4?

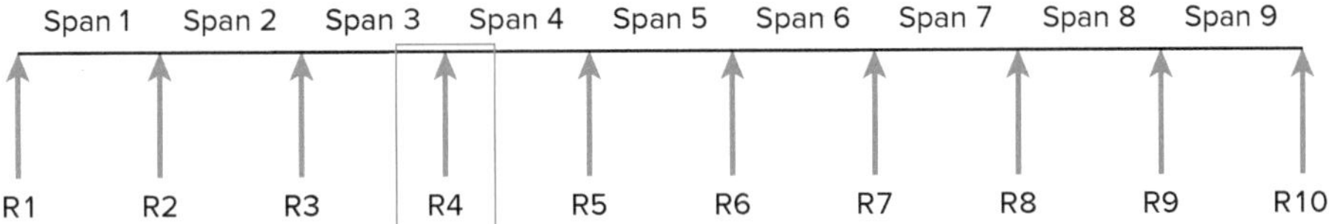

Solution

It is too difficult to assume many span variations.

1) All spans
2) Odd spans
3) Even spans
4) Third spans
5) 1st, 5th, 6th spans

Influence lines for a reaction at an interior support involve several steps:

1. Remove the support being investigated.
2. Displace the beam vertically upward or downward at a distance of one unit. The resulting beam shape will be the shape of the reaction influence diagram.
3. For maximum effect, load the portions that displace vertically in the direction of the initial displacement from step 2.

The diagram below shows how the deflected shape indicates the correct solution.

What is the approximate shape of the reaction influence diagram for reaction 4 for the 4-span continuous structure shown in Figure 4.9?

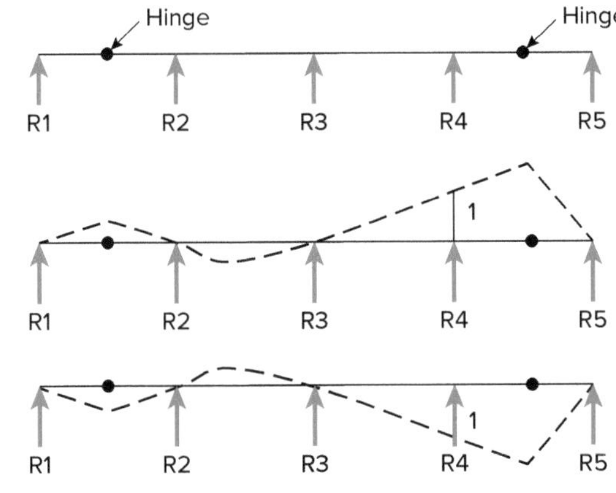

FIGURE 4.9 Influence Diagram for Reaction

The maximum reaction at R4 will occur when spans 1, 3, and 4 are loaded.

The influence line for beam shear involves several steps:

1. Replace the point of interest with a link element of nominal length. If the point of interest is a reaction, place a hinge at that point and lift the hinge upward a unit distance.
2. Push the two ends of the beam together at the link until it is vertical. The total beam length has not changed, and the slope of the sections to the left or right are the same.
3. Determine the ratio of δ_{left} and δ_{right}. The larger amount will indicate which side of the beam to load.

The use of a link element in Figure 4.10 shows how the total length of the beam does not change.

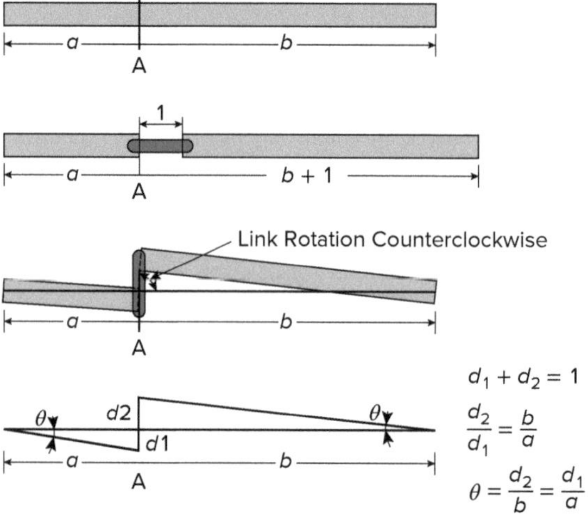

FIGURE 4.10 Influence Line for Shear—Link Element Usage

From this example, we know that $d_2 > d_1$. Therefore, place the load on the d_2 or positive side. If the link rotation was clockwise instead of counterclockwise, then the d_2 value would be below the horizontal axis and the sections below the axis would be loaded for maximum effect. The diagram below provides an example of a three-span structure.

Where should a uniformly distributed load be placed on the beam to maximize the shear at section A in Figure 4.11?

The influence line for the shear solution (Fig. 4.11):

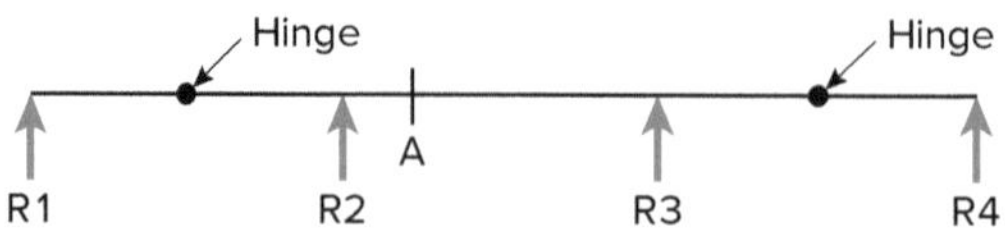

FIGURE 4.11 Influence Line for Shear–Three-Span Structure

For counterclockwise link rotation (Fig. 4.12):

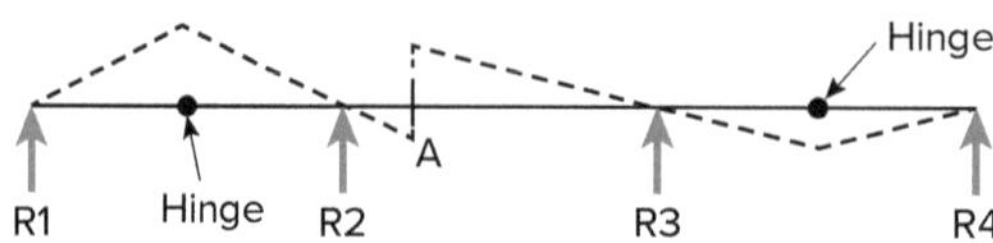

FIGURE 4.12 Solution Influence Line for Shear with Counterclockwise Rotation of Link Element

Note that for the counterclockwise link rotation shown, the larger distance at the point of interest is above the horizontal axis. Therefore, the sections located above the horizontal axis will be loaded for maximum effect.

For clockwise link rotation (Fig. 4.13):

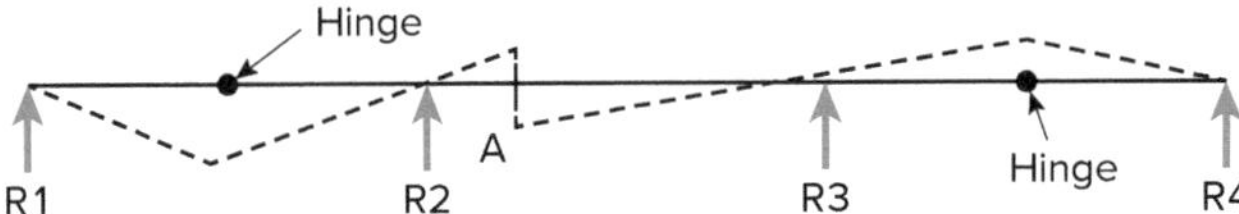

FIGURE 4.13 Solution Influence Line for Shear with Clockwise Rotation of Link Element

The influence line for the beam moment involves several steps:

1. Replace the point of interest with an imaginary hinge.
2. Rotate the beam one unit rotation by applying equal but opposite moments to each of the two beam sections.
3. The deformed shape will indicate which portion of the beam to load for maximum effect.

Note that when a point of interest is a support, this unit rotation can be achieved by simply lifting the point at the hinge location. Figure 4.14 indicates the use of the influence line for moment.

Where should a uniformly distributed load be placed on the following beam to maximize the moment at point A (Fig. 4.14)?

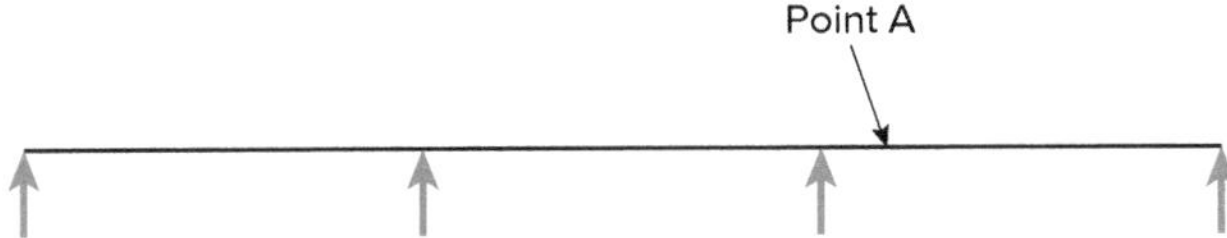

FIGURE 4.14 Influence Line for Moment–Three-Span Example

Assume a joint is located at point A and rotate it counterclockwise (Fig. 4.15).

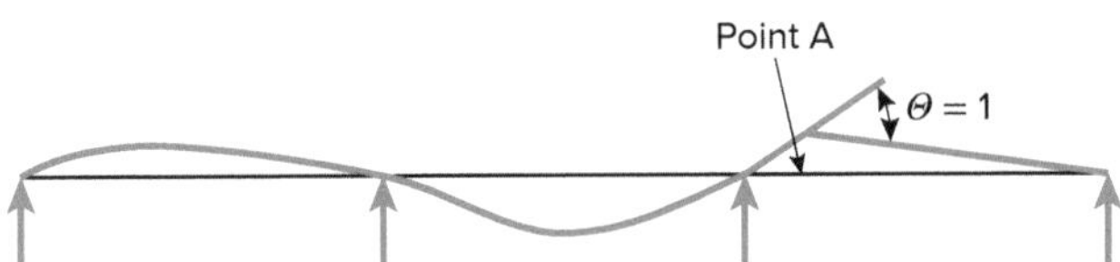

FIGURE 4.15 Solution Influence Line for Moment with Counterclockwise Rotation

Note that the beam deflection is zero at each support. Since point A is located above the initial beam location, load all of the spans located above the undeformed shape.

If the rotation was clockwise, the deflection would be as shown in Figure 4.16.

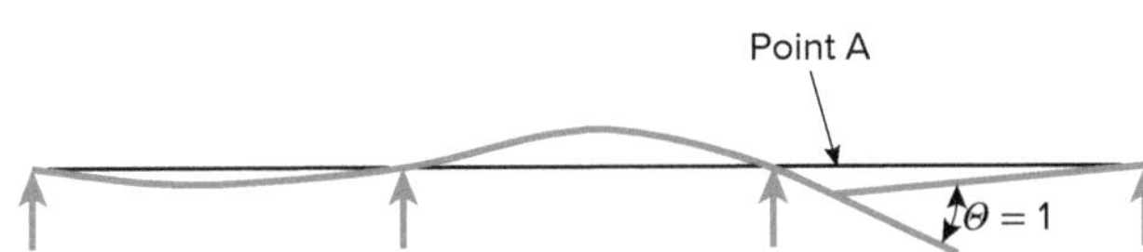

FIGURE 4.16 Solution Influence Line for Moment with Clockwise Rotation

In this case, load all the spans below the undeformed beam.

4.3.6 Three-Dimensional Analysis

For analysis types that do not conform to the traditional two-dimensional environment, such as the one shown in Example 4.8, the analysis follows from the two-dimensional in each direction independently.

Example 4.8: Analysis of a Plate

What is the reaction at $L3$? Each side of the plate is 42 ft.

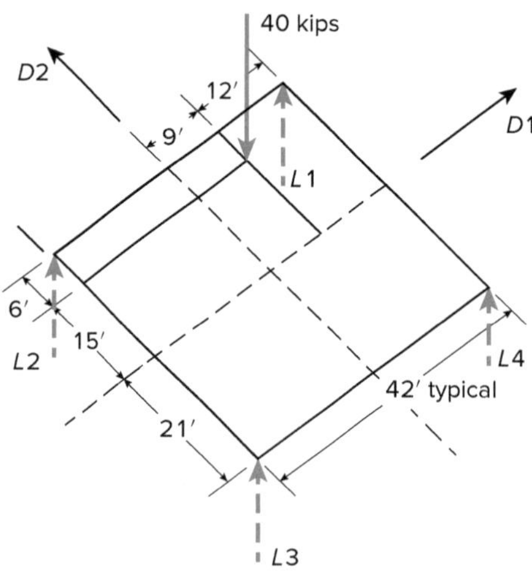

Solution

Assume the $D2$ direction first:

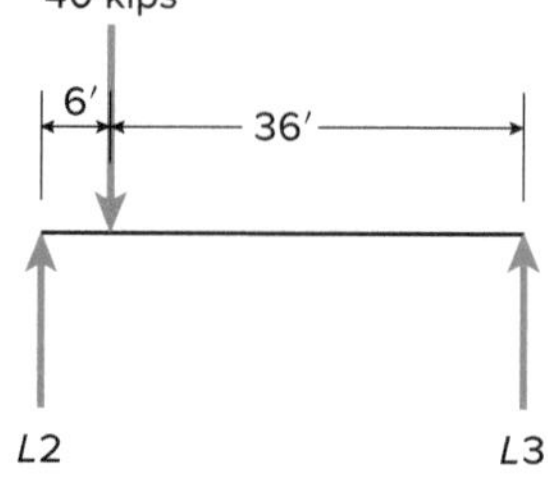

$$L3 = 40\text{ kips}\left(\frac{6\text{ ft}}{42\text{ ft}}\right) = 5.71\text{ kips}$$

Assume the $D1$ direction next:

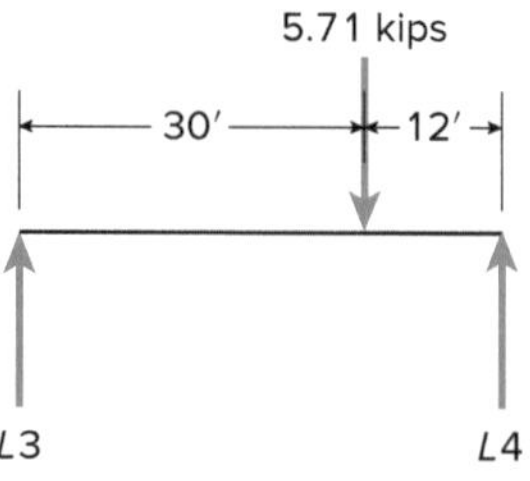

$$L3 = 5.71\text{ kips}\left(\frac{12\text{ ft}}{42\text{ ft}}\right) = 1.63\text{ kips}$$

The first step is to solve for the reactions in one direction; start with the reaction in the $D1$ or $D2$ direction only.

The second step is to solve for the reactions in the orthogonal direction as indicated in the diagram to the right.

4.4 TRUSS ANALYSIS

Truss analysis is based on two primary solution techniques: method of joints and method of sections. Truss joints are assumed to have a pin support between members, which indicates that there are two unknowns at each truss joint.

4.4.1 Two-Dimensional Truss Reactions

The calculation of truss joint forces follows from the three equations of equilibrium used in beam analysis. However, if the loads are placed symmetrically about the span, and the geometry is likewise symmetrical (typically the case for trusses), then the reactions on the left and right sides of the truss are simply half of the total applied load.

Example 4.9: Truss Reactions Due to Symmetry

Calculate the reaction of components due to the symmetrically applied load.

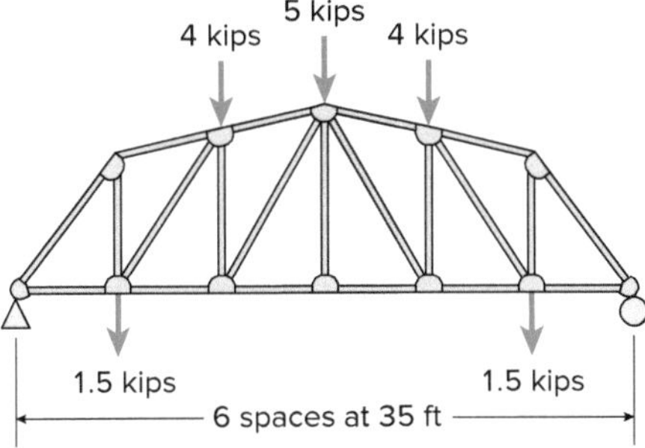

Solution

$$\text{Reactions} = \frac{\text{total load}}{2}$$

$$= \frac{(1.5 \text{ kips})(2) + 5 \text{ kips} + (4 \text{ kips})(2)}{2}$$

$$= 8 \text{ kips left/right}$$

4.4.2 Truss Member Forces

4.4.2.1 Method of Joints

Unknown member forces that use the method of joints require a consistent sign convention when evaluating if members are in tension or compression. Assume that every truss member at a joint is in tension, with the force acting away from the joint. If a member is indeed a compression member, then the solution will result in a negative number for that member. However, the assumption of all members in tension at the start will remove any guessing as to the actual sense of the stress.

Decomposing inclined members is a required part of the method of joints. Recall the trigonometric functions relative to a common angle in Figure 4.17.

The method of joints is illustrated in Example 4.10.

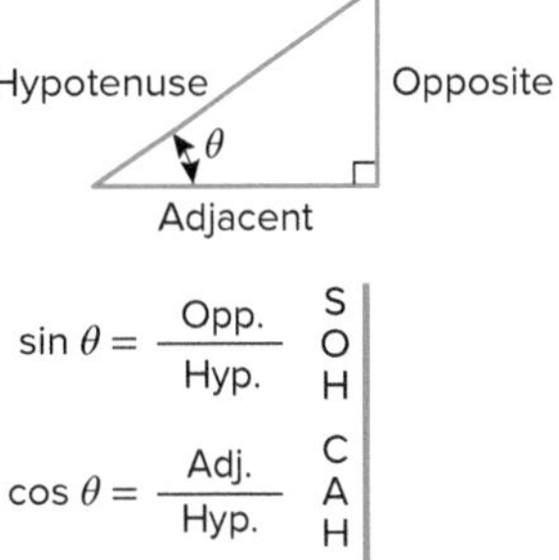

$$\sin\theta = \frac{\text{Opp.}}{\text{Hyp.}} \quad \text{SOH}$$

$$\cos\theta = \frac{\text{Adj.}}{\text{Hyp.}} \quad \text{CAH}$$

$$\tan\theta = \frac{\text{Opp.}}{\text{Adj.}} \quad \text{TOA}$$

FIGURE 4.17 Trigonometric Functions Defined

Example 4.10: Truss Member Forces—Method of Joints

Find the force in member GH.

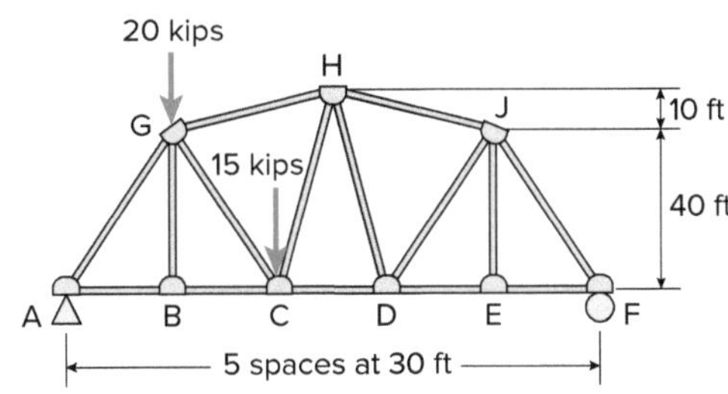

Example 4.10 *(continued)*

Solution

Find reactions. There is no symmetry of loading, so use the equation of equilibrium.

$\curvearrowleft + \Sigma M_A = 0 \Rightarrow -(20 \text{ kips})(30 \text{ ft}) - (15 \text{ kips})(60 \text{ ft}) + f_F(150 \text{ ft}) = 0$

$f_F = 10 \text{ kips}\uparrow$

$+\uparrow \Sigma F y = 0 \Rightarrow F_{A_y} - 20 \text{ kips} - 15 \text{ kips} + 10 \text{ kips} = 0$

$F_{A_y} = 25 \text{ kips}\uparrow$

$F_{A_y} = \phi$ by inspection

Joint A

$\theta = \tan^{-1}\left(\dfrac{40 \text{ ft}}{30 \text{ ft}}\right) = 53.13°$

$+\uparrow \Sigma f y = 0 \Rightarrow F_{AG} \times \sin(53.13°) + 25 \text{ kips} = 0$

$\Rightarrow F_{AG} = -31.25 \text{ kips (compression)}$

$\Sigma F_x = 0 \Rightarrow F_{AG} \times \cos(53.13°) + F_{AB} = 0$

$F_{AB} = 31.25 \text{ kips } (\cos(53.13°)) = 18.75 \text{ kips (tension)}$

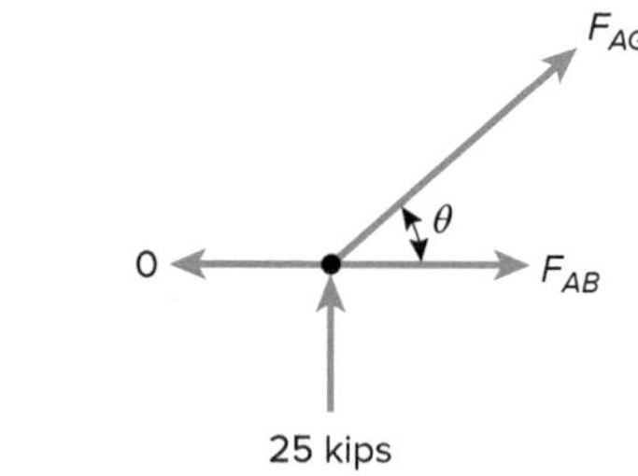

Joint B

$+\uparrow \Sigma F y = 0 \Rightarrow F_{BG} = \phi$

$+\rightarrow \Sigma F x = 0 \Rightarrow F_{BC} = F_{AB} = 18.75 \text{ kips (tension)}$

Joint B

F_{BG}

F_{AB} F_{BC}

Joint C

$\theta_2 = \tan^{-1}\left(\dfrac{10}{45}\right) = 12.53°$

$\theta_3 = \tan^{-1}\left(\dfrac{30 \text{ ft}}{40 \text{ ft}}\right) = 36.87°$

$+\rightarrow \Sigma F x = 0 \Rightarrow 31.25 \text{ kips} \times \cos(53.13°)$

$= F_{GH} \cos(12.53°)$

$+F_{GC}\sin(36.87°) = 0$

$\Rightarrow 18.75 \text{ kips} + 0.98\, F_{GH} + 0.6\, F_{GC} = 0$ **(1)**

Joint C

20 kips

θ_2 F_{GH}

$53.13° = \theta_1$

$-31.25 \text{ kips} = F_{AG}$ θ_3

F_{GC}

$F_{BG} = 0$

$+\rightarrow \Sigma F x = 0 \Rightarrow 31.25 \text{ kips} \times \sin(53.13°) + F_{GH}(12.53°)$

$-F_{GC} \times \cos(36.87°) - 20 \text{ kips} = 0$

$\Rightarrow 25 \text{ kips} + 0.22\, F_{GH} - 0.80\, F_{GC} - 20 \text{ kips} = 0$ **(2)**

There are two equations with two unknowns. Solve Equation (1) for F_{GC} and plug the answer into Equation (2).

$F_{GC} = \dfrac{-18.75}{0.6} - \dfrac{0.98}{0.6} F_{GH}$

$= -31.25 \text{ kips} - 1.63\, F_{GH}$

Plug 1.63 F_{GH} into Equation (2).

$25 \text{ kips} + 0.22 F_{GH} - 0.80(-31.25 \text{ kips} - 1.63 F_{GH}) - 20 \text{ kips} = 0$

$25 \text{ kips} + 0.22 F_{GH} + 25 \text{ kips} + 1.30 F_{GH} - 20 \text{ kips} = 0$

$1.52 F_{GH} = -30 \text{ kips}$

$F_{GH} = 19.74 \text{ kips (compression)}$

The method of joints for truss member forces may be time prohibitive if the member of interest is located toward the center of the truss. In general, the method of joints is attractive if the joint is at the support locations or possibly one joint adjacent. The rationale for this in the PE Civil exam is purely time related. The same solution in the method of joints for a member near the center of the truss would take longer to solve than utilizing the method of sections shown in the next part.

4.4.2.2 Method of Sections

The method of sections is similar to the free-body diagrams that were shown for beam analysis. The general rule of thumb is to provide a section cut through the truss while only intersecting three members; this coincides with the number of equilibrium equations that carries over from beam analysis. There are rare instances when cutting through more members will still result in a solution. The direction of the section cut is arbitrary. Example 4.11 shows the method of sections for a typical truss.

Example 4.11: Truss Member Forces—Method of Sections

Find the force in member GH. (Use information from Example 4.10.)

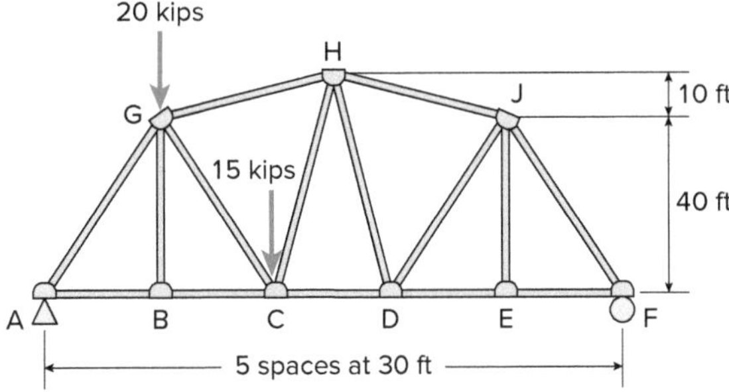

Solution

$\rightarrow F_{A_y} = 25 \text{ kips}\uparrow$

$F_{F_Y} = 10 \text{ kips}\uparrow$

Take a cut on the left side:

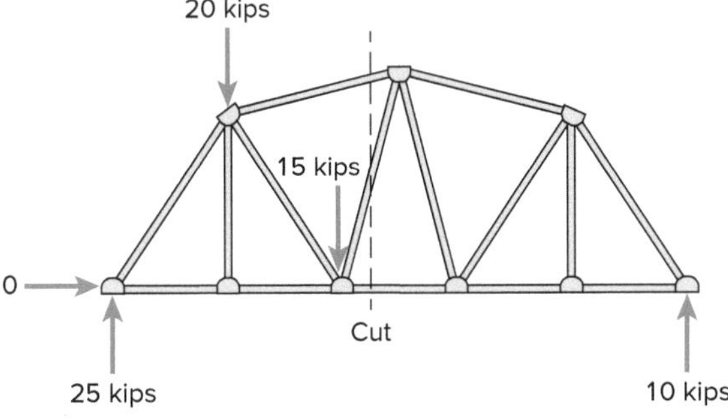

$$\theta_1 = \tan^{-1}\left(\frac{10 \text{ ft}}{45 \text{ ft}}\right) = 12.53°$$

$$\theta_2 = \tan^{-1}\left(\frac{50 \text{ ft}}{15 \text{ ft}}\right) = 73.3°$$

Example 4.11 (continued)

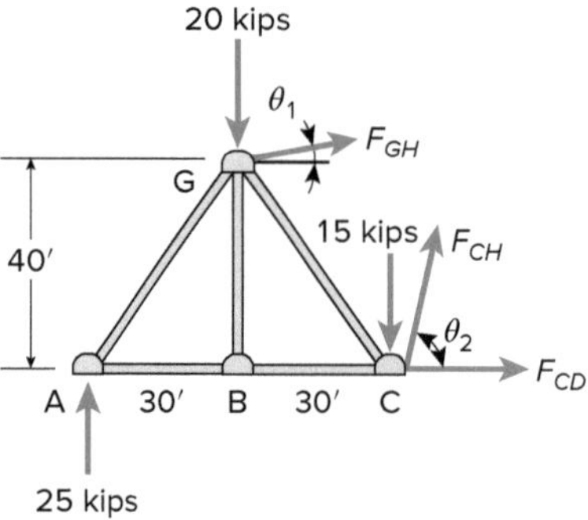

$$\curvearrowleft +\Sigma M_c = 0 \Rightarrow 20\text{ kips}(30\text{ ft}) - F_{GH} \times \cos(12.53°)\,40\text{ ft} - F_{GH} \times \sin(12.53°) \times 30\text{ ft}$$
$$-25\text{ kips}(60\text{ ft}) = 0$$
$$\Rightarrow 600\text{ kips-ft} - 39.05\text{ ft } F_{GH} - 6.51\text{ ft } F_{GH} - 1{,}500\text{ kips-ft} = 0$$
$$\Rightarrow F_{GH} = -\frac{900\text{ kips-ft}}{45.56} = 19.75\text{ kips-ft (compression)}$$

4.4.3 Zero-Force Members

Zero-force members in trusses indicate that for the loading condition shown, some truss members will not take force. It is critical that zero-force members remain in the truss for overall stability. Zero-force members are simply a function of the location and orientation of the applied loads shown for a specific problem. If the load changes location or direction, the truss must be analyzed a second time. There are four rules that aid in the identification of zero-force members.

Rule 1. If a joint has only two members with no force applied at the joint, then both members are zero-force members. The two members at the joint must be noncollinear (that is, not in the same plane or line of action). See Figure 4.18.

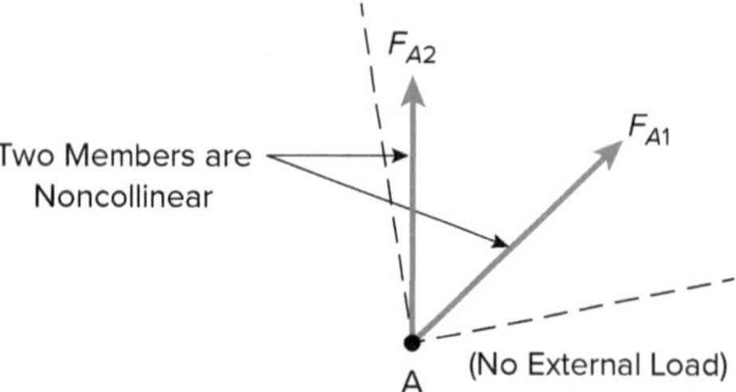

FIGURE 4.18 Zero-Force Members–Rule 1 Joint

Rule 2. If a joint has only three members, two of which are along the same line of action, with no force applied at the joint, then the third member is the zero-force member. The two collinear members do take load. See Figure 4.19.

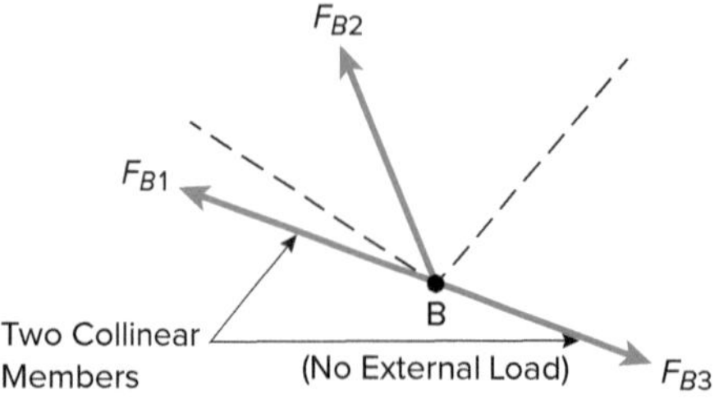

FIGURE 4.19 Zero-Force Members–Rule 2 Joint

Rule 3. Once a zero-force member is identified (using the first two rules), assume that the member does not take a load and can be eliminated from the truss. The resulting layout of the truss can be reviewed again and there could be other zero-force members. It is important to note that removing members of the truss leads to stability issues. Thus, rule 3 should only be used if the problem explicitly involves zero-force members in the analysis; it should not be used for stability calculation purposes.

Rule 4. The geometry at the supports can result in additional zero-force members. For example, a roller support does not provide horizontal reactions. If there is only one horizontal member connected to a roller with no diagonals, then the horizontal member must be a zero-force member. A quick illustration of the free-body diagram of a support joint is useful for zero-force member determination, as shown in Example 4.12.

Example 4.12: Zero-Force Member at Roller Support

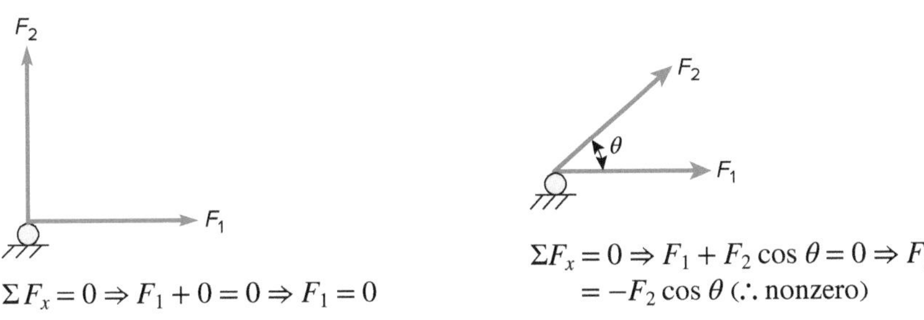

$\Sigma F_x = 0 \Rightarrow F_1 + 0 = 0 \Rightarrow F_1 = 0$

$\Sigma F_x = 0 \Rightarrow F_1 + F_2 \cos\theta = 0 \Rightarrow F_1 = -F_2 \cos\theta$ ($\therefore$ nonzero)

Example 4.13: Truss for Determination of Zero-Force Members

How many zero-force members are in the truss below?

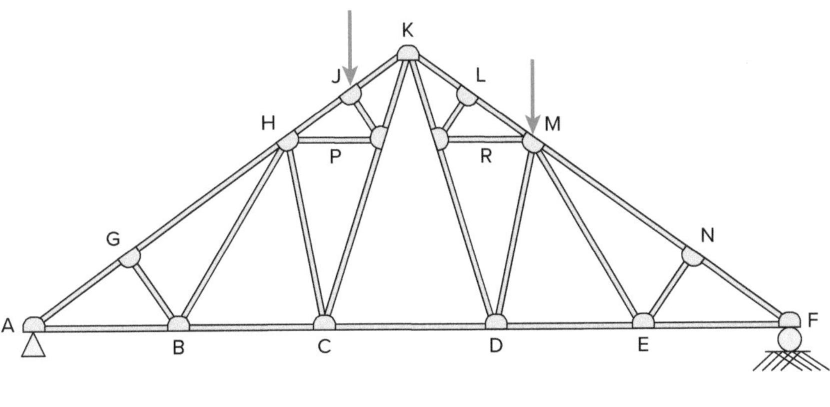

A. 4
B. 5
C. 6
D. 8

Solution

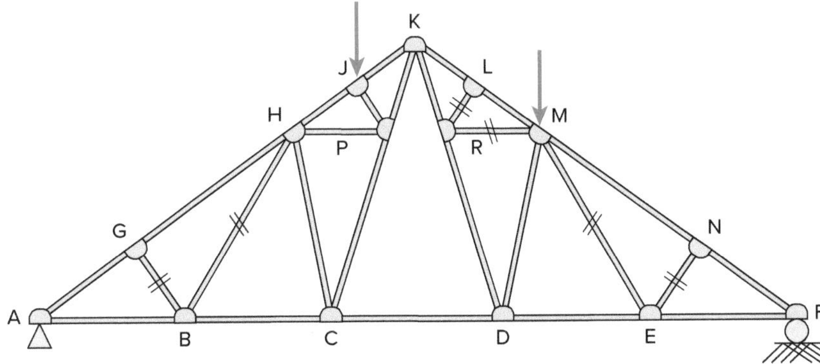

Example 4.13 *(continued)*

Moving from left to right for the truss shown, joint G is the first joint to investigate. This joint meets rule 2; the collinear members are AG and GH. Thus, member GB is a zero-force member. Joint L is the next joint to investigate. This joint meets rule 2; the collinear members are KL and LM. Thus, member LR is a zero-force member. The next joint of interest is joint N. This joint meets rule 2; the collinear members are MN and NF. Thus, member NE is a zero-force member. Once these three members have been identified, look at the truss a second time, with the elements removed. Again, these elements of the truss are always needed for stability purposes. They are only removed when calculating zero-force members in this problem. The first joint of interest is joint B. It meets rule 2; the collinear members are AB and BC. Thus, member BH is a zero-force member. The next joint to look at is joint R. This meets rule 2; the collinear members are KR and RD. Thus, member RL is a zero-force member. Joint E is the last joint to assess. This meets rule 2, with collinear elements DE and EF. Thus, member ME is a zero-force member.

Answer: C

4.5 STRESSES ON A CROSS SECTION

The following sections will investigate axial, shear, and flexural stresses on a cross section.

4.5.1 Axial Stress

Axial stress is defined as the applied axial force divided by the cross-sectional area, P/A. Axial stress is also caused by a bending moment.

4.5.1.1 Tension Members

The capacity of tension members is defined in Section J4.1 of the American Institute of Steel Construction's *Steel Construction Manual (14*th *ed.)*, hereafter referred to as the *Steel Manual* [4], as the smaller of tensile yielding of the gross cross section (away from connections, with the area of the element unchanged) or the fracture of the net cross section (at a connection, taking into account holes in the element for bolts). The equations are shown below for yielding of the gross and fracture of the net cross sections (Equations 4-14 and 4-15, respectively).

$R_n = F_y \times A_g$ (*Steel Manual*, Sec. J4-1) **Equation 4-14**

F_y = yield strength of steel (ksi)

A_g = gross area (in^2)

$\phi = 0.90$ (LRFD) $\Omega = 1.67$ (ASD)

$R_n = F_u \times A_e$ (*Steel Manual*, Sec. J4-2) **Equation 4-15**

$\phi = 0.75$ (LRFD) $\Omega = 2.00$ (ASD)

F_u = ultimate strength of steel (ksi)

A_e = effective net area; for bolted splice plates, $A_e = A_n \leq 0.85 \times A_g$

Note that shear lag (D3, page 16.1-27 in the *Steel Manual*) is accounted for in the A_e calculation.

4.5.1.2 Compression Members

Compression members encompass columns of various shapes—whether they are made of steel, concrete, masonry, wood, or iron. The capacity of compression members is governed by the Euler buckling load, which is covered in more detail in Section 4.8.1.

4.5.2 Shear Stress

Shear stresses are the largest at the supports for a typical beam and represent the vertical force over the cross-sectional area.

4.5.2.1 Maximum Shear Stress

The general equation for the shear distribution over the depth of a cross section is shown in Equation 4-16.

$$\nu = \frac{VQ}{Ib}$$ **Equation 4-16**

V = applied shear force at the section of interest

Q = first moment of the area above the neutral axis, identified for a rectangular cross section of width b and height h in Equation 4-17

$$Q = \frac{bh^2}{8}$$ **Equation 4-17**

I = moment of inertia (identified for a rectangular cross section of width b and height h in Equation 4-18)

$$I = \frac{bh^3}{12}$$ **Equation 4-18**

b = width of a rectangular cross section

The general equation for shear stress, indicated in Equation 4-16, does not provide an indication of the shape of the stress over the depth. The shear stress is the minimum at the extreme fiber of the cross section and the maximum at mid-height. The distribution follows a parabolic shape, as shown in Figure 4.20.

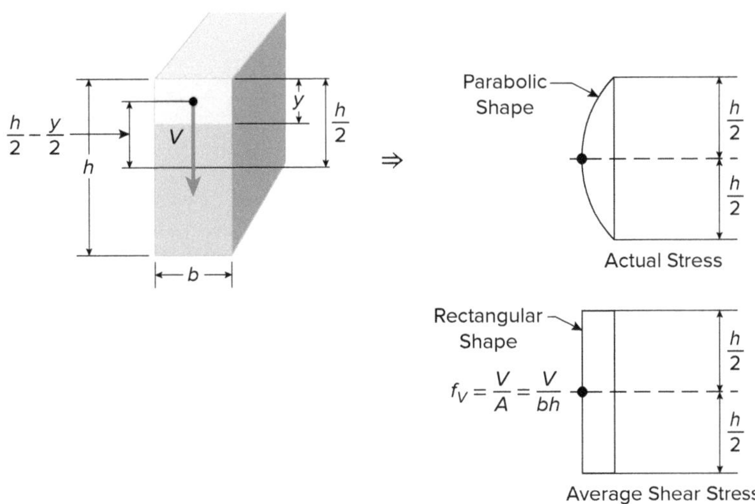

FIGURE 4.20 Shear Distribution Over Depth

4.5.2.2 Average Shear Stress

The general equation for shear stress is not used in design. The average shear stress is used, which simplifies the stress over the depth as a constant. This stress distribution is applicable for rectangular cross sections only. For I-shaped sections, the total area cannot be used, as the flange area at the extreme fiber does not provide any resistance to shear, with respect to the parabolic shape. In other words, an I-shape has the largest area where the demand for shear is the lowest. It would be aggressive to assume that the flange areas resist the shear at the extreme fibers. Thus, for I-shapes, the resistance to shear is defined as the total depth d times the web width t_w. This is shown in Figure 4.21.

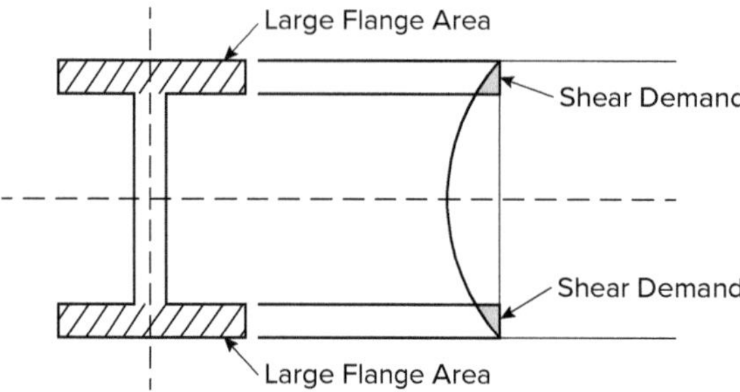

The largest steel area corresponds to the lowest shear demand. Therefore, it would be aggressive to assume that the flanges help resist shear force.

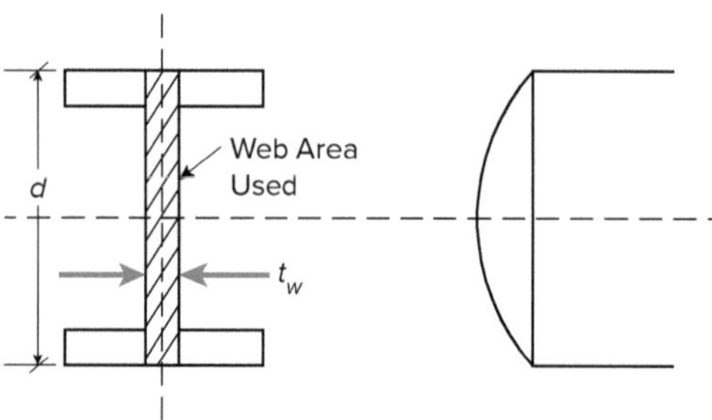

Web Area for I-Shapes in Shear = $d \times t_w$

FIGURE 4.21 I-Shape Cross Section—Shear Area

4.5.3 Bending Stress

Bending or flexural stresses occur along the entire beam length, but are maximum near midlength for simply supported beams. The general equation for flexural stress is shown in Equation 4-19.

$$f_b = \frac{My}{I}$$ **Equation 4-19**

f_b = flexural stress in cross section

M = externally applied moment on cross section

y = distance from the neutral axis, to any point on the cross section

I = moment of inertia

There is an important distinction for the y value in Equation 4-19, in that it is a completely arbitrary distance from the neutral axis. It could be 1, 2, or 5 inches from the neutral axis of the cross section. When the distance y reaches the extreme fiber of the cross section above or below the neutral axis, that results in the maximum bending stress in the cross section. This results in Equation 4-20, in which the y value has been replaced with the c value. M and I were previously defined.

$$f_{b_{max}} = \frac{Mc}{I}$$ **Equation 4-20**

A simplification of the maximum bending stress equation is shown in Equation 4-21, which introduces a new term, the section modulus, *S*. Equation 4-22 is also used to define the maximum flexural stress:

$$f_{b_{max}} = \frac{M}{S}$$ **Equation 4-21**

$$S = \frac{bh^2}{6}$$ **Equation 4-22**

4.5.4 Combined Stresses on a Cross Section

The analysis of a compression member that is not loaded through the geometric centroid of the cross section induces not only pure uniaxial compression, but also a bending moment into the column. The capacity of such a member requires an envelope solution, such that the applied moment and compression stresses are within the limits of the cross section. One such envelope solution is shown in Figure 4.22.

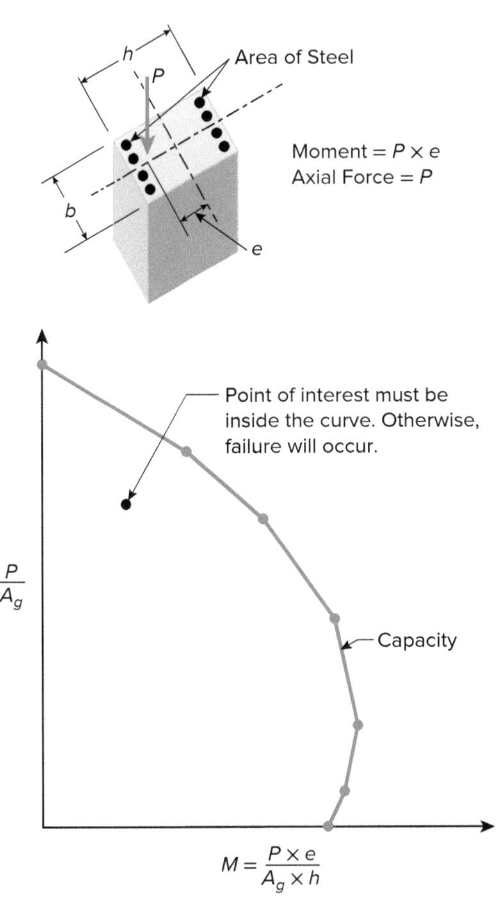

FIGURE 4.22 Compression Member Subjected to Eccentric Loading—Envelope Capacity

4.5.5 Eccentric Loading of Columns

Subjecting column sections to a load through the geometric centroid results in an axial stress of *P*/*A*. However, in some cases, the application of load is not possible through the geometric centroid, and the applied force is offset in one or more directions. The first case, as shown in Figure 4.23, investigates results in an offset, e, in one direction for a column section.

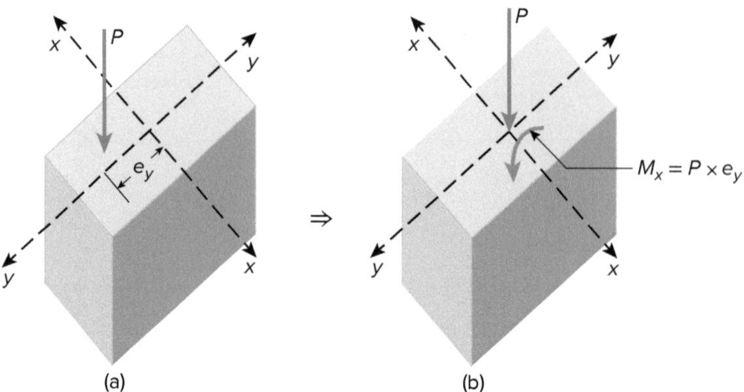

FIGURE 4.23 Eccentrically Loaded Column in One Direction

The applied load P can apply tension or compression to the portions of the column section about the x or y axis, as applicable. This results in additional compression where the P load is applied, and additional tension where it is not applied, as shown in Figure 4.24.

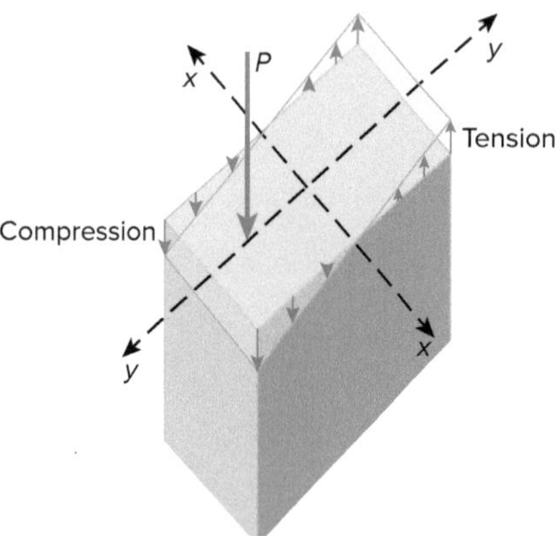

FIGURE 4.24 Tension and Compression for Eccentrically Loaded Column in One Direction

Sign convention plays an important role when calculating the resulting stress at various points on the cross section. The compressive stress due to the applied load P must be accounted for as the more traditional P/A component. In addition, the stress due to the applied moment, $P \times e$, must be accounted for as well. The equation needed for this stress depends on the point of interest on the cross section. For instance, if the point of interest is not at the top or bottom of the cross section (that is, extreme fiber), then Equation 4-19 (previously shown and included below for emphasis) would be used.

$$f_b = \frac{My}{I}$$ **Equation 4-19** (p. 134)

The reason for using Equation 4-19 in this case is that the distance y is arbitrary from the neutral axis to any point on the cross section. However, if the point of interest on the cross section is at the extreme fiber, then Equation 4-20 or 4-21 (shown previously but identified again for clarity) would be more applicable.

$$f_{b_{max}} = \frac{Mc}{I}$$ **Equation 4-20** (p. 134)

$$f_{b_{\max}} = \frac{M}{S}$$ **Equation 4-21** (p. 135)

Thus, the total stress would be:

$$\sigma = \pm\frac{P}{A} \pm \frac{My}{I}$$ **Equation 4-23**

Note that the sign of the stress in the above equation has been left somewhat generic, as compression stresses can be defined as either positive or negative when the point of interest on the cross section is an arbitrary distance. For a point at the extreme fiber of the cross section:

$$\sigma = \pm\frac{P}{A} \pm \frac{Mc}{I}$$ **Equation 4-24**

or

$$\sigma = \pm\frac{P}{A} \pm \frac{M}{S}$$ **Equation 4-25**

Column sections can also be subjected to an offset e in two directions, as shown in Figure 4.25.

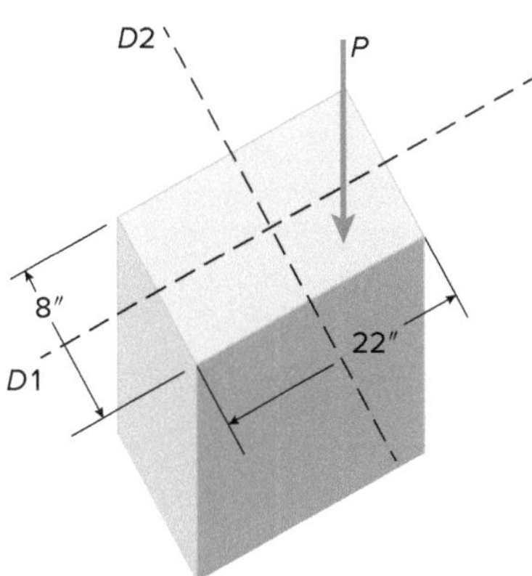

FIGURE 4.25 Eccentrically Loaded Column in Two Directions

The solution to determine the resulting stress follows from the previous example, except that an additional term is required for bending about the second axis. Also, the height and width of the cross section will depend on the axis where bending is taking place. For example, in Figure 4.25, the bending about the $D1$ axis would result in a height dimension of 8 in, while the width dimension would be 22 in. When determining the moment of inertia or section modulus about the $D1$-axis, these values would be used. However, when determining the moment of inertia or section modulus about the orthogonal $D2$ axis, the values would flip. For example, for $D2$-axis bending, the height dimension would now be 22 in, while the width dimension would be 8 in. Providing the correct value in the denominator term of Equations 4-23 through 4-25 (as applicable) is critical in determining the stress at the point of interest. As for the sense of the stress at various locations in the cross section, it again depends where the point of interest is relative to the applied load P. If the point of interest is in the quadrant where P is located, it would add compression about both axes. However, if the point of interest is in one of the other three quadrants, it would result in compression about one axis and tension on the other, or possibly tension on both axes. This is shown in Figure 4.26.

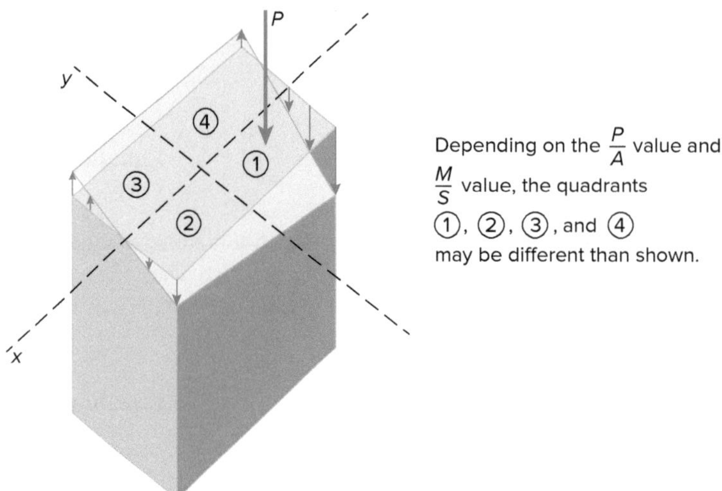

FIGURE 4.26 **Location of Resultant Tension or Compression Due to Eccentric Moment About a Cross Section**

The resulting stress follows from the previous discussion, but with one additional term. For eccentric loading about both axes, the total stress equation would be:

$$\sigma = \pm\frac{P}{A} \pm \frac{M_x y}{I_x} \pm \frac{M_y y}{I_y}$$ **Equation 4-26**

Note that the sign of the stress in the above equation has been left somewhat generic, as compression stresses can be defined as either positive or negative when the point of interest on the cross section is an arbitrary distance. For a point at the extreme fiber of the cross section, the maximum total stress would be:

$$\sigma = \pm\frac{P}{A} \pm \frac{M_x c}{I_x} \pm \frac{M_y c}{I_y}$$ **Equation 4-27**

or

$$\sigma = \pm\frac{P}{A} \pm \frac{M_x}{S_x} \pm \frac{M_y}{S_y}$$ **Equation 4-28**

4.6 BEAM DEFLECTION

Beam deflections were previously shown in Figures 4.8–4.10. Regardless of the boundary conditions for the beam of interest, the numerator term involves loading and beam length. The denominator defines the stiffness of the beam in terms of the section modulus value E and the geometric term, moment of inertia, I. Units must be consistent for all deflection equations. The applied load must agree with the modulus value—that is, if the loading is shown in lb or lb/ft, then the modulus value must also be in lb/in^2; however, if the loading is shown in kips or kips/ft, then the modulus value must be in kips/in^2. In a similar sense, the beam length must be converted to inches, as it must agree with the moment of inertia term, which is in in^4.

4.7 BEAM DESIGN

4.7.1 Concrete Beam Design

Concrete beam design conforms to ACI 318-14 [3].

4.7.1.1 Rectangular Singly Reinforced Beam Design for Flexure

Singly reinforced beams have reinforcing steel located on the tension face only. This results in the reinforcing steel taking the tension stress, and the concrete bearing the compression stress. Prior to this condition, when the beam has been cast and left to cure, there are no initial cracks in the beam. For a bridge application, once the beams have been subjected to live load traffic, if the loading is relatively light, the beam will remain uncracked, until a tensile stress known as the modulus of rupture is reached. The modulus of rupture is defined as the stress that causes concrete to crack at the extreme fiber in tension. It is defined by ACI in Equation 4-29.

$$f_t = 7.5\sqrt{f'_c}$$ **Equation 4-29**

f_t = tensile stress at the extreme fiber of an uncracked concrete section that will cause a crack to form

f'_c = concrete compressive strength at 28 days

A modification of the modulus of rupture is the cracking moment in a beam section, as outlined in Example 4.14.

Example 4.14: Concrete Beam Cracking Moment

Determine the cracking moment for the singly reinforced concrete beam shown.

$f'_c = 4{,}000$ psi

$f_y = 60{,}000$ psi

flexural reinforcement = 2 #9 bars

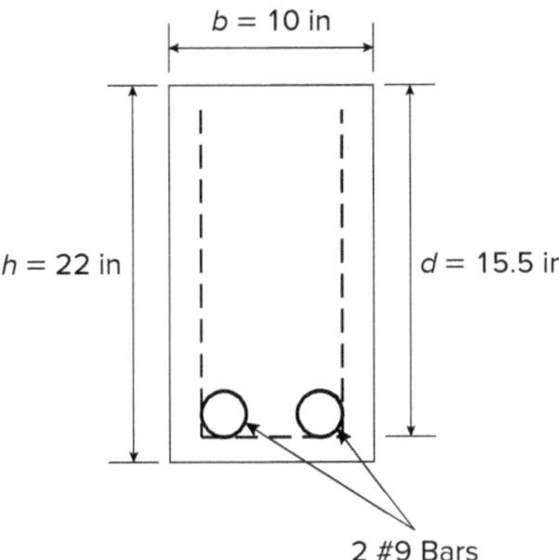

Solution

The stress that will cause cracking at the extreme fiber is the modulus of rupture, f_t.

$$f_t = 7.5 \times \sqrt{(4{,}000\,\text{psi})} = 474.3\text{ psi}$$

Recall the relationship between moment and maximum flexural stress:

$$f_{b_{max}} = \frac{Mc}{I}$$

Rearrange this equation for the moment:

$$M = \frac{f_{b_{max}} I}{c}$$

$$I = \frac{bh^3}{12}$$

$$I = \frac{(10\text{ in})(22\text{ in})^3}{12} = 8{,}873\text{ in}^4$$

The distance c is from the neutral axis to the extreme fiber. For a rectangular section, $c = \frac{h}{2}$.

Example 4.14 *(continued)*

$$c = \frac{h}{2} = \frac{22 \text{ in}}{2} = 11 \text{ in}$$

$$M_{\text{cracking}} = \frac{474.3 \text{ psi}(8{,}873 \text{ in}^4)}{11 \text{ in}} = 382{,}587 \text{ lb/in} = 31.88 \text{ kips/ft}$$

Flexural capacity as per ACI 318-14 is based on the material capacity of concrete and steel, f'_c and f_y, respectively. Thus, the stresses over the depth for the design flexural capacity are shown in Figure 4.27.

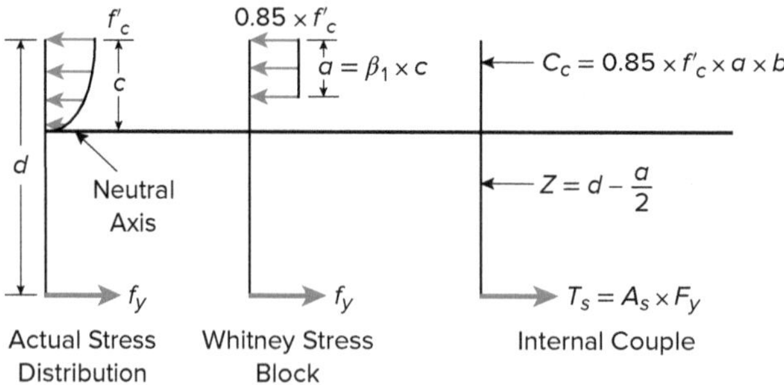

FIGURE 4.27 Singly Reinforced Concrete Beam Stresses for Flexural Capacity

The stresses shown in the left portion of Figure 4.27 can be simplified due to the presence of the parabolic shape. Thus, the illustration in the center indicates the Whitney stress block. It simplifies the area under the parabolic shape into an equivalent rectangle. In this case, the horizontal distance has been reduced from f'_c to $0.85 \times f'_c$, and the vertical distance has been reduced from c to a value of a by the factor β_1. The β_1 factor depends on f_c and will essentially come out in the next step when calculating the depth of the compression block. The right illustration from Figure 4.27 converts the stress in the center illustration to a force.

$$T_{\text{steel}} = A_s \times F_y \text{ and } C_{\text{concrete}} = 0.85 \times f'_c \times b \times a \qquad \textbf{Equation 4-30}$$

The tension in the steel T and compression in the concrete C form what is known as a force couple—essentially two equal but opposite forces separated by a perpendicular distance. Since the forces are equal, the a value can be solved by setting the two values equal to each other. Thus, the depth of the compression block a is shown in Equation 4-30.

$$a = \frac{A_s \times f_y}{0.85 \times f'_c \times b} \qquad \textbf{Equation 4-31}$$

Once the compression block is known, the nominal moment capacity of the section can be determined by summing moments about either the tension in the steel T or compression in the concrete C from Figure 4.27. Convention holds to sum forces about the compression in the concrete C_c term. This results in the nominal moment capacity equation (Equation 4-32).

$$M_n = A_s \times f_y\left(d - \frac{a}{2}\right) \qquad \textbf{Equation 4-32}$$

The design moment capacity includes the strength reduction factor, ϕ, which varies for flexural design depending on the tensile strain in the steel at failure. The reduction factor as per ACI 318-14 varies from 0.65 to 0.90 and is dependent on which material fails first at ultimate load. Steel has a large ductility, which is defined as the change in length divided by the original length. Steel strains beyond yield from flexure stresses result in significant deformations prior to failure, providing warning that a failure is likely. Concrete, on the other hand, fails in a brittle manner with no impending warning. Thus, the range of reduction factors takes this into account such that a tension-controlled failure results in a reduction factor of 0.90, while a compression-controlled failure results in a reduction factor of 0.65.

The determination of the strength reduction factor depends on the strain in the steel at failure. Concrete reaches its strain limit at 0.003 at failure. Once that limit has been reached, the strain in the steel is investigated. When the strain in the steel $\varepsilon_t < 0.002$, the steel was not utilized or stressed significantly when the concrete reached its limit. If the strain in the steel $\varepsilon_t > 0.005$, then the steel was stressed a significant amount prior to the concrete reaching its limit; thus, it is a tension-controlled failure. When the strain in the steel is between the two limits of 0.002 and 0.005, the strength reduction factor varies linearly from 0.65 to 0.90.

4.7.1.1.1 Placement of Steel

Common dimensions for reinforced concrete beams are shown in Figure 4.28. The *d* value is the vertical distance from the compression face to the centroid of the reinforcing steel, also known as the effective depth of the beam. The width of the beam is defined as *b*. The concrete cover shown is measured from the bottom of the cross section to the bottom of the reinforcing bar. The cover varies depending on application. For instance, casting against earth requires a larger cover than a typical interior cast-in-place application. The rationale for casting against earth is that there is a higher chance of variation compared with casting against plywood formwork. For a precast application, the tolerance can reduce further, as there is more quality control in a plant environment. For the same beam being cast in the three different locations previously noted, the overall height would vary as the concrete cover changes. However, the effective depth *d* would be constant. For this reason, in many design equations, the effective depth *d* is used, not the total height of the beam.

One Row of Steel (a)

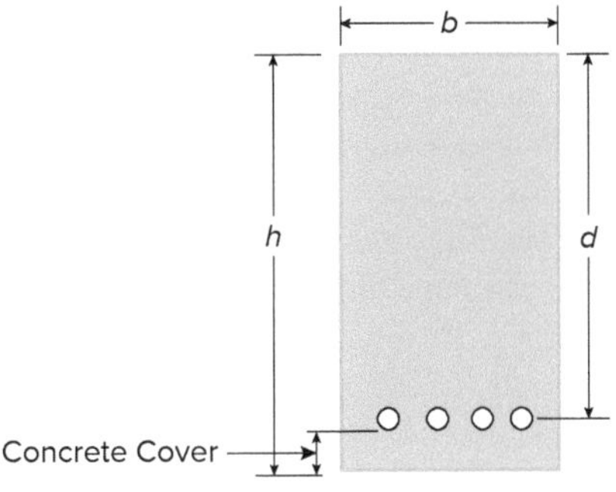

Multiple Rows of Steel (b)

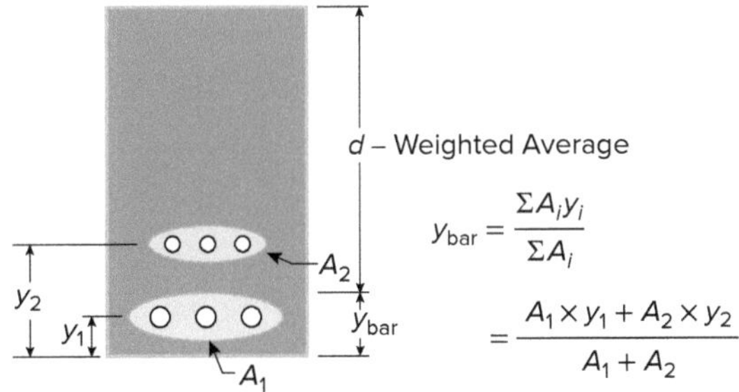

FIGURE 4.28 Dimensions for Rectangular Concrete Beams

The effective depth *d* is measured to the centroid of the reinforcing steel. Beams with more than one row of steel require calculation of a centroidal distance y_{bar} from a common data point, similar to what was noted in the moment of inertia calculation for combined shapes. If the area of steel is exactly the same in each row, then the centroid will be half the distance between the two rows.

4.7.1.1.1.1 Development of Reinforcement

Reinforcing steel must meet certain deformation rib height and spacing requirements that have been set by an ASTM standard, which in turn provide engagement of the reinforcing steel under load. However, full capacity of a bar is not available at the start of its length. The bar requires a certain distance to engage with the surrounding concrete in order to ensure that the full capacity of the reinforcing steel can be achieved. This is referred to as the development length of the steel. There are several parameters that will increase or decrease the development length, based on experimental data. The process begins with the basic development length, as per ACI 318-14, Section 25.4.2.1. The development length l_d for deformed bars and wires in tension shall be a minimum of 12 in, but may be larger depending on concrete cover and bar spacing requirements. The additional two equations to check are likely beyond the scope of a typical exam problem. Note that development lengths are also used in masonry for building design.

Modifications for the development length relate to the possible use of lightweight concrete, the use of an epoxy coating on the steel, the bar size, and the location where the bars are placed during casting (at the top of the section or other locations). These are shown in ACI 318-14, Table 25.4.2.4.

In some cases, a bar cannot be developed for a given length due to space limitations. For example, if the development length for a No. 5 bar is 18 in, but only 12 in is available, then an alternative must be sought. The solution would be to provide a bent bar, which provides development in a shorter distance, but exhibits either a 90° or 180° bend near the terminal end of the bar. ACI also accounts for this in Section 25.4.3, "Development of Standard Hooks in Tension." The development length values, l_{dh}, are the greater of 6 in, $8 \times d_{\text{bar}}$, and a third equation that relates to the modification factors previously noted. These include lightweight concrete, epoxy-coated bars, concrete cover, and possible confining reinforcement.

4.7.1.1.1.2 Lap Length of Reinforcement

The development of reinforcing steel in columns, continuous beams, and foundation applications results in reinforcing steel that is truncated relative to the element's total length for practical construction. This is shown schematically in Figure 4.29 for a spread-footing application that includes a long column section above. Casting the footing section with shorter bars near the top that project beyond the concrete limits of the footing itself requires less effort by construction personnel. Once the concrete in the footing has cured, then the bars that are protruding from the footing column section are lapped with new steel bars that continue for a given height. For building construction, there is an upper limit to the length of reinforcing steel that can be safely transported to a job site. Thus, if a column section is designed to span a vertical height of 80 ft, it may not be possible to use a single reinforcing steel bar for that application. If the maximum transportable length is 30 ft, then multiple bars must be used.

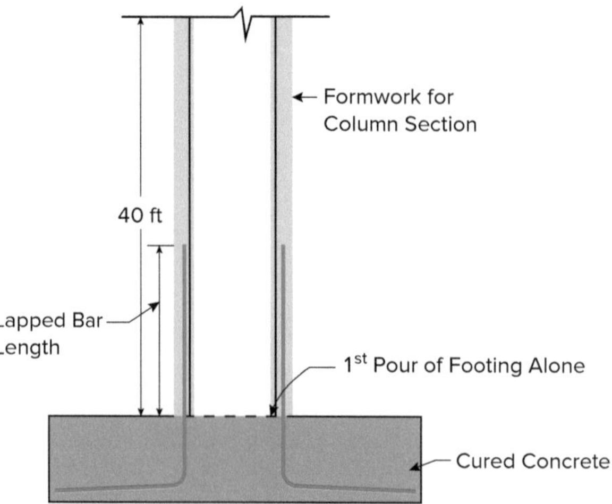

FIGURE 4.29 Spread Footing Example of Lapped Bars

The restriction in available length of bars requires an overlap distance for the reinforcing steel in certain applications. This length is referred to as a lap length. It is calculated similarly to what was identified previously for development lengths in terms of lightweight reinforcement and cover requirements for steel. The basic lap length and associated modification factors are indicated in ACI 318-14, Section 25.5.2, "Lap Splice Lengths of Deformed Bars and Deformed Wires in Tension."

4.7.1.2 Rectangular Singly Reinforced Beam Design for Shear

The **factored shear force** V_u is typically the largest at the support locations. At the end supports for a simply supported beam, the critical shear force is taken at a distance (d) away from the face of the support. The rationale for this is based on laboratory tests that indicate the inclination of a shear crack to be approximately 45°, as shown in Figure 4.30. If a concentrated load was applied to the beam near the support, then the first location that would cause the crack to form would be a horizontal distance equal to the vertical dimension (d) from flexural design.

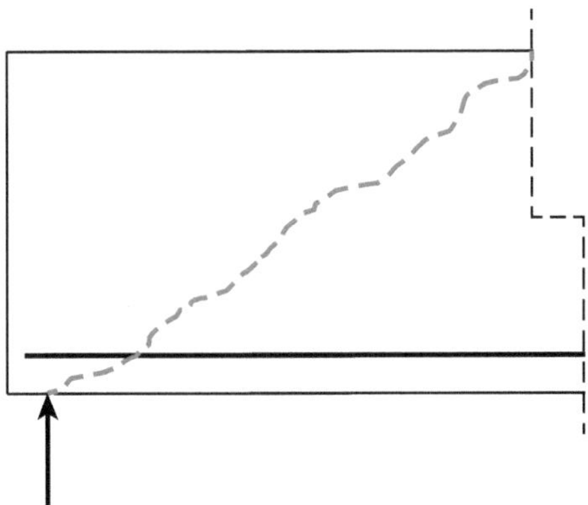

FIGURE 4.30 Inclination of Shear Crack

The strength reduction factor, ϕ, is equal to 0.75 for shear, and it does not vary, as was the case for flexural capacity. The concrete shear capacity is shown in Equation 4-33.

$$V_c = 2\sqrt{f'_c}\, b \times d \qquad \textbf{Equation 4-33}$$

> **TIP**
>
> Note that the value of f'_c in the square root must be entered in lb/in^2 or psi; using kips/in^2 or ksi will result in an incorrect value.

Note that the concrete shear capacity equations used in the AASHTO *Bridge Design Specifications* [5] require f'_c to be in ksi; thus, it is important to be cognizant of the problem statement with regard to the equation that is required.

In cases where the shear demand is relatively small, the ACI design code permits the concrete to take all of the factored shear force (V_u); otherwise, steel stirrups are required to increase the capacity of the cross section.

If $V_u \le \dfrac{\phi V_c}{2}$, then no stirrups are required.

If $V_u > \dfrac{\phi V_c}{2}$, then determine the spacing of the stirrups.

In most cases, steel stirrups are required; thus, the shear capacity is a combination of concrete and steel, V_c and V_s, respectively. The general design equation for shear is shown in Equation 4-34.

$$\phi V_n = \phi(V_c + V_s) \ge V_u \qquad \textbf{Equation 4-34}$$

Steel stirrups are shown in cross section and elevation in Figures 4.31 and 4.32.

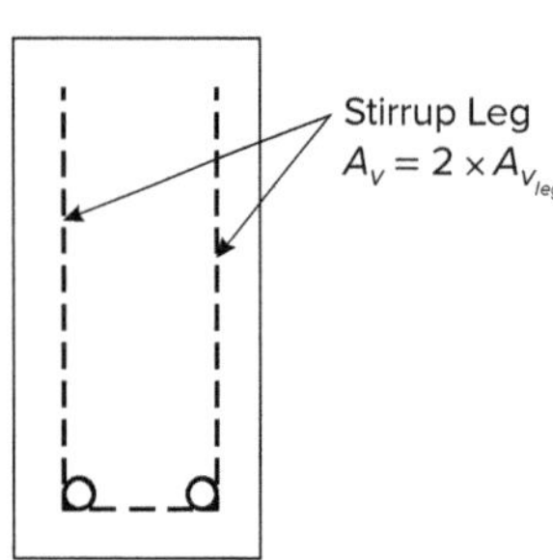

FIGURE 4.31 Cross Section of Concrete Beam—Showing Steel Stirrups

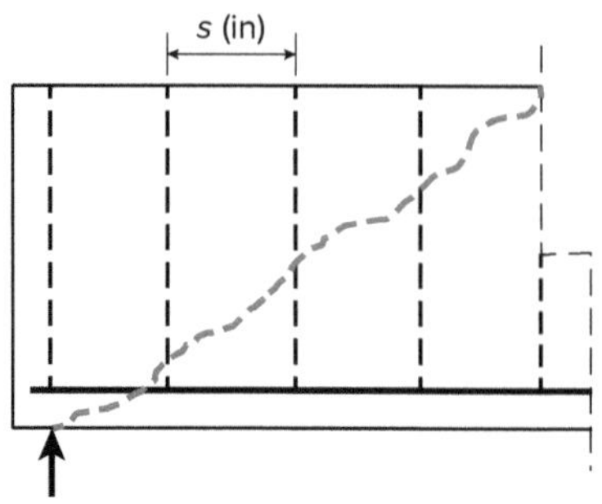

FIGURE 4.32 Elevation of Concrete Beam—Showing Steel Stirrup Spacing, *s*

Rearranging Equation 4-34 for the steel contribution V_s results in the following relationship:

$$V_s \geq \frac{V_u}{\phi} - V_c$$ **Equation 4-35**

The equation for the steel contribution to shear strength is defined in Equation 4-36.

$$V_s = \frac{A_v \times f_y \times d}{s}$$ **Equation 4-36**

d = effective depth (in)

f_y = yield strength of reinforcing steel (ksi)

A_v = shear area determined as the individual bar area times the number of vertical stirrup legs (in^2)

s = longitudinal spacing of stirrups (in)

Setting Equations 4-35 and 4-36 equal to each other, the spacing of the stirrups along the length of the span can be determined from the factored shear force V_u, as shown in Equation 4-37.

$$s_{\text{req}_{\text{strength}}} = \frac{A_v \times f_y \times d}{V_s}$$ **Equation 4-37**

The stirrup spacing s also must be checked based on the demand on the stirrups.

If $V_{s_{\text{required}}} \leq 2 \times V_c = 4 \times b \times d \times \sqrt{f'_c}$

s must be the smaller of $d/2$ or 24 in.

If $2 \times V_c < V_{s_{\text{required}}} \leq 4 \times V_c = 8 \times b \times d \times \sqrt{f'_c}$

s must be the smaller of $d/4$ or 12 in.

A final check for the stirrup spacing is based on the cross-sectional area and stirrup size, from Equation 4-38.

$$s_{\text{req}_{\text{width}}} = \text{smaller}\left(\frac{A_v \times f_y}{50 \times b_w}, \frac{A_v \times f_y}{0.75 \times b_w \sqrt{f'_c}}\right)$$ **Equation 4-38**

However, the final check is only required if the factored shear force meets the requirements of Equation 4-39.

$$\frac{\phi V_c}{2} < V_u \leq \phi V_c$$ **Equation 4-39**

4.7.1.3 Reinforced Concrete T-Beams

T-beams represent a rectangular beam with a walking or driving surface above, as illustrated in Figure 4.33. There are two types of T-beams: **monolithic (or built-in) T-beams** and **isolated T-beams**. The monolithic or built-in T-beams are cast at one time and include the entire cross section with several T-beams adjacent to each other. Isolated T-beams have a truncated flange on each side of the web and look like the capital letter T in section view.

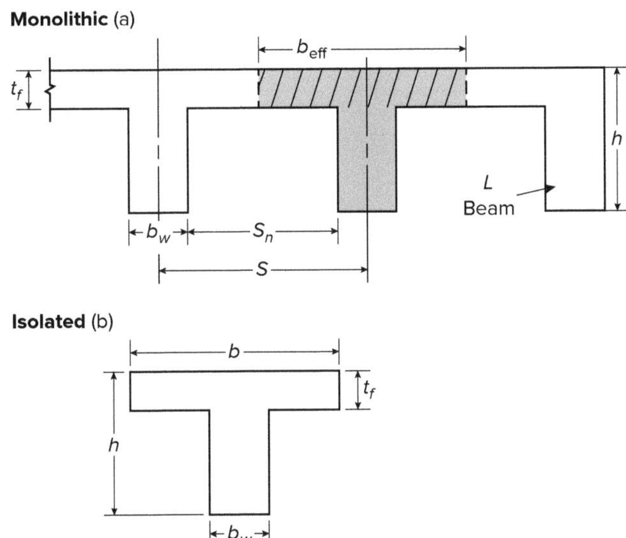

FIGURE 4.33 T-Beam Dimensions

The main difference between looking at the flexural capacity of a T-beam and a rectangular section is that the portion of the flange that resists the applied load must be determined. This is known as the effective flange width of the T-beam. For built-in or monolithic T-beams, ACI limits the effective width for interior T-beams to the smallest of the three values shown below:

1. $b_w + 16h_f$
2. $b_w + S_n$
3. $b_w + L_n/4$

} Smallest of the three values (ACI 318-14, Sec. 6.3.2.1)

Note that the values shown in the above inequality are identified in Figure 4.34.

For exterior T-beams, the limit is reduced by half on one side of the web because of the front or rear façade of a typical building.

Isolated T-beams are also restricted on the effective flange width. First, the web width b_w must not exceed $2 \times h_f$. If this occurs, the web width is reduced, so that the effective width is no more than four times this value. This is shown in Equation 4-40.

$$b_e \leq 4 \times b_w \leq 8 \times h_f \qquad \textbf{Equation 4-40}$$

Once the effective flange width is known, the depth of the compression block a must be determined. The equation is similar to that shown in the rectangular beam analysis, with the modification in the denominator of the b_e term noted above, as shown in Equation 4-41.

$$a = \frac{A_s \times f_y}{0.85 \times f'_c \times b_e} \qquad \textbf{Equation 4-41}$$

Once the depth of the compression block a has been determined, it must be compared to the height or thickness of the slab h_f. This results in one of two cases.

If the depth of the compression block a is less than or equal to the height of the slab, that is, $a \leq h_f$, then the T-beam may be treated as a rectangular beam, and the design equation, shown in Equation 4-32 (discussed earlier), is used.

$$M_n = A_s \times f_y \times \left(d - \frac{a}{2}\right)$$ **Equation 4-32** (p. 140)

If the depth of the compression block a is greater than the height of the slab, that is, $a > h_f$, then the calculated a value initially determined is no longer valid as the compression block has shifted into the web section. The analysis follows from what was shown in the rectangular beam case, except that the compression force is now in two pieces: one from the flange and one from the web, as illustrated in Figure 4.34.

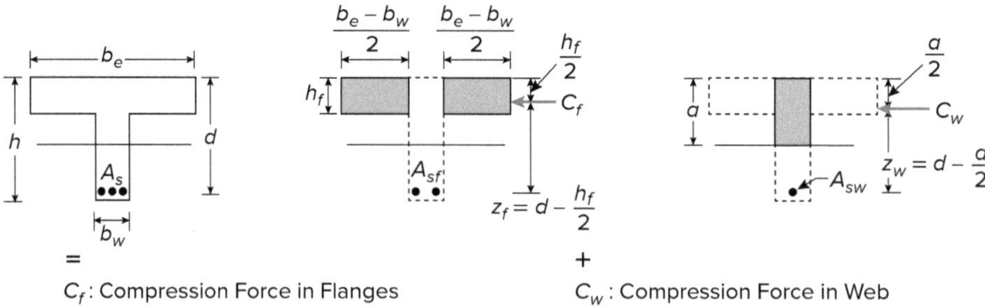

FIGURE 4.34 T-Beam Case 2 Compression Forces

The flange and web compression forces are shown in Equations 4-42 and 4-43, respectively.

$$C_f = 0.85 \times f'_c \times h_f \times (b_e - b_w)$$ **Equation 4-42**

$$C_w = 0.85 \times f'_c \times a \times b_w$$ **Equation 4-43**

Tension is still in equilibrium with the two compression forces. Thus, the new depth of the compression block within the web section b_w can be determined by setting tension equal to compression, as was done in the rectangular beam analysis. The new a value is shown in Equation 4-44.

$$a = \frac{(A_s \times f_y) - (0.85 \times f'_c \times h_f) \times (b_e - b_w)}{0.85 \times f'_c \times b_w}$$ **Equation 4-44**

The nominal flexural capacity follows from what was done in the rectangular beam case, namely, summing moments about the application point of either the tension or compression forces. For the T-beam analysis, the moment is summed about the tension force, such that the capacity is in terms of the two compression forces, as shown in Equation 4-45.

$$M_n = 0.85 \times f'_c \left[h_f \times (b_e - b_w) \times \left(d - \frac{h_f}{2}\right) + b_w \times a \times \left(d - \frac{a}{2}\right)\right]$$ **Equation 4-45**

Example 4.15: T-Beams

What is the effective width of the T-beam shown above? The span length, L, is equal to 36 ft.

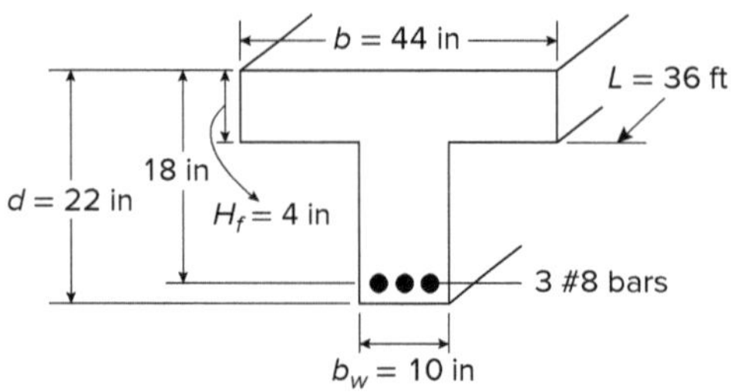

Example 4.15 *(continued)*

A. 32 in

B. 40 in

C. 44 in

D. 29 in

Solution

The web width, b_w, must not exceed $2h_f$. The isolated T-beam, b_w, is first checked with respect to h_f.

$b_w \leq 2\,h_f \rightarrow b_w \leq 2(4\text{ in}) = 8\text{ in}$

The actual b_w is 10 in, but as per the h_f condition, the maximum can be 8 in. Therefore, use $b_w = 8$ in.

The effective $b_e \leq 4\,b_w \rightarrow b_e \leq 4(8\text{ in}) = 32$ in.

Answer: A

4.7.2 Steel Beam Design

Steel I-beams are efficient structural shapes for flexure, providing the largest amount of area located the furthest distance from the neutral axis. However, the I-shape results in two items that must be checked for flexural capacity. The first item of capacity relates to the individual thickness of the flange and web sections and is referred to as a check on the compactness of the cross section related to local buckling of the web and/or flange section. If the compact section criteria have been met, the second item of capacity relates to a global failure of the I-beam and is referred to as lateral torsional buckling. This relates to the compression flange on an I-shape and is discussed in further detail in Section 4.7.2.2.

4.7.2.1 Compact Section Criteria

The majority of the rolled steel shapes in the *Steel Manual* do not have any issue related to local buckling of the web or flange prior to the I-beam obtaining its full plastic capacity, M_p. There are ten outlying shapes that do not meet the requirement for compact flanges for $F_y = 50$ ksi, including W21×48, W14×99, W14×90, W12×65, W10×12, W8×31, W8×10, W6×15, W6x9, and W6×8.5.

Girders or beams that are not fabricated to form rolled steel shapes from the *Steel Manual* are referred to as plate girders. Plate girders can be in many different depths with varying flange widths and thicknesses to accommodate continuous spans and to economize on the amount of steel used on a project. As such, the *Steel Manual* has two equations to check that the flange and web meet the minimum thickness to ensure that no local buckling issues are encountered. The equations for flange and web compactness are shown in Equations 4-46 and 4-47.

$$\lambda_f = \frac{b_f}{2t_f} \leq \lambda_{pf} = 0.38\sqrt{\frac{E}{F_y}}$$ **Equation 4-46**

$$\lambda_w = \frac{h}{t_w} \leq \lambda_{pw} = 3.76\sqrt{\frac{E}{F_y}}$$ **Equation 4-47**

b_f = flange width (in)

t_f = flange thickness (in)

h = beam height (in)

t_w = web width (in)

For the more traditional yield strength, $F_y = 50$ ksi, the previous equations reduce down to:

$$\frac{b_f}{2t_f} \leq 9.15$$ **Equation 4-48**

$$\frac{h}{t_w} \leq 90.6$$ **Equation 4-49**

4.7.2.2 Lateral Torsional Buckling

A steel I-shape will ultimately fail in flexure from the compression flange shifting out of plane to the left or right of the vertical orientation. This is referred to as lateral torsional buckling. An example of a simply supported beam subjected to a uniformly distributed load helps illustrate the failure. As the top compression flange begins to shift to the left or right in the lateral direction, the ends of the beam are still fairly vertical. This results in a twisting or torsional stress in the beam.

There are two methods to reduce the effects of lateral torsional buckling (LTB). For a top compression flange, the most effective method is to provide a reinforced concrete deck that is integrally attached to the steel beam via shear studs. The shear studs are welded to the top flange of the steel beam and then embedded into the concrete deck, resulting in a continuous positive connection between the two elements. The second method is to provide bracing of the top compression flange at discrete distances along the length of the span. The distance between brace points on the compression flange is referred to as the unbraced length, L_b. If no bracing is provided, then L_b is simply equal to the beam length. Of course, the bracing can be at any arbitrary distance: A beam could be braced at midspan, at third points, at quarter points, and so forth.

The unbraced length of the compression flange L_b is compared with two other values: L_p and L_r. L_p is the maximum unbraced length of the compression flange that results in the plastic moment for the beam, $M_p = F_y \times Z_x$. L_r is defined as the maximum unbraced length of the compression flange that results in first yield with residual stresses, that is, $M_r = 0.7 \times F_y \times S_x$. The available capacity of the beam relative to the unbraced length of the compression flange is shown in AISC Table 3-10, and is shown schematically in Figure 4.35.

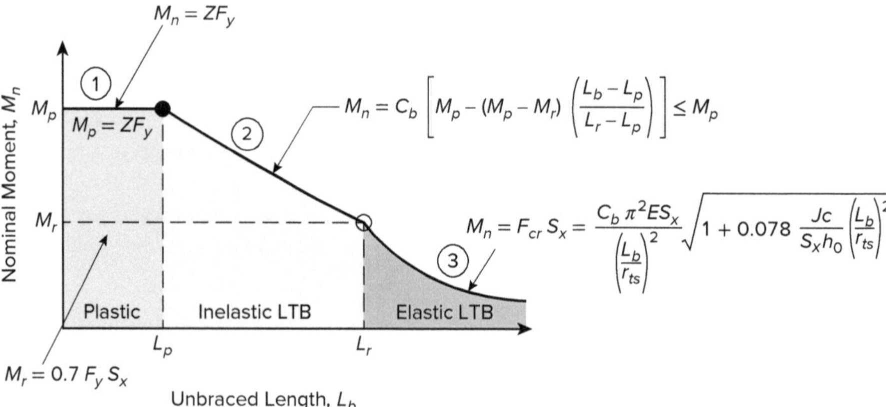

FIGURE 4.35 Moment Capacity as a Function of the Unbraced Length of the Compression Flange

For I-shapes, the selection of the appropriate beam for the given unbraced length and design moment is taken from Table 3-10 in the *Steel Manual*. The two key points noted previously, L_p and L_r, are denoted in AISC Table 3-10 as closed and open dots, respectively. A sample page from the *Steel Manual* for Table 3-10 is shown in Figure 4.36.

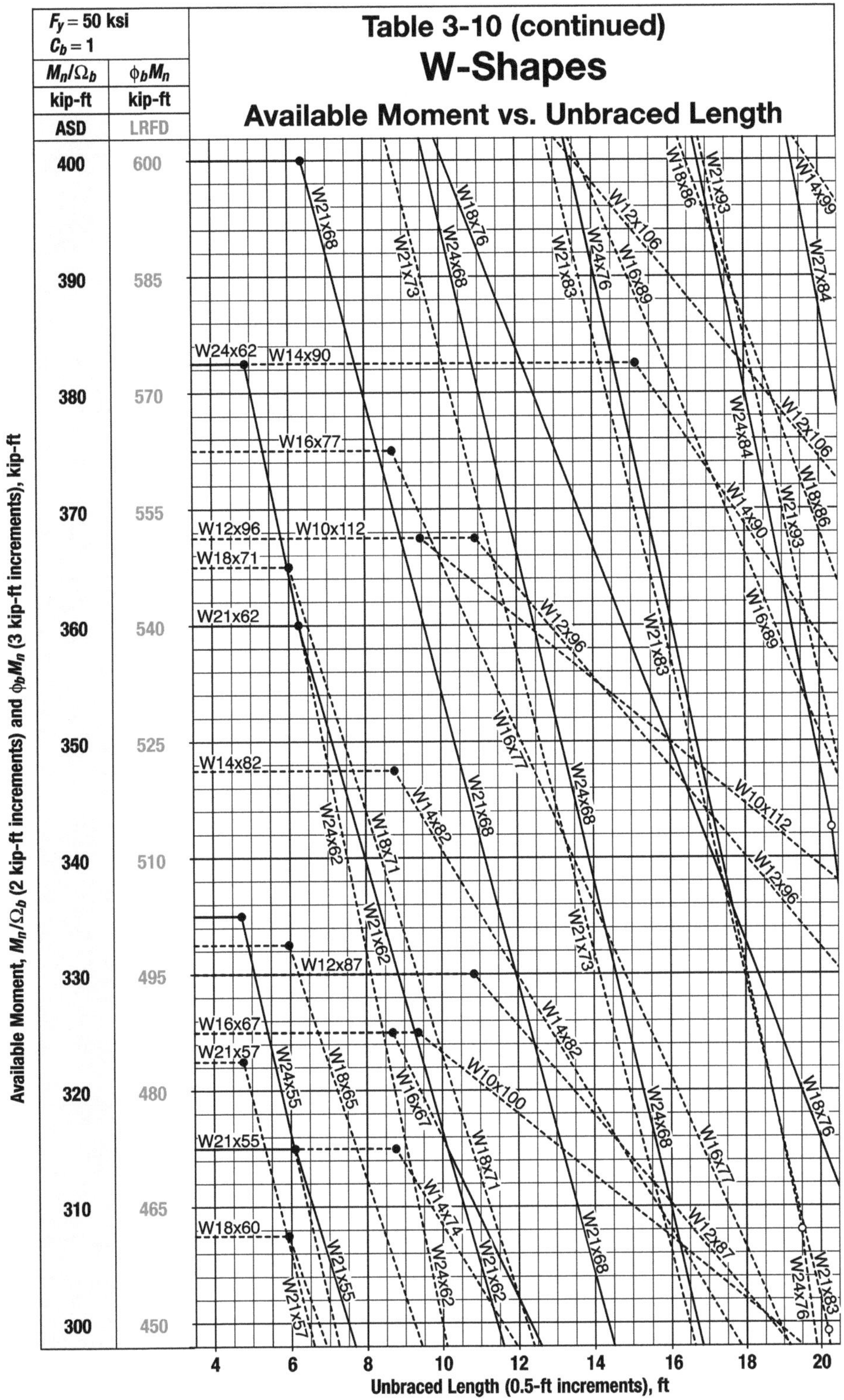

FIGURE 4.36 Table 3-10 Sample from the AISC *Steel Manual*

In order to use this table, two items are needed: the design moment of the beam in ASD or LRFD and the unbraced length of the compression flange. The first beam to the right of the intersection will meet the design requirements. However, there are two types of lines in Table 3-10: solid and dashed lines. Solid lines are the lightest by weight, and thus the least expensive to purchase. In general, a solid line is the goal of most designs unless other restraints come into play. For instance, if the first solid line encountered is a W20 beam, but the project of interest can only accommodate a W18 beam, then a dashed line can be selected. Alternatively, steel availability may force selection from a solid line beam to a dashed line beam. Dashed line beams will always meet the requirements; however, they pay a small weight penalty for this option.

Example 4.16: Steel Beams

A steel fabricator has a stock of W14×90 beams with a nominal design moment (ϕM_n) of approximately 574 kip-ft. If the beam is unbraced over its length, what is the largest span this beam can be subjected to in order to obtain this design moment?

A. 18 ft
B. 14 ft
C. 12 ft
D. 8 ft

Solution

From Figure 4.36 (p. 149), beam W14×90 is located at approximately 574 kip-ft of capacity on the vertical axis. Following the end of the flat portion of the curve down to the horizontal axis, the maximum unbraced length is approximately 15 ft. Thus, the largest span without exceeding this length would be 14 ft.

Answer: B

4.7.3 Shear Capacity

In general, shear capacity of I-shaped sections is not a governing design factor unless there is a concentrated load adjacent to a support. Still, the average shear stress for an I-shape does increase substantially when compared to a rectangular shape of the same height. Recall from Section 4.5.2.2 that the average shear stress for an I-shape section is defined as the total depth d times the web width t_w. The design shear stresses from the *Steel Manual* for ASD and LRFD are shown in Equations 4-50 and 4-51.

$$F_{v_{\text{allowable}}} = 0.6 \times \frac{F_y}{\Omega}$$ **Equation 4-50**

$$F_{v_{\text{ultimate}}} = \phi \times 0.6 \times F_y$$ **Equation 4-51**

The reduction factors for ASD and LRFD are 1.5 and 1.0, respectively. Thus, the allowable shear stresses become:

$$F_{v_{\text{allowable}}} = 0.4 \times F_y$$ **Equation 4-52**

$$F_{v_{\text{ultimate}}} = 0.6 \times F_y$$ **Equation 4-53**

Alternatively, if the allowable value being considered is average shear force, not shear stress, the equations above would be modified by multiplying through by the shear area $d \times t_w$.

$$V_{\text{allowable}} = 0.4 \times F_y \times d \times t_w$$ **Equation 4-54**

$$V_n = 0.6 \times F_y \times d \times t_w$$ **Equation 4-55**

4.8 COLUMN DESIGN

Column elements generally fall into two categories at ultimate loading: short or stub columns and long or slender columns. Short columns will crush at ultimate load and will not deform laterally, whereas long or slender column sections will fail due to buckling.

4.8.1 Euler Buckling Load

Long columns will fail due to buckling as defined by the Euler buckling load. The Euler buckling load and buckling stress equations follow.

$$P_{cr} = \frac{\pi^2 \times E \times A}{\left(\frac{KL}{r}\right)^2} \qquad \textbf{Equation 4-56}$$

$$F_{cr} = \frac{\pi^2 \times E}{\left(\frac{KL}{r}\right)^2} \qquad \textbf{Equation 4-57}$$

E = modulus of elasticity of material (ksi)

K = effective length factor (dimensionless)

L = unbraced length of column (in)

r = radius of gyration (in)

The K value relates to the boundary conditions of the column at the top or bottom, as well as a theoretical or recommended design value. Typical values are shown in Figure 4.37.

Column	Theoretical *K*	Design *K*
Fixed / Fixed (Deflected Shape)	0.5	0.65 to 0.90
Pinned / Fixed (Deflected Shape)	0.707	0.80 to 0.90
Pinned / Pinned (Deflected Shape)	1	1
Free / Fixed (Deflected Shape)	2	2.0 to 2.1
Free to Translate Only / Fixed (Deflected Shape)	1	1.2
Free to Translate Only / Pinned (Deflected Shape)	2	2

FIGURE 4.37 Effective Length Factors

Example 4.17: Euler Buckling Load

What is the theoretical resistance to buckling for the column below? The boundary conditions are fixed at the base and free to rotate at the top, and the assumed height is L.

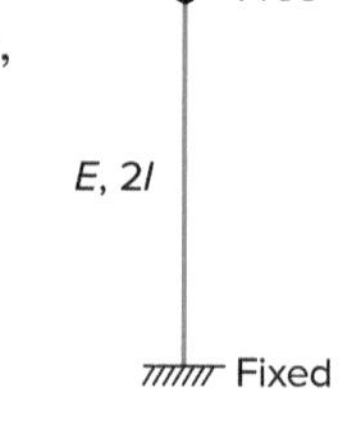

A. $P_{cr} = \dfrac{\pi^2 4EI}{(L)^2}$

B. $P_{cr} = \dfrac{\pi^2 E2I}{2(L)^2}$

C. $P_{cr} = \dfrac{\pi^2 EI}{4(L)^2}$

D. $P_{cr} = \dfrac{\pi^2 EI}{2(L)^2}$

Solution:

The Euler formula for long columns, critical force (the applied force at which the column will buckle, Equation 4-55, p. 151) is:

$$P_{cr} = \frac{\pi^2 \times EA}{\left(\frac{KL}{r}\right)^2} \qquad \text{Substitute } r = \sqrt{\frac{I}{A}}$$

$$= \frac{\pi^2 EA}{\left(\frac{KL}{\sqrt{\frac{I}{A}}}\right)^2} = \frac{\pi^2 EA}{(KL)^2} \times \frac{I}{A} = \frac{\pi^2 EI}{(KL)^2}$$

Given:

$E = E$

$I = 2I$

$K = 2.0$ (Recall the K values from Figure 4.37. For boundary conditions of fixed-free, the theoretical K value is 2.0.)

$$P_{cr} = \frac{\pi^2 E(2I)}{(2L)^2} = \frac{\pi^2 E(2I)}{4(L)^2} = \frac{\pi^2 EI}{2(L)^2}$$

Answer: D

The unbraced length value is easier to visualize for two cases. The first is for a building example where the column length would be the floor-to-floor distance. The second case is typical of transmission towers where a single element may be braced at several locations along its length. See Figure 4.38.

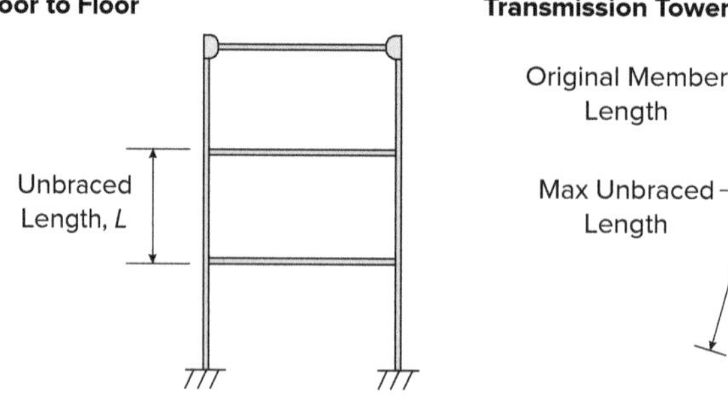

FIGURE 4.38 Unbraced Length Examples

The radius of gyration is a geometric parameter that varies depending on the cross section being investigated. It can be calculated as shown in Equation 4-58, and will generally have a strong and a weak axis for a rectangular or I-shaped section.

$$r = \sqrt{\frac{I}{A}}$$ **Equation 4-58**

4.8.2 Slenderness Ratio

The squared term in the denominator in Equation 4-57 (*KL*/*r*, p. 151) is referred to as the slenderness ratio for a cross section. For a strong and a weak axis, the larger value of the slenderness ratio will indicate the weaker direction, and thus it will buckle first. For example, if a column section has the same length about the strong and weak axes, and all other values are maintained, it will fail about the weak axis. If the same column was braced at midlength about the weak axis, it is possible that the strong axis direction may govern.

4.8.3 Concrete Columns

The type of concrete column—short or long—will be discussed with respect to two types of building construction: sway and nonsway frames.

Nonsway or braced frames resist lateral movement due to rigid elements in the form of shear walls or cross-bracing. Under axial loading, the column sections for nonsway or braced frames deflect laterally along their length, but the tops of the columns do not translate. Sway or unbraced frames result in lateral translation of the top of the column sections under applied vertical loading. The main differences are shown in Figure 4.39.

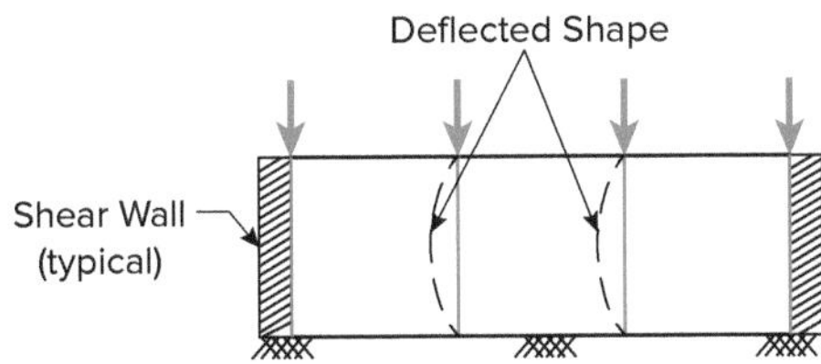

Braced Frames (b)

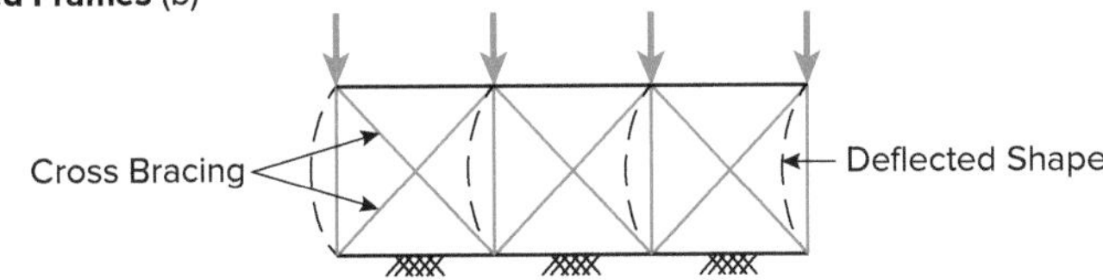

Sway/Unbraced Frames (c)

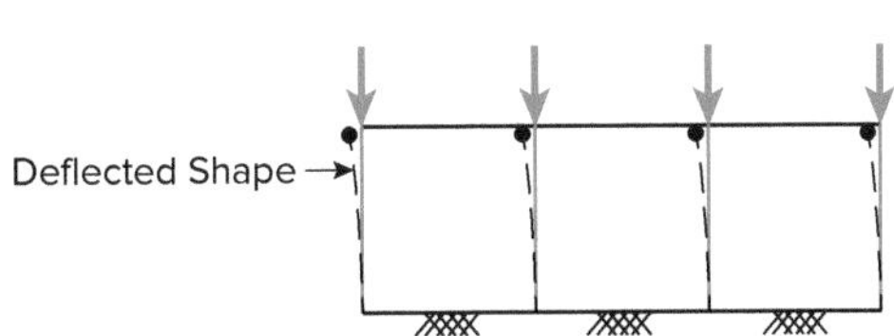

FIGURE 4.39 Nonsway or Braced Frames and Sway or Unbraced Frames Under Axial Loading

The definition of a short column for both types of construction is based on the slenderness ratio of the column. For the nonsway or braced frames, the slenderness ratio limit is shown in Equation 4-59.

$$\frac{k_b \times L_u}{r} \leq 34 - 12\left(\frac{M_1}{M_2}\right) \leq 40$$ **Equation 4-59**

k_b = effective length factor, which can be conservatively taken as equal to 1.0

L_u = unbraced length of the column (in)

r = radius of gyration (in)

M_1 = absolute value of the smaller moment at the top or bottom of the column

M_2 = absolute value of the larger moment at the top or bottom of the column

Note that the ratio of M_1/M_2 is positive if the column is in single curvature due to opposing moments (that is, one moment is clockwise and the other is counterclockwise), whereas the ratio is negative if the column is in double curvature and subjected to the same direction moments.

For sway or unbraced frames, the slenderness ratio is limited to a value of 22, as shown in Equation 4-60.

TIP

For exam purposes, the k_b value for nonsway or braced frames will be provided.

$$\frac{k_b \times L_u}{r} \leq 22$$ **Equation 4-60**

The main difference between the nonsway or braced frame and the sway or unbraced frame is the k_b value in the numerator. The previous value in Equation 4-59 was limited to 1.0; for the sway or unbraced frame, the k_b value is ≥ 1.0.

4.8.3.1 Concrete Column Reinforcing Steel

4.8.3.1.1 Minimum Longitudinal Steel

ACI 318-14 identifies the minimum longitudinal steel in Section 10.6.1.1 to be between $0.01 \times A_g$ and $0.08 \times A_g$, where A_g is defined as the gross cross-sectional area of the column. If the area of steel calculated falls below the minimum, the cross-sectional area is too large and must be reduced. For instance, if the column is circular, this means the diameter must be reduced. A similar reduction would be needed for a square or rectangular column cross section. Alternately, if the area of steel required is greater than 8%, the overall column size is too small and must be increased in size to get under the 8% maximum.

There is also a requirement for the number of longitudinal bars in a cross section, as identified in ACI 318-14, Section 10.7.3.1:

- Minimum of three longitudinal bars for a triangular cross section
- Minimum of four longitudinal bars for a rectangular or circular transverse steel
- Minimum of six longitudinal bars for columns composed of helical or spiral steel or of special moment frames

4.8.3.1.2 Minimum Spiral/Hoop Steel

The minimum shear reinforcement is outlined in ACI 318-14, Section 10.6.2.2 as the greater of Equations 4-61 and 4.62:

$$0.75 \times \sqrt{f'_c} \times \frac{b_w s}{f_{yt}}$$ **Equation 4-61**

$$50 \times \frac{b_w s}{f_{yt}}$$ **Equation 4-62**

b_w = web width or circular diameter (in)

s = center-to-center spacing of reinforcement (in)

f_{yt} = specified yield strength of transverse reinforcement (psi)

Example 4.18: Concrete Columns

Determine the axial design strength of a short column that is 16 in × 16 in with ties. The concrete compressive strength is $f'_c = 5$ ksi and the steel reinforcing bars are Grade 60 ($f_y = 60$ ksi). The column is reinforced with 8 #6 longitudinal bars. The normalized eccentricity $\dfrac{M_u}{P_u h} < 0.1$.

A. 41 kips
B. 213 kips
C. 667 kips
D. 812 kips

Since the statement indicates it is a short column, there is no need to calculate the slenderness ratio. The problem statement indicates that the normalized eccentricity is $\dfrac{M_u}{P_u h} < 0.1$; thus, the effects of flexural moment can be neglected. In addition, it is a column with ties. Therefore, the following equation applies:

$$\phi P_n = 0.80\ \phi\ [0.85 f'_c\ (A_g - A_{st}) + f_y A_{st}] \geq P_u$$

$$\phi = 0.65, f'_c = 5 \text{ ksi}, A_g = 16 \text{ in} \times 16 \text{ in}, A_{st} = 8 \times 0.44 \text{ in}^2 \text{ for \#6 bars}, f_y = 60 \text{ ksi}$$

$$\phi P_n = 0.80 \times 0.65\ [0.85 \times 5 \text{ ksi}(16 \text{ in} \times 16 \text{ in} - 8 \times 0.44 \text{ in}^2) + 60 \text{ ksi}(8 \times 0.44 \text{ in}^2)]$$

$$\phi P_n = 667 \text{ kips}$$

$$\rho_g = \frac{A_{st}}{A_g} = \frac{8 \text{ bars} \times 0.44 \text{ in}^2 \text{ per \#6 bar}}{16'' \times 16''} = 0.0138 \approx 0.1 \qquad (0.01 \leq \rho_g \leq 0.08), \text{ OK}$$

Answer: C

4.8.4 Steel Column Design

The reduction factors for column design as per the *Steel Manual* are:

$\phi_c = 0.90$ (LRFD)

$\Omega_c = 1.67$ (ASD)

The *Steel Manual* differentiates steel columns into those without slenderness and those with slenderness. The limit between the two, as outlined in Section B4.1 of the *Steel Manual*, relates to the width-to-thickness ratio, λ_r from Table B4.1a. If the limit in the table is exceeded, then the section is a slender element. Thus, the limit for W-shapes is:

$$\frac{0.5 \times b_f}{t_f} \leq 0.56 \times \sqrt{\frac{E}{F_y}} \qquad \textbf{Equation 4-63}$$

If we focus on elements without slenderness, then the capacity is the smaller of two governing limit states: flexural buckling as outlined in the *Steel Manual*, Section E3 or torsional and flexural-torsional buckling as outlined in Section E4. The latter of the two is a bit more involved and likely beyond the scope of the exam. For the flexural buckling from Section E3, the critical stress, F_{cr}, is determined as a function of the slenderness ratio noted previously for concrete columns, KL/r. There are two cases, as illustrated in Equations 4-64 and 4-65.

$$\text{If } \frac{KL}{r} \leq 4.71\sqrt{\frac{E}{F_y}}, \text{ then } F_{cr} = \left[0.658^{\frac{F_y}{F_e}}\right] \times F_y \qquad \textbf{Equation 4-64}$$

$$\text{If } \frac{KL}{r} > 4.71\sqrt{\frac{E}{F_y}}, \text{ then } F_{cr} = 0.877 \times F_e \qquad \textbf{Equation 4-65}$$

F_e = elastic buckling stress from Equation 4-66

$$F_e = \frac{\pi^2 E}{\left(\frac{KL}{r}\right)^2}$$ **Equation 4-66**

As an alternative to calculating the capacity with respect to the governing radius of gyration, there are two tables in the *Steel Manual* that provide the critical stress in tabular format. Table 4.1 provides available axial compression for W-shapes with respect to the smaller radius of gyration, r_y. Alternatively, if the smaller radius of gyration does not govern for a W-shape, Table 4.22 provides the available critical stress relative to the slenderness ratio, KL/r. If Table 4.22 is used, the available force must be multiplied by the area of the cross section, as stress is the value obtained from the table.

4.9 SLAB DESIGN

4.9.1 Concrete Slab Design

The design of reinforced concrete slabs is typical for the construction of retail and commercial applications. Flexural and shear capacity follow from rectangular beam design, with a few modifications noted below.

4.9.1.1 One-Way Flexural Slab Design

Slabs are identified as either being one-way slabs (that is, loaded in one direction only) or two-way slabs (loaded in two directions). The demarcation of the two types of slabs is the ratio of the length to the width. If the length is greater than or equal to two times the width, then it can be treated as a one-way slab. If the length is less than the width, it can be treated as a two-way slab. Also, the width of the slab for design purposes is reduced to 12 in. This aids in calculating the required area of steel and ultimately the spacing of the steel along the length of the slab. The total area of steel for the 12-in slab width relates to the bar spacing shown for a specific problem. For instance, if a No. 6 bar was selected with a spacing of 8 in, then the total area of steel for a 12-in width would simply be the single bar area of 0.44 in^2 times the ratio of the bar spacing to the total width of 12 in. Thus, the total area would be:

TIP

In all design equations, the width b is set at 12 in.

$$0.44\ \text{in}^2\left(\frac{12\ \text{in}}{8\ \text{in}}\right) = 0.66\ \text{in}^2$$

Maximum spacing of the main longitudinal steel is the smaller of $3 \times h$ or 18 in. The rationale for this requirement is to provide smaller bars at smaller spacings as opposed to larger bars that are more widely spaced.

Temperature and shrinkage steel is required to be orthogonal to the main reinforcing steel, as shown in Figure 4.40. This serviceability requirement reduces the chance of the concrete cracking due to seasonal changes in temperature and humidity. For reinforcing steel with a yield strength of 60 ksi, the required area of temperature and shrinkage steel is shown in Equation 4-67.

$$\rho_g = \frac{A_s}{bh} = 0.0018$$ **Equation 4-67**

Once the area of steel is determined, the analysis for flexural capacity follows from that of the rectangular beams; that is, determine the depth of the compression block a, and then determine the design moment capacity:

$$a = \frac{A_s \times f_y}{0.85 \times f'_c \times b}$$ **Equation 4-31** (p. 140)

$$M_n = A_s \times f_y\left(d - \frac{a}{2}\right)$$ **Equation 4-32** (p. 140)

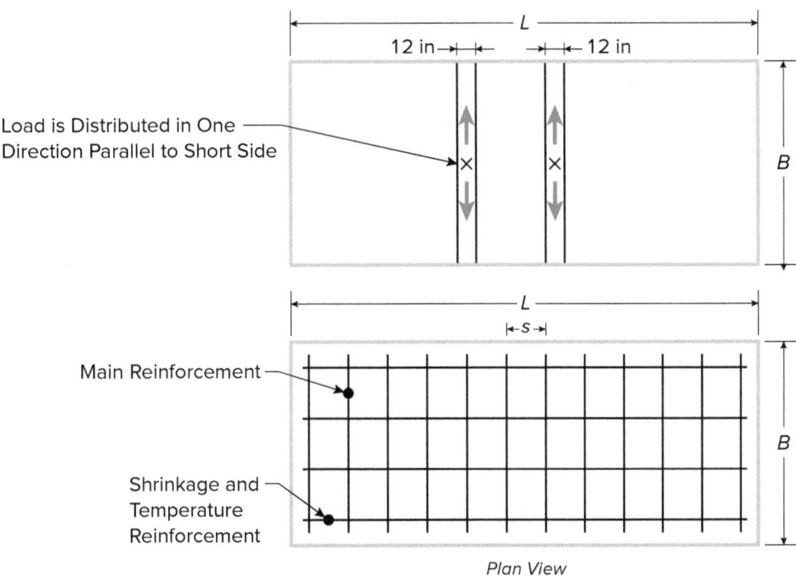

FIGURE 4.40 One-Way Slab Design for Flexure

4.9.1.2 One-Way Shear Design

One-way slab design for shear is simpler than what was previously shown for rectangular beams. Specifically, the capacity is reduced from that of the concrete and steel, V_c and V_s, to account for the concrete portion, V_C, in terms of capacity. If the footing is not sufficient in shear for the thickness shown, then the thickness of the slab is increased until it meets the shear requirement. The flexural capacity is then adjusted as needed for the modified slab thickness or height.

Example 4.19: Structural Slabs

A structural slab is to be constructed for a warehouse. The following properties of the slab are given:

$f'_c = 5{,}500$ psi

$f_y = 60{,}000$ psi

$h = 15$ in

$d = 12$ in

$\varphi = 0.9$

Flexural steel #7 bars spaced at 14 in on center

The nominal moment capacity of the slab is most nearly:

A. 27.12 kip-ft
B. 35.04 kip-ft
C. 30.13 kip-ft
D. 31.53 kip-ft

Solution

For structural slabs, the width is set to 12 in; thus, $b = 12$ in.

The area of steel for a single #7 bar is 0.60 in^2. The bars are spaced at 14 in on center. The total area of steel for a 12-in width is then:

$$A_{s_{total}} = 0.60\text{ in}^2 \frac{12\text{ in}}{14\text{ in}} = 0.514\text{ in}^2$$

The depth of the compression block:

$$a = \frac{A_s \times f_y}{0.85 \times f'_c \times b}$$ **Equation 4-31** (p. 140)

Example 4.19 *(continued)*

$f'_c = 5{,}500 \text{ psi} = 5.5 \text{ ksi}$

$f_y = 60{,}000 \text{ psi} = 60 \text{ ksi}$

$$a = \frac{0.514 \text{ in}^2 (60 \text{ ksi})}{(0.85)(5.5 \text{ ksi})(12 \text{ in})} = 0.55 \text{ in}$$

The nominal moment capacity:

$$M_n = A_s \times f_y \left(d - \frac{a}{2}\right)$$ **Equation 4-32** (p. 140)

$$M_n = 0.514 \text{ in}^2 \times 60 \text{ ksi} \left(12 \text{ in} - \frac{0.55 \text{ in}}{2}\right) = 361.60 \text{ kip-in}$$

Convert to kip-ft: $\dfrac{361.60 \text{ kip-in}}{12 \text{ in}} = 30.13 \text{ kip-ft}$

Answer: C

4.10 FOOTING DESIGN

4.10.1 Allowable Bearing Capacity for Shallow Footings

> **TIP**
>
> Note that if there is nothing about gross or net given in a problem statement, then assume a net bearing capacity.

The allowable bearing capacity of a soil can be provided as either a net or gross allowable from a geotechnical engineering standpoint. The main difference is that the gross allowable is generally given at the existing ground line for a project site. The location of the base of the footing, as shown in Figure 4.41, is not at the existing ground line, to account for minimum embedment for frost or other considerations. In the instance of gross allowable soil pressure, the weight of the footing, the column section below grade, and the soil overburden must be accounted for by subtracting these values from the gross allowable soil capacity. The net allowable soil capacity is provided at the vertical distance from the existing ground line to the proposed bottom of the new footing, and thus it does not require any additional modifications.

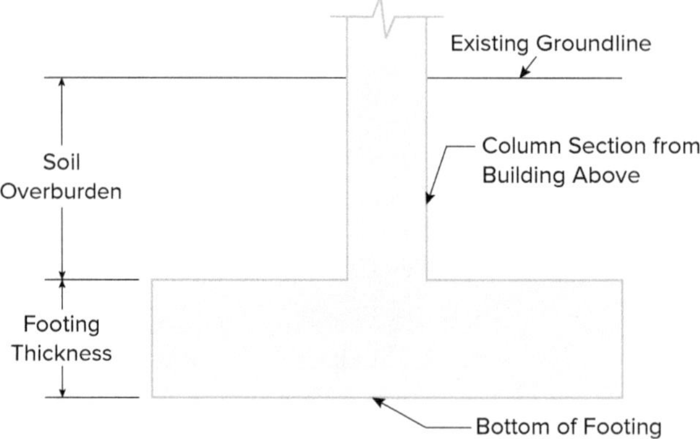

FIGURE 4.41 Spread Footing Located Below Grade

4.10.2 Sizing of Footings from Structural Loads

Once the allowable bearing capacity has been determined as either a net or gross allowable, then the overall dimensions of the footing can be determined as a *P*/*A* stress calculation. The one caveat is that the applied load from the column above must be used at the service level with no load factors. This is because the design philosophy for geotechnical engineering relates to ASD with service loads and factor of safety. The allowable bearing capacity that has been provided in a given problem has already been reduced by an appropriate factor of safety. It would be inconsistent to use load factors in this instance, as it would be mixing LRFD and ASD design principles. Once the total load has been determined, the footing area is limited by Equation 4-68.

$$A_{max} = \frac{P_{Total}}{q_{allowable}}$$ **Equation 4-68**

P_{Total} = total applied load from the building above (that is, dead plus live load, for example)

$q_{allowable}$ = bearing capacity determined by a geotechnical engineer

4.10.3 Footing Structural Design

After the size of the footing has been determined, the design of the footing follows the LRFD design principles previously noted. This requires load factors for dead, live, and any other applicable loads for a given problem. Thus, the total load will be greater than the value shown in the numerator of Equation 4-68. By dividing this new total load by the area of the footing, the factored area load q_u is determined. At first glance, this may be disconcerting, as the factored area load q_u will exceed the allowable bearing capacity in the previous section. However, this is not the case, as the footing dimensions were determined using ASD and the allowable bearing capacity. The design of the footing uses load factors from LRFD design procedures.

4.10.3.1 One-Way Shear

The critical section for one-way shear is shown in Figure 4.42. For simplicity, only the right side of the footing has been identified as critical—in reality either side may fail. The distance x in the figure can be determined from the overall dimensions of the footing by working from the center of the footing. For the square footing shown in Figure 4.42, the x distance is calculated as $b/2 - c/2 - d$, where c is defined as the column width in the direction of interest.

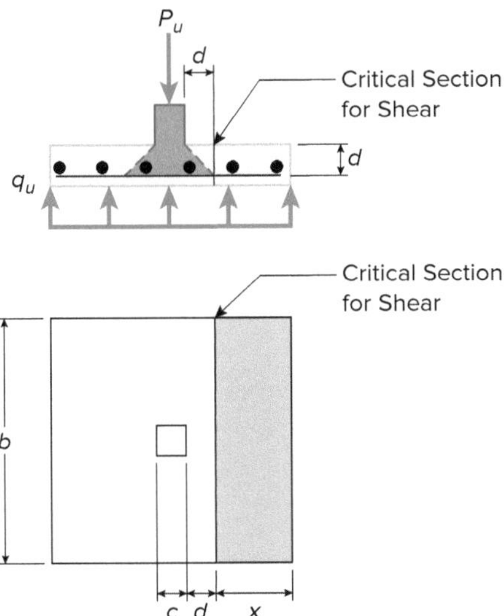

FIGURE 4.42 Critical Location for One-Way Shear in Shallow Footings

The shear capacity follows from what was shown in the rectangular beams, but the steel contribution goes to zero as was the case for shear in slabs. Thus, the equation for capacity is just that of the concrete. Taking into account the strength reduction factor, ϕ, Equation 4-69 must be met for one-way shear capacity.

$$\varphi V_c \geq V_u$$ **Equation 4-69**

Substituting the values for V_c and using the $\phi = 0.75$ result in Equation 4-70.

$$(0.75)2bd\sqrt{f'_c} \geq q_u \times bx$$ **Equation 4-70**

q_u = factored total load divided by the footing area

School of PE

Structural Mechanics

4.10.3.2 Two-Way Shear

Two-way or punching shear is an issue when the applied axial load is relatively large and the footing is thin in comparison. Picture a three-hole punch on a piece of paper, where the punch represents the column load and the paper represents the footing thickness. This is shown in Figure 4.43.

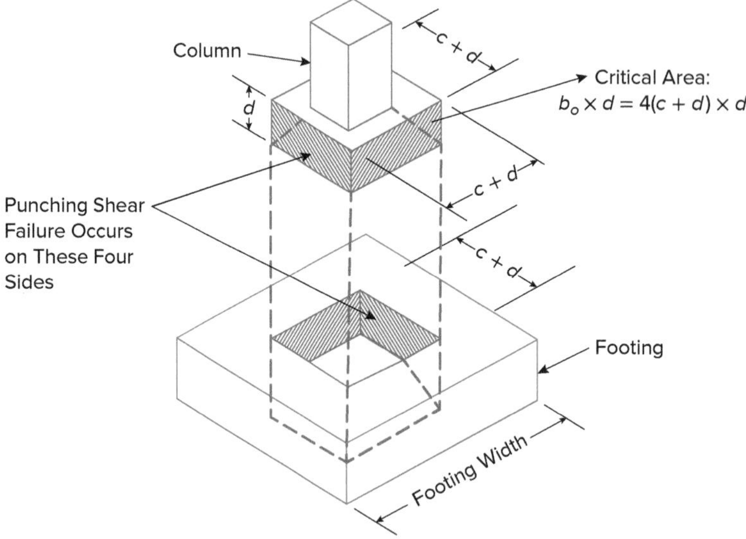

FIGURE 4.43 Critical Location of Two-Way Punching Shear for Footings

Capacity comes from ACI 318-14, Section 22.6.5, as shown in Equation 4-71.

$$\varphi V_n = \varphi(V_c) \qquad \textbf{Equation 4-71}$$

The V_c component is the smallest of the following three values:

$$V_c = 4\lambda\sqrt{f'_c}\, b_0 d \qquad \textbf{Equation 4-72}$$

$$V_c = \left(2 + \frac{4}{\beta}\right)\lambda\sqrt{f'_c}\, b_0 d \qquad \textbf{Equation 4-73}$$

$$V_c = \left(2 + \frac{\alpha_s d}{b_0}\right)\lambda\sqrt{f'_c}\, b_0 d \qquad \textbf{Equation 4-74}$$

λ = lightweight concrete factor ($\lambda = 1.0$ for normal-weight concrete)

$b_0 = 4(c + d)$

β = ratio of long to short side of column section

α_s is shown in Figure 4.44, with values of:

α_s = 40 for interior columns

= 30 for edge columns

= 20 for corner columns

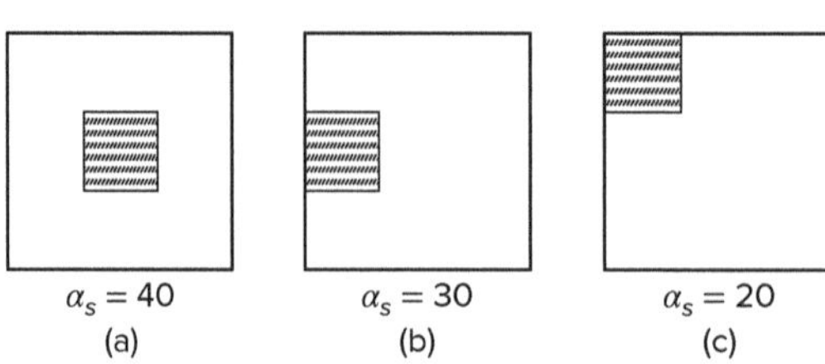

FIGURE 4.44 Column Location for Two-Way Shear Capacity

4.10.3.3 Flexure

The critical section for flexure is at the face of the column section. For a rectangular footing, this results in two moments that must be checked to determine which is the largest. This is shown in Figure 4.45. Once the moment is found, the capacity must be determined from the rectangular beam analysis previously introduced.

$M_{A-A} = q_u \times b \times f \times (f/2)$
$M_{B-B} = q_u \times k \times g \times (g/2)$

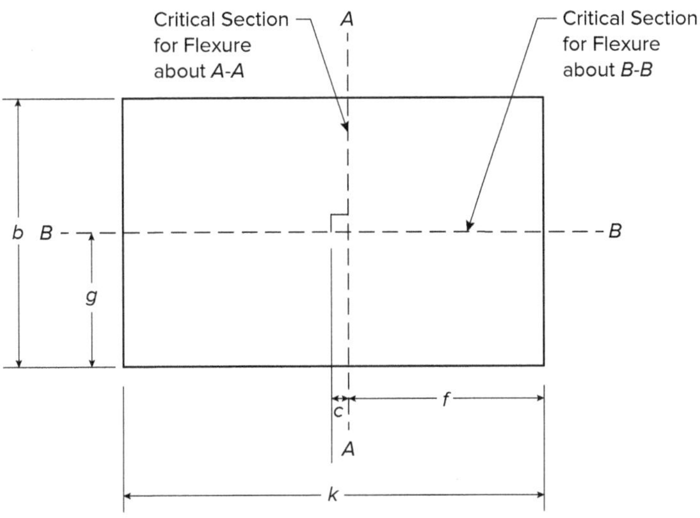

FIGURE 4.45 Bending Moment for Shallow Footings

4.11 RETAINING WALL DESIGN

4.11.1 Lateral Earth Coefficients

There are three lateral earth coefficients that relate to the movement of a wall section relative to the soil mass behind that it is supporting: at-rest, active, and passive.

4.11.1.1 At-Rest Coefficient

The at-rest condition is for rigid walls, in which the deflection at the top of the wall is essentially zero. In this condition, no failure wedge is formed behind the wall, which results in the smallest force on the wall. The calculation for the at-rest coefficient is shown in Equation 4-75. In this and the remaining definitions:

ϕ = friction angle of the soil

K = ratio of lateral pressure to vertical pressure (σ_h/σ_v)

$K_0 = 1 - \sin(\phi)$ **Equation 4-75**

4.11.1.2 Active Coefficient

The active lateral earth coefficient is applicable when the top of the wall section moves relative to the soil such that a potential failure wedge is formed and the soil will provide additional force on the wall. The active coefficient is shown in Equation 4-76.

$K_a = \tan^2\left(45 - \frac{\phi}{2}\right) = \frac{(1-\sin(\phi))}{(1+\sin(\phi))} = \frac{\sigma_a}{\sigma_v}$ **Equation 4-76**

4.11.1.3 Passive Coefficient

The passive lateral earth coefficient comes into play when the wall pushes into the soil mass. A visual of this occurs on the wing walls of a bridge abutment under seismic loading, where the walls are retaining the road bed. When seismic loading perpendicular to the bridge is experienced, the wing walls will attempt to push the soil, as shown in Figure 4.46. The passive coefficient is shown in Equation 4-77.

$$K_p = \tan^2\left(45 + \frac{\phi}{2}\right) = \frac{(1 + \sin(\phi))}{(1 - \sin(\phi))} = \frac{\sigma_p}{\sigma_v}$$ **Equation 4-77**

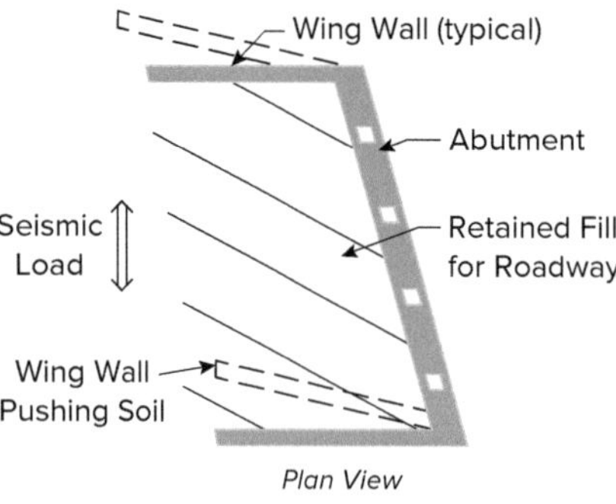

FIGURE 4.46 Passive Earth Pressure Coefficient Example

The resulting horizontal pressure or stress at a depth h can then be calculated as the appropriate K value times the depth h times the unit weight of the soil, as shown in Equation 4-78.

$$\sigma_h = K \times \gamma_{soil} \times h$$ **Equation 4-78**

4.11.2 Bearing, Sliding, and Overturning Calculations

Example 4.20 illustrates a check for a retaining wall for bearing, sliding, and overturning under service loading. In each case, the calculation for the wall is simplified by taking the length in or out of the page equal to 1 ft. Thus, the only other parameters needed are the height and width of the footing or stem and the height of the soil for the bearing comparison. The unit weight of the concrete is fixed at 150 lb/ft^3, unless a different concrete mix has been specified in the problem statement. The unit weight of the soil will be provided.

TIP

An active condition is generally assumed in practice, so pay attention to the given information in a problem to identify which type is presented.

Sliding calculations include the forces noted above with the addition of the lateral earth pressure. From the previous discussion, the K value of the soil must be determined. The force due to lateral earth pressure over a given depth h is shown in Equation 4-79. Note that the height will be from the ground line at the top of the wall to the bottom of the footing.

$$P_{Lateral} = \frac{1}{2} \times \gamma_{soil} \times K \times h^2$$ **Equation 4-79**

The final item needed for a sliding check is the coefficient of friction between the base of the footing and the soil, which would be provided in a given problem statement on the exam.

Overturning calculations use the values previously noted, with the addition of the individual moment arms of the forces. The moment arms for the concrete stem and footing follow from the given geometry, as does the moment arm for the soil on the heel of the footing. The lateral earth pressure is a triangular distribution, and thus the moment arm from the base of the footing is a distance of one-third of the wall height.

Example 4.20: Bearing, Sliding, and Overturning for a Cantilever Retaining Wall

Check that the following retaining wall is adequate for bearing, sliding, and overturning.

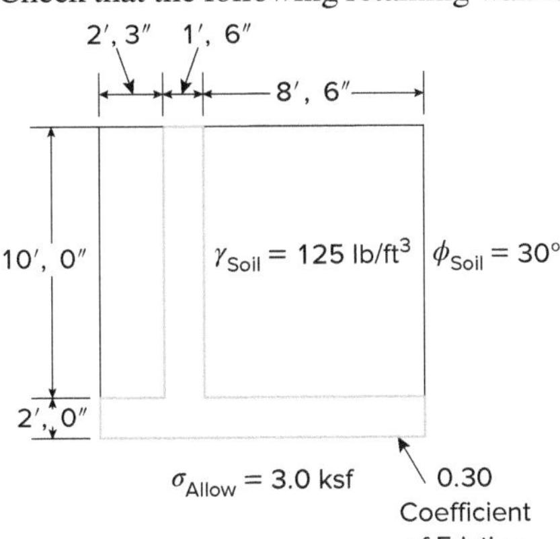

Bearing calculation:

Concrete area (2 ft, 3 in + 1 ft, 6 in + 8 ft, 6 in)(2 ft) + (1 ft, 6 in)(10 ft) = 39.5 ft^2

Soil area = (10 ft)(8.5 ft) = 85 ft^2

Concrete weight = (39.5 ft^2)(1 ft)(0.150 ksf) = 5.93 kips

Soil weight = (85 ft^2)(1 ft)(0.125 ksf) = 10.63 kips

$$\sigma_{\text{applied}} = \frac{(5.93 \text{ kips} + 10.63 \text{ kips})}{(12.25 \text{ ft})(1 \text{ ft})} = 1.35 \text{ ksf}$$

1.35 ksf < 3.0 ksf

Allowable (OK)

Sliding calculation: Assume an active K value.

Total weight = 5.93 kips + 10.63 kips = 16.56 kips

Resistance = (16.56 kips)(0.30) = 4.97 kips

$$\text{Driving force} = K_a = \frac{1 - \sin 30}{1 + \sin 30} = 0.333$$

$$\text{Force} = \frac{1}{2}\gamma_{\text{Soil}}\, k_a \text{h}^2 = \frac{1}{2}(0.125 \text{ kcf})(12 \text{ ft})^2(0.333) = 3 \text{ kips}$$

4.97 kips > 3 kips

Overturning calculation:

Assume rotation about the lower left joint.

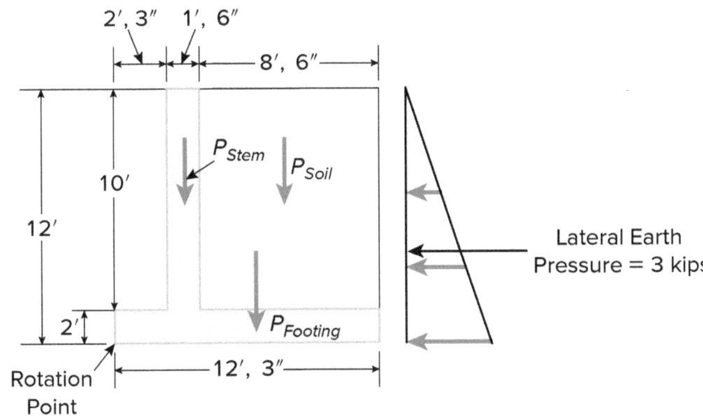

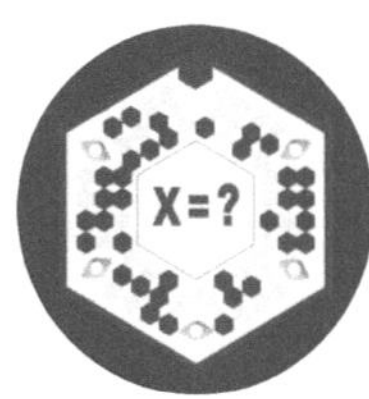

Example 4.20 (continued)

$P_{soil} = 10.63$ kips

$$\text{Moment Arm } (2 \text{ ft}, 3 \text{ in} + 1 \text{ ft}, 6 \text{ in}) + \frac{8 \text{ ft}, 6 \text{ in}}{2} = 8 \text{ ft}$$

$P_{stem} = (1 \text{ ft}, 6 \text{ in})(10 \text{ ft})(0.150 \text{ kcf}) = 2.25$ kips

$$\text{Moment Arm } (2 \text{ ft}, 3 \text{ in}) + \frac{1 \text{ ft}, 6 \text{ in}}{2} = 3 \text{ ft}$$

$P_{footing} = (2 \text{ ft})(12 \text{ ft}, 3 \text{ in})(0.150 \text{ kcf}) = 3.67$ kips

$$\text{Moment Arm} = \frac{12.25 \text{ ft}}{2} = 6.125 \text{ ft}$$

$$\text{Driving Moment} = (3 \text{ kips})(12 \text{ ft})\left(\frac{1}{3}\right) = 12 \text{ kips/ft}$$

Resistance Moment (10.63 kips)(8 ft) + (2.25 kips)(3 ft) + (3.67 kips)(6.125 ft)
= 114.27 kips/ft

114.27 kips/ft >> 12 kips/ft

Footing is acceptable for overturning.

4.11.3 Location of Steel for Tension Forces

The location of flexural steel for a retaining wall will be based on what face of the wall stem and footing will be placed in tension, as shown in Figure 4.47.

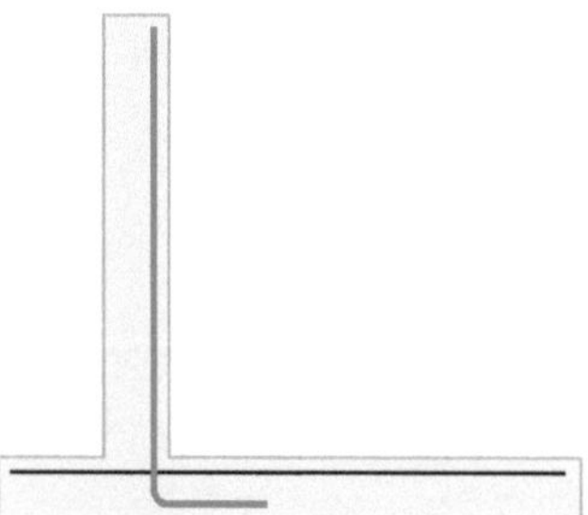

FIGURE 4.47 Retaining Wall Location of Tension Steel

REFERENCES

1. American Society of Civil Engineers (ASCE). 2010. ASCE 7-10: *Minimum Design Loads and Associated Criteria for Buildings and Other Structures*. Reston, VA: ASCE.
2. International Code Council (ICC). 2018. *International Building Code*. Washington, DC: ICC. https://codes.iccsafe.org/content/IBC2018?site_type=public.
3. American Concrete Institute (ACI). 2014. ACI 318-14: *Building Code Requirements for Structural Concrete and Commentary*. Farmington Hills, MI: ACI.
4. American Institute of Steel Construction (AISC). 2011. *Steel Construction Manual (14th ed.)*. Chicago, IL: AISC.
5. American Association of State Highway and Transportation Officials (AASHTO). 2017. *AASHTO LRFD Bridge Design Specifications (8th ed.)*. Washington, DC: AASHTO.

Hydraulics and Hydrology

5

Jeffrey S. MacKay, PE

CONTENTS

CONTENTS (*continued*)

CONTENTS (*continued*)

EXAM GUIDE

V. Hydraulics and Hydrology

A. Open-channel flow
B. Stormwater collection and drainage (e.g., culvert, stormwater inlets, gutter flow, street flow, storm-sewer pipes)
C. Storm characteristics (e.g., storm frequency, rainfall measurement, and distribution)
D. Runoff analysis (e.g., rational and SCS/NRCS methods, hydrographic application, runoff time of concentration)
E. Detention/retention ponds
F. Pressure conduit (e.g., single pipe, force mains, Hazen-Williams, Darcy-Weisbach, major and minor losses)
G. Energy and/or continuity equation (e.g., Bernoulli)

Approximate Number of Questions on Exam: 7

NCEES Principles & Practice of Engineering Examination, Civil Breadth Exam Specifications

COMMONLY USED ABBREVIATIONS

A	area
b	channel width
B	weir or channel width
C	rational coefficient
CN	curve number
D	pipe diameter
d	flow depth
E	specific energy
f	Darcy friction factor
Fr	Froude number
g	gravitational acceleration
H	total energy head
HW	headwater
h	energy head
I	rainfall intensity
I_a	initial abstraction
K	conveyance
K_L	minor loss coefficient
L	length
N	number of end contractions
n	Manning's roughness coefficient
p	pressure
S	storage capacity
t_c	time of concentration
R	hydraulic radius
R_e	Reynolds number
Q	flow rate
S	slope of energy grade line
SG	specific gravity
t	time
T	spread or width at surface
TW	tailwater
v	velocity

V	volume
Y	weir height
z	elevation

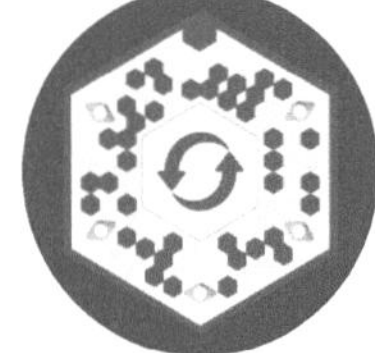

COMMONLY USED SYMBOLS

θ	angle
γ	specific weight
ρ	density
μ	absolute viscosity
ν	kinematic viscosity
ϵ	specific roughness

INTRODUCTION

This chapter contains the relevant PE Civil Breadth exam material on hydraulics and hydrology aspects of water resources engineering.

5.1 OPEN-CHANNEL FLOW

In water resources, open-channel flow refers to the conveyance of water with a free surface; that is, the water is at the same pressure as the atmosphere around it. Another way of describing it is flow that is not entirely confined within rigid boundaries, meaning that a part of the flow is in contact with nothing but air. Water is normally confined in some type of channel, such as a stream, pipe, flume, swale, or gutter. Gravity is the driving force in open-channel flow (Fig. 5.1).

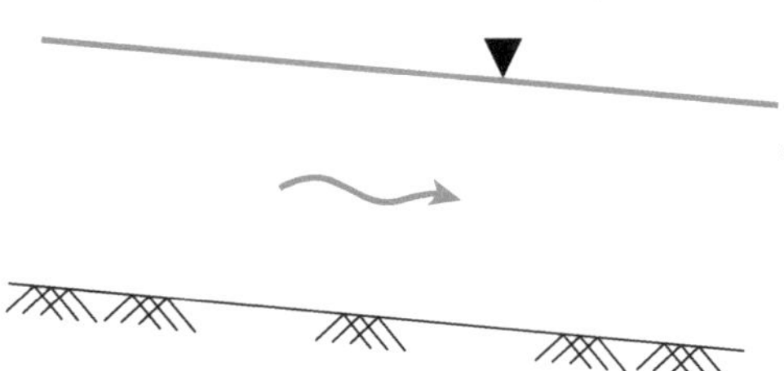

FIGURE 5.1 Open-Channel Flow

A channel in which the cross-sectional shape, size, and bottom slope are constant is a prismatic channel. Most of the man-made, or artificial, channels that are used in civil engineering applications are prismatic channels. The rectangle, trapezoid, triangle, and circle are commonly used man-made channel shapes. For example:

- Rectangular: concrete spillways, wooden flumes
- Trapezoidal: lined and unlined earthen channels
- Triangular: roadside swales
- Circular: storm and sanitary pipes

Natural channels generally have varying cross sections and are nonprismatic.

5.1.1 Basic Flow Parameters

The flow depth (d) in a channel is generally measured from the channel bottom to the free water surface. Several hydraulic parameters affect flow depth d and velocity v in open-channel flow. Since gravity is the driving force, much like a ball rolling down a hill, the

> **TIP**
>
> Basically, more contact with the channel surface means more resistance.

steepness of the channel, or slope (S), is a significant factor in water movement. Friction resists motion; therefore, the roughness or smoothness of the channel also affects flow velocity. Finally, related to friction is the hydraulic radius (R), which is the ratio of the flow area (A) to the wetted perimeter (P).

5.1.1.1 Hydraulic Radius

R(ft, m) = A, area (ft^2, m^2)/P, wetted perimeter (ft, m) ($R = A/P$) **Equation 5-1**

$R = D/4$ when a circular pipe is full or half-full

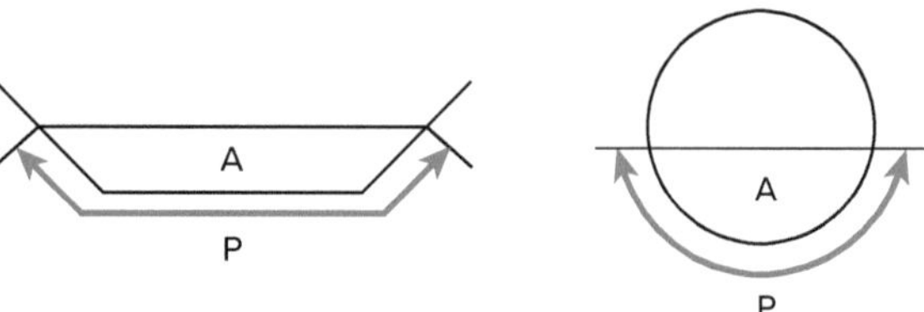

FIGURE 5.2 Hydraulic Radius Parameters

Figure 5.3 contains formulas for area, wetted perimeter, and hydraulic radius for basic channel sections.

Shape of Cross Section	b, d	T, θ, d	T, θ, d, b	θ, d, D
Area (A)	db	$\frac{d^2}{\tan\theta}$	$\left(b + \frac{d}{\tan\theta}\right)d$	$\frac{1}{8}(\theta - \sin\theta)D^2$ θ in Radians
Wetted Perimeter (P)	$2d + b$	$\frac{2d}{\sin\theta}$	$b + 2\left(\frac{d}{\sin\theta}\right)$	$\frac{1}{2}\theta D$ θ in Radians
Hydraulic Radius (R)	$\frac{db}{b + 2d}$	$\frac{d\cos\theta}{2}$	$\frac{bd\sin\theta + d^2\cos\theta}{b\sin\theta + 2d}$	$\frac{1}{4}\left(1 - \frac{\sin\theta}{\theta}\right)D$ θ in Radians

FIGURE 5.3 Hydraulic Parameters of Common Cross Sections

5.1.2 Types of Flow

Time as the criterion:

- **Steady flow:** depth of flow does not change over time
- **Unsteady flow:** depth of flow changes with time

Space as the criterion:

- **Uniform flow:** depth of flow is the same at every section of the channel
- **Varied flow:** depth of flow changes along the length of the channel; can be further classified as either rapidly or gradually varied (Fig. 5.4)

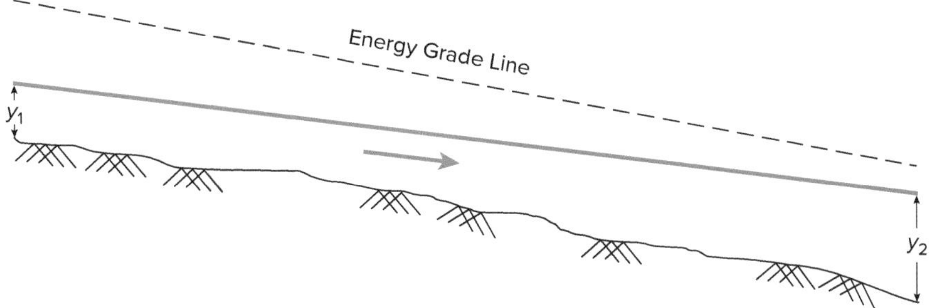

FIGURE 5.4 Non-Uniform Flow

5.1.3 Critical Depth

Critical depth, d_c, is the normal flow depth in a channel at critical flow. Critical flow is an unstable and turbulent condition where a slight change in energy can cause an abrupt rise or fall in flow depth. Flow will pass through critical depth on its way to a normal depth either above (subcritical) or below (supercritical) critical depth. Critical depth often occurs at significant changes in channel slope (for example, steep to mild), transitions from unrestricted flow to restricted flow (for instance, culverts), downstream of sluice gates, and at weirs and spillways (Fig. 5.5). For a given channel, the critical depth corresponds to the minimum energy needed to convey a given flow rate, Q. Critical depth is independent of channel slope.

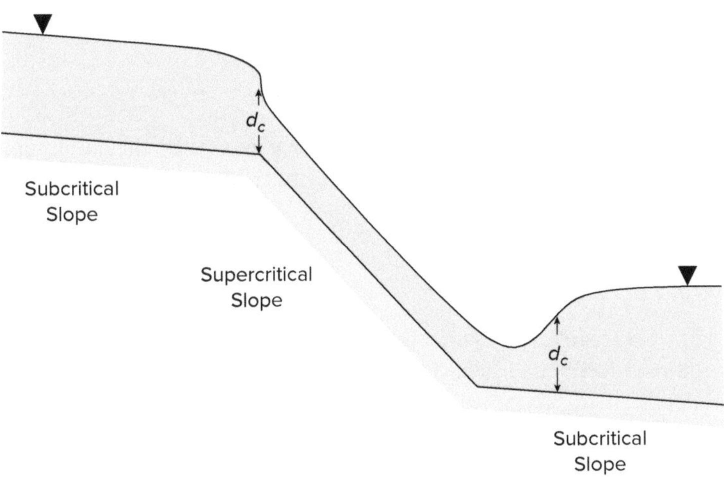

FIGURE 5.5 Examples of Occurrences of Critical Depth

Critical depth can be calculated for various channel shapes using Equations 5-2 through 5-5.

Rectangular channel:

$$d_c = \left[\frac{C_1 Q}{b}\right]^{\frac{2}{3}}$$ **Equation 5-2**

$C_1 = 0.176$ (USCS), 0.319 (SI)

Triangular channel:

$$d_c = C_2\left[\frac{Q}{z_1 + z_2}\right]^{\frac{2}{5}}$$ **Equation 5-3**

$C_2 = 0.757$ (USCS), 0.96 (SI)

Other shapes (including trapezoidal and circular):

$$Q = \left[\frac{gA^3}{T}\right]^{\frac{1}{2}}$$ **Equation 5-4**

This requires a trial-and-error solution.

Circular channels (approximate):

$$d_c = C_3 \frac{Q^{\frac{1}{2}}}{D^{\frac{1}{4}}}$$ **Equation 5-5**

$C_3 = 0.42$ (USCS), 0.562 (SI)

5.1.4 Froude Number

The Froude number is a dimensionless number that is used to determine flow regime (Fig. 5.6).

$$Fr = \frac{v}{\sqrt{gD_h}}$$ **Equation 5-6**

v = velocity (ft/s)

g = gravitational constant (32.2 ft/s^2)

D_h = hydraulic depth (ft) = area (ft^2)/top width (ft)

Critical flow ($Fr = 1$): the transition or control flow that possesses the minimum possible energy for that flow rate. Flow depth occurs at critical depth.

Supercritical flow ($Fr > 1$): fast, rapid, and typically shallow. Supercritical flow is dominated by inertial forces and behaves as rapid or unstable flow. Flow depth is less than critical depth.

Subcritical flow ($Fr < 1$): slow, tranquil, and typically deep. Subcritical flow is dominated by gravitational forces and behaves in a slow or stable way. Flow depth is greater than critical depth.

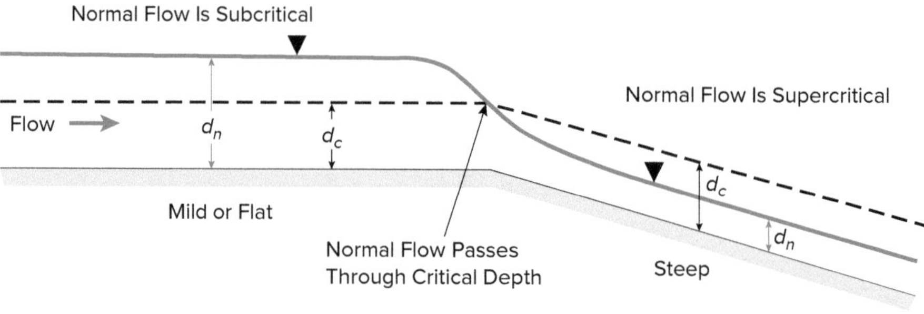

FIGURE 5.6 Flow Regimes

5.1.5 Hydraulic Jump

A hydraulic jump occurs when there is a rapid increase in depth creating an abrupt rise of the water surface. Supercritical flow transitions to subcritical flow through a hydraulic jump, representing a high energy loss with erosive potential. They can occur downstream of sluice gates, at the base of spillways, or where a steep channel slope suddenly becomes much flatter (Fig. 5.7).

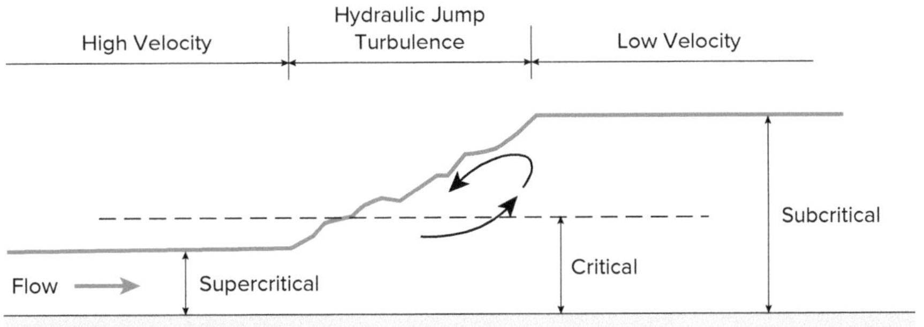

FIGURE 5.7 Hydraulic Jump

Labeled as d_1 and d_2 in Figure 5.8, **conjugate depths** occur on either side of the hydraulic jump. The rise in flow depth from immediately upstream to immediately downstream of the jump is referred to as the height of the hydraulic jump: $h = d_2 - d_1$.

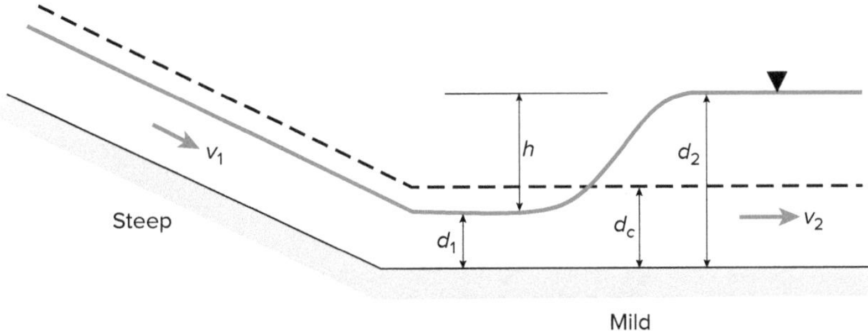

FIGURE 5.8 Hydraulic Jump Parameters

For rectangular channels, conjugate depths may be used in the equations below to solve for conditions upstream or downstream of the jump.

$$d_1 = -\frac{1}{2}d_2 + \left[\frac{2v_2^2 d_2}{g} + \frac{d_2^2}{4}\right]^{\frac{1}{2}}$$ **Equation 5-7**

$$v_1^2 = \left(\frac{gd_2}{2d_1}\right)(d_1 + d_2)$$ **Equation 5-8**

d_1, d_2 = upstream and downstream depth, respectively (ft)

v_1, v_2 = upstream and downstream velocity, respectively (ft/s)

5.1.6 Manning's Equation

Manning's equation says that flow velocity is proportional to the hydraulic radius and slope of the energy grade line (EGL) and is inversely proportional to channel roughness (Fig. 5.9). Channel roughness in Manning's equation is expressed as n, which is called Manning's coefficient. Manning's coefficient represents the friction applied to the flow by the channel. Tables of n values for various types of channel lining can be found in many references. Examples of n values for common storm-sewer materials and earthen ditch linings are provided in Table 5.1.

TIP

If needed to solve a problem on the PE Civil exam, Manning's n values would likely be given.

$$v = \frac{1.49}{n}R^{2/3}S^{1/2}$$ **Equation 5-9**

Recall that flow rate is computed by multiplying flow velocity by flow area, or $Q = vA$. When A is inserted into the right side of the equation, Manning's equation then becomes a solution for Q.

$$Q = \frac{1.49}{n}AR^{2/3}S^{1/2}$$ **Equation 5-10**

v = velocity (ft/s)

Q = flow rate (ft^3/s)

n = Manning's coefficient

A = flow area (ft^2)

R = hydraulic radius (ft)

S = slope of EGL (ft/ft) = channel slope, S_0, for uniform flow

$S_0 = \Delta y/\Delta x$

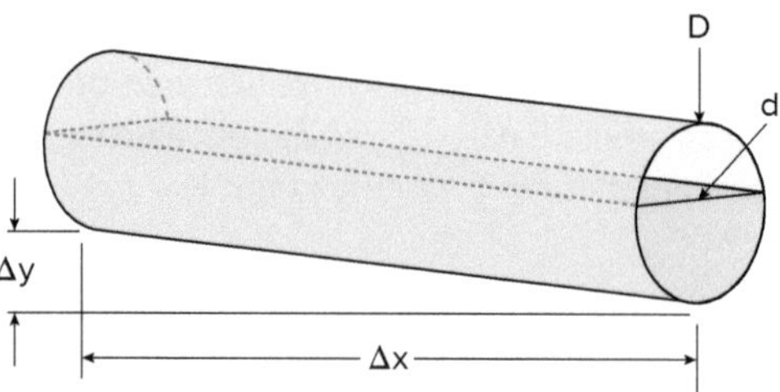

FIGURE 5.9 Manning's Equation Parameters

TABLE 5.1 Typical Range of Manning's Coefficients

CHANNEL TYPE/MATERIAL	MANNING'S n
Concrete Pipe	0.010 to 0.015
Corrugated Metal Pipe	0.011 to 0.037
Plastic Pipe, Smooth	0.009 to 0.015
Plastic Pipe, Corrugated	0.018 to 0.025
Pavement/Gutter Section	0.012 to 0.016
Concrete Channel	0.011 to 0.015
Riprap Channel	0.020 to 0.035
Vegetated Channel	0.020 to 0.035

Example 5.1: Manning's Equation

Flow is constant and continuous through a 1,000-ft-long, 3-ft-wide rectangular flume. The flume is wooden and has a Manning's n of 0.020. A 12-in flow depth is measured in an upstream segment of the channel, which has a 1% longitudinal slope. If at the same time the flow depth in a downstream channel segment is measured as 9.4 in, then what is the approximate velocity in the downstream segment in feet per second?

Solution

The conditions in the upstream channel segment are used to calculate the flow rate. The flow rate is in turn used to calculate the velocity in the downstream channel segment.

$$Q = \frac{1.49}{n} A_1 R_1^{2/3} S_1^{1/2}$$

Flow area and hydraulic radius for the upstream channel segment, where d is flow depth and b is channel bottom width:

$$A_1 = d_1 b = \frac{(12 \text{ in})}{(12 \text{ in/ft})}(3 \text{ ft}) = 3 \text{ ft}^2$$

$$P_1 = 2d_1 + b = 2\frac{(12 \text{ in})}{(12 \text{ in/ft})} + 3 \text{ ft} = 5 \text{ ft}$$

Example 5.1 *(continued)*

$$R_1 = A_1/P_1 = \frac{(3\ \text{ft}^2)}{(5\ \text{ft})} = 0.6\ \text{ft}$$

$$Q = \frac{1.49}{0.02}(3\ \text{ft}^2)(0.6\ \text{ft})^{2/3}(0.01)^{1/2} = 15.9\ \text{ft}^3/\text{s}$$

The velocity in the downstream channel segment is found by dividing the flow rate, which is constant, by the flow area.

$$v_2 = Q/A_2$$

$$A_2 = d_2 b = (9.4\ \text{in})(3.0\ \text{ft})/(12\ \text{in/ft}) = 2.4\ \text{ft}^2$$

$$v_2 = (15.9\ \text{ft}^3/\text{s})/(2.4\ \text{ft}^2) = 6.8\ \text{ft/s}$$

5.1.7 Friction Loss

The total friction loss (or energy used) along a channel is $h_f = LS$.

$$h_f = \frac{Ln^2v^2}{2.208R^{4/3}}\ (\text{USCS})$$ **Equation 5-11**

h_f = friction head loss (ft)

L = channel length (ft)

5.1.8 Most Efficient Cross Sections

A channel section is considered most efficient when it conveys a maximum discharge for a given cross-section area (A), flow resistance (Manning's n), and bottom slope (S_0). The most efficient cross section has a minimum wetted perimeter (P) for a given cross-section area (A), thus the hydraulic radius (R) is maximized. The optimum bottom width (b) to flow depth (d) ratio for a trapezoidal channel with side slopes (z) and a rectangular channel ($z = 0$) is:

$$\frac{b}{d} = 2\left[\left(z^2 + 1\right)^{\frac{1}{2}} - z\right]$$ **Equation 5-12**

z = channel side slope, horizontal to vertical (=1) ratio

A circle has the least wetted perimeter (P) for a given cross-section area (A) of any geometric shape. The most efficient circular cross section is a semicircle, which is the most efficient cross section of all. All cross sections that satisfy Equation 5-12 are such that a semicircle can be inscribed in them, as shown in Figure 5.10. The most efficient trapezoidal cross section is half of a hexagon. The most efficient rectangular cross section is half of a square.

Semicircular and circular shapes are practical for pipes, but earthen channels must be trapezoidal. The trapezoidal section determined from Equation 5-12 will be the most economical section to build as far as excavation and channel lining are concerned. Triangular (or v-shaped) channels are not very efficient, in general.

TABLE 5.2 Optimum Channel Dimensions

CROSS SECTION	WIDTH, *b*	AREA, *A*	WETTED PERIMETER, *P*
Semicircle	$2d$	$(\pi/2)d^2$	πd
Trapezoidal	$2d/\sqrt{3}$	$\sqrt{3}d^2$	$\sqrt{3}(2d)$
Rectangular	$2d$	$2d^2$	$4d$

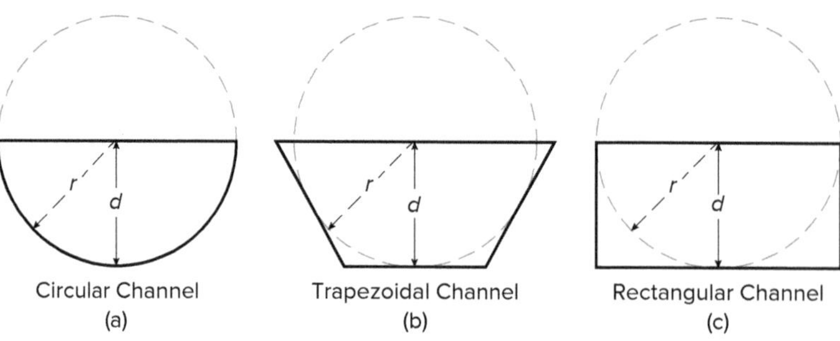

FIGURE 5.10 Most Efficient Cross Sections

Example 5.2: Most Efficient Cross Sections

A circular metal storm pipe under a road is deformed due to excessive vehicle loads. The deformed shape resembles an ellipse. Has the pipe's capacity increased, decreased, or remained constant? Can you tell?

Solution

The deformed pipe has less capacity than the original circular pipe. A circle is the most efficient cross section. A deformed pipe maintains the same wetted perimeter, but the flow area is reduced. A smaller hydraulic radius results in less flow capacity.

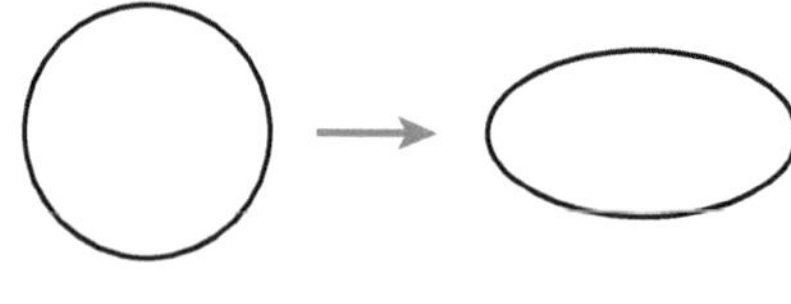

5.1.9 Flow Measurement

Weirs

Weirs are useful hydraulic tools that allow engineers to control water height and velocity and can be used to calculate discharge. The discharge rate is a function of the head over the weir and the width of the weir crest. Weirs are also flow controls in that they force flow to pass through critical depth—from subcritical flow upstream to supercritical flow downstream. Figure 5.11 illustrates different weir types in profile view.

> **TIP**
>
> The main types of weirs used by engineers are sharp-crested and broad-crested weirs.

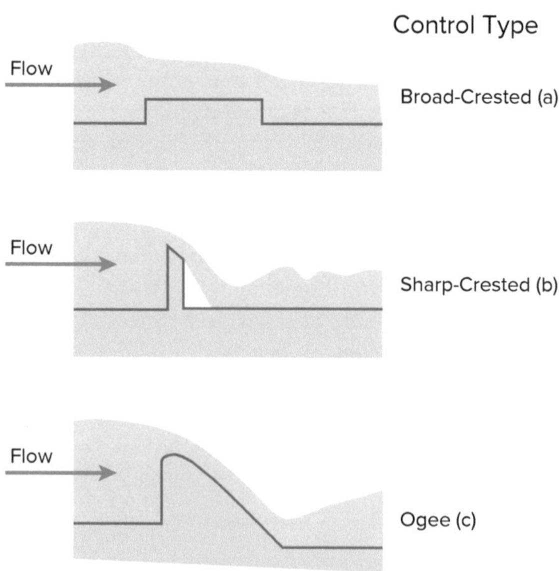

FIGURE 5.11 Flow Controls—Weirs

5.1.10 Sharp-Crested Weir

Sharp-crested weirs are used for measuring smaller flows (small rivers and canals). As depicted in Figure 5.12, the detached water surface falling away from the downstream edge of the weir is known as a free-falling nappe.

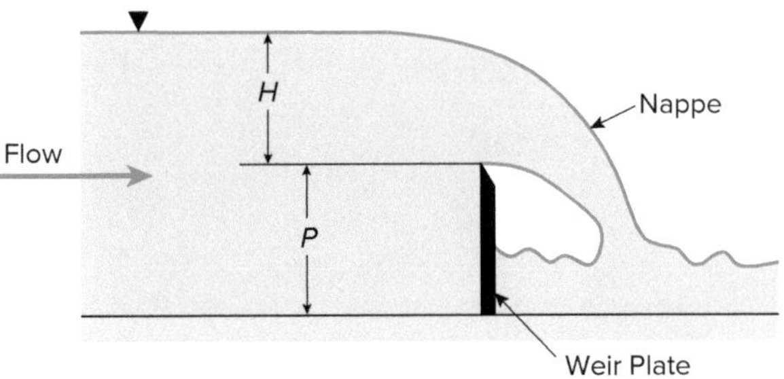

FIGURE 5.12 Sharp-Crested Weir

5.1.11 Contracted (Rectangular) and Suppressed Weirs

$$Q = \frac{2}{3} C_D B\left(H^{\frac{3}{2}}\right)\sqrt{2g}$$ **Equation 5-13**

Q = volumetric flow rate over the weir (ft^3/s)

g = acceleration of gravity (32.2 ft/s^2)

C_D = discharge coefficient, usually ranging from 0.60 to 0.62

$$C_D = 0.602 + 0.083(H/P)$$ **Equation 5-14**

H = head over the weir, from the weir crest to the upstream water surface (ft)

P = height of the weir plate (ft)

B = width of the contracted notch (rectangular), or the width of the channel (suppressed) (ft)

In the case of a contracted weir, the effective width of the weir crest, $B_{\text{effective}}$, should be calculated to account for the flow contraction transition from just upstream of the weir to over the weir crest (Fig. 5.13).

$$B_{\text{effective}} = B_{\text{actual}} - 0.1NH$$ **Equation 5-15**

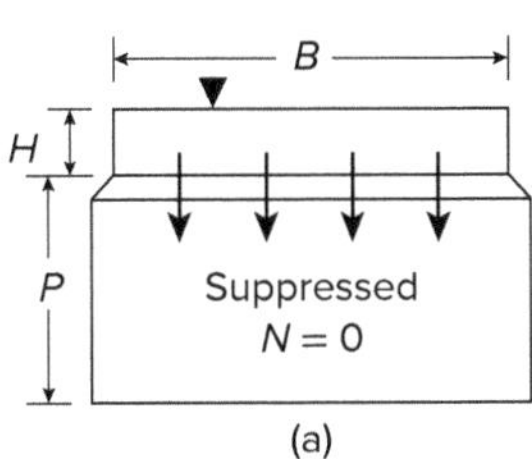

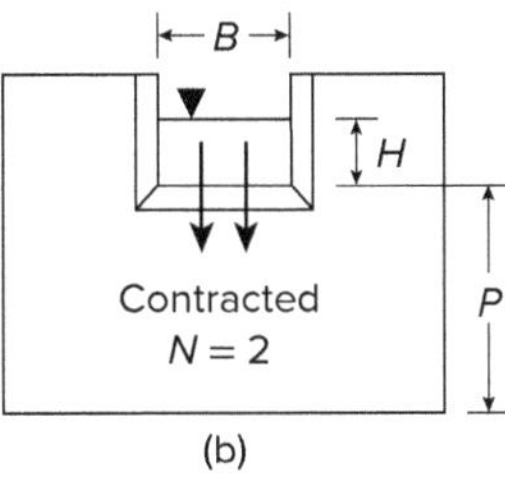

FIGURE 5.13 Suppressed and Contracted Weirs

Example 5.3: Contracted and Suppressed Weirs

A sharp-crested rectangular weir is located in a 10-ft-wide rectangular concrete channel. The crest of the weir, which is contracted 5 ft on one side, is 6 ft above the bottom of the channel. What is the approximate flow rate over the weir, in cubic feet per second, if the head is measured as 1.4 ft? Use Equation 5-14 to find the discharge coefficient.

Solution

$$Q = \frac{2}{3} C_D B\left(H^{\frac{3}{2}}\right)\sqrt{2g}$$

Example 5.3 *(continued)*

The width of the contracted notch (B) is 10 ft − 5 ft = 5 ft.

Since only one side is contracted, $N = 1$.

$$B_{\text{effective}} = B_{\text{actual}} - 0.1NH = 5\text{ ft} - 0.1(1)(1.4\text{ ft}) = 4.9\text{ ft}$$

$$C_D = 0.602 + 0.083(H/P) = 0.602 + 0.083(1.4\text{ ft}/6\text{ ft}) = 0.621$$

$$Q = \frac{2}{3}(0.621)(4.9\text{ ft})\left(1.4\text{ ft}^{\frac{3}{2}}\right)\sqrt{2(32.2)} = 27.0\text{ ft}^3/\text{s}$$

5.1.12 V-Notch (Triangular) Weirs—most accurate for small discharges

$$Q = \frac{8}{15}C_e\left(\tan\frac{\theta}{2}\right)\left(h_e^{\frac{5}{2}}\right)\sqrt{2g} \qquad \textbf{Equation 5-16}$$

Q = volumetric flow rate over the weir (ft^3/s)

g = acceleration of gravity (32.2 ft/s^2)

C_e = discharge coefficient, typically 0.58 to 0.61; can be found using Figure 5.14(a)

$h_e = h_u + k_h$

h_u = head flowing through the notch (ft)

k_h (ft) can be found using Figure 5.14(b)

θ = notch angle, degrees

When $\theta = 90°$, this equation can be simplified to:

$$Q = 2.49h_e^{2.48} \qquad \textbf{Equation 5-17}$$

for 0.2 ft < h_e < 1.25 ft

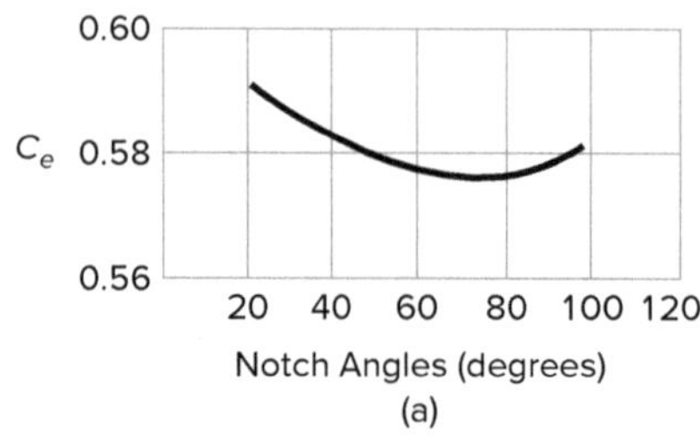

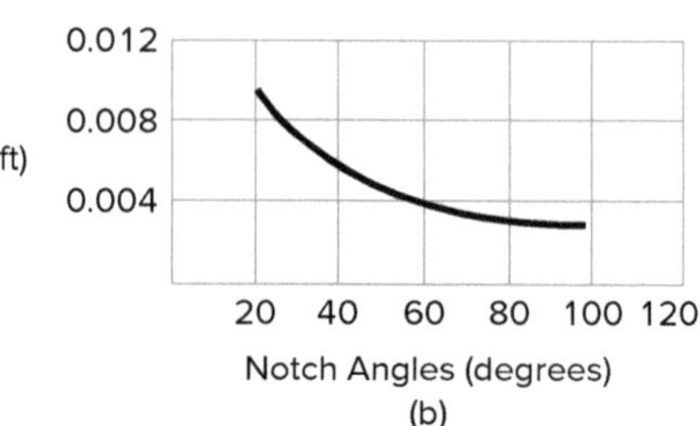

FIGURE 5.14 Coefficients for V-Notch Weirs

A diagram of a V-notch weir is shown in Figure 5.15.

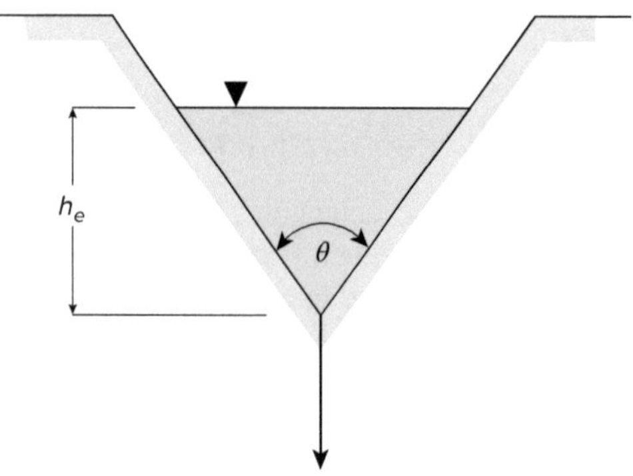

FIGURE 5.15 V-Notch (Triangular) Weir

5.1.13 Cippoletti (Trapezoidal) Weirs—used when discharge is too great for a rectangular weir (Fig. 5.16)

$$Q = 3.36LH^{3/2}$$ **Equation 5-18**

Q = volumetric flow rate over the weir (ft^3/s)

L = bottom length of the weir crest (ft)

H = head over the weir, from the weir crest to the upstream water surface (ft)

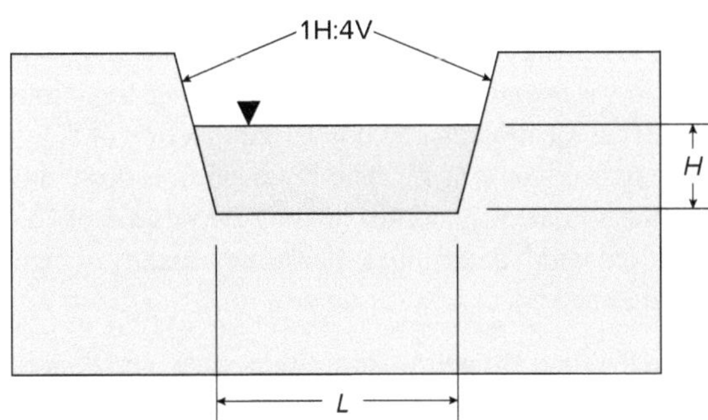

FIGURE 5.16 Cippoletti Weir

5.1.14 Broad-Crested Weirs

Broad-crested weirs can be used for measuring flows in medium-to-large rivers and canals. A weir may be considered broad crested if the weir thickness (direction of flow) is greater than half of the hydraulic head. An overflow spillway is used to provide a stable, controlled release of flow from dams. Spillways may have a shape known as an ogee, which closely approximates the underside of a nappe from a sharp-crested weir (Fig. 5.17). The ogee shape prevents the water surface from breaking contact with the spillway.

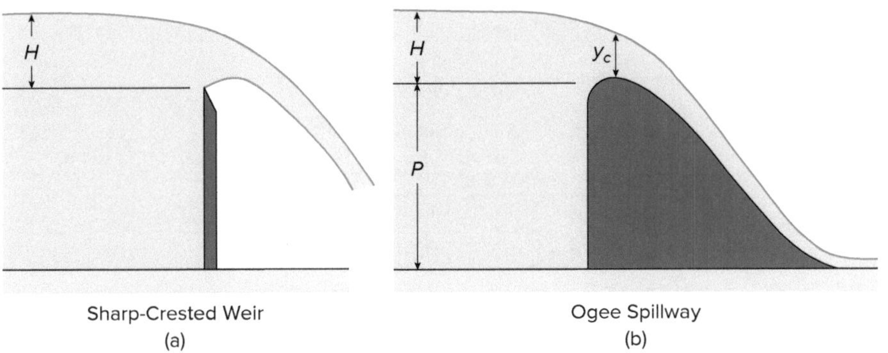

FIGURE 5.17 Ogee Spillway

The general broad-crested weir equation is:

$$Q = \frac{2}{3}C_1L\left(H^{3/2}\right)\sqrt{2g}$$ **Equation 5-19**

H = upstream hydraulic head (ft)

C_1 = weir coefficient, dimensionless (0.60 to 0.75 for ogee spillways; 0.50 to 0.57 for broad-crested weirs)

H is assumed to be the depth of water above the weir crest, measured upstream of the weir crest.

The Horton equation can also be used for broad-crested weirs and spillways:

$$Q = C_sLH^{3/2}$$ **Equation 5-20**

C_s = spillway coefficient
ogee spillways = 3.30 to 3.98
broad-crested weirs = 2.63 to 3.33

If the velocity of the approach is significant, the velocity head is added to H.

$$Q = C_s L\left(H + \frac{v^2}{2g}\right)^{3/2}$$ **Equation 5-21**

5.1.15 Parshall Flume

The Parshall flume is a fixed hydraulic structure used to measure flow rate in industrial discharges and influent/effluent flows in wastewater treatment plants. Under free-flow conditions, the depth of water at a specified location can be converted to a rate of flow. The single, primary point of measurement is denoted as H_a. The H_a location is upstream of the throat. From the beginning of the throat, it is approximately two-thirds of the length of the converging section. The throat width determines the flume capacity, which ranges from 0.005 ft^3/s (1 in) to 3,280 ft^3/s (50 ft).

Water enters the converging section of a Parshall flume in a subcritical state, and then is accelerated through the throat as supercritical flow. The diverging section forces it back to subcritical flow (Fig. 5.18).

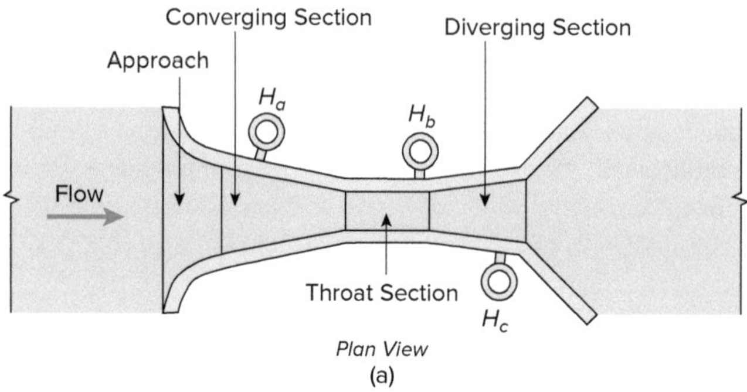

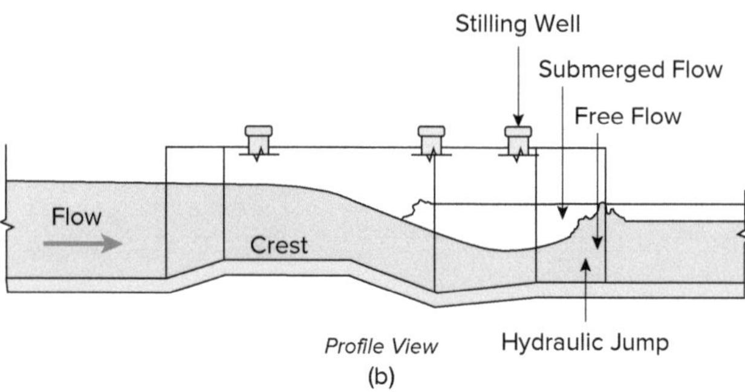

FIGURE 5.18 Parshall Flume

Two flow conditions can occur in a Parshall flume: **free flow** and **submerged flow**. Only one head measurement (H_a) is needed for free-flow conditions, whereas a secondary head measurement (H_b) is required to determine submerged conditions. When a free-flow condition exists, a hydraulic jump occurs downstream of the flume. When a submerged condition exists, flow out of the flume is more restricted and the hydraulic jump disappears. Calculations for free-flow conditions are as follows:

$$Q = CH_a^n$$ **Equation 5-22**

Q = flow rate (ft^3/s)

C is the free-flow coefficient for the flume; see Table 5.3.

H_a = head at the primary point of measurement (ft)

n varies with flume size; see Table 5.3.

TABLE 5.3 Parshall Flume Coefficients

THROAT WIDTH	COEFFICIENT (*C*)	EXPONENT (*n*)	THROAT WIDTH	COEFFICIENT (*C*)	EXPONENT (*n*)
1 in	0.338	1.55	6 ft	24.00	1.59
2 in	0.676	1.55	7 ft	28.00	1.60
3 in	0.992	1.55	8 ft	32.00	1.61
6 in	2.06	1.58	10 ft	39.38	1.60
9 in	3.07	1.53	12 ft	46.75	1.60
1 ft	3.95	1.55	15 ft	57.81	1.60
1.5 ft	6.00	1.54	20 ft	76.25	1.60
2 ft	8.00	1.55	25 ft	94.69	1.60
3 ft	12.00	1.57	30 ft	113.13	1.60
4 ft	16.00	1.58	40 ft	150.00	1.60
5 ft	20.00	1.59	50 ft	186.88	1.60

Source: *Water Measurement Manual (3rd ed.)*, US Department of the Interior/Bureau of Reclamation, 2001.

Example 5.4: Parshall Flume

The effluent flow from a wastewater treatment plant is monitored with the aid of a 1.5-ft-wide Parshall flume. Depths in the converging, throat, and diverging sections of the flume were observed to be 3.2 ft, 2.5 ft, and 1.5 ft, respectively. What is the approximate effluent flow rate, in ft^3/s?

Solution

$$Q = CH_a^n$$

Table 5.3 can be used to find C and n. The depth measurement in the converging section should be used.

$$Q = (6.0)(3.2\text{ ft})^{1.54} = 36.0\text{ ft}^3/\text{s}$$

5.2 STORMWATER COLLECTION AND DRAINAGE

5.2.1 Sizing Circular Pipes

Since the hydraulic radius and flow area of a circular pipe flowing full are both functions of only pipe diameter (that is, flow depth equals pipe diameter), Manning's equation can be rewritten in terms of D. For example, to calculate the minimum pipe diameter to convey a given flow rate at full flow, one can use:

$$D = 1.335\left(\frac{nQ}{\sqrt{S}}\right)^{3/8}$$ **Equation 5-23**

To calculate the flow rate or velocity of a pipe flowing full, use:

$$Q = \frac{0.463D^{\frac{8}{3}}\sqrt{S}}{n}$$ **Equation 5-24**

$$v = \frac{0.591D^{\frac{2}{3}}\sqrt{S}}{n}$$ **Equation 5-25**

In all of the above equations, the variables and units are defined as follows:

D = pipe diameter (ft)

Q = volumetric flow rate (ft^3/s)

v = flow velocity (ft/s)

S = pipe slope (ft/ft)

n = Manning's n

Note that the flow rate of a half-full circular channel is 50% of the same channel when it is full. See Figures 5.19 and 5.20.

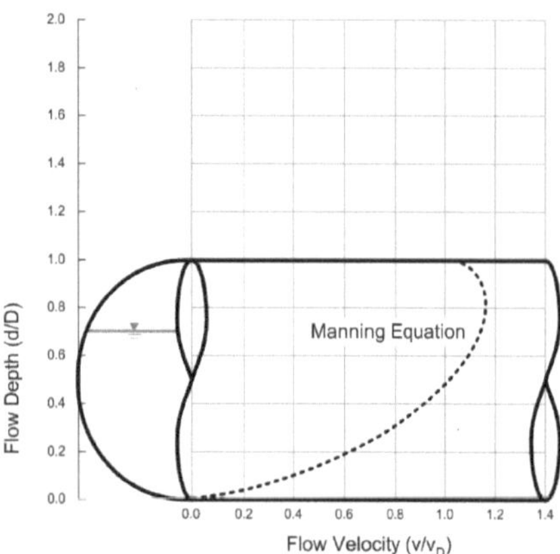

FIGURE 5.19 Hydraulic Relationship of the Manning Equation

Used by permission from ADS Environmental Services.

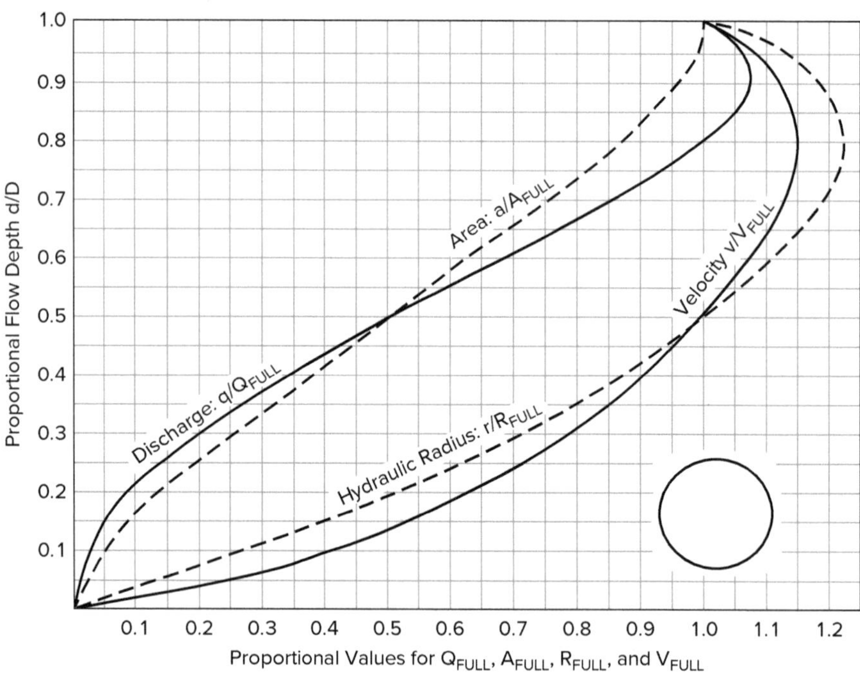

FIGURE 5.20 Hydraulic Elements Chart for Circular Channels

Example 5.5: Pipe Discharge

What is the discharge, in ft^3/s, in a 30-in inside diameter pipe, with a Manning's n of 0.024 and a flow depth of 7.5 in, with a 0.50% downward slope?

Example 5.5 *(continued)*

Solution

$d/D = (7.5 \text{ in})/(30 \text{ in}) = 0.25$

Per the hydraulic elements chart, $Q/Q_{\text{full}} = 0.14$

Therefore, $Q = 0.14Q_{\text{full}}$

$$Q_{\text{full}} = \frac{0.463D^{8/3}S^{1/2}}{n} = \frac{0.463(2.5 \text{ ft})^{8/3}(0.005)^{1/2}}{0.024} = 15.7 \text{ ft}^3\text{/s}$$

$Q = 0.14(15.7 \text{ ft}^3\text{/s}) = 2.2 \text{ ft}^3\text{/s}$

5.2.2 Gutter Flow

Gutter flow calculations are used to relate the flow rate Q in a curbed channel to the spread of water T from the curb into the roadway section (Fig. 5.21). The portion of the road directly adjacent to the curb is sometimes depressed, or at a steeper slope than the normal crown of the roadway. This portion is called the gutter, and it may be formed with concrete when the curb is installed. Calculating the flow capacity of the gutter can be done using the first equation below, where T is simply the gutter width. The minimum gutter width T needed to convey a given flow rate Q can be calculated using the second equation.

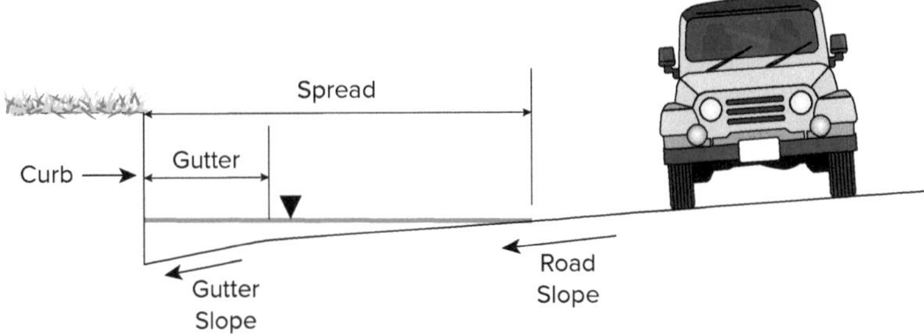

FIGURE 5.21 Gutter Flow and Spread

$$Q = \frac{0.56}{n} S_x^{1.67} S^{0.5} T^{2.67}$$ **Equation 5-26**

$$T = \frac{(1.79Qn)^{\frac{3}{8}}}{S_x^{\frac{5}{8}} S^{\frac{3}{16}}}$$ **Equation 5-27**

Q = flow rate in gutter ($\text{ft}^3\text{/s}$)

T = spread from curb (ft)

n = Manning's n

S_x = cross slope (ft/ft)

S = longitudinal (direction of flow) slope (ft/ft)

When water spreads beyond the gutter into the travel lane, the problem becomes much more difficult because of the compound slope.

Take caution with the terminology used in this type of problem. When the problem involves calculating the "travel lane encroachment" or the "spread into the travel lane," the gutter (or shoulder if defined) width needs to be subtracted from the computed

> **TIP**
>
> Problems on the PE Civil exam will most likely be limited to a single cross slope.

spread T. Figure 5.22 illustrates the difference between spread, which is always the distance water extends from the curb, and encroachment, which depends on how the roadway cross section is defined.

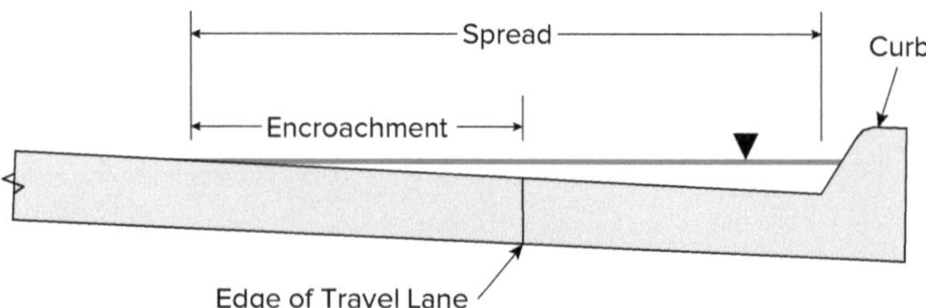

FIGURE 5.22 Spread into a Roadway Section

Example 5.6: Gutter Flow

What is the maximum flow that can be conveyed without encroaching on the travel lane? The profile grade of the road is 3.0% and $n = 0.011$.

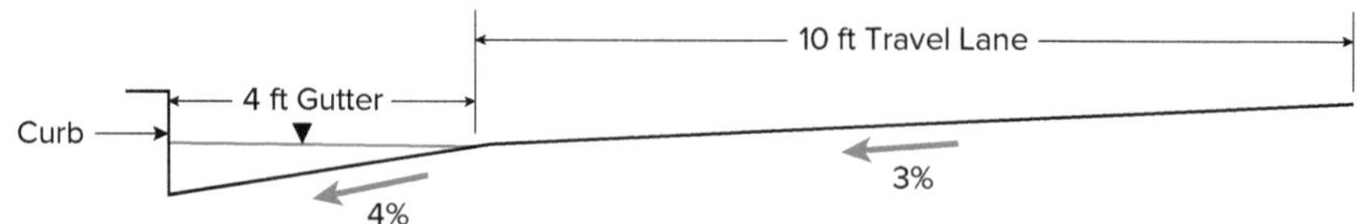

Solution

S_x = cross slope = 4% = 0.04

S = longitudinal (direction of flow) slope = 3% = 0.03

T = spread from curb = 4 ft

$$Q = \frac{0.56}{n} S_x^{1.67} S^{0.5} T^{2.67}$$

$$= \frac{0.56}{0.011} (0.04 \text{ ft/ft})^{1.67} (0.03 \text{ ft/ft})^{0.5} (4 \text{ ft})^{2.67}$$

$Q = 1.7 \text{ ft}^3\text{/s}$

5.2.3 Sizing Other Channel Shapes

Channel design parameters for rectangular, trapezoidal, and V-shaped channels can be solved quickly using Tables 7-10 and 7-11 from the *Handbook of Hydraulics* [1] (excerpted here in Tables 5.4 and 5.5). The basic section is a trapezoid, as shown in Figure 5.23. Vertical sides are a rectangular section. V-shaped channels have zero bottom width (b). The first column in the tables is the ratio of depth to bottom width, D/b. Values of D/b range from 0.01 to 2.25. The remaining columns each correspond to a different side slope, expressed as horizontal to vertical (H:V).

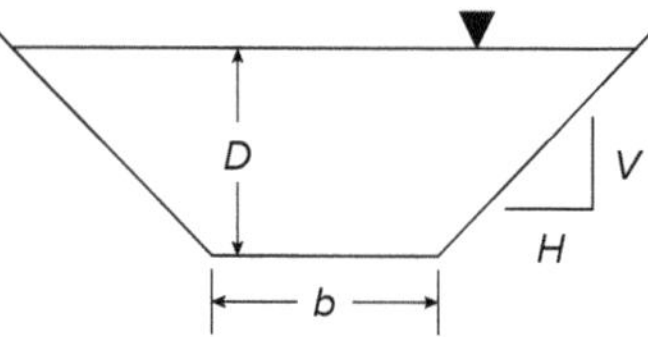

FIGURE 5.23 Basic Trapezoidal Channel Section

> **TIP**
>
> Tables 7-10 and 7-11 from the *Handbook of Hydraulics* are available in Appendix A.11: Conveyance Factors in Modified Manning's Equation.

Tables 5.4 and 5.5 are tabulations of K and K', respectively. K and K' are variables in a modified Manning's formula. The procedures for using these tables follow.

Use Table 5.4 when D is known to find Q or b.

$$Q = \frac{K}{n} D^{\frac{8}{3}} \sqrt{S} \Rightarrow K = \frac{Qn}{D^{\frac{8}{3}} \sqrt{S}} \qquad \textbf{Equation 5-28}$$

TABLE 5.4 Excerpt of Table 7-10 from the *Handbook of Hydraulics* [1]

	SIDE SLOPES OF CHANNEL, RATIO OF HORIZONTAL TO VERTICAL									
D/b	**VERTICAL**	**¼-1**	**½-1**	**¾-1**	**1-1**	**1½-1**	**2-1**	**2½-1**	**3-1**	**4-1**
.01	146.7	147.2	147.6	148.0	148.3	148.8	149.2	149.5	149.9	150.5
.02	72.4	72.9	73.4	73.7	74.0	74.5	74.9	75.3	75.6	76.3
.03	47.6	48.2	48.6	49.0	49.3	49.8	50.2	50.6	50.9	51.6
↓	↓	↓	↓	↓	↓	↓	↓	↓	↓	↓
2.25	.212	.439	.700	.973	1.24	1.77	2.28	2.77	3.26	4.22
∞	.000	.091	.274	.500	.743	1.24	1.74	2.23	2.71	3.67

Source: King, H. W. *Handbook of Hydraulics (5th ed.)*. McGraw-Hill, 1963. Used with permission.

The first five rows in Table 5.4 are shown above. Note that the last row in Table 5.4 is for V-shaped channels. When $b = 0$, the ratio D/b becomes ∞.

Use Table 5.5 when b is known to find Q or D.

$$Q = \frac{K'}{n} b^{\frac{8}{3}} \sqrt{S} \Rightarrow K' = \frac{Qn}{b^{\frac{8}{3}} \sqrt{S}}$$ **Equation 5-29**

TABLE 5.5 Excerpt of Table 7-11 from the *Handbook of Hydraulics* [1]

	SIDE SLOPES OF CHANNEL, RATIO OF HORIZONTAL TO VERTICAL									
D/b	**VERTICAL**	**¼-1**	**½-1**	**¾-1**	**1-1**	**1½-1**	**2-1**	**2½-1**	**3-1**	**4-1**
.01	.00068	.00068	.00069	.00069	.00069	.00069	.00069	.00069	.00070	.00070
.02	.00213	.00215	.00216	.00217	.00218	.00220	.00221	.00222	.00223	.00225
.03	.00414	.00419	.00423	.00426	.00428	.00433	.00436	.00439	.00443	.00449
↓	↓	↓	↓	↓	↓	↓	↓	↓	↓	↓
2.25	1.84	3.81	6.09	8.46	10.8	15.4	19.8/	24.1	28.4	36.7

Source: King, H. W. *Handbook of Hydraulics (5th ed.)*. McGraw-Hill, 1963. Used with permission.

If Q, n, D or b, and S are given in the problem:

1. Calculate K or K' using the above equations.
2. Locate the nearest K or K' in Table 5.5 under the side slope column.
3. Note the corresponding D/b value in the left-hand column.
4. Multiply D/b by b or D given in the problem.

If D and b are given in the problem:

1. Calculate D/b and select K or K' under the side slope column.
2. Plug K or K' into the equations above and solve for the unknown.

When the side slopes of the channel section are not equal, the side slopes can be averaged to use the tables. For example, 2H:1V and 4H:1V side slopes on a V-shaped channel can be averaged to use the 3H:1V side slope column in the tables.

Example 5.7: Sizing a Trapezoidal Channel

What is the normal depth, in feet, in a trapezoidal drainage ditch with a 2-ft bottom width, 2H:1V side slopes, Manning's n of 0.05, and 1% channel slope, conveying 13 ft^3/s?

Example 5.7 *(continued)*

Solution

Since b is known, use Table 5.5:

1) $K' = \dfrac{Qn}{b^{8/3}\sqrt{S}} = \dfrac{(13\ \text{ft}^3/\text{s})(0.05)}{(2\ \text{ft})^{8/3}\sqrt{.01}} = 1.02$

2) Column for 2H:1V side slopes ► nearest value is $K' = 0.990$

	SIDE SLOPES OF CHANNEL, RATIO OF HORIZONTAL TO VERTICAL									
D/b	**VERTICAL**	**¼-1**	**½-1**	**¾-1**	**1-1**	**1½-1**	**2-1**	**2½-1**	**3-1**	**4-1**
.56	.343	.422	.497	.566	.630	.748	.857	.963	1.07	1.27
.57	.351	.433	.511	.584	.651	.775	.889	1.000	1.11	1.32
.58	.359	.445	.526	.602	.673	.802	.922	1.038	1.15	1.37
.59	.367	.456	.542	.621	.694	.830	.956	1.077	1.20	1.43
.60	.375	.468	.557	.640	.717	.858	.990	1.117	1.24	1.49

Source: King, H. W. *Handbook of Hydraulics (5th ed.)*. McGraw-Hill, 1963. Used with permission.

3) $d/b = 0.60$

4) $d = 0.60b = 0.60(2\ \text{ft}) = 1.2\ \text{ft}$

5.2.4 Culverts

Culverts are enclosed channels that run underneath embankments that allow water to pass from one side to the other. They serve an important function in railroad and roadway design, in particular, as these linear features cut across numerous low-lying areas where water collects and is conveyed. Culverts can be narrow or wide, long or short, and can have a variety of cross-sectional shapes (as shown in Figure 5.24). All culverts have an entrance (upstream end), an outlet (downstream end), and a barrel.

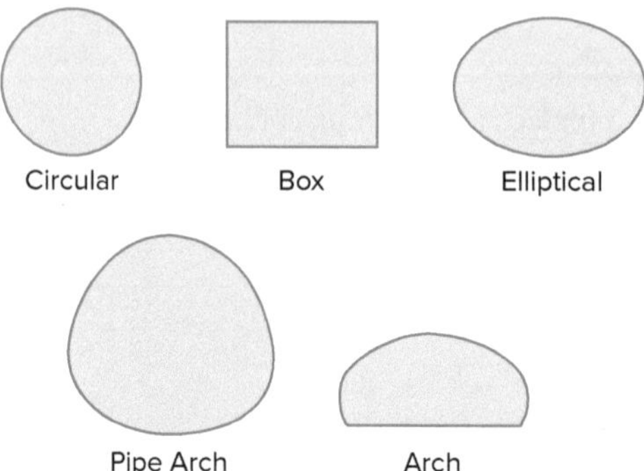

FIGURE 5.24 Commonly Used Culvert Shapes

A culvert is normally sized to convey a design discharge with a maximum depth of water at the culvert entrance (Fig. 5.25). This depth, called headwater (HW), is often required to be at least 1 to 2 ft lower than the edge of the roadway elevation to limit the frequency of flooding. Well-established design methodologies are used to estimate HW, which will correspond to a given discharge (Q). Since energy is needed to force flow into and through the culvert, a higher HW will typically correspond with a higher Q.

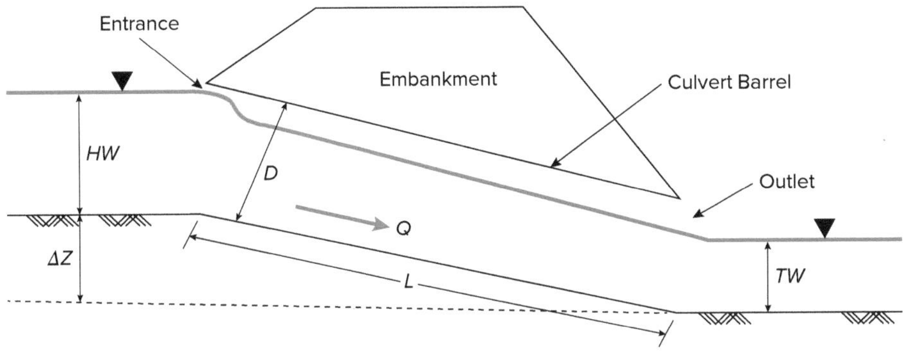

FIGURE 5.25 Culvert Geometric Parameters

Tailwater (TW) is defined as the depth of water at the downstream end of the culvert. Research sponsored by the Federal Highway Administration (FHWA) provides a wealth of information about how culvert hydraulics work. One of two conditions can occur: inlet control or outlet control. The characteristics and computations used to solve culvert flow for each condition are explained as follows.

5.2.4.1 Inlet Control

Inlet control means that culvert barrel conditions allow more flow to pass through the culvert than can be accepted through the entrance (or inlet). The inlet is the choke point that restricts the passage of water into the barrel. Inlet control can be summarized as follows:

- **Flow capacity** is controlled by the HW depth, cross-sectional area, and type of inlet edge.
- The **culvert barrel** is always partially full.
- **Flow** passes through critical depth at the inlet and remains supercritical through the culvert.
- The culvert barrel slope is usually steep.

HW is influenced by the size and shape of the culvert opening and the entrance conditions. Smoother transitions into the culvert barrel improve efficiency and result in lower HW for a given discharge. Examples of improved conditions are the end mitered to the embankment slope, a beveled or chamfered edge, or wingwalls placed at an angle from the culvert barrel to help funnel flow into the culvert.

Many scenarios exist for a culvert operating in inlet control. For example, when a mildly sloping channel upstream of the culvert transitions to a steep culvert slope, the culvert inlet will act as a flow control. The flow regime will transition from subcritical to supercritical flow, with critical depth occurring at or near the culvert entrance. This scenario is depicted in Figure 5.26.

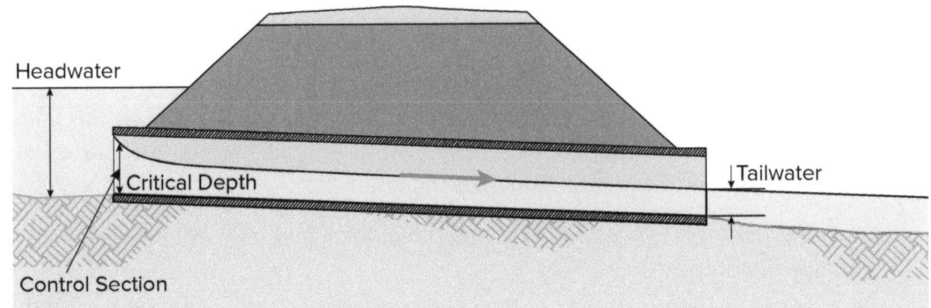

FIGURE 5.26 Culvert in Inlet Control

Source: *Hydraulic Design of Highway Culverts (3rd ed.) (FHWA-HIF-12-026, HDS 5)*, 2012 [2].

FHWA's Hydraulic Design Series No. 5 (HDS-5), *Hydraulic Design of Highway Culverts* [2], contains methodologies for modeling culverts operating in inlet control. When the culvert entrance is unsubmerged, the culvert entrance behaves like a weir. This condition occurs at lower flow rates. At higher flow rates, the culvert entrance may be submerged. When this occurs, the culvert entrance behaves like an orifice. The publication contains numerous nomographs that can be used to quickly calculate HW depth for different culvert shapes and sizes in inlet control. Figure 5.27 shows an example of an inlet control culvert nomograph. For this nomograph, values on two of the scales are needed to find a value on the third scale. For example, if the culvert diameter D and design discharge Q are known, the HW/D ratio can be found.

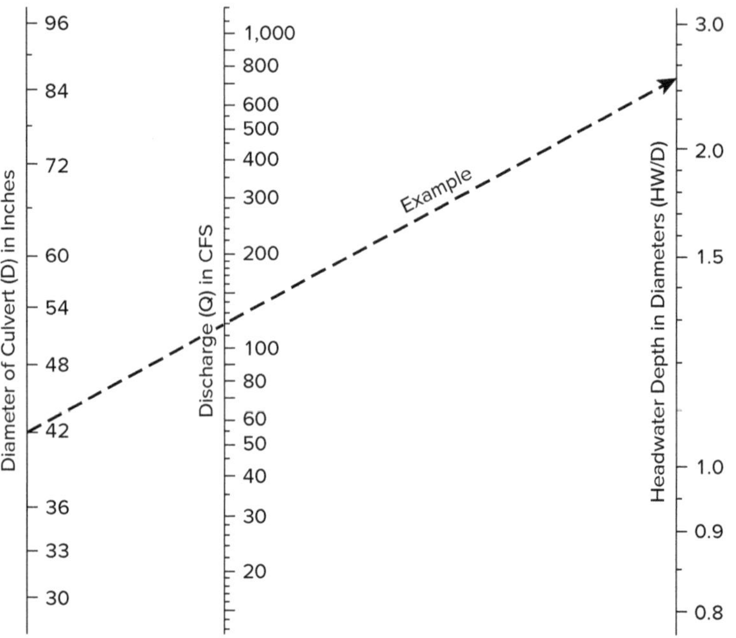

FIGURE 5.27 **Portion of an Inlet Control Nomograph for Concrete Pipe Culverts**

Source: *Hydraulic Design of Highway Culverts (3rd ed.) (FHWA-HIF-12-026, HDS 5)*, 2012 [2].

TIP

For similar problems on the exam, a nomograph will be provided.

Example 5.8: Inlet Control

Using the nomograph in Figure 5.27, determine the HW depth for a 3.5-ft-diameter concrete pipe culvert for a design flow of 120 ft^3/s.

Solution

Drawing a straight line through the two points on the scales for D (3.5 ft × 12 in = 42 in) and Q (120 ft^3/s), HW/D is 2.5.

$$\text{HW} = 2.5D = 2.5(3.5 \text{ ft}) = 8.75 \text{ ft}$$

5.2.4.2 Outlet Control

Outlet control means that the culvert barrel cannot accept as much flow as the inlet allows. This may occur with a high TW, a long culvert with a rough interior, or a flat culvert slope. Outlet control can be summarized as follows:

- Inlet control factors (HW depth, cross-sectional area, and type of inlet edge) also affect outlet control.
- Culvert barrel characteristics (roughness, area, shape, length, and slope) and TW depth are also factors.

- The culvert barrel may flow full or partially full.
- Flow remains subcritical through the culvert.
- The culvert barrel slope is usually flat or mild.

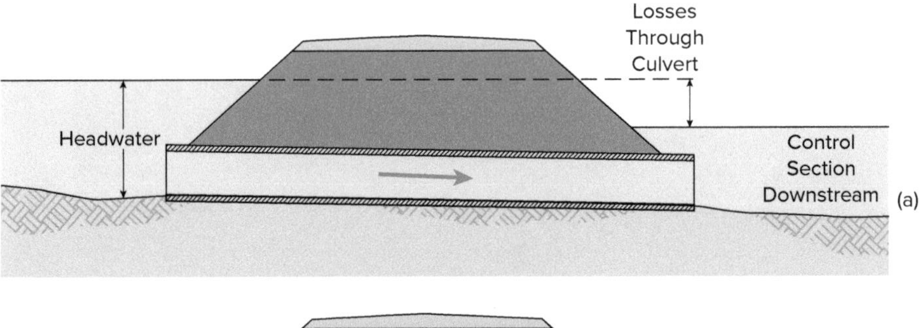

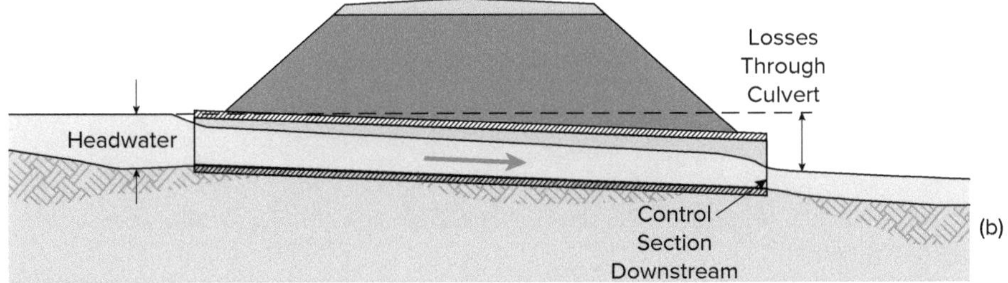

FIGURE 5.28 **Culvert in Outlet Control with Submerged (a) and Unsubmerged (b) Inlet**

Source: *Hydraulic Design of Highway Culverts (3rd ed.) (FHWA-HIF-12-026, HDS 5)*, 2012 [2].

Full-flow outlet control conditions, as depicted in the top panel in Figure 5.28, can be calculated by performing an energy balance between the ends of the culvert. The total energy (head loss, H_L) that is needed (or used) to convey flow through the culvert includes energy loss at the entrance (H_e), friction losses through the barrel (H_f), and energy loss at the outlet (H_o). All three of these losses are a function of the velocity head ($v^2/2g$).

> **TIP**
>
> The entrance loss coefficient, k_e, that is used to calculate H_e would normally be given on the exam.

$$H_L = H_e + H_f + H_o$$ **Equation 5-30**

$$H_e = k_e(v^2/2g)$$ **Equation 5-31**

$$H_f = (K_u n^2 L R^{-1.33})(v^2/2g)$$ **Equation 5-32**

$$H_o = v^2/2g$$ **Equation 5-33**

K_u = 29 (USCS); 19.63 (SI)
k_e = entrance loss coefficient (dimensionless)
n = Manning's roughness coefficient
L = length of culvert barrel (ft, m)
R = hydraulic radius of the full culvert barrel (ft, m)
v = velocity in the barrel (ft/s, m/s)

Figure 5.29 also depicts a culvert operating in outlet control and flowing full. Here, the EGL and the hydraulic grade line (HGL) are plotted. The energy losses that occur are evident by the drops in the EGL and HGL at the entrance and outlet, as well as the gradual slopes along the culvert barrel.

> **TIP**
>
> EGL and HGL are covered in depth in Section 5.7.5.

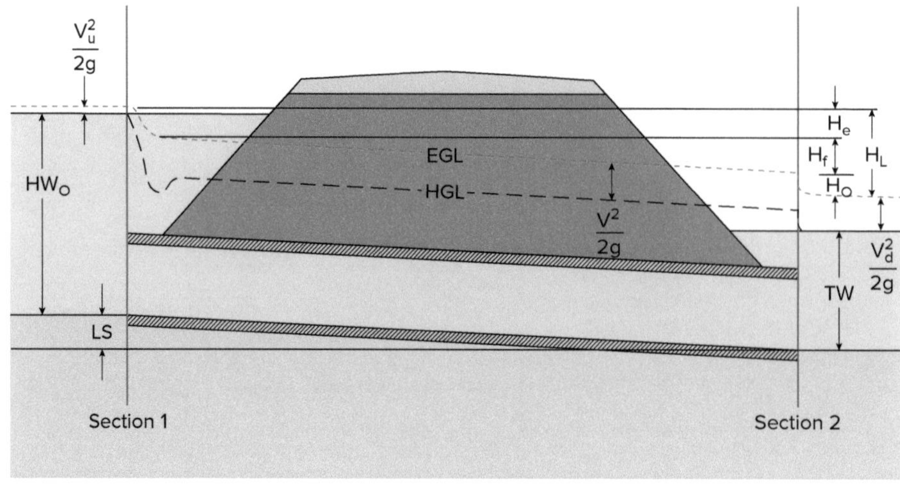

FIGURE 5.29 Full-Flow Energy and Hydraulic Grade Lines

Source: *Hydraulic Design of Highway Culverts (3rd ed.) (FHWA-HIF-12-026, HDS 5)*, 2012 [2].

If an energy balance is performed at the HW and TW ends of the culvert in Figure 5.29, it can be written as:

$$HW_o + LS + (v_u^2/2g) = TW + (v_d^2/2g) + H_L \quad \textbf{Equation 5-34}$$

HW_o = headwater depth above the entrance invert (ft, m)

v_u = approach velocity (ft/s, m/s)

v_d = downstream velocity (ft/s, m/s)

TW = tailwater depth about the outlet invert (ft, m)

H_L = sum of all losses (ft, m)

The velocity head upstream of the culvert entrance is neglected when the flow velocity in the approach (v_u) is low. This is often the case in culvert hydraulics. Likewise, the velocity head resulting from the velocity immediately downstream of the culvert (v_d) is usually neglected. Equation 5-35 is the result of both approach and downstream velocities being neglected. H_L is the difference in HW and TW elevations.

$$HW_o = TW + H_L - LS \quad \textbf{Equation 5-35}$$

The FHWA culvert design methodology involves first computing the HW assuming an inlet control condition. The HW is then also found assuming an outlet control condition. The higher of the two HW values for a given flow rate is selected as the basis of the culvert design.

5.3 STORM CHARACTERISTICS

5.3.1 Hydrology

5.3.1.1 Hydrologic Cycle

Water is continuously recirculating between the atmosphere, land, and oceans. Energy from the sun causes water to evaporate from the ocean surface. Evaporated water joins the atmosphere and eventually moves inland as clouds. Atmospheric conditions act to condense and precipitate water onto the land surface. Gravitational forces move surface and subsurface water to rivers and streams, many of which eventually lead back to the ocean. This hydrologic cycle is summarized in Figure 5.30. The actual process is more complex and contains many subcycles.

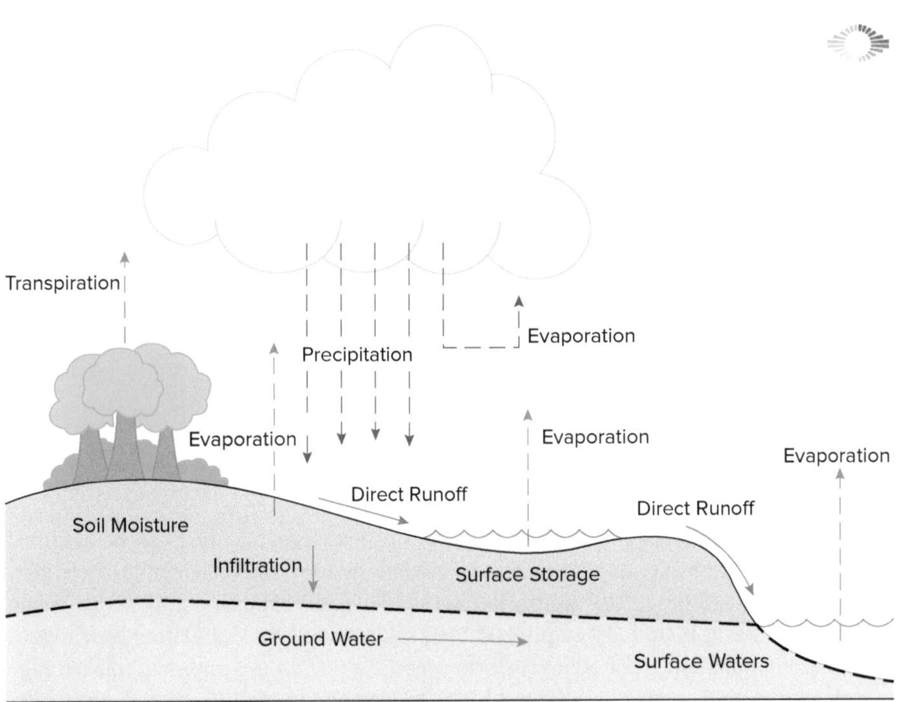

FIGURE 5.30 Schematic of the Hydrologic Cycle

5.3.1.2 Engineering Hydrology

Engineering hydrology views the **hydrologic cycle** quantitatively. It uses hydrologic principles to solve engineering problems arising from our use of the earth's water resources. Water resource engineers play an important role in the planning, analysis, design, construction, and operation of projects to control, utilize, and manage water resources. Hydrology in general can be divided into two primary branches: surface water and groundwater.

> **TIP**
>
> Surface water hydrology is the focus of the PE Civil Breadth exam.

5.3.2 Water Balance Problems

Quantification of the hydrologic cycle can be performed using a simple mass balance equation, where the change in storage (or stored water) (ΔS) is equal to inflows (I) minus outflows (O). This is a basic hydrologic principle that may be applied on a global, regional, or local scale. The water-holding elements of the hydrological cycle are the atmosphere (clouds), vegetation, snow packs, land surface, soil, waterbodies (streams, lakes, rivers, oceans), and aquifers. The hydrologic processes include precipitation, evaporation, transpiration, infiltration, overland flow, surface runoff, and groundwater outflow.

A water balance (or water budget) problem for the purposes of the PE Civil Breadth exam involves analyzing the change in water stored in a lake, stormwater basin, reservoir, and the like. The net of inflow (positive) and outflow (negative) volumes of one of these storage features results in the change in stored water.

Change in stored water (ΔS) = inflows − outflows **Equation 5-36**

Common input (or inflow) and output (or outflow) components for a storage area are as follows:

Inflow:

- Precipitation
- Water channeled into a given area (for example, from surface runoff)
- Groundwater inflow from adjoining areas

Outflow:

- Surface runoff outflow (for example, through a detention outlet structure)
- Water channeled out of the same area
- Evaporation
- Transpiration
- Infiltration or exfiltration

Change in storage:

This occurs as a change in:

- Groundwater
- Soil moisture
- Surface reservoir water and depression storage
- Detention storage

TIP

A water balance problem involving a stormwater detention basin is a good example of an application for the PE Civil Breadth exam.

Normally, for these types of problems, the sides of the storage feature can be assumed to be vertical. Therefore, depth is equal to volume divided by surface area. Detention basin problems can be greatly simplified into a few water balance components. When a stormwater basin is initially empty and the problem involves calculating how long it takes to drain completely following a storm event, $\Delta S = 0$. In this case, the total storage volume and rate of outflow are the minimum parameters needed to solve the problem. A problem may instead involve determining the time it takes for the storage feature to fill, which requires knowing the total storage volume and the **net rate of inflow** (inflow minus outflow rate). The following problem is an example of this application of the water balance concept.

Example 5.9: Water Balance Problems

A stormwater basin covers approximately 0.25 acres on a commercial site. The basin is empty before a cloud burst creates a short, intense rainstorm. The average rate of runoff into the basin during the storm is 10 ft^3/s, and the average discharge rate is 5 ft^3/s. If the basin is 5 ft deep and the inflow and outflow rates are assumed to be constant, how many hours before the basin will fill up?

Solution

$\Delta S = Q_{in} - Q_{out}$, where ΔS = the total basin volume

$V_{full} = (Q_{in} - Q_{out})t$, where Q is in ft^3/s, t is in seconds

$t = V_{full}/(Q_{in} - Q_{out})$

$V_{full} = (0.25 \text{ acre})\,(43{,}560 \text{ ft}^2/\text{acre})\,(5 \text{ ft}) = 54{,}450 \text{ ft}^3$

$t = (54{,}450 \text{ ft}^3) / (10 \text{ ft}^3/\text{s} - 5 \text{ ft}^3/\text{s}) = 10{,}890 \text{ s}$

$t = (10{,}890 \text{ s}) / (3{,}600 \text{ s/hr}) = 3 \text{ hr}$

5.3.3 Precipitation

Precipitation is any type of water that forms in the earth's atmosphere and then drops onto the surface of the earth. Precipitation can take the form of rain, snow, hail, or sleet. There are many uses for precipitation data in civil engineering, including runoff estimation analysis, water balance studies, flood analysis for design of hydraulic structures (for example, culverts and bridges), flood forecasting, and low flow studies.

Precipitation depth (in, mm) is often quantified at a point (for example, a rain gauge) or over an area (such as a watershed). The duration (hr) of a precipitation event is the period of time during which precipitation occurred. The depth of precipitation that accumulates per unit time is the intensity (in/hr, mm/hr). A **rainfall (or storm) hyetograph** is a plot of rainfall depth or intensity versus time. A cumulative rainfall hyetograph is called a **rainfall mass curve**. Both are shown in Figure 5.31.

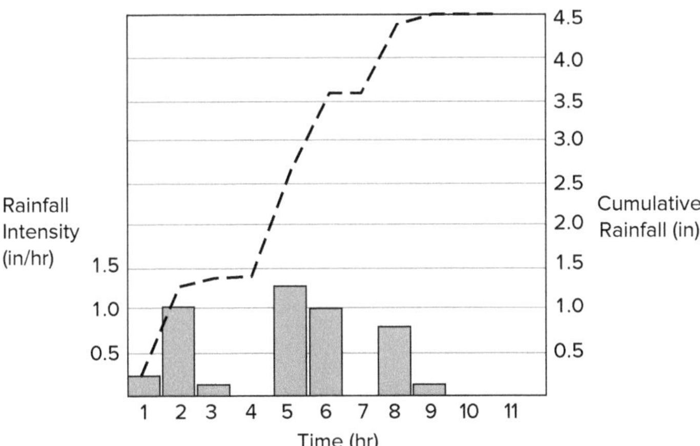

FIGURE 5.31 Example of a Storm Hyetograph with a Cumulative (Mass) Rainfall Curve

5.3.3.1 Rainfall Measurement

Rainfall events are measured using **rain gauges**. Nonrecording gauges provide only a measure of the cumulative amount at the end of a time period (usually 24 hours). The standard type is a large cylinder with a funnel and a plastic measuring tube inside the cylinder. Recording gauges automatically record the amount of rainfall reaching the surface as a function of time throughout the duration of a storm. Three types in common use are weighing, tipping bucket, and float (Fig. 5.32).

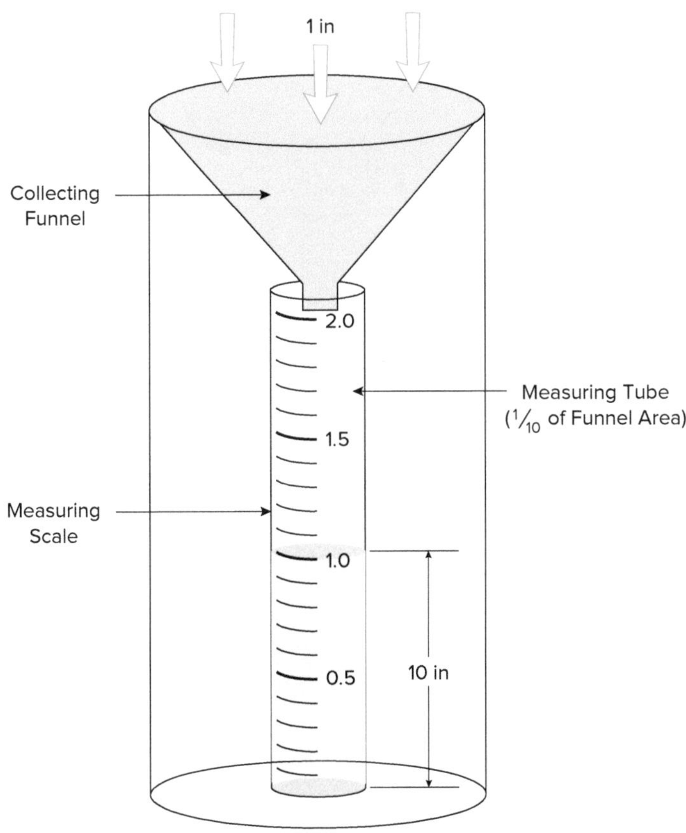

FIGURE 5.32 Standard Nonrecording Rain Gauge

Recording gauge data can look like the chart in Figure 5.33. It is a cumulative mass (depth) curve. Notice that right around the two-hour mark, when the curve has reached the maximum value in the chart (10 mm), the curve suddenly drops back down to zero. This is because the gauge emptied itself to make room for more rain. To determine the total rainfall accumulation, the hydrologist would add the amount before it emptied (10 mm) to the amount that accumulated after (roughly 7 mm).

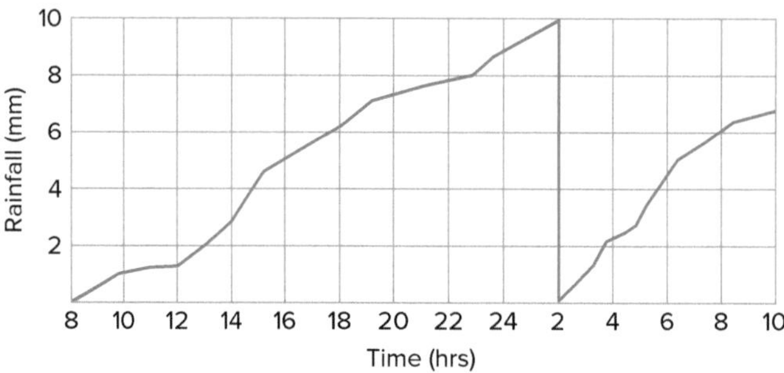

FIGURE 5.33 Sample Rain Gauge Record

5.3.3.2 Mean Areal Precipitation

A single-point precipitation measurement from a rain gauge on a 100-acre site is sufficient for estimating the total volume of precipitation falling over that area. A single point is less representative of the volume as the catchment area increases due to rainfall variability. The accuracy of mean and total precipitation estimates is increased by utilizing a dense network of point measurements. These measurements are converted to areal estimates using a variety of techniques, including the arithmetic mean method and the **Thiessen polygon method**.

The **arithmetic mean method** calculates areal precipitation using the arithmetic mean of all the point measurements considered in the analysis. Each gauge is assigned the same weight. To improve accuracy, gauges should be uniformly distributed over a relatively flat region. The formula that is used is:

$$\overline{P} = \sum \frac{P_i}{n}$$ **Equation 5-37**

$\overline{P}$ = mean precipitation (in, mm)

P_i = precipitation depth in gauge i (in, mm)

n = number of gauges used in the analysis

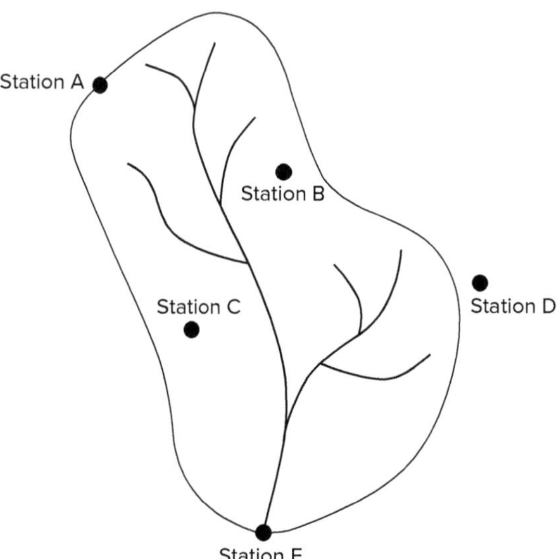

FIGURE 5.34 Rain Gauge Layout in a River Basin

Oftentimes, only the gauges that are physically located within the study boundary are included in the analysis. However, if a gauge that is located outside of the study boundary is the closest to a significant portion of the study area (such as Station D in Figure 5.34), one should consider including it in the analysis.

The Thiessen polygon method is a common method used to determine average precipitation over an area when there is more than one measurement. The watershed is divided into polygons that are centered on each measurement point (gauge), and then a weighted average of the measurements is calculated based on the relative sizes of the polygons. For example, a measurement within a large polygon is given more weight than a measurement within a small polygon. The weighted average is calculated by:

$$\overline{P} = \frac{\sum_{i=1}^{n} P_i A_i}{\sum_{i=1}^{n} A_i} = \frac{P_1A_1 + P_2A_2 + P_3A_3 + \cdots + P_nA_n}{A_1 + A_2 + A_3 + \cdots + A_n}$$ **Equation 5-38**

$\overline{P}$ = mean precipitation (in, mm)

P_i = precipitation depth in gauge i (in, mm)

A_i = area of polygon i

The following steps and Figures 5.35 and 5.36 illustrate how Thiessen polygons are made:

1. Connect all of the measurement points (gauges) with dashed lines.
2. Perpendicularly bisect each of the connecting lines and extend the bisecting lines until they either intersect the study area boundary or another bisecting line.
3. Calculate the area for each subdivided portion of the study area. When a grid is superimposed on the study area, use squares as the unit of measure (that is, count the number of squares).
4. Use the weighted formula to find the mean precipitation.

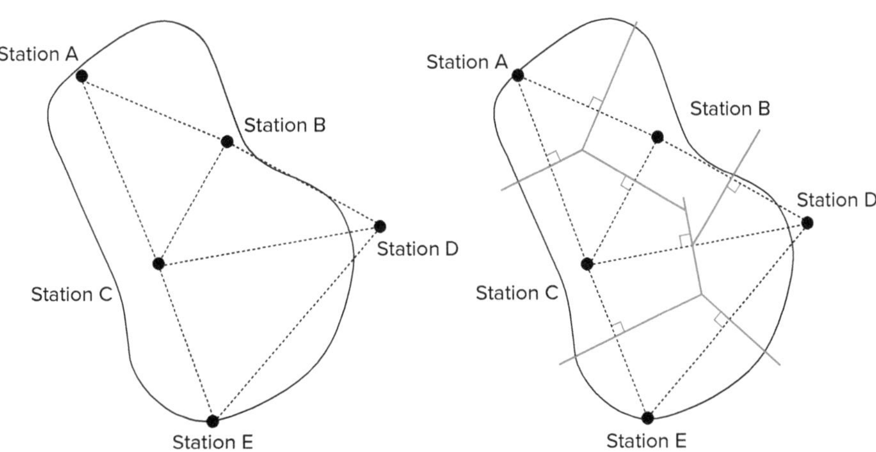

FIGURE 5.35 Step 1 (left) and Step 2 (right) of Thiessen Polygon Method

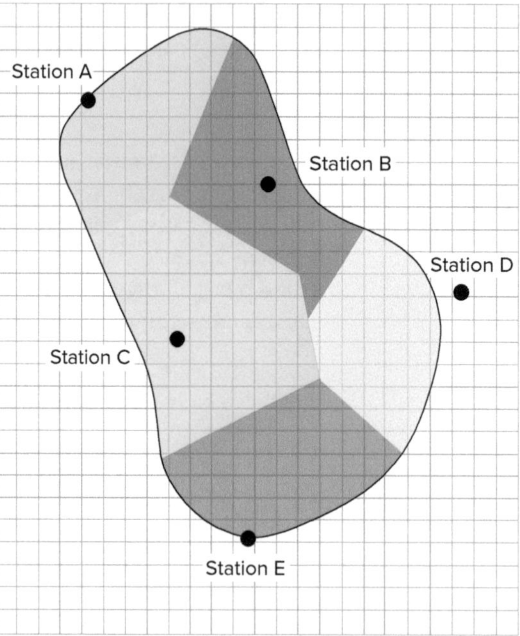

FIGURE 5.36 Step 3 of Thiessen Polygon Method

> **TIP**
>
> Streamflow event frequency is based only on annual peak flow magnitude.

5.3.4 Design Storms and Floods

Frequency (or recurrence interval) is used to describe the probability of a given precipitation or peak streamflow event occurring. Rainfall event frequency is based on the magnitude and the duration of a rainfall event. A minimum length of record (typically ten years) is usually required to perform a frequency analysis. More years of historical data result in better estimates.

The **recurrence interval** (or frequency, F) is the average number of years between rainfall or streamflow events of a defined magnitude. Design storms and floods in civil engineering practice are often specified by their recurrence interval (for example, a 100-year storm).

Probability (p) is often specified as a percentage (for example, 1% flood). It is the chance that a storm or flood of a defined magnitude will be exceeded in any given year. A 1% annual exceedance probability (AEP) is the same as a 100-year return period.

> **TIP**
>
> The AEP is always a fraction of one.

For example, a 0.04 AEP flood has a 4% chance of occurring in any given year. This corresponds to a 25-year frequency flood. A comparison of flood events is more easily understood using recurrence interval terminology (for example, a 100-year flood is greater than a 25-year flood). The AEP, on the other hand, reminds us that when a large flood event occurs, it does not reduce the chances of another one occurring within a short period of time (see Table 5.6).

$$p = \frac{1}{F}$$ **Equation 5-39**

The probability of a storm or flood with a given frequency occurring in n years is:

$$p = 1 - \left(1 - \frac{1}{F}\right)^n$$ **Equation 5-40**

The probability of a storm or flood occurring in m consecutive years is:

$$p = p^m$$ **Equation 5-41**

The probability of a storm or flood not occurring is:

$$p_{\text{not}} = 1 - p$$ **Equation 5-42**

TABLE 5.6 Recurrence Intervals and Probabilities of Occurrences

RECURRENCE INTERVAL (YEARS)	PROBABILITY OF OCCURRENCE IN ANY GIVEN YEAR	% CHANCE OF OCCURRENCE IN ANY GIVEN YEAR
100	1 in 100	1
50	1 in 50	2
25	1 in 25	4
10	1 in 10	10
5	1 in 5	20
2	1 in 2	50

Source: US Geological Survey, Department of the Interior/USGS.

Example 5.10: Probability of Flood

An undersized culvert is due to be replaced in 10 years, but the highway department is concerned about the frequency of the road flooding. A 25-year or greater magnitude event causes the road to flood. What is the probability of the road flooding before the culvert is replaced?

Solution

$$p = 1 - \left(1 - \frac{1}{F}\right)^n \blacktriangleright p = 1 - \left(1 - \frac{1}{25}\right)^{10}$$

$$p = 0.335 = 34\%$$

5.4 RUNOFF ANALYSIS

5.4.1 Hydrographs

Stormwater runoff entering a watercourse as surface runoff produces a hydrograph, a plot of discharge (in cubic feet per second) over time. Hydrographs show how a stream or river responds to precipitation events within its drainage basin. Discharge can be calculated from fieldwork data by multiplying the cross-sectional area by average velocity. The data are mainly collected from stream gauges located at different sites along the watercourse. Figure 5.37 illustrates the three phases of a hydrograph: rising limb, peak, and falling or recession limb.

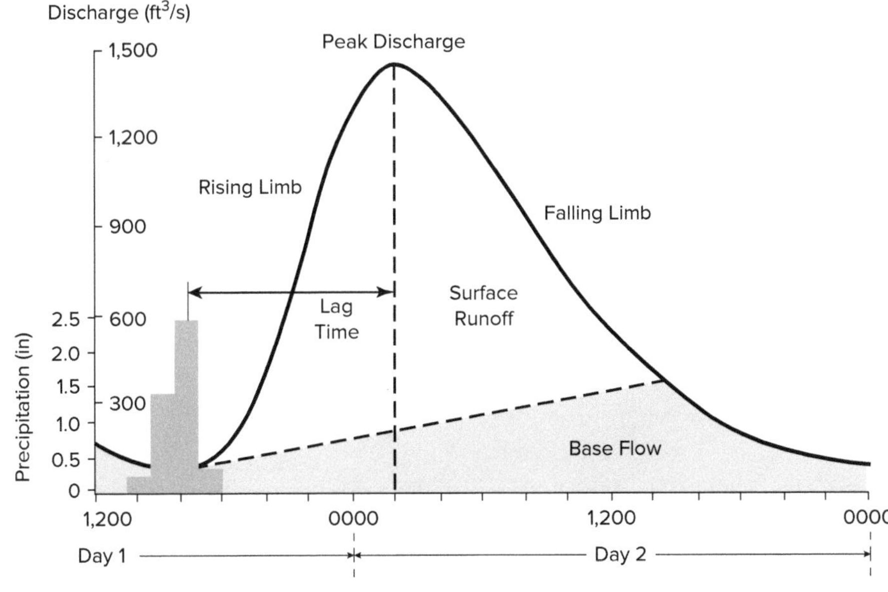

FIGURE 5.37 River Hydrograph

River hydrographs are composed of two flow components: surface runoff and base flow. **Surface runoff** is storm-induced flow that reaches the stream shortly after it falls as rain and is discharged from the basin within one or two days. Base flow is sustained stream flow from deep subsurface flow and delayed shallow subsurface flow. Base flow may gradually increase following a storm as groundwater is recharged. Methods have been developed from gauge data to separate base flow from surface runoff.

Three primary types of hydrographs are used in watershed hydrology:

- **Natural hydrograph:** obtained directly from flow records of a gauged stream
- **Synthetic hydrograph:** obtained by using watershed parameters and storm characteristics to simulate a natural hydrograph
- **Unit hydrograph:** a discharge hydrograph resulting from one inch of runoff distributed uniformly over the watershed resulting from rainfall of a specified duration

5.4.2 Time of Concentration

The time of concentration, or t_c, is defined as the time needed for water to travel from the hydraulically most distant point in a drainage area to the point of interest (Fig. 5.38). The phrase "hydraulically most distant" is important because it means the longest path in terms of time. The travel time, t (seconds), of surface runoff is a function of the length of the flow path and the flow velocity. In general:

$$t = \frac{L}{v}$$ **Equation 5-43**

L = length of flow path (ft)

v = flow velocity (ft/s)

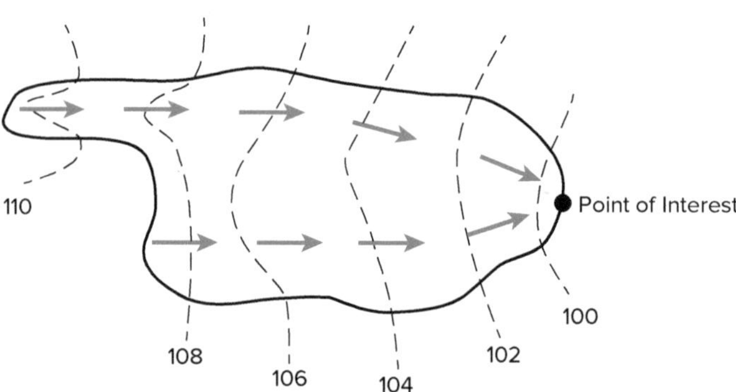

FIGURE 5.38 Time of Concentration Flow Paths

Various methods of calculating velocity are used depending on flow type. Shallow, slow-moving flow over a parking lot can be described as **sheet flow**. This type of flow is very different from flow conveyed through a storm sewer, which is described as **channel flow**. Whereas channel flow velocity can be calculated using Manning's equation, sheet flow requires a different formula to solve. Likewise, surface flow that does not resemble sheet flow or channel flow is referred to as shallow concentrated flow.

In many cases, surface runoff progresses from sheet flow to concentrated channel flow as it makes its way from the upper part of a drainage area to the outlet. When this happens, the travel time associated with each flow segment is calculated, and the total travel times are added together to arrive at the time of concentration. The segmental method is illustrated in Figure 5.39. The National Resources Conservation Service (NRCS) segmental method does just this:

$$t_c = t_{\text{sheet}} + t_{\text{shallow}} + t_{\text{channel}}$$ **Equation 5-44**

Any unit of time can be used in Equation 5-44, as long as it is consistent. Channel flow velocity would simply be calculated from Manning's equation.

TIP

The equations or charts needed to calculate sheet and shallow flow would be provided on the exam.

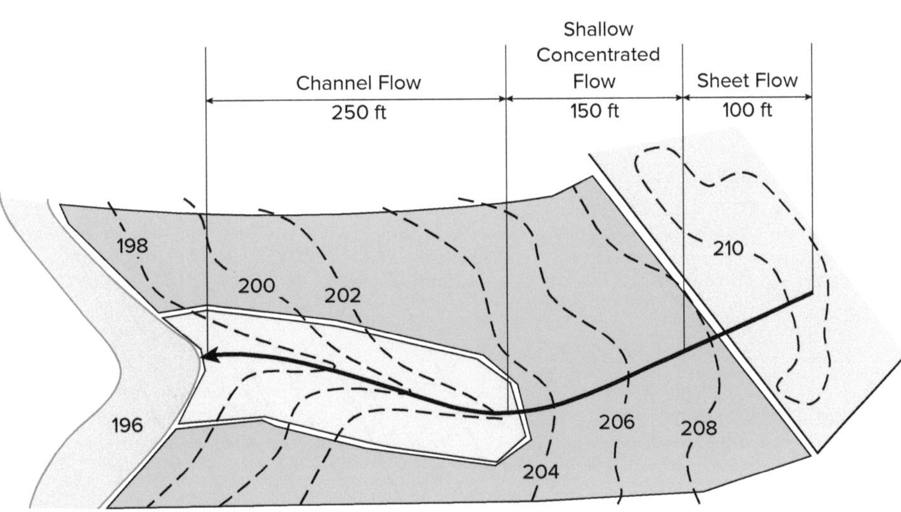

FIGURE 5.39 Segmental Method Flow Components

Conceptually then, the following can be said about time of concentration:

- Longer flow path length = longer t_c
- Rougher surfaces/channels (for example, woods, heavy brush) = longer t_c
- Flatter surfaces/channels = longer t_c

Example 5.11: Time of Concentration

A drainage basin has a total length from the high point to the low point of 1,300 ft. Storm runoff takes the path shown in the figure. Sheet and overland flow velocities over different surfaces can be found using the chart below. What is most nearly the time of concentration for the drainage basin, in minutes?

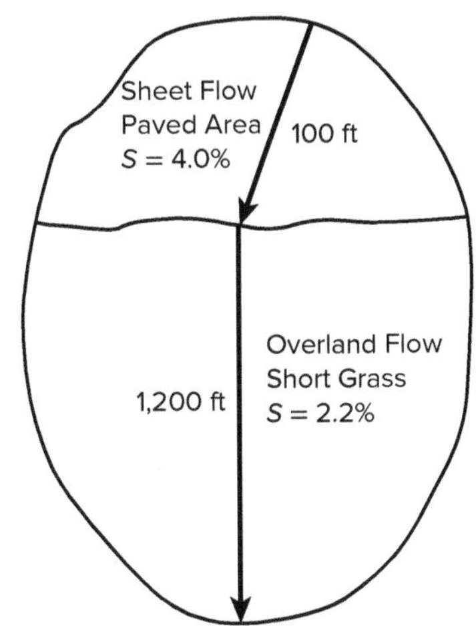

Example 5.11 *(continued)*

Solution

$$t_c = \left(\frac{L_1}{v_1} + \frac{L_2}{v_2}\right)$$

$L_1 = 100$ ft, $L_2 = 1{,}200$ ft

From the chart: $v_1 = 4.0$ ft/s, $v_2 = 1.0$ ft/s

$$t_c = \left(\frac{100 \text{ ft}}{4.0 \text{ ft/s}} + \frac{1{,}200 \text{ ft}}{1.0 \text{ ft/s}}\right)/(60 \text{ s/min}) = 20.4 \text{ min}$$

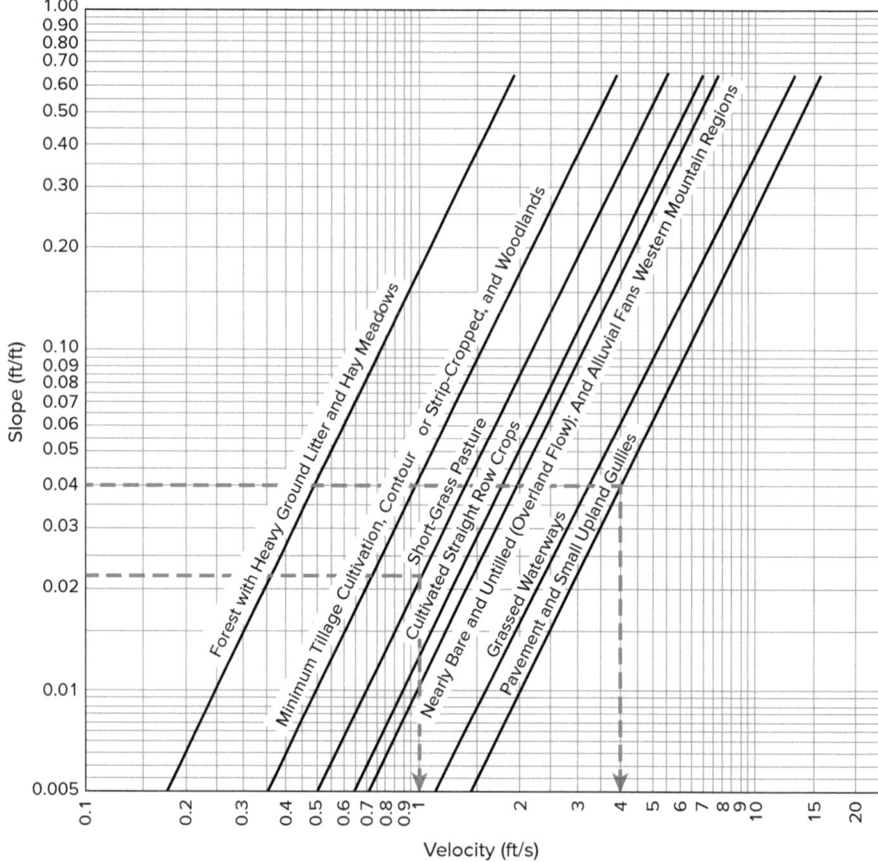

Source: *Part 630 Hydrology: National Engineering Handbook (210-VI-NEH)*, Natural Resources Conservation Service (USDA), 2010 [3].

5.4.3 Drainage Area Characteristics—Effects on Runoff

The characteristics of a drainage basin explain how a stream or river responds to a rainfall event. As depicted in Figure 5.40, water enters the river through surface runoff, throughflow, and groundwater flow. Compared to these sources, precipitation falling directly into the river is negligible. Surface runoff reaches the river orders of magnitude faster than subsurface flow. As a result, the drainage basin characteristics that increase surface runoff produce a hydrograph with a more pronounced peak.

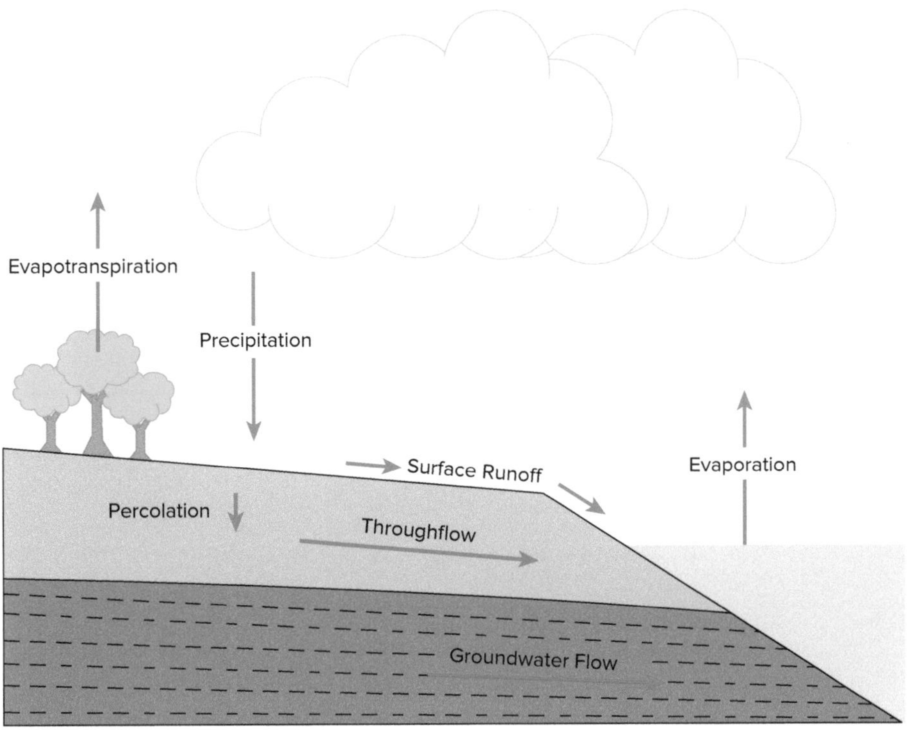

FIGURE 5.40 Water Entry into Rivers

- **Slope.** The size, shape, and slopes (or relief) of a drainage basin can influence the shape of the hydrograph (Fig. 5.41). Gentle-to-flat slopes make it more likely that rainfall will pool on the surface and provide more time for infiltration to occur. Conversely, steep slopes are more likely to convert rainfall to surface runoff. Water runs off more quickly on steep gradients and causes a steep rising limb on the hydrograph. Gentle and flat slopes will cause the rising limb to be flatter.

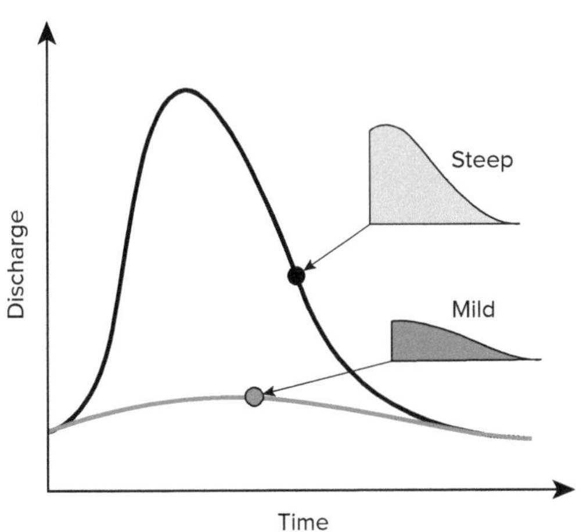

FIGURE 5.41 Effect of Drainage Basin Steepness on Runoff Hydrographs

- **Geology.** Soils and geology are also important factors in shaping a hydrograph. Impermeable rock will encourage throughflow and surface runoff, particularly where there are shallow compact soils. Permeable soils (for example, sand) and rock (for example, limestone) allow water to drain through, leading to higher groundwater storage and flow. Discharge is lower and the lag time more extended in these areas.

- **Drainage Density.** Drainage density refers to the presence of stream channels within the drainage basin. A greater density of channels allows surface runoff to reach a channel more quickly, thereby increasing discharge and reducing lag time.
- **Land Use.** The way in which the land is used will also have an influence on the hydrograph. Vegetation intercepts precipitation and allows evaporation to take place directly into the atmosphere. As a result, less surface runoff occurs. Contrast that with urban areas, where impermeable surfaces allow nearly all of the rain that reaches the surface to become runoff.
- **Size and Shape.** The size and shape of the drainage basin influences the shape of the hydrograph in many ways (Fig. 5.42). A stream or river will be slower to respond to a larger drainage basin due to the distance that water has to travel to reach the channel. The shape of the drainage basin is a factor because it can affect the timing of runoff concentrating at the basin outlet. Circular-shaped drainage basins have more rapid responses compared to long, narrow drainage basins, which take longer to arrive at the outlet.

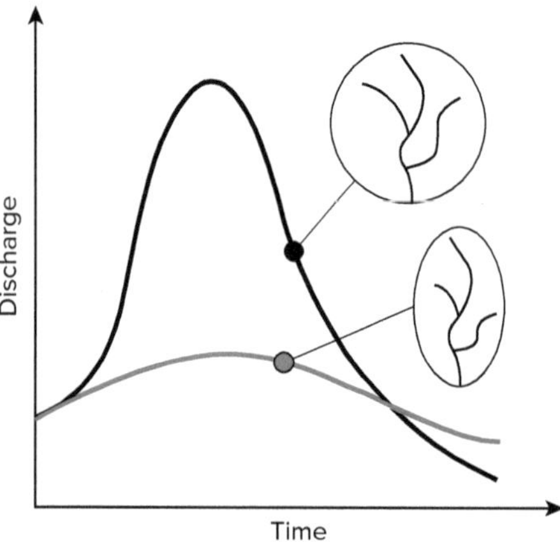

FIGURE 5.42 Effect of Drainage Basin Shape on Runoff Hydrographs

5.4.4 Rational Method

The rational method was developed more than 100 years ago for the purpose of predicting peak flow rates for small, urban watersheds. It is a valid hydrologic design tool for urban (or largely impervious areas) watersheds. Most references recommend limiting its application to 200 acres or less. The rational method cannot be used to compute runoff volume, and it does not by itself provide information pertaining to the runoff hydrograph shape. The peak flow rate, Q_{peak} (ft^3/s), is calculated as:

$$Q_{peak} = (C)(I)(A)$$ **Equation 5-45**

C = runoff coefficient (dimensionless)

I = rainfall intensity (in/hr)

A = drainage area (acres)

Conversion note: 1 acre-in/hr = 1.008 ft^3/s

> **TIP**
>
> Runoff coefficients vary widely by source and, as such, are provided on the PE Civil exam.

The runoff coefficient (C) is a dimensionless ratio intended to indicate the amount of rainfall converted to runoff by a drainage area (Fig. 5.43). Values of C are always less than one, with higher values indicating more impervious, steeper, or finer-textured soils (for example, clays). Lower values of C indicate areas that are more pervious, flatter, or contain coarser-textured soils (for example, sands).

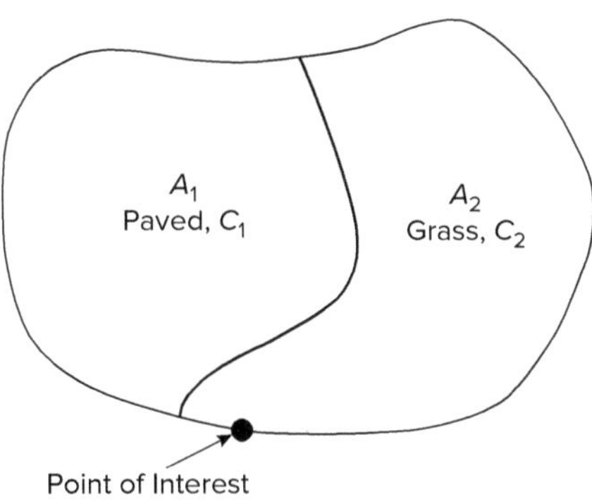

FIGURE 5.43 Drainage Area with Multiple Cover Types

When the watershed contains multiple runoff coefficients, a weighted value (C_w) must be calculated.

$$C_w = \Sigma C_i A_i / A_{total}$$ **Equation 5-46**

$$C_w = [C_1A_1 + C_2A_2]/[A_1 + A_2]$$

Storm intensity (I) is a function of geographic location, design frequency (or return period), and storm duration. By definition, the storm duration is equal to the time of concentration when using the rational method. The longer the length of the storm, the lower the storm intensity. The relation among storm duration, storm intensity, and storm return period is represented by a family of curves called the **intensity-duration-frequency (IDF) curves**. An example of an IDF curve is shown in Figure 5.44. The x-axis represents the storm duration; by definition, this would be the time of concentration for the drainage area. Note that both axes are log scale.

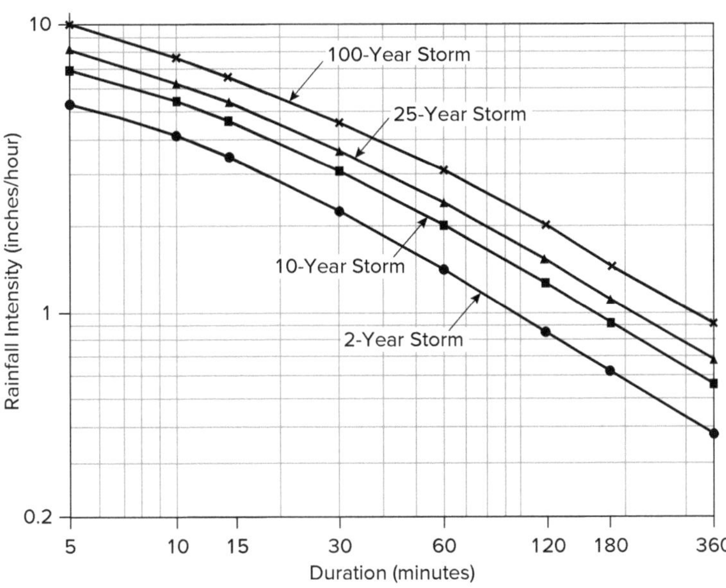

FIGURE 5.44 IDF Curves

The **drainage area** (A) in the rational method is simply the total area that contributes surface runoff to the point of interest. If multiple discrete areas drain to a point of interest, the areas must be combined and treated as one. As in the figure below, drainage areas A_1 and A_2 would be added together and a single peak flow rate computed at the point of interest (Fig. 5.45). Of course, if the drainage areas contain different runoff coefficients, a weighted coefficient (C_w) would also need to be calculated.

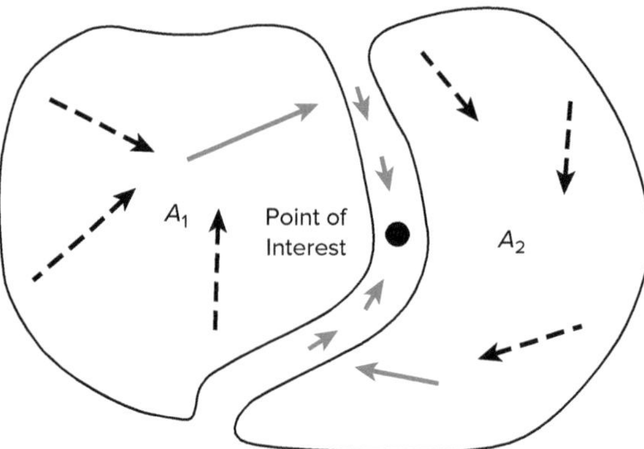

FIGURE 5.45 Drainage Areas Combining at a Single Point of Interest

Delineating drainage areas to a point of interest can be simple or complex. Generally speaking, a single watershed, free of obstructions and storage features, is a simple exercise in following the ridge of the topographic contours on a map or plan. A delineation of this type is shown in Figure 5.46.

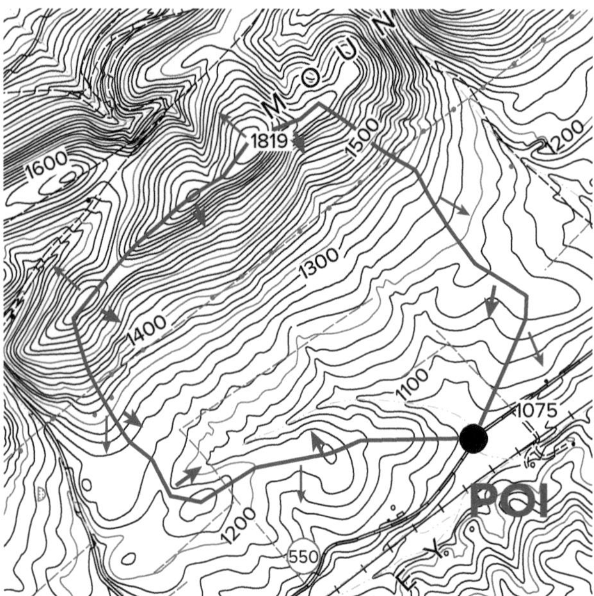

FIGURE 5.46 Drainage Area Delineation to a Point of Interest

5.4.5 Storm-Sewer Systems

Complexities are introduced when the point of interest is downstream of topographic or man-made features that constrict or redirect flow. Closer inspection of drainage area features may be necessary to determine if any parts of the drainage area are attenuated (for example, a pond or depression), lost (for example, a sinkhole), or directed away from the point of interest. When the point of interest is part of a storm-sewer system, the assumptions of the rational method must be respected to properly compute peak flows for various system components (for example, inlet spacing, pipe sizing). To size a pipe in a storm-sewer system, all subareas upstream of the pipe must be combined into a single drainage area, which has a corresponding C_w.

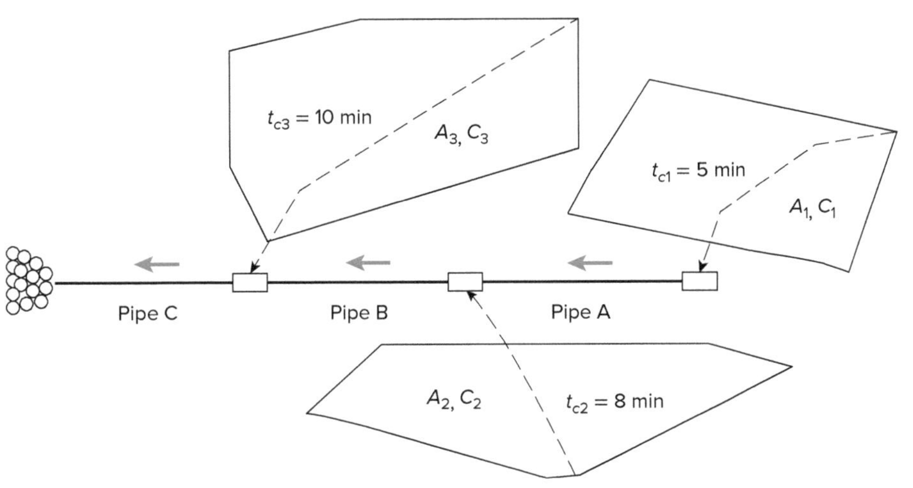

FIGURE 5.47 Storm-Sewer System

Suppose you need to size pipe C in the storm-sewer system shown in Figure 5.47. In order to do so, all areas contributing flow to the system upstream of pipe C (A_1, A_2, and A_3) are added together. The corresponding C_w can be calculated using the weighted formula. Rainfall intensity (I) is normally found using regional IDF curves or equations. Assuming that the design storm frequency is given, only the storm duration is needed to find I using the IDF information. The storm duration and time of concentration are the same for the rational method. Thus, the time it takes all parts of the contributing drainage area to reach the point of interest (pipe C) defines t_c. Assuming that travel time in the pipes is negligible, the t_c that would be used is 10 minutes.

Example 5.12: Storm-Sewer Systems

A storm-sewer system has the following drainage area characteristics:

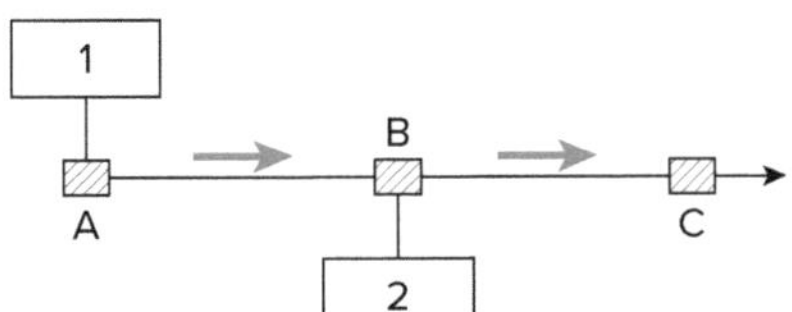

AREA	SIZE (ac)	t_c (min)	C
1	3.0	10	0.3
2	1.5	5	0.7

The travel time from storm inlet A to inlet B is negligible.

Rainfall intensity is given by: $I\text{ (in/hr)} = \dfrac{105}{t_c(\text{min}) + 15}$

What peak rate of discharge, in cubic feet per second, should be used to design the pipe between inlets B and C?

Solution

Refer to Equations 5-45 and 5-46.

$Q = C_w I A_{\text{total}}$

Calculate C_w at inlet B because the pipe being sized begins there.

$$C_w = \frac{(C_1A_1 + C_2A_2)}{(A_1 + A_2)}$$

$$C_w = \frac{(0.3 \times 3.0) + (0.7 \times 1.5)}{(3.0 + 1.5)}$$

Example 5.12 *(continued)*

$$C_w = \frac{0.9 + 1.05}{4.5}$$

$$C_w = 0.43$$

Rainfall intensity is based on the time it takes *all* contributing areas to reach inlet B.

Area 1: 10 min + 0 min (A to B) = 10 min
Area 2: 5 min

$$I\ (\text{in/hr}) = \frac{105}{10\ \text{min} + 15\ \text{min}} = 4.2\ \text{in/hr}$$

$$Q = 0.43(4.2\ \text{in/hr})(3.0\ \text{acres} + 1.5\ \text{acres}) = 8.1\ \text{ft}^3/\text{s}$$

5.4.6 NRCS Curve Number Method

The NRCS runoff curve number method is a method of estimating runoff (rainfall excess) from a single rainfall event. The curve number method is described in detail in the NRCS *National Engineering Handbook, Part 630, Hydrology* [3]. Originally developed for agricultural watersheds in the midwestern US, it has since been adapted for other developed and undeveloped watersheds.

The curve number method utilizes an empirical relationship between rainfall and retention (rainfall not converted into runoff) and the runoff properties of a watershed (for example, soils and land use). It assumes that a certain amount of rainfall is abstracted at the beginning of a storm. Rainfall interception (for example, tree canopies), depression storage, and soil infiltration are the primary means of abstraction. Runoff is generated when the total rainfall (P) exceeds the initial abstraction (I_a). The potential maximum runoff (that is, the rainfall that is available for runoff) is calculated as $P - I_a$. For a given storm, the maximum loss of rainfall is calculated as $S + I_a$. The portion of the rainfall that becomes runoff (Q) is calculated by:

$$Q = \frac{(P - I_a)^2}{(P - I_a) + S}$$ **Equation 5-47**

The potential maximum retention, S, is a function of only the curve number (CN).

$$S = \frac{1{,}000}{\text{CN}} - 10$$ **Equation 5-48**

Initial abstraction is assumed to be 20% of the potential maximum retention.

$$I_a = 0.2S$$ **Equation 5-49**

Substituting out I_a, the equation becomes:

$$Q = \frac{(P - 0.2S)^2}{(P + 0.8S)}$$ **Equation 5-50**

Q = runoff depth (in)
P = storm precipitation depth (in)
S = potential maximum retention after runoff begins (storage capacity) (in)
CN = curve number (dimensionless)

Figure 5.48 provides a graphical solution for runoff, Q.

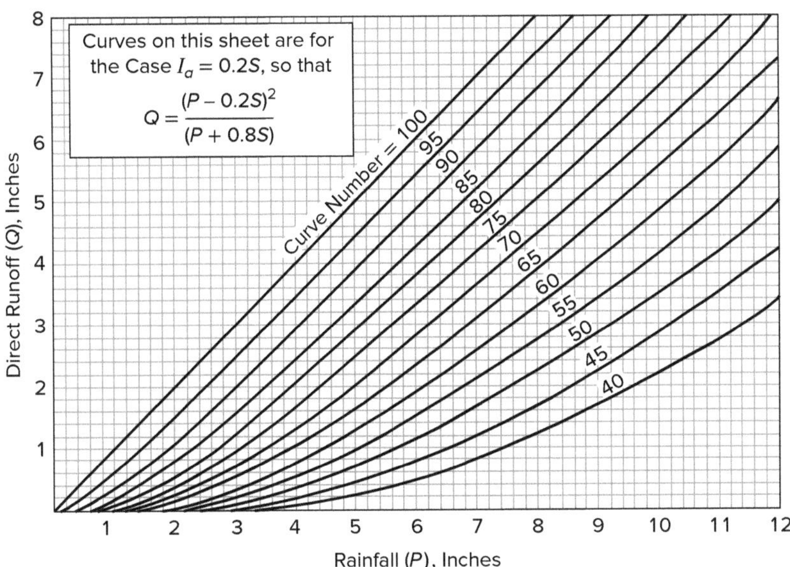

FIGURE 5.48 Graphical Solution for the NRCS Runoff Curve Number Method

Source: *Urban Hydrology for Small Watersheds* TR-55, *Natural* Resources Conservation Service (USDA), 1986.

The curve number is a function of a variety of factors, some of which may not be applicable for a given scenario.

- Hydrologic soil group (HSG)
- Cover type (for example, grass, impervious)
- Treatment (for agricultural land only)
- Hydrologic condition
- Antecedent runoff condition
- Connectivity of impervious areas

The infiltration rates of soils are affected by subsurface permeability and surface intake rates. Soils are classified into four HSGs (A, B, C, and D) according to their minimum infiltration rate. A comparison of the HSGs is shown in Table 5.7.

TABLE 5.7 Hydrologic Soil Groups

	GROUP A	GROUP B	GROUP C	GROUP D
Runoff Potential	Low	Moderate	Moderate	High
Infiltration Rates (in/hr)	High > 0.30	Moderate 0.15 to 0.30	Moderate 0.05 to 0.15	Low 0 to 0.05
Soils/Texture	Sand or Gravel	Moderately Coarse	Moderately Fine to Fine	Clay

Curve numbers from the *National Engineering Handbook* [3] are provided in Table 5.8.

TABLE 5.8 Sample Runoff Curve Number Values[1]

COVER TYPE	HYDROLOGIC SOIL GROUP			
	A	B	C	D
Meadow	30	58	71	78
Woods[2]	30	55	70	77
Impervious	98	98	98	98
Open Space[2]	39	61	74	80
Row Crops, Straight[2]	72	81	88	91

[1]Antecedent runoff condition II
[2]Good hydrologic condition
Modified from *Part 630 Hydrology: National Engineering Handbook (210-VI-NEH)*, Natural Resources Conservation Service (USDA), 2010 [3].

When the drainage area being analyzed has multiple land uses, either a weighted CN is calculated for the entire area, or runoff is calculated for each sub-area (with its own unique CN). In general, whenever pervious areas and impervious areas can be reasonably separated for analyses, runoff should be calculated for each and then added together for a total runoff volume.

Curve number weighting is normally only done when using curve numbers for calculating peak flows (that is, procedures in TR-55 [4]).

Example 5.13: Curve Number Method

A 30-acre rolling meadow with generally clayey soils receives 4 inches of rainfall over a 24-hour period. What is the approximate volume of runoff from the area, in acre-feet?

Solution

From Table 5.8, the CN is 78.

$$S = \frac{1{,}000}{\text{CN}} - 10 = \frac{1{,}000}{78} - 10 = 2.82 \text{ in}$$

$$Q = \frac{(P - 0.2S)^2}{(P + 0.8S)} = \frac{(4 \text{ in} - 0.2 \times 2.82 \text{ in})^2}{(4 \text{ in} + 0.8 \times 2.82 \text{ in})} = 1.89 \text{ in}$$

$$V_{\text{runoff}} = Q_{\text{depth}} \times A$$

$$V_{\text{runoff}} = (1.89 \text{ in.})(1 \text{ ft}/12 \text{ in.})(30 \text{ acres}) = 4.725 \text{ acre-ft}$$

5.5 DETENTION AND RETENTION PONDS

5.5.1 Stormwater Management

Land development affects the quantity and quality of stormwater runoff that naturally discharges from a site. Construction normally includes removal of vegetation, bulk earthwork (grading), soil compaction, and building impervious surfaces. Each of these activities affects what happens to precipitation falling on the site.

Preparing a site for grading involves clearing trees, shrubs, and other vegetation that stabilize the soil beneath it. Vegetation intercepts and slows rainfall, reducing its erosive energy, reducing overland flow of runoff, and allowing infiltration to occur. The root systems of plants provide pathways for downward water movement into the soil mantle.

The topsoil layer is usually removed and stockpiled prior to grading a site. **Topsoil** is the upper, outermost layer of soil (usually the top 2 to 8 in) containing a mix of decaying organic matter and deposited eroded material. Topsoil is spongy and contains an abundance of nutrients that support vegetative growth, microorganisms that aid in breaking down pollutants, and pore space that is conducive to infiltrating runoff. When topsoil is not replaced at the same depth as it was prior to site grading, the soil storage capacity is significantly reduced. Bulk grading of soil often replaces small, subtle depressions in the landscape with smooth, uniform surfaces that are incapable of retaining water.

Macropores, or small openings in soil, provide a mechanism for water to move through the soil. When soil is disturbed (for example, by grading, stockpiling, or heavy equipment traffic), the soil is compacted, macropores are smashed, and the soil permeability characteristics are substantially reduced.

When soil is replaced by impervious cover, nearly all of the rainfall that occurs over the area becomes surface runoff. This causes additional hydraulic and pollutant loading on the adjacent vegetated areas. Reduced infiltration over a large area, such as urban areas, can result in decreased base flows and increased peak flows in nearby streams.

Urbanization is a general term for the clustered transition of undeveloped areas to developed areas. Figure 5.49 illustrates a typical water balance before and after urbanization occurs. Figure 5.50 shows the resultant hydrographs.

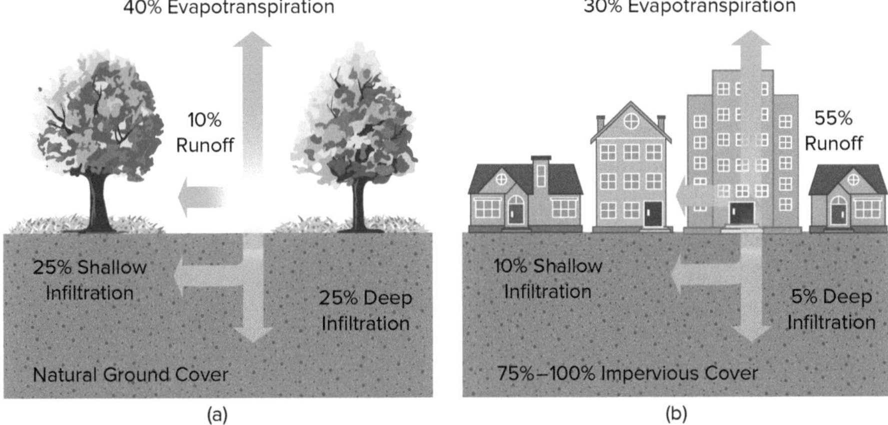

FIGURE 5.49 Effects of Development on Runoff

Source: United States Department of Agriculture.

The US Army Corps of Engineers defines stormwater management as follows:

> **Stormwater management** is the mechanism for controlling stormwater runoff for the purposes of reducing downstream erosion, water quality degradation, and flooding and mitigating the adverse effects of changes in land use on the aquatic environment. [5]

Stormwater management is aimed at controlling the volume and timing of runoff (for example, for flood control and water supplies) and minimizing the release of potential contaminants to water resources. Therefore, stormwater from developed areas needs to be managed in a way so as to not pollute or degrade water resources.

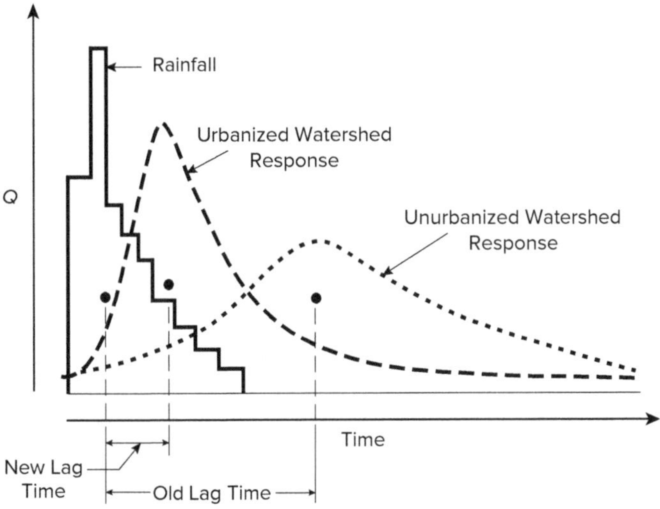

FIGURE 5.50 Effects of Urbanization on Runoff Hydrographs

5.5.2 Stormwater Controls

Stormwater controls are used to reduce the effects of development on receiving waterbodies by removing pollutants and attempting to mimic the natural hydrologic balance. They are designed to provide one or more of these key stormwater functions:

- **Peak Rate of Runoff.** Peak rates are typically controlled using an aboveground or underground storage area coupled with an outlet control structure. Detention storage reduces peak discharge rates by attenuating inflow and releasing it at a controlled rate.
- **Volume of Runoff, Infiltration, and Evapotranspiration.** Volume control is accomplished passively or actively through structural and nonstructural stormwater controls. Passive volume control occurs when stormwater is discharged over vegetated areas. Active volume control includes engineered stormwater controls that capture and retain runoff. Evapotranspiration and infiltration are the processes by which the retained runoff is permanently removed.
- **Water Quality of Runoff.** The removal of pollutants from stormwater runoff is accomplished by a number of physical phenomena. Sedimentation removes suspended solids by gravity; filtration removes particles (including those dissolved via ion exchange) by passing runoff through a fine media; and infiltration eliminates pollutants entirely by removing runoff from surface flow.

Examples of terms used to describe stormwater controls are: best management practice, stormwater control measure (SCM), and stormwater management facility.

Types of structural SCMs (not all inclusive):

- Basins: wet (retention), dry (detention), infiltration
- Vegetated swale
- Vegetated filter strip
- Filters: sand filter, bioretention (rain garden), green roof, manufactured
- Infiltrators: trench, dry well, permeable pavement

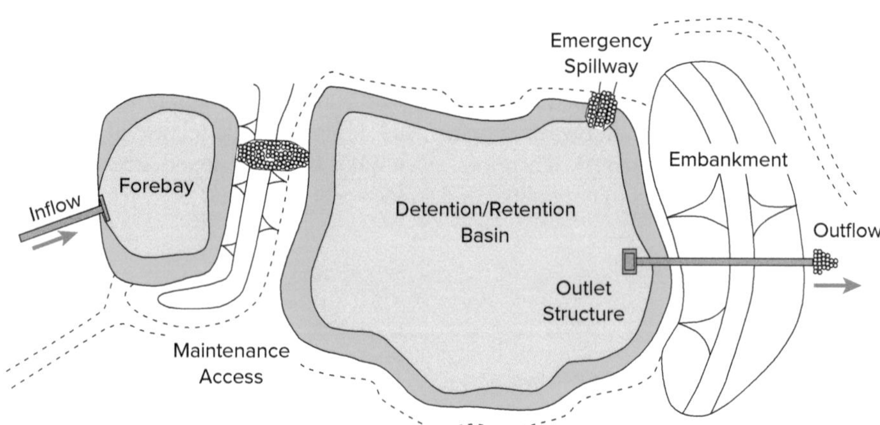

FIGURE 5.51 Stormwater Detention/Retention Basin

Types of source controls (not all inclusive):

- Elimination or disconnection of impervious surfaces
- Rainwater harvesting
- Soil and vegetative restoration
- Street sweeping

5.5.2.1 Detention Versus Volume Reduction

Stormwater detention basins have historically been used to control peak discharge rates of flow from an improved site. While they are effective in controlling peak rates, they cannot significantly reduce total stormwater volumes discharged. Detention ponds thus create a tradeoff in which the peak flow rate is reduced (compared to no stormwater controls), but the duration of higher-volume stormwater discharges is extended (see Figure 5.52). An unintended consequence of this is when a subwatershed has multiple detention basins, discharges can combine to create erosive flood conditions.

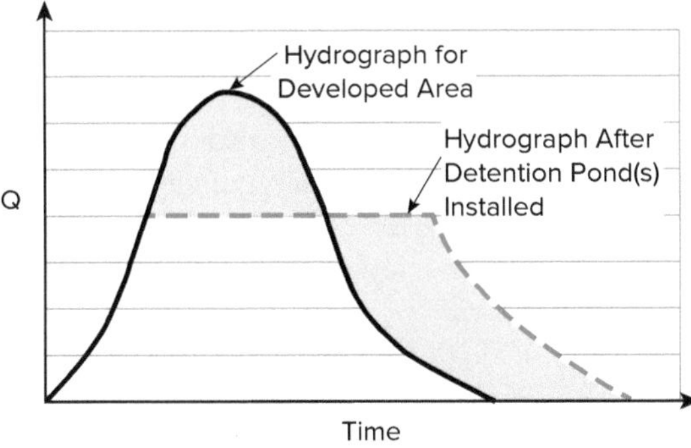

FIGURE 5.52 Effect of Stormwater Detention Basin on an Outflow Hydrograph

5.5.3 Sizing Stormwater Facilities

Detention basins manage peak flow rates by storing and slowly releasing stormwater runoff directed into the basin. The engineer sizes the basin and designs the outlet structure, which controls the effluent flow rate from the basin. A **stage-storage curve** indicates the volume of storage available at incremental elevations within the basin. The stage-discharge curve indicates the total discharge rate from the basin at those same incremental elevations. Figure 5.53 is an elevation view of a stormwater detention basin. An elevation scale is shown in the middle of the basin: elevation 0 is at the bottom, and elevation 5 is at the top. Also shown is the basin outlet structure. Notice that the invert of the lowest opening in the outlet structure is at elevation 1. This means that a portion of the basin's capacity is reserved for either a permanent pool or infiltration.

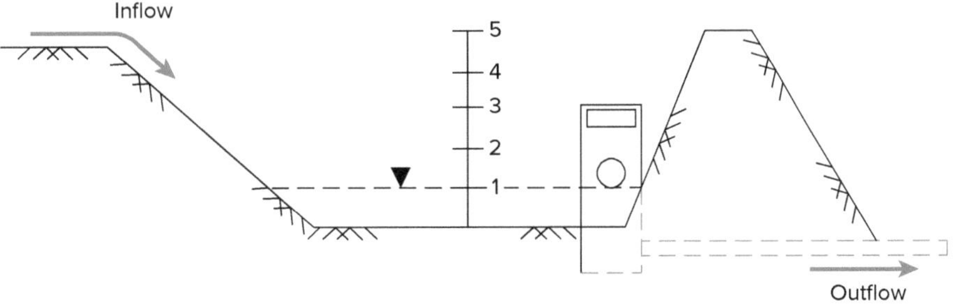

FIGURE 5.53 Elevation View of a Detention Basin

The stage-storage data for this basin is developed using the topographic contour lines for each incremental elevation. Figure 5.54 is a sample contour grading plan for a stormwater basin. The volume of storage available at each elevation within the basin is computed and tabulated, as shown in Table 5.9.

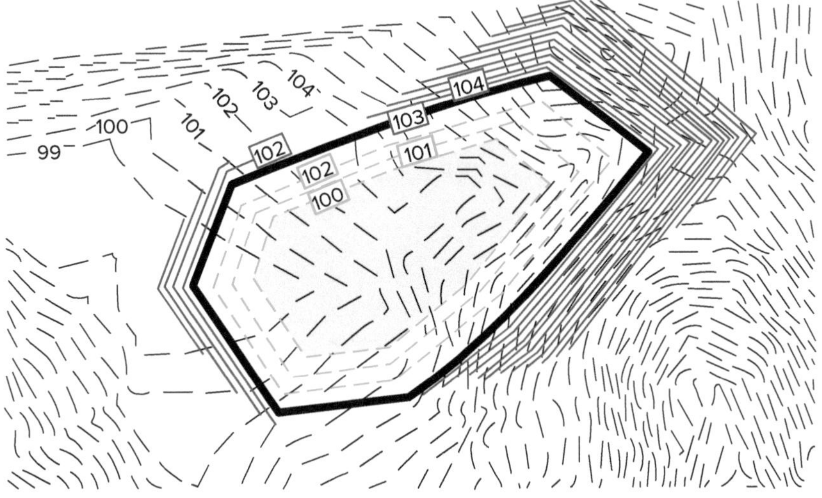

FIGURE 5.54 Contour Grading Plan of a Detention Basin

The stage-discharge data are determined by computing the flow through the outlet structure at various elevations. As the water surface elevation within the basin increases, the flow through the openings in the structure increases (as hydraulic head increases). Sample stage-discharge data are tabulated in Table 5.9. Notice that there is zero discharge until the water surface is above elevation 1. That is because of the location of the lowest opening in the outlet structure, which is above the basin bottom.

TABLE 5.9 Stage-Storage and Stage-Discharge Curve Data

STAGE (ft)	STORAGE (ft^3)	DISCHARGE (ft^3/s)
0	0	0
1	200	0
2	450	5
3	700	12
4	950	22
5	1,200	30

The inflow hydrograph of a detention basin will look different from the outflow hydrograph because of the effects of storage and managed release. When designed properly, the peak discharge rate of the outflow hydrograph should be no greater (and usually less) than that of the inflow hydrograph. When you superimpose the hydrographs, as shown in Figure 5.55, the intersection of the curves is meaningful. As indicated in the plot of basin storage (also in Fig. 5.55), the peak storage used (which is also the peak water surface elevation) in the basin is reached when the inflow and outflow hydrographs intersect, or when the outflow rate begins to exceed the inflow rate.

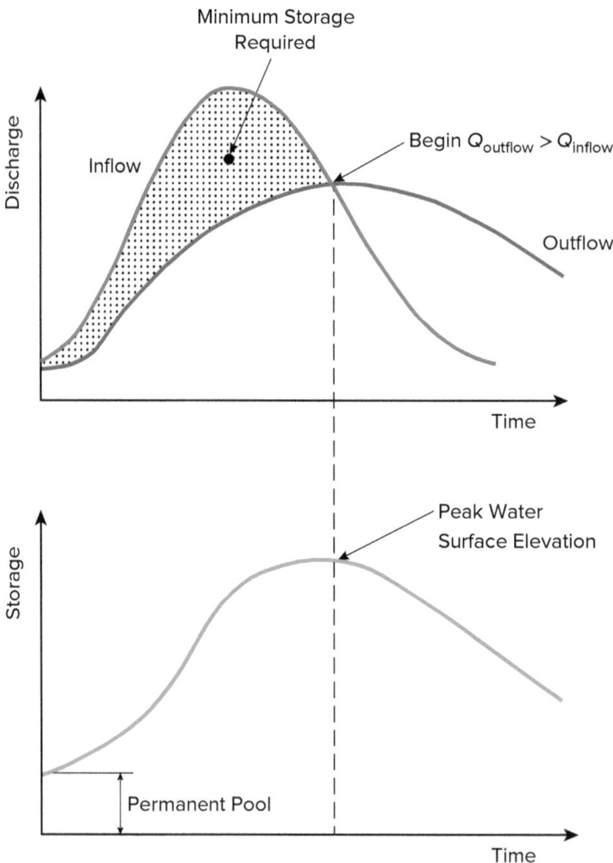

FIGURE 5.55 Inflow and Outflow Hydrographs from a Detention Basin

Another piece of information may be derived from the inflow and outflow hydrographs. The shaded area in Figure 5.55 indicates the minimum volume of storage that would be required for that basin. From the beginning of the hydrographs to the intersection point, the inflow rate exceeds the outflow rate. Thus, the water level in the basin is rising, and storage is needed to contain the increasing volume of water.

Example 5.14: Sizing Stormwater Facilities

Inflow and outflow hydrographs from a given storm event for a stormwater basin are shown below. The inflow hydrograph is a solid red line; the outflow hydrograph is a dashed blue line. The stormwater basin is designed to hold and infiltrate a portion of the stored volume, while the remaining volume is temporarily stored and released through an engineered outlet structure.

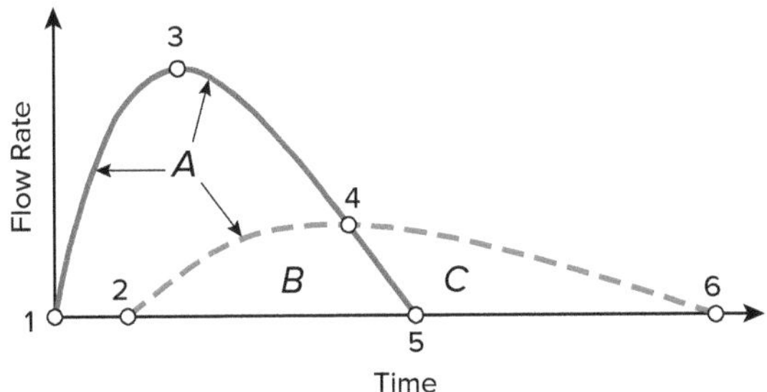

1. How would the volume of water retained in the basin (for infiltration) be calculated?
2. How would you determine the volume of storage used for this event?
3. At what point does the water level in the basin reach a maximum?

Solution

1. The volume of water retained in a basin is found by subtracting the entire area of the outflow hydrograph from the inflow hydrograph. In other words, total inflow minus total outflow is the volume retained. In this case: inflow $(A + B)$ – outflow $(B + C)$ = volume retained $(A - C)$.
2. The volume of storage used is found by subtracting the outflow volume that occurs during inflow from the total inflow. In this figure, the volume of storage is represented by area A.
3. The water level in the basin reaches a maximum when the outflow rate begins to exceed the inflow rate. This is represented by point 4 in the figure.

5.6 PRESSURE CONDUIT

5.6.1 Basic Water Properties

Water is an incompressible liquid because its volume does not change significantly with changing pressure. **Density** is the ratio of the mass to the volume of a substance. Specific weight, or unit weight, is the ratio of the weight (force due to gravity) to the volume of a substance. In the US customary system (USCS), because the unit of mass (slug) is rarely used in practice, pound-mass per cubic foot (lbm/ft^3) is the conventional unit for density. The unit for specific weight is pound-force per cubic foot (lbf/ft^3). Both density and **specific weight** have a numeric value of approximately 62.4 at 60 degrees Fahrenheit (°F). There are 7.48 gallons in a cubic foot, with each gallon weighing approximately 8.34 pounds.

In the metric system, the density of water is often cited as 1.0 gram per cubic centimeter (g/cm^3) or 1,000 kilograms per cubic meter (kg/m^3) at 4 degrees Celsius (°C). The term for weight is called a newton (N), and the specific weight of water is 9,800 newtons per cubic meter (9,800 N/m^3). This is expressed as 9.8 kilonewtons per cubic meter (9.8 kN/m^3), where the prefix kilo stands for 1,000.

The specific gravity (SG) of a liquid is the ratio of its density compared to that of pure water. Therefore, the specific gravity of pure water is 1. Pure water is at its highest density at 39°F (4°C).

Viscosity is an important fluid property when analyzing fluid motion near a solid boundary. It is a measure of the fluid's resistance to gradual deformation by shear or tensile stress. Shear resistance is caused by intermolecular friction when layers of fluid attempt to slide by one another. In relative terms, molasses is highly viscous, water is medium viscous, and oil is low viscous. There are two related measures of fluid viscosity: dynamic (or absolute) and kinematic. **Absolute viscosity** is a measure of internal resistance; **kinematic viscosity** is a ratio of absolute viscosity to density ($\nu = \mu/\rho$). Water viscosity units and values at various temperatures are provided in the Tables 5.10A and 5.10B.

TABLE 5.10A Basic Properties of Water (USCS)

TEMP. T (°F)	DENSITY ρ (slug/ft^3)	SPECIFIC WEIGHT γ (lbf/ft^3)	DYNAMIC VISCOSITY μ ($\times 10^{-5}$ lb-s/ft^2)	KINEMATIC VISCOSITY ν ($\times 10^{-5}$ ft^2/s)	VAPOR PRESSURE HEAD h_ν (ft)
32	1.940	62.42	3.746	1.931	0.20
40	1.940	62.43	3.229	1.664	0.28
50	1.940	62.41	2.735	1.410	0.41
60	1.938	62.37	2.359	1.217	0.59
70	1.936	62.30	2.050	1.059	0.84
80	1.934	62.22	1.799	0.930	1.17
90	1.931	62.11	1.595	0.826	1.62
100	1.927	62.00	1.424	0.739	2.21
110	1.923	61.86	1.284	0.667	2.97
120	1.918	61.71	1.168	0.609	3.96
130	1.913	61.55	1.069	0.558	5.21
140	1.908	61.38	0.981	0.514	6.78
150	1.902	61.20	0.905	0.476	8.76
160	1.896	61.00	0.838	0.442	11.21
170	1.890	60.80	0.780	0.413	14.20
180	1.883	60.58	0.726	0.385	17.87
190	1.876	60.36	0.678	0.362	22.29
200	1.868	60.12	0.637	0.341	27.61
212	1.860	59.83	0.593	0.319	35.38

TABLE 5.10B Basic Properties of Water (SI)

TEMP. T (°C)	DENSITY ρ (kg/m^3)	SPECIFIC WEIGHT γ (kN/m^3)	DYNAMIC VISCOSITY μ ($\times 10^{-3}$ kg/m-s)	KINEMATIC VISCOSITY ν ($\times 10^{-6}$ m^2/s)	VAPOR PRESSURE P_ν (kN/m^2)
0	999.8	9.805	1.781	1.785	0.61
5	1,000.0	9.807	1.518	1.519	0.87
10	999.7	9.804	1.307	1.306	1.23
15	999.1	9.798	1.139	1.139	1.70
20	998.2	9.789	1.002	1.003	2.34
25	997.0	9.777	0.890	0.893	3.17
30	995.7	9.764	0.798	0.800	4.24
40	992.2	9.730	0.653	0.658	7.38
50	988.0	9.689	0.547	0.553	12.33
60	983.2	9.642	0.466	0.474	19.92
70	977.8	9.589	0.404	0.413	31.16
80	971.8	9.530	0.354	0.364	47.34
90	965.3	9.466	0.315	0.326	70.10
100	958.4	9.399	0.282	0.294	101.33

5.6.2 Pressure

The definition of pressure is the amount of force acting on a surface per unit area, as shown in Figure 5.56.

Simply stated, pressure = force/area.

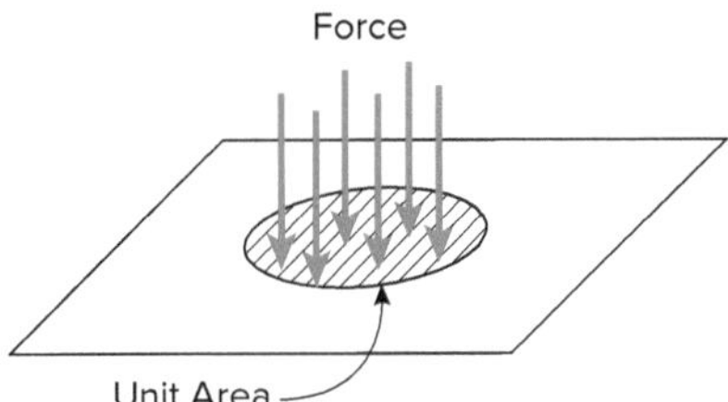

FIGURE 5.56 Pressure Definition

Water, whether it is stored in a tank or flowing through a pipe, exerts force (and pressure) against the walls of its container. Pressure is usually expressed in pounds per square inch (lbf/in^2 or psi) in US units; in SI metric units, it is newtons per square meter (N/m^2), which is also called a Pascal (Pa). Kilopascal (kPa) is used in most hydraulics applications (1 kPa = 1,000 Pa = 0.145 psi) because 1 Pa is relatively small (1 Pa = 0.000145 psi). Pressure can also be expressed in atmosphere, bar, kilopascal, and inches (or mm) of mercury or water.

Hydrostatic pressure is the pressure exerted by a fluid due to the force of gravity. In water systems, hydrostatic pressure increases with depth measured from the surface because of the increasing weight of the water above. It depends only upon the depth of water, the density of water, and the acceleration of gravity.

$$P_{\text{hydrostatic}} = \frac{\text{weight}}{\text{area}} = \frac{\text{mg}}{A} = \frac{\rho V g}{A} = \rho g h$$ **Equation 5-51**

Water pressure and depth can be converted back and forth using simple expressions. For example, 1.0 psi of pressure is equivalent to 2.31 ft of water height. Similarly, 1.0 ft of water height is approximately equal to 0.434 psi of pressure. These are called fluid height equivalents for pressure. The following expressions may be useful in pressure flow calculations.

US unit: P (psi) = 0.43 × h (ft) **Equation 5-52**

SI metric units: P (kPa) = 9.8 × h (m) **Equation 5-53**

Absolute pressure, P_{absolute}, is the total pressure on an object in a body of water (for example, a lake or an ocean) or a container when the container is open to the atmosphere above. The pressure is created from the combined weight of the water and atmospheric molecules. Absolute pressure is said to be measured relative to zero pressure (a vacuum). Gauge pressure, P_{gauge}, is the pressure measured relative to the ambient air (atmospheric) pressure, $P_{\text{atmospheric}}$. At sea level, atmospheric pressure is 14.7 psi (or 1 atm).

$$P_{\text{absolute}} = P_{\text{atmospheric}} + P_{\text{gauge}}$$ **Equation 5-54**

5.6.2.1 Measuring Devices

Total (stagnation) pressure is the sum of the static and dynamic pressures in a piping system. **Static pressure** is measured with gauges attached to the side of a pipe or tank wall; it represents the pressure felt at rest. **Dynamic pressure** represents the kinetic energy of water in motion and is a function of flow velocity and density. Computationally, it is the difference between the total pressure and the static pressure. Figure 5.57 depicts the differences between total, static, and dynamic pressure.

$$P_{\text{total}} = P_{\text{static}} + P_{\text{dynamic}}$$ **Equation 5-55**

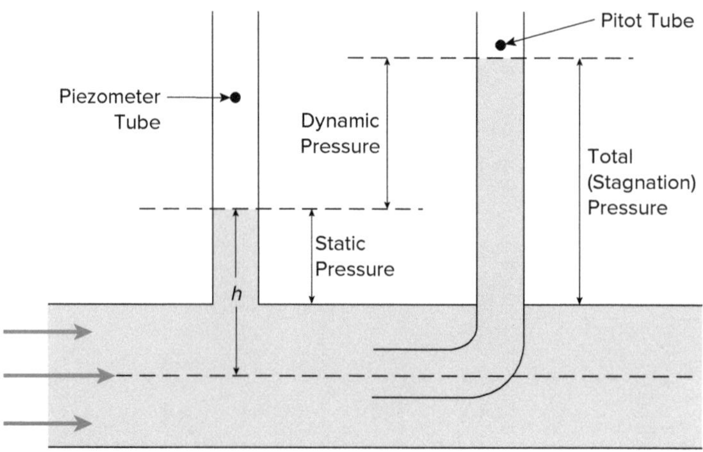

FIGURE 5.57 Pressure-Measuring Devices

5.6.3 Pipe Materials and Sizes

Pipes used to distribute drinking water are made of plastic, metal, or concrete. The most commonly used materials for piping carrying drinking-water supply are galvanized steel or iron, copper, unplasticized polyvinylchloride (PVC), chlorinated polyvinylchloride (CPVC), and polyethylene (PE). All of them have advantages and disadvantages; the properties of each pipe material should fulfill some specified requirements. When in contact with water or soil, the material should be resistant (corrosion-proof) to possible chemical reactions, and the material should not allow toxic substances to be released into the water. Pipes also have to be resistant against specified internal and external pressures.

Galvanized iron and steel were once commonly used as conduits for drinking water, but their use is declining. Internal corrosion causes rust discoloration and can impart an undesirable taste and smell. They are still commonly used for fire protection applications. Nominal diameters for pressure applications range from ⅛ in to 30 in.

PVC and CPVC pipes are much lighter than iron and steel pipes and do not corrode. Plastic pipes, in general, are more susceptible to damage when exposed above ground, and they become brittle when exposed to ultraviolet light. Nominal diameters for pressure applications range from ⅛ in to 24 in.

PE pipes are grouped into high-, medium-, and low-density PE. Medium- and low-density PE pipes have thinner walls and are more flexible than high-density PE pipes. PE is a preferred material for long-distance water supply piping.

Copper tubing is smaller in overall diameter than the equivalent galvanized steel pipes and fittings. Due to its thinner wall section, it is relatively light to handle and is available in coil form or straight lengths as required. Copper tube or pipe is also particularly useful for hot-water supply systems.

Water service pipe is normally no smaller than ¾ in. Most codes stipulate a minimum ½-in pipe diameter for each type of fixture. Most codes also require residential static water pressure to be between 40 and 80 psi. If the municipal pressure is higher than 80 psi, codes normally require the installation of a pressure-reducing valve to knock the pressure down to 80 psi. A booster pump and pressure tank are installed in areas where the municipal pressure is too low.

Certain pipe materials, such as steel and PVC, have nominal pipe sizes, which often do not equal the inside or outside diameter of the pipe. For example, steel pipes with nominal sizes between ⅛ in and 12 in have inside and outside diameters that are different from the nominal diameter. The outside diameter is equal to the nominal diameter for 14-in pipes and larger.

One should use only the inside pipe diameter when performing hydraulic calculations.

TIP

Due to the large variety of pipe materials and sizes, relevant pipe dimensions are normally given on the PE Civil exam.

5.7 CONTINUITY AND ENERGY EQUATIONS

5.7.1 Continuity Equation

A fundamental principle used in analyzing uniform flow is **flow continuity**. Mass is always conserved in water systems regardless of the pipeline configuration and flow direction. **Constant flow** in a channel means that the product of the area and velocity will be the same for any two cross sections within that channel. Equation 5-56 and Figure 5.58 illustrate this principle.

$$Q = A_1 v_1 = A_2 v_2$$ **Equation 5-56**

A = cross-sectional area (ft^2) = $\pi D^2/4$ for a circular pipe
D = pipe diameter (ft)
v = velocity (ft/s)
Q = volumetric flow rate (ft^3/s)
Unit conversion: 1 ft^3/s = 448.83 gallons per minute (gpm)

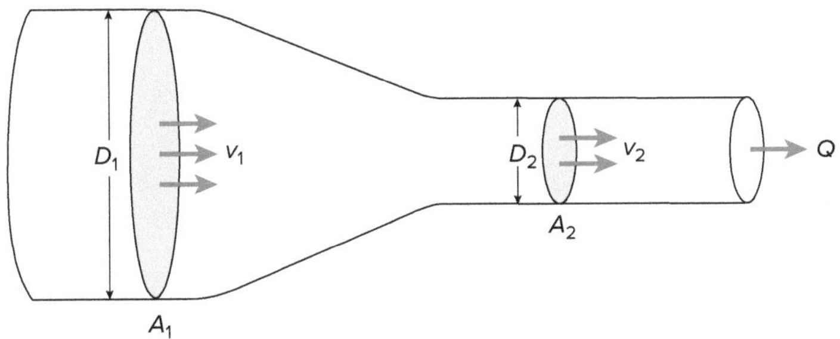

FIGURE 5.58 Continuity Principle

Example 5.15: Continuity

Ten cubic feet per second flow through a pipe of changing dimension, as shown below. What does the diameter of the second pipe section need to be (in feet) to double the velocity compared to the diameter of the first pipe section?

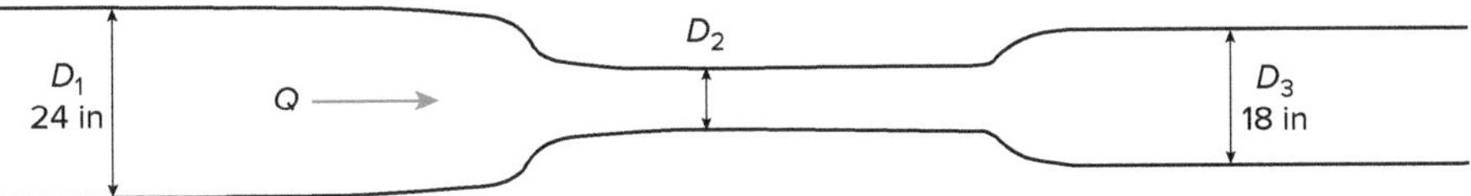

Solution

First, solve for the velocity in the first pipe section.

$$v = \frac{Q}{A}$$

$$Q = 10 \text{ ft}^3/\text{s}$$

$$A_1 = \frac{\pi D_1^2}{4} = \frac{\pi (2.0 \text{ ft})^2}{4} = 3.14 \text{ ft}^2$$

$$v_1 = (10 \text{ ft}^3/\text{s})(3.14 \text{ ft}^2) = 3.2 \text{ ft/s}$$

Per the problem statement, the velocity in the second pipe section is twice that of the first:

$$v_2 = 2v_1 = 2(3.2 \text{ ft/s}) = 6.4 \text{ ft/s}$$

$$A_2 = \left(\frac{\pi}{4}\right) D_2^2$$

$$10 \text{ ft}^3/\text{s} = (6.4 \text{ ft/s})\left(\frac{\pi}{4}\right) D_2^2$$

$$D_2^2 = 2.0 \blacktriangleright D_2 = 1.4 \text{ ft}$$

5.7.2 Fluid Energy (Bernoulli Equation)

Water in any piping system possesses energy. The **total energy** is the sum of kinetic, potential, and pressure energies. The Bernoulli equation says that in a frictionless environment, energy is conserved. Thus, throughout a water piping system, as elevations and pipe dimensions change, energy converts from one form to another, while the total energy remains constant.

Energy head can be thought of as the height of a water column. The total energy head, H (ft), at any point in a water supply and distribution system is the sum of three types of energy heads.

$$H = h_v + h_p + h_z$$ **Equation 5-57**

A moving mass of water possesses more energy than a stationary one—the difference is kinetic energy. In the Bernoulli equation, this is called **velocity head**, h_v.

$$h_v = v^2/2g$$ **Equation 5-58**

v = velocity (ft/s)

$g = 32.2\ \text{ft/s}^2$

A mass of water at a high pressure possesses more energy than one at a lower pressure—the difference is pressure energy. In the Bernoulli equation, this is called **pressure head**, h_p.

$$h_p = p/\gamma$$ **Equation 5-59**

p = pressure (lbf/ft^2)

γ = specific weight (lbf/ft^3)

Water at a higher elevation possesses more energy than water at a lower elevation—the difference is potential energy. In the Bernoulli equation, this is called **elevation head**, h_z.

$$h_z = z$$ **Equation 5-60**

z = elevation (ft)

Problems often involve calculating a pressure, elevation, or velocity at a point in the system. When the unknown parameter is at a different elevation (z) or it is in a pipe of a different diameter (D) compared to the point it is being compared to, it is a Bernoulli equation energy balance problem. For example, for the diagram in Figure 5.59, $H_A + h_A = H_B + h_f$. The total energy in reservoir A, H_A, plus the energy head added by the pump, h_A, is equal to the total energy in reservoir B, H_B, plus the friction head in the pipelines between the reservoirs, h_f. Pump energy head and friction head are discussed in the sections that follow.

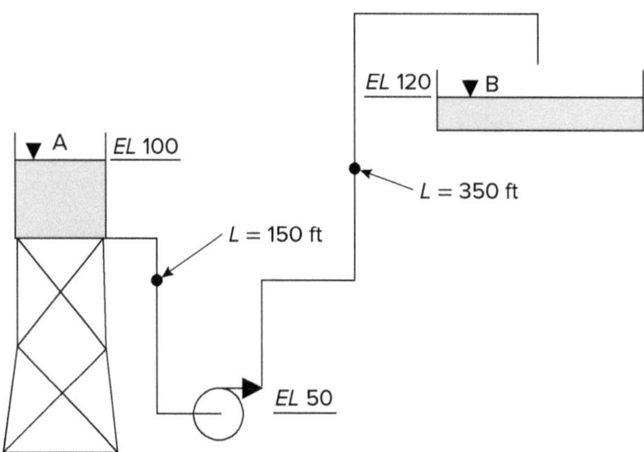

FIGURE 5.59 Problem Requiring a Bernoulli Energy Balance

5.7.3 Friction Losses

When water flows through a pipe, the pressure (and pressure head) continuously drops in the streamwise direction because of friction along the walls of the pipe, changes in pipe geometry, and various components connected to the system. Several equations have been developed to calculate the head loss due to friction. The applicability of each method depends, in part, on the flow characteristics, such as whether laminar or turbulent flow exists. The **Reynolds number** is a dimensionless parameter that represents the ratio of inertial forces to viscous forces.

$$R_e = \frac{D_h v}{\nu}$$ **Equation 5-61**

D_h = hydraulic diameter (ft)

v = velocity (ft/s)

ν = kinematic viscosity (ft^2/s)

Laminar flow occurs when the Reynolds number is less than 2,300. Velocities are low, and fluid particles move parallel to the overall flow direction. Laminar flow is rare in practice in water systems. **Turbulent flow** occurs when the Reynolds number is greater than 4,000. Velocities are higher, and fluid particle paths are completely irregular. It is what normally occurs in water systems. By definition, **transitional flow** is what occurs between Reynolds numbers of 2,300 and 4,000. Transitional flow has no practical application. Figure 5.60 depicts the three types of flow.

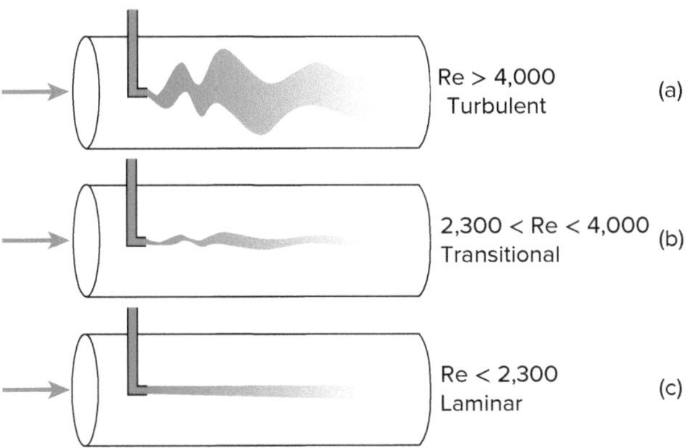

FIGURE 5.60 Flow Types in Fluid Dynamics

Example 5.16: Turbulent Flow

What is the minimum velocity, in feet per second, in a 6-in-diameter pipe carrying water at 50°F to be considered turbulent flow?

Solution

Rearrange Equation 5-61.

$$v = \frac{\nu R_e}{D_h}$$

Kinematic viscosity of water at 50°F = 1.410×10^{-5} ft^2/s

$$v = \frac{(1.410 \times 10^{-5}\ \text{ft}^2/\text{s})(4{,}000)}{(0.5\ \text{ft})} = 0.11\ \text{ft/s}$$

5.7.3.1 Major Losses

Major losses, denoted by h_f (or sometimes h_{major}), are due to frictional losses through sections of constant-diameter pipe, in which the flow is fully developed.

5.7.3.1.1 Darcy-Weisbach Equation

Friction head, h_f (ft), is the energy required to overcome resistance to flow in pipes.

$$h_f = \frac{fLv^2}{2Dg} \qquad \textbf{Equation 5-62}$$

f = Darcy's friction factor (dimensionless)

L = pipe length (ft)

v = fluid velocity (ft/s)

D = inside pipe diameter (ft)

g = gravitational constant = 32.2 ft/s^2

Darcy's friction factor, f, is found using the following for circular pipes:

Laminar flow:

$$f = 64/Re \qquad \textbf{Equation 5-63}$$

> **TIP**
>
> If needed to solve a problem on the PE Civil exam, f would likely be given.

Turbulent flow:

Moody diagram (shown in Figure 5.61; it can be found in most hydraulics references)

- Need Re and ε/D (relative roughness)
- ε is specific roughness, a pipe material property; see Table 5.11.

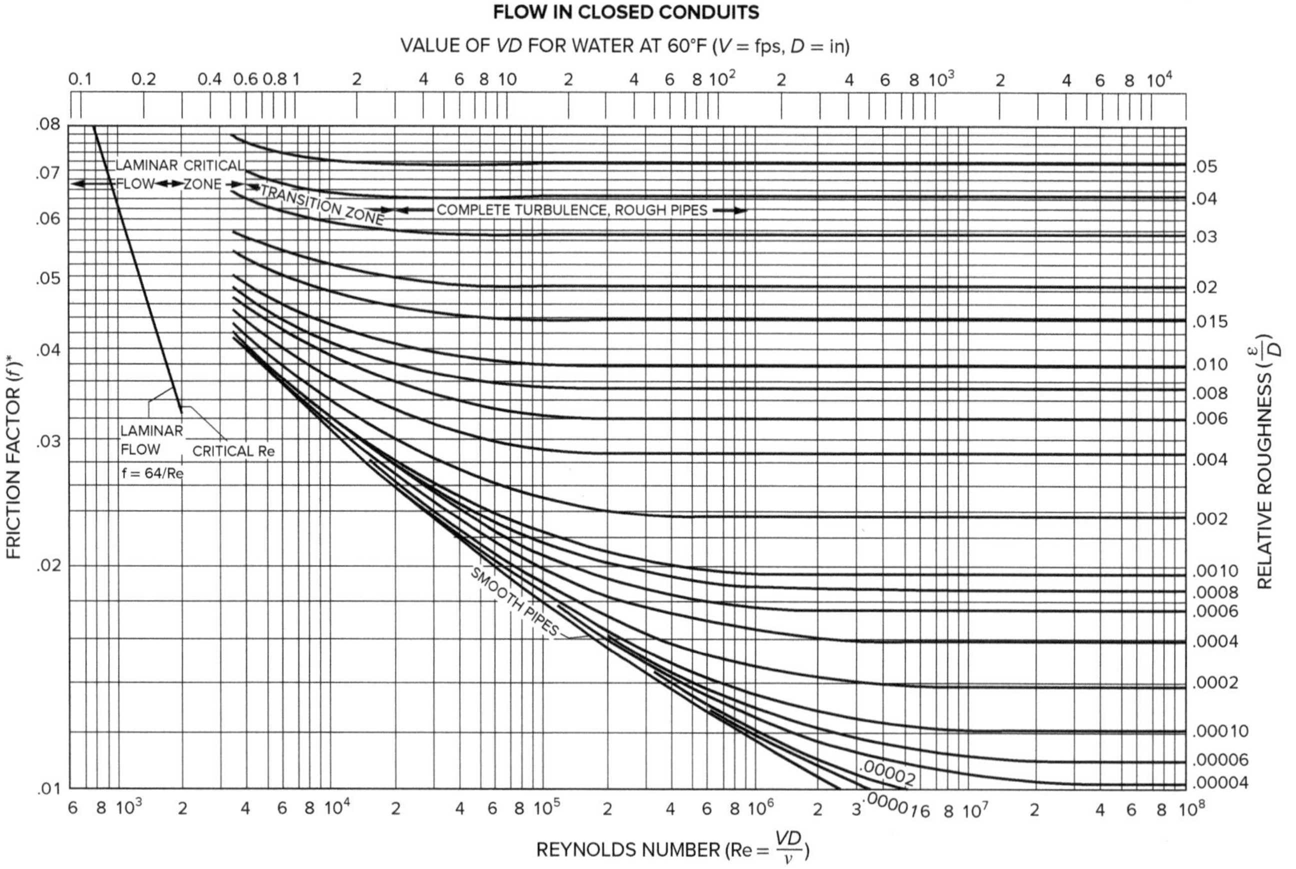

FLOW IN CLOSED CONDUITS

VALUE OF VD FOR WATER AT 60°F (V = fps, D = in)

*The Fanning Friction is this factor divided by 4.

FIGURE 5.61 Moody Diagram for Finding Darcy Friction Factor

TABLE 5.11 Specific Roughness for Select Pipe Materials

	SPECIFIC ROUGHNESS, ε	
MATERIAL	**ft**	**mm**
Plastic (Smooth)	0.000005	0.0015
Commercial Steel	0.00015	0.045
Cast Iron	0.00085	0.26
Concrete	0.001 to 0.01	0.3 to 3.0

When solving for velocity, the Darcy-Weisbach equation becomes:

$$v = \sqrt{\frac{2Dgh_f}{fL}}$$ **Equation 5-64**

$$h_f = \frac{\Delta p}{\gamma}$$ **Equation 5-65**

Note: Equation 5-65 for Δp is applicable only when the changes in elevation and velocity heads are negligible. It is the pressure head loss (in feet) due to pipe friction.

Example 5.17: Darcy-Weisbach Equation

Water is pumped from a lake at 240 ft above mean sea level to a water treatment plant at 300 ft above mean sea level. Water is conveyed through 3,000 ft of 6-in inside diameter galvanized steel pipe. Minor losses are negligible. The velocity and friction factor are 5 ft/s and 0.017, respectively. If 60 psig of pressure is needed at the water treatment plant, what pressure, in psig, is required in the pipeline at the lake?

Solution

Set up energy balance because there is a change in elevation between the ends of the pipe and there is friction loss.

$$H_A = H_B + h_f$$

$$h_{pA} + h_{vA} + h_{zA} = h_{pB} + h_{vB} + h_{zB} + h_f$$

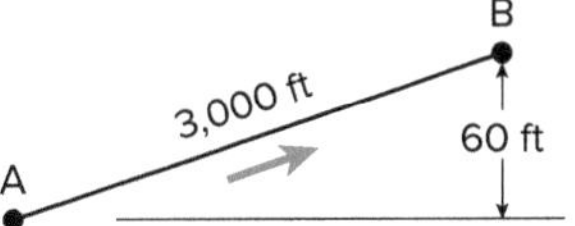

Velocity terms cancel because the pipe size is constant.

Set the beginning of the pipeline at $EL = 0$, such that the end is $EL = 60$ ft.

$$h_{pA} = h_{pB} + 60 + h_f$$

$$h_f = \frac{fLv^2}{2Dg} = \frac{0.017(3{,}000\text{ ft})(5\text{ ft/s})^2}{2(0.50\text{ ft})(32.2\text{ ft/s}^2)} = 40\text{ ft}$$

$$h_{pB} = 60\text{ psi} \times 2.31\text{ ft/psi} = 139\text{ ft}$$

$$h_{pA} = 139\text{ ft} + 60\text{ ft} + 40\text{ ft} = 239\text{ ft}$$

$$P_A = 239\text{ ft}/2.31\text{ ft/psi} = 103\text{ psi}$$

5.7.3.1.2 Hazen-Williams Equation

Friction head, h_f, can also be calculated with the Hazen-Williams equation. This equation is only applicable to turbulent flow and liquids.

$$h_f = \frac{3.022v^{1.85}L}{C^{1.85}D^{1.17}}$$ **Equation 5-66**

$$h_f = \frac{10.44Q^{1.85}L}{C^{1.85}d^{4.87}}$$ **Equation 5-67**

v = fluid velocity (ft/s)

Q = flow rate (gpm)

L = pipe length (ft)

C = friction coefficient; see Table 5.12

D = inside pipe diameter (ft)

d = inside pipe diameter (in)

TIP

If needed to solve a problem on the PE Civil exam, *C* would likely be given.

TABLE 5.12 Hazen-Williams Coefficients for Select Pipes

VALUES OF HAZEN-WILLIAMS COEFFICIENT *C*	
Pipe Material	*C*
Ductile Iron	140
Concrete (regardless of age)	130
Cast Iron:	
New	130
5 yr old	120
20 yr old	100
Welded Steel, New	120
Wood Stave (regardless of age)	120
Vitrified Clay	110
Riveted Steel, New	110
Brick Sewers	100
Asbestos-Cement	140
Plastic	150

Source: *NCEES Reference Handbook, v 9.5*, 2013.

The Hazen-Williams equation can be rearranged to solve for other variables.

$$v = \left(\frac{h_f C^{1.85} D^{1.17}}{3.022L}\right)^{0.54}$$ **Equation 5-68**

$$L = \frac{h_f C^{1.85} D^{1.17}}{3.022 v^{1.85}}$$ **Equation 5-69**

$$D = \left(\frac{3.022 v^{1.85} L}{C^{1.85} h_f}\right)^{0.855}$$ **Equation 5-70**

Example 5.18: Hazen-Williams Equation

Two pressure gauges indicate a loss of 10 psig of pressure across 250 ft of 3-in inside diameter PVC pipe (C = 120). What is the approximate velocity of water in the pipe in feet per second?

Solution

In this problem, there is only a pressure loss, so a Bernoulli energy balance is not necessary. The friction loss equation alone can be used to solve the problem.

$$v = \left(\frac{h_f C^{1.85} D^{1.17}}{3.022L}\right)^{0.54}$$

$$h_f = \Delta h_p = (10\text{ psi})(2.31\text{ ft/psi}) = 23.1\text{ ft}$$

$$v = \left[\frac{(23.1\text{ ft})(120)^{1.85}(0.25\text{ ft})^{1.17}}{3.022(250\text{ ft})}\right]^{0.54} = 7.6\text{ ft/s}$$

5.7.3.2 Minor Losses

Minor (or local) losses, denoted by h_{minor}, are those caused by changes in velocity direction or magnitude from devices in the piping system (that is, anything other than straight, constant-diameter pipe sections). These include entrance and exit transitions (to/from a reservoir), expansion and contraction, and pipe connections and fittings (Fig. 5.62). These devices induce flow separation and mixing, which cause additional head losses. These losses are called minor because they are small compared to the friction head (h_f) in a typical water distribution system with long pipes. However, as illustrated below, in systems with many pipe fittings over a short distance, the minor losses may actually be greater than the major (friction) losses. Also, a head loss for a valve may be very small if it is completely open, but a partially closed valve can result in a large head loss.

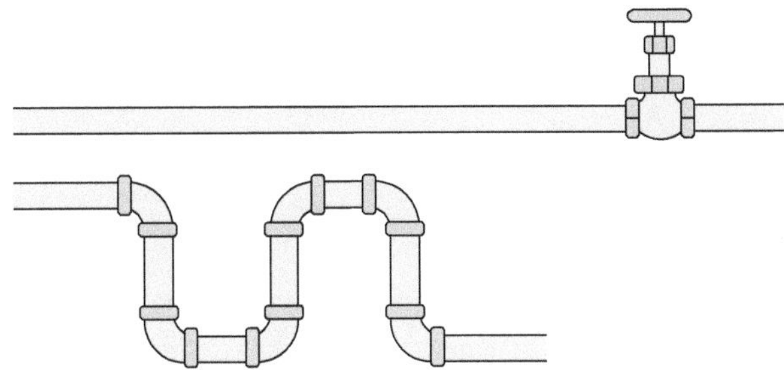

FIGURE 5.62 Fittings in Water Lines

Two approaches can be used to calculate minor head losses in piping systems: **loss coefficients** and **equivalent lengths**. While both approaches are used in practice, the use of loss coefficients is more common.

5.7.3.2.1 Loss Coefficients

The general equation for minor losses in pipes with the same diameter and velocity both upstream and downstream of the non-uniformity is:

$$h_{\text{minor}} = K_L \frac{v^2}{2g}$$ **Equation 5-71**

> **TIP**
>
> If needed to solve a problem on the PE Civil exam, values of K_L would likely be given.

K_L = loss coefficient (dimensionless)

v = flow velocity (ft/s)

Loss coefficients are experimentally derived. Examples are given in Table 5.13.

TABLE 5.13 K_L Factors for Commonly Used Components

VALVE OR FITTING		*K* FACTOR
Globe Valve	Wide open	10
	1/2 open	12.5
Gate Valve	Wide open	0.20
	3/4 open	0.90
	1/2 open	4.5
	1/4 open	24
Return Bend		2.2
Standard Tee		1.8
Standard Elbow		0.90
45° Elbow		0.42
90° Elbow		0.75
Ball Check Valve		4
Union Socket		0.04

5.7.3.2.2 Equivalent Lengths

The head loss through a system component can be converted to a length of pipe equivalent, L_e, needed for the same head loss:

$$h_m = f \frac{L_e}{D} \frac{v^2}{2g}$$ **Equation 5-72**

To find L_e for a pipeline with a constant diameter, the k values can be added together and multiplied by D/f:

$$L_e = \frac{kD}{f}$$ **Equation 5-73**

$$L_{\text{total}} = L_{\text{actual}} + \Sigma L_e$$ **Equation 5-74**

Example 5.19: Equivalent Lengths

The minor loss coefficients add up to $k = 5$ in 1,000 ft of a 12-in-diameter pipe. If $f = 0.020$, what is the equivalent length of the minor losses in feet?

Solution

$$L_e = \frac{kD}{f} = \frac{5(12\text{ in})\left(\dfrac{1\text{ ft}}{12\text{ in}}\right)}{0.02} = 250\text{ ft}$$

Conceptually, the friction loss from 250 ft of 12-in pipe is the same as the combined minor head losses in this 1,000 ft of 12-in pipe.

5.7.3.3 Total Friction Head Loss

The total head loss due to friction (h_L) is the sum of all major (h_f or h_{major}) and minor (h_{minor}) losses in a system. For a system where pipe diameter is constant, the following equation can be applied to find all friction-related losses:

$$\Delta h_L = \Sigma h_{\text{major}} + \Sigma h_{\text{minor}} = \frac{v^2}{2g}\left(\frac{fL}{D} + \Sigma K_L\right) \qquad \textbf{Equation 5-75}$$

5.7.4 Pipe Networks

There are two basic pipe configurations in a pipe network: series and parallel. When pipes are connected in series, volumetric flow rate is constant, and head loss is the sum of the head loss in the components (Figs. 5.63 and 5.64). When pipes are configured in parallel, volumetric flow rate is the sum of the flow in the components, and head loss across all branches is the same.

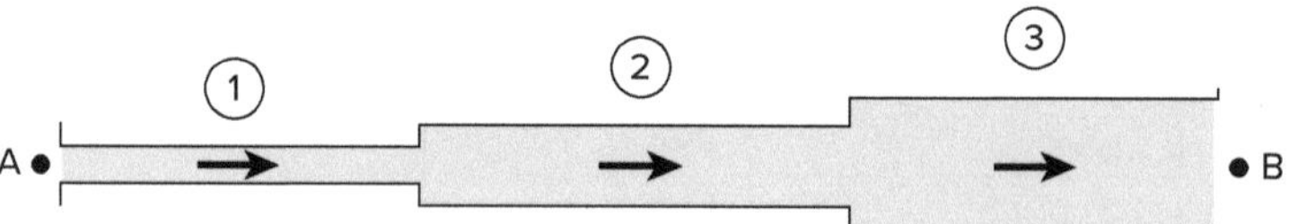

FIGURE 5.63 Pipes in Series

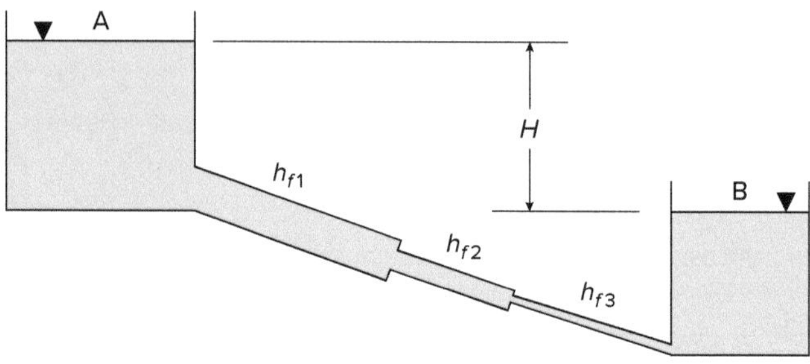

FIGURE 5.64 Pipes in Series Between Reservoirs

5.7.4.1 Pipes in Series

Friction losses are added together for pipes connected in series.

$$h_{f\text{total}} = h_{fa} + h_{fb} \qquad \textbf{Equation 5-76}$$

5.7.4.2 Pipes in Parallel

The three basic principles for analyzing pipes in parallel are shown in Figure 5.65 and explained in the list following.

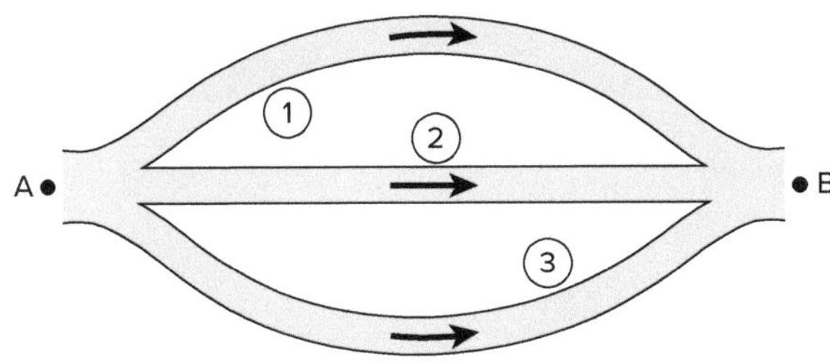

FIGURE 5.65 Flow in Parallel Branches

1. Flow divides so that friction loss is equal in the parallel pipes.

$$h_{f1} = h_{f2} = h_{f3} \quad \textbf{Equation 5-77}$$

2. Head loss between junctions can be calculated using any branch.

$$h_{fA-B} = h_{f1} = h_{f2} = h_{f3} \quad \textbf{Equation 5-78}$$

3. Total flow between junctions is equal to the sum of flow in the branches.

$$Q_A = Q_1 + Q_2 + Q_3 = Q_B \quad \textbf{Equation 5-79}$$

$$Q_A = \frac{\pi}{4}\left(D_1^2 v_1 + D_2^2 v_2 + D_3^2 v_3\right) = Q_B \quad \textbf{Equation 5-80}$$

If the system has only two parallel branches, the unknown branch flows can be determined by solving Equations 5-77 through 5-80 simultaneously. See Fig. 5.66.

The primary hydraulic properties of a pipe are roughness (C), diameter (D), and length (L). Suppose two parallel branches have identical properties—one would expect the flow rate in the pipes to be equal. What would happen if one of the properties were changed for one of the pipes?

- The shortest one (L) will carry more flow.
- The smoothest one (C) will carry more flow.
- The largest one (D) will carry more flow.

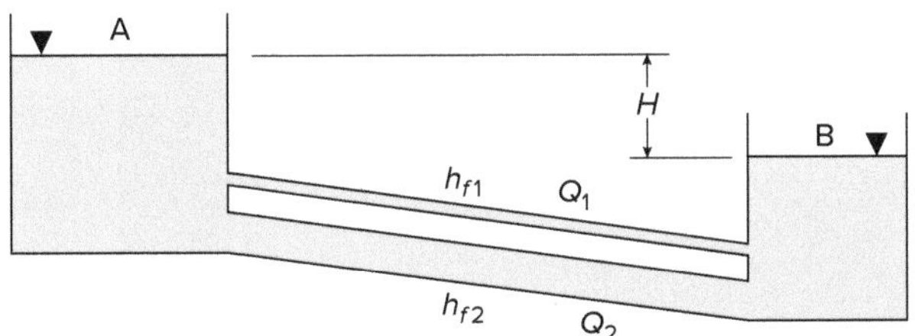

FIGURE 5.66 Pipes in Parallel Between Reservoirs

Example 5.20: Pipes in Parallel

A pressure gauge at the junction of pipe 1 with two pipe branches—a and b—reads 30 psig, while the junction with pipe 2 reads 25 psig. Assuming there are no significant elevation differences, what is the velocity in branch a, in feet per second?

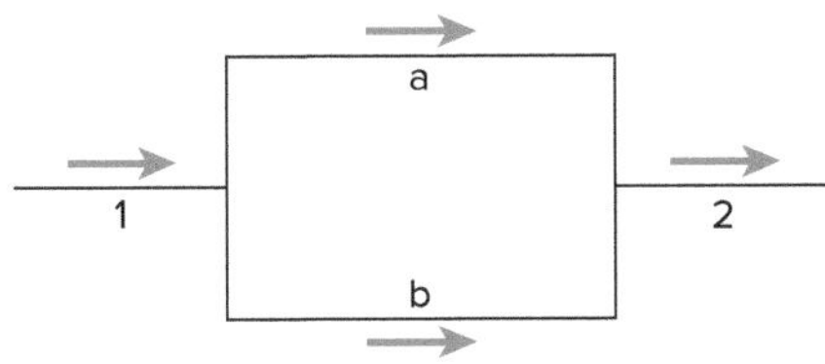

Example 5.20 (continued)

Branch a: $L = 400$ ft, $D = 12$ in, $C = 100$

Branch b: $L = 300$ ft, $D = 12$ in, $C = 120$

Solution

$$v = \left(\frac{h_f C^{1.85} D^{1.17}}{3.022L}\right)^{0.541}$$

$h_f(\text{ft}) = \Delta P(\text{psi}) \times (2.31\ \text{ft/psi}) = 5 \times 2.31 = 11.55$ ft

$$v = \left[\frac{(11.55\ \text{ft})(100)^{1.85}(1.0\ \text{ft})^{1.17}}{3.022(400\ \text{ft})}\right]^{0.541} = 8.1\ \text{ft/s}$$

TIP

As noted in previous sections, all energy head terms are in units of feet or meters.

5.7.5 Hydraulic and Energy Grade Lines

The EGL is a linear plot of the total energy head (H) along a closed system.

$$\text{EGL} = h_p + h_v + h_z \qquad \textbf{Equation 5-81}$$

The HGL does not take into account kinetic energy (velocity head) and plots the elevation that one would find using a piezometer (to measure static energy).

$$\text{HGL} = h_p + h_z \qquad \textbf{Equation 5-82}$$

As shown in Figure 5.67, for a straight horizontal pipe, the change in the EGL from section 1 to section 2 is the head loss due to friction (h_f). The friction slope S can be found by dividing h_f by the pipe length L.

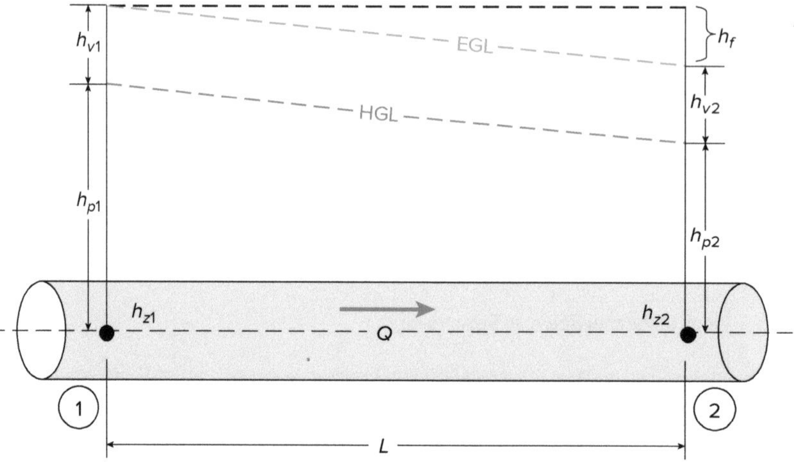

FIGURE 5.67 Hydraulic and Energy Grade Lines

5.7.6 Pumps

Pumps are hydraulic machines used to move or raise water and generate the delivery force (pressure). Pumps transfer water by converting mechanical energy into pressure energy. The pressure applied to the water forces the water to flow at the required discharge rate and to overcome friction and local (or minor) head losses in the pipeline.

5.7.6.1 Types of Pumps

Pumps can be divided into two families based on their application and capabilities (Fig. 5.68). Dynamic (centrifugal) pumps and (positive) displacement pumps are the two major groups.

- **Dynamic.** Kinetic energy is continuously added to increase the fluid velocity within the pump, which in turn is converted to static pressure energy. Centrifugal pumps are most commonly used in water distribution systems.
- **Positive displacement.** Energy is periodically added by application of force, or mechanical displacement. These pumps have a piston (or equivalent) moving in a closely fitting cylinder, and forces are exerted on the fluid by motion of the piston.

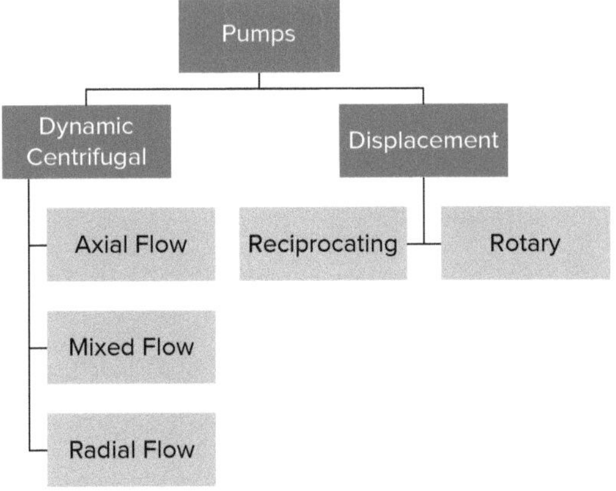

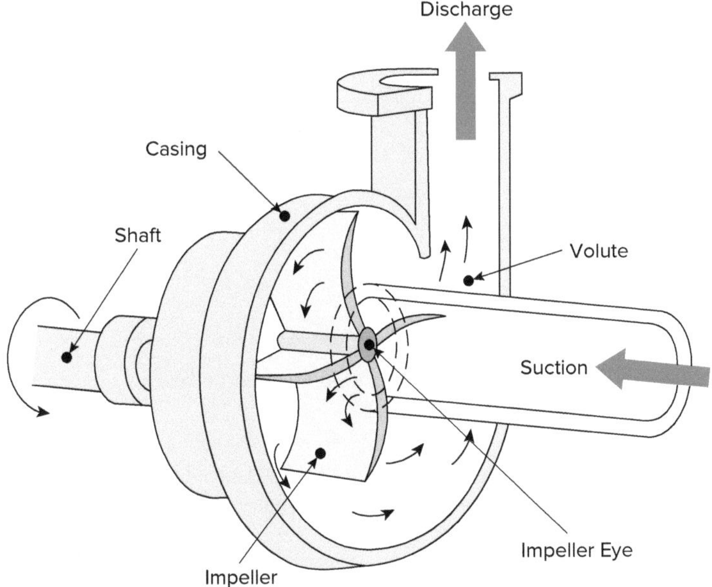

FIGURE 5.68 Types of Pumps and Centrifugal Pump Illustration

5.7.6.2 Total Dynamic Head

When a pump is lifting or pumping water, total dynamic head (TDH) is the vertical distance (in feet or meters) from the elevation of the EGL on the suction side of the pump to the elevation of the EGL on the discharge side of the pump (Fig. 5.69). TDH is the difference between the operating suction (upstream side) pressure and discharge (downstream side) pressure of the pump, and it is the sum of the total static head and pipe friction losses.

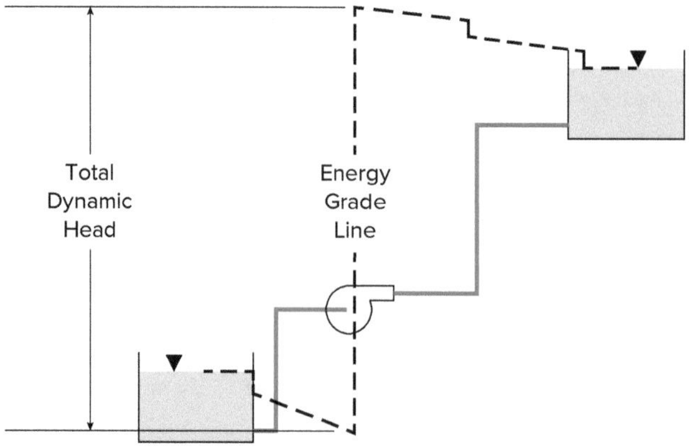

Total Dynamic Head

Energy Grade Line

FIGURE 5.69 Total Dynamic Head

5.7.6.3 System Curve

System curves are developed by engineers to indicate the energy head needed to overcome static and friction energy losses in the hydraulic system at different flow rates (Fig. 5.70). A system curve is a plot of TDH, or head that must be added (h_A) by the pump, versus flow rate.

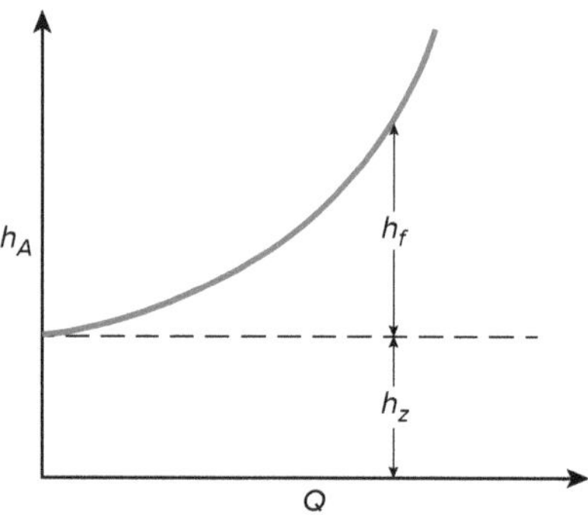

FIGURE 5.70 System Curve

h_A = total static head (h_z) + total friction head (h_f)

h_z = static discharge head − static suction head

h_z = static discharge head + static suction lift

h_f = friction head in the suction line + friction head in the discharge line

5.7.6.4 Pump Performance Curve

A pump performance curve (or pump curve) indicates the TDH provided by the pump at different flow rates (Fig. 5.71). Pump curves are supplied by the manufacturer of the pump. The TDH at zero flow is called the shutoff head.

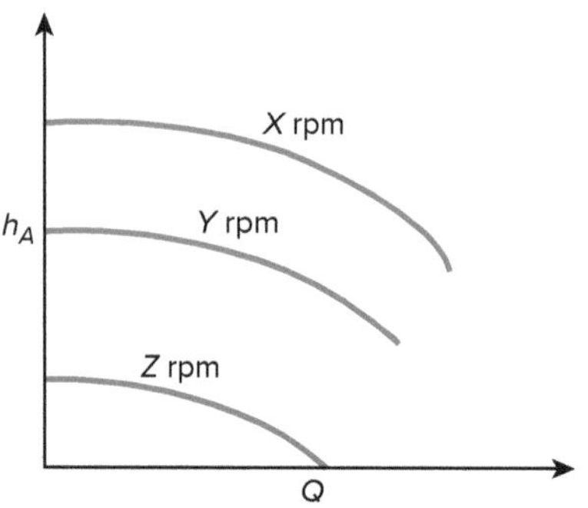

FIGURE 5.71 Pump Performance Curve

5.7.6.5 Operating Point

The intersection of a pump (performance) curve and a system curve is called the operating point. It indicates the discharge the pump will deliver and the TDH the pump will generate (Fig. 5.72).

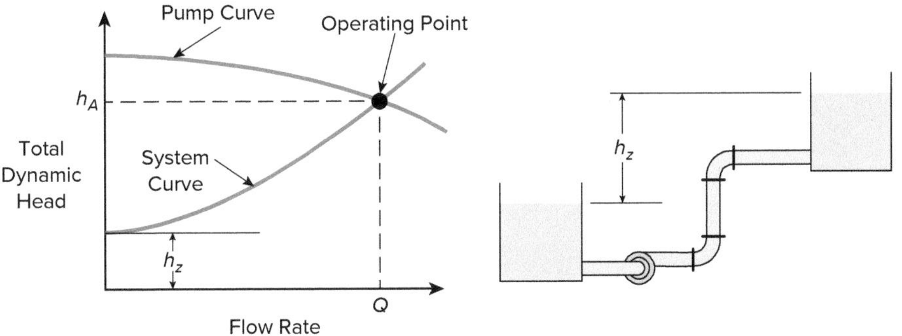

FIGURE 5.72 Pump Operating Point

Example 5.21: Operating Point

What is the expected flow rate, in gallons per minute, for the pump application in the following chart?

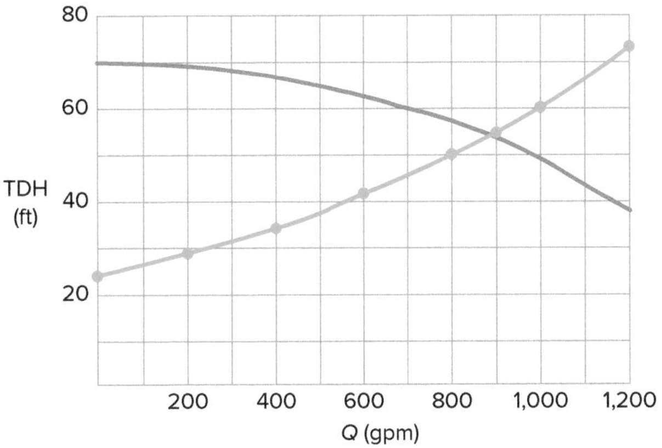

Solution

The intersection of the pump performance curve (blue) and system curve (green) indicates the system operating point. The operating point defines the system head and flow rate. In this case, the flow rate is approximately 880 gpm.

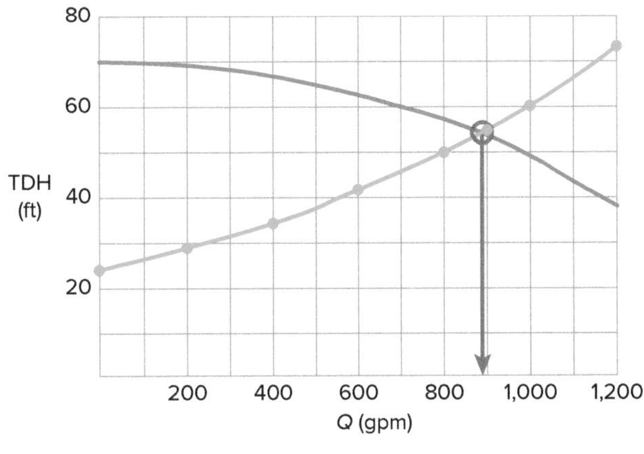

5.7.7 Orifice Discharge

5.7.7.1 Free and Submerged Discharge

The velocity (ft/s, m/s) of a free jet from an orifice in a tank is expressed by:

$$v_o = C_v\sqrt{2gh} \qquad \textbf{Equation 5-83}$$

The flow rate $\dot{V}$(ft^3/s, m^3/s) is then:

$$\dot{V} = C_d A_o \sqrt{2gh} \qquad \textbf{Equation 5-84}$$

C_v = coefficient of velocity, see Table 5.14

C or C_d = coefficient of discharge, see Table 5.14

A_o = orifice cross-sectional area (ft^2, m^2)

h = water surface to orifice center (ft, m)

If the orifice is submerged, h is the difference between water levels on either side of the orifice (see Figure 5.73b). If the orifice is in a tank, and the tank is pressurized, replace h with $h + p/\gamma$. See Figure 5.73a.

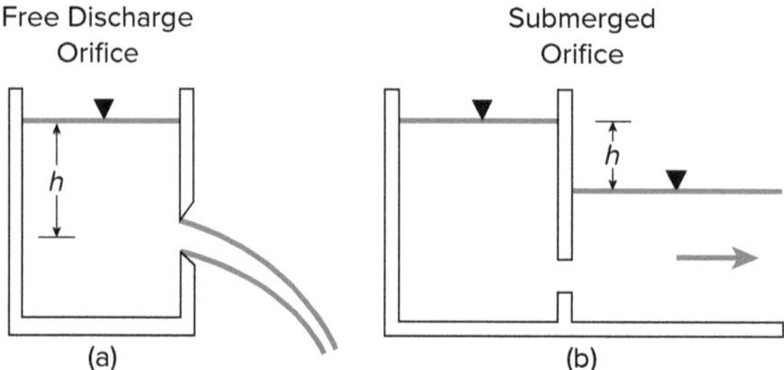

FIGURE 5.73 Free Discharge and Submerged Orifices

TABLE 5.14 Orifice Coefficients

	SHARP EDGED	ROUNDED	SHORT TUBE	BORDA
C	0.61	0.98	0.80	0.51
C_c	0.62	1.00	1.00	0.52
C_v	0.98	0.98	0.80	0.98

Source: Bober, W.; Kenyon, R. A. *Fluid Mechanics*. Wiley, 1980. Used with permission.

5.7.7.2 Time to Empty a Tank

The time it takes for water to drain from a tank is a function of both the discharge rate from an opening in the tank and the volume of water to be drained (Fig. 5.74). The following equation can be used to calculate the time for the water surface in a tank to drop from z_1 to z_2:

$$t(s) = \frac{2A_t(\sqrt{z_1} - \sqrt{z_2})}{C_d A_o \sqrt{2g}} \qquad \textbf{Equation 5-85}$$

A_t = tank cross-section area (ft^2, m^2)

z_1 and z_2 (ft, m) are measured from the center of the orifice, which is $z = 0$.

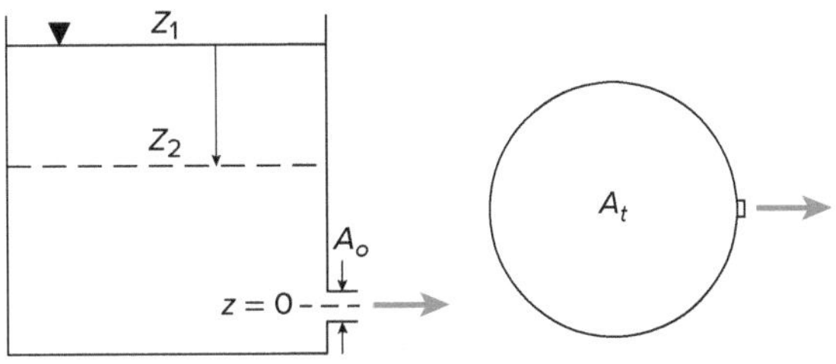

FIGURE 5.74 Discharge from a Tank

Example 5.22: Discharge from a Tank

A 20-ft-diameter cylindrical tank has a 3-in-diameter sharp-edged orifice at its base. If the tank is initially filled with 50 ft of water and is drained for 1 hour, what is the final depth, in feet?

Solution

$$t(s) = \frac{2A_t(\sqrt{z_1} - \sqrt{z_2})}{C_d A_o \sqrt{2g}}$$

Calculate the cross-sectional area of the tank:

$$A_t = \pi(20^2)\text{ft}/4 = 314 \text{ ft}^2$$

Calculate the cross-sectional area of the orifice:

$$A_o = \frac{\pi[(3 \text{ in}/12 \text{ in})^2]}{4} = 0.0491 \text{ ft}^2$$

$$t = (1 \text{ hr})(60 \text{ min/hr})(60 \text{ s/min}) = 3{,}600 \text{ sec}$$

Per Table 5.14, $C_d = 0.61$

$$3{,}600 \text{ s} = \frac{2(314 \text{ ft}^2)\left(\sqrt{50 \text{ ft}} - \sqrt{z_2}\right)}{(0.61)(0.049 \text{ ft})\sqrt{2(32.2)}}$$

$$1.40 = \sqrt{50 \text{ ft}} - \sqrt{z_2}$$

$$z_2 = 32 \text{ ft}$$

5.7.8 Nozzles

A nozzle is a short tube with a taper or constriction used (as on a hose) to speed up or direct fluid flow. Discharge through a nozzle is often a simple conversion from pressure head to velocity or flow rate.

Example 5.23: Nozzles

A 6-in-diameter pipe under 75 psi of pressure discharges through a 2-in-diameter nozzle. Neglecting energy losses, what is the flow rate through the nozzle in cubic feet per second?

Example 5.23 *(continued)*

Solution

$v_o = C_v\sqrt{2gh}$

$C_v = 1.0$ since the problem states to neglect energy losses

$h = \text{P (psi)} \times 2.31 \text{ (ft/psi)} = (75 \text{ psi})(2.31 \text{ ft/psi}) = 173.25 \text{ ft}$

$v = (2 \times 32.2 \text{ ft/s}^2 \times 173.25 \text{ ft})^{1/2} = 105.6 \text{ ft/s}$

$Q = vA$

$$A = \pi D^2/4 = \frac{\pi(2/12)^2}{4} = 0.022 \text{ ft}^2$$

$Q = (105.6 \text{ ft/s})(0.022 \text{ ft}^2) = 2.3 \text{ ft}^3/\text{s}$

5.7.9 Water Hammer

Water hammer is a dynamic pressure that is created in a piping system when a valve is closed suddenly. The fluid momentum is stopped abruptly and a pressure wave travels through the piping system. It is caused by the sudden transformation of kinetic energy to pressure energy. The pressure spikes can exceed five to ten times the working pressure of the system, which puts stress on joints, pipe walls, and pipe support systems. The pressure wave will travel back and forth between the valve and the source (for example, a reservoir) until friction losses in the pipes settle the wave (Fig. 5.75). Water hammer will sound like a banging or hammering in the pipes, which is caused by the pressure wave hitting a closed valve or joints at high force.

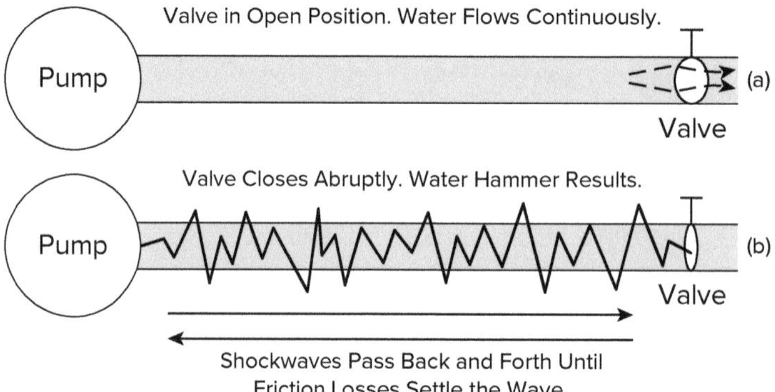

FIGURE 5.75 **Illustration of Water Hammer**

When piping systems are properly engineered, the likelihood of water hammer occurring is greatly reduced. Examples of good design practices are:

- Knowing optimal valve locations within a piping system
- Using slow-closing valves
- Using silent or non-slam check valves instead of swing check valves

One way to mitigate the effects of water hammer is to install arrestors, which provide a point of relief for pressure spikes caused by water hammer by acting like a shock absorber (Fig. 5.76). Examples of water hammer arrestors include air chambers and expansion tanks. Pressure-reducing valves and pressure-relief valves can also be effective safety measures. In large systems, training and educating operators to open and close manual or actuated valves properly (for example, quarter-turn valves such as ball valves, butterfly valves, and plug valves) is critical to safe operation.

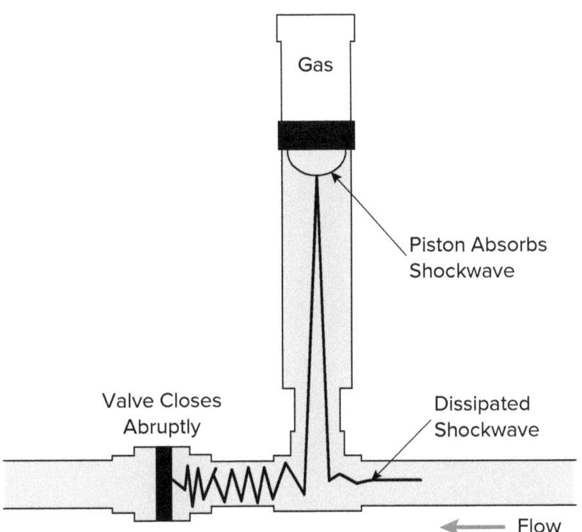

FIGURE 5.76 Water Hammer Arrestor

5.7.10 Siphons

A siphon is a liquid reservoir with an inverted U-shaped tube. A siphon is designed to drain liquid from a source (for instance, a pond or reservoir) by liquid flow such that it reaches an elevation above the source before discharging the liquid at an elevation below the source (Fig. 5.77). Flow will continue without the input of external energy (for example, a pump) until the source elevation is below the discharge elevation. The siphon conduit must be filled with water first, which may require pumping to prime the siphon tube.

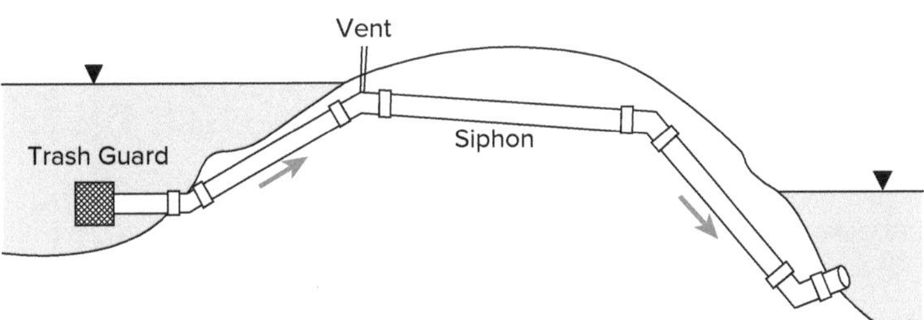

FIGURE 5.77 Siphon Through an Embankment

5.7.11 Buoyancy

Buoyant force, F_b, is an upward force that acts on an object partially or completely submerged in a fluid. The buoyant force on an object is equal to the weight of the displaced fluid.

$$F_b = \gamma V$$ **Equation 5-86**

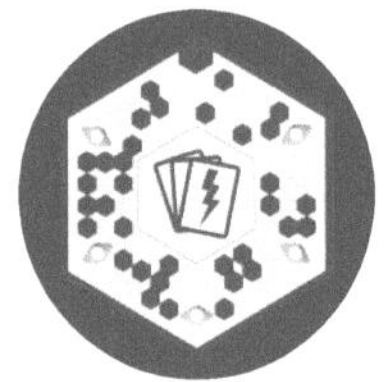

F_b = buoyant force (lb)

γ = unit weight of water (lbf/ft^3)

V = volume of water displaced (ft^3)

Example 5.24: Buoyancy

A floating object displaces liquid equal in weight to its own weight.

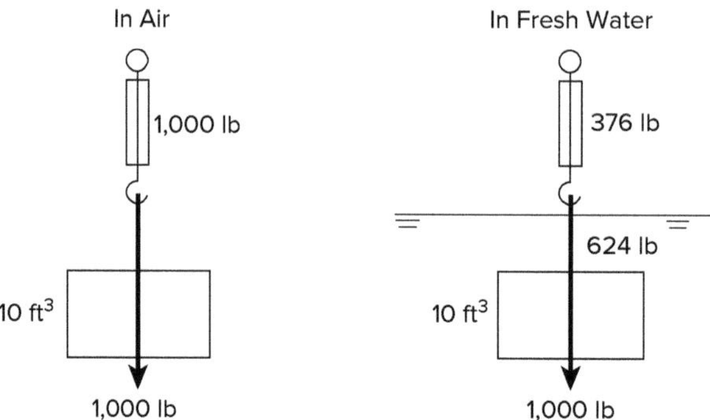

What is the specific gravity of the object in the figure above?

Solution

SG = weight of object in air divided by weight of water displaced

The weight of water displaced is found by multiplying 10 ft^3 of water by 62.4 lbf/ft^3 = 624 lb.

By definition, this is also the buoyant force on the object.

SG = (1,000 lb)/(624 lb) = 1.60

5.7.12 Buoyant Forces

Barges are flat-bottomed boats used to carry freight along rivers and canals. Buoyant forces keep the barge deck above water, even while carrying massive freight.

Example 5.25: Buoyant Forces

A towboat pulls a barge up a river carrying 100 tons of concrete beams for a bridge. The approximate dimensions of the barge are 195 ft long × 35 ft wide × 12 ft deep. If the barge weighs 280 tons when it is not carrying freight and 3 ft of freeboard must be maintained, how much additional weight, in tons, can the barge carry in freshwater? Assume 1 ton equals 2,000 pounds (short ton).

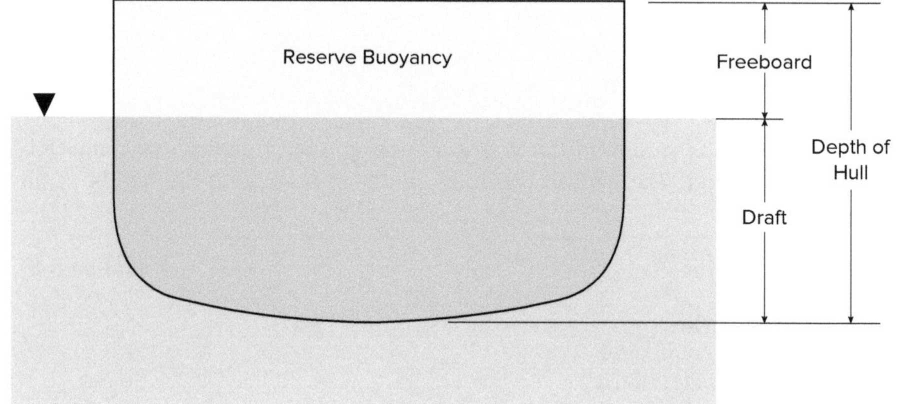

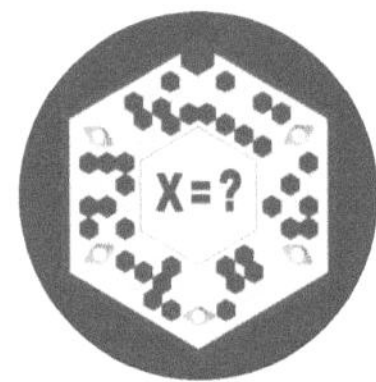

Example 5.25 *(continued)*

Solution

Additional weight the barge can carry = weight of water displaced – barge weight – freight weight

Volume displaced = (195 ft)(35 ft)(12 ft – 3 ft) = 61,425 ft^3

Weight = volume × unit weight

Weight of displaced water = 61,425 ft^3 × 62.4 lbf/ft^3 = 3,832,920 lbf

Weight of displaced water (in tons) = (3,832,920 lbf)/(2,000 lbf/ton) = 1,916 tons

Weight barge can carry = 1,916 tons – 280 tons – 100 tons = 1,536 tons

REFERENCES

1. King, H. W.; Brater, E. F. 1963. *Handbook of Hydraulics (5th ed.)*. New York: McGraw-Hill.
2. Schall, J. D.; Thompson, P. L.; Zerges, S. M.; Kilgore, R. T.; Morris, J. L. 2012. *Hydraulic Design of Highway Culverts (3rd ed.) (FHWA-HIF-12-026, HDS 5)*. Washington, DC: US Dept. of Transportation, Federal Highway Administration.
3. Natural Resources Conservation Service (NRCS). 1997. *National Engineering Handbook, Part 630: Hydrology*. Washington, DC: NRCS.
4. US Department of Agriculture (USDA). 1986. TR-55: *Urban Hydrology for Small Watersheds*. Washington, DC: USDA.
5. US Army Corps of Engineers (USACE), Omaha District. 2013. "Nationwide Permit Definitions." USACE, Dec. 26, 2013. https://www.nwo.usace.army.mil/Media/Fact-Sheets/Fact-Sheet-Article-View/Article/487703/nationwide-permit-definitions/

CHAPTER

Geometrics

6

Nicholas G. Siegl, PE, ENV SP

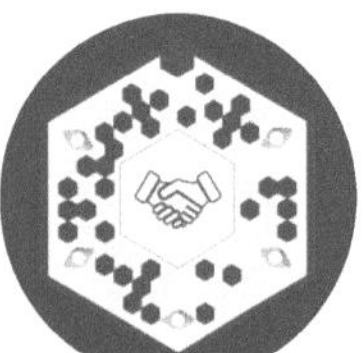

CONTENTS

CONTENTS (*continued*)

CONTENTS (*continued*)

Geometrics

EXAM GUIDE

VI. Geometrics

A. Basic circular curve elements (e.g., middle ordinate, length, chord, radius)
B. Basic vertical curve elements
C. Traffic volume (e.g., vehicle mix, flow, and speed)

Approximate Number of Questions on Exam: 3

NCEES Principles & Practice of Engineering Examination, Civil Breadth Exam Specifications

INTRODUCTION

The geometric design of highways and streets involves the development of horizontal and vertical alignments. Geometric design is affected by many factors, including design speed, terrain, adjacent land uses, and the volume and composition of traffic, safety, driver expectancy, and sight distance. Geometric elements, particularly horizontal curves, should be highly visible to drivers because they directly affect driver expectancy and vehicle operations, including braking, acceleration, and maximum speed. As part of the driving task, drivers continuously monitor, evaluate, and react to curves and other geometric elements.

When designing a horizontal curve, roadway designers should consider the types of vehicles that will be using the roadway. Semitrailers are much larger and heavier than passenger cars and therefore need more space for turning movements and have difficulty maintaining high speed on steep grades. All geometric elements should be designed to meet the design speed of the facility and create a safe, balanced, and overall pleasing design.

6.1 HORIZONTAL ALIGNMENTS

Commonly Used Abbreviations

AASHTO American Association of State Highway and Transportation Officials
c length of subchord
D degree of curvature (arc definition and chord definition)
d angle of subchord
E external distance
e superelevation
ft foot
g gravitational acceleration (32.2 ft/s^2)
HSO horizontal sightline offset
I intersection (or central angle) of a horizontal curve; also called Δ
L length of curve (from PC to PT)
l curve length for subchord
LC length of long chord
M middle ordinate
PI point of intersection
PC point of curve
PCC point of compound curvature
POC point on curve
POT point on tangent
PRC point of reverse curvature
PT point of tangent
R radius
s second

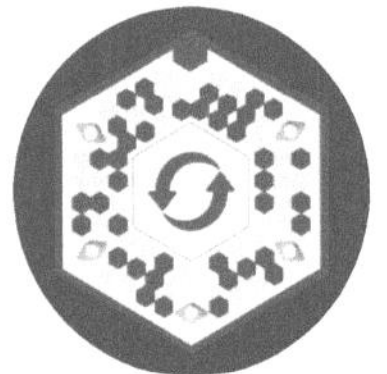

Commonly Used Symbols

α horizontal angle between the back tangent and long chord of a circular curve
β horizontal central angle associated with a chord length
Δ intersection angle (also called I) between two tangents
θ horizontal angle for layout of circular curves

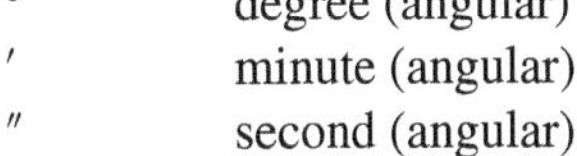

° degree (angular)
′ minute (angular)
″ second (angular)

6.1.1 Elements of a Straight Line

6.1.1.1 Direction

Most fixed-alignment transportation facilities are essentially linear systems that connect points (for example, A and B) that are located some distance apart. Geometrically, the shortest distance between two points is a straight line. Most alignments consist of one or more straight lines connected by horizontal curves. Therefore, the straight line is a fundamental component of transportation system alignment and deserves careful consideration. The most basic characteristics of a straight line are its length, its direction, and the locations of its endpoints. The section discusses methods of specifying the direction of a line and determining its length using coordinates.

One can think of the meridian as the y-axis in an x-y coordinate system. The angle between the meridian and the line is specified as either an azimuth or a bearing.

Also, the direction (or heading) of a line is always measured with reference to some starting point on the line. If that point is not an endpoint, one can start at that point and head along the line in one of two possible directions—either toward one endpoint or toward the other endpoint. Because the starting point and the two endpoints are on the same line, the two possible directions are direct opposites, exactly 180° apart. Therefore, the direction of a line is measured with reference to the meridian and from a starting point.

TIP

The direction of a straight line is typically expressed as an angle measured with reference to the meridian, which, for the purpose of the PE Civil exam, is simply a straight line that runs due north-south on a two-dimensional plane.

6.1.1.2 Azimuths

Azimuths are used extensively in route surveying and aviation. An azimuth is a horizontal angle turned clockwise from a reference heading to the line under consideration.

Every line has two azimuths. A line has one azimuth when heading forward and a second one when heading back. Their values differ by exactly 180°. Therefore, it is very important to call out the endpoints of a line in the correct order. The azimuth of a line heading from point A to point B is very different from the azimuth of a line heading from point B to point A.

When a line is heading due north, its azimuth is called out as "Az 0°" (or "Az 360°"). Both calls indicate exactly the same direction. Therefore, the azimuth of a line may be 0° or 360°, or it may be any value in between, but it can never be greater than 360°.

While azimuths are not used as much as bearings, there are some distinct advantages to using them. If you know the azimuths of two intersecting lines, you can quickly find the deflection angle between them by simply taking the difference in the azimuths. This is very simple and easy because all azimuths are equal to or between 0° and 360°.

TIP

Although there are situations where azimuths may be measured from due south, for the purpose of the exam, they are typically measured from due north in a clockwise direction.

Example 6.1: Azimuths

Suppose you're standing at point A facing due east. Your heading is Az 90°. You walk exactly 100 ft to point B, and then turn 35° to the right (clockwise). What is your new heading?

Example 6.1 *(continued)*

Solution

As shown in the figure, when heading from B to C, your heading is Az 90° + 35° = Az 125°. However, when heading from C to B, your heading is Az 125° + 180° = Az 305°.

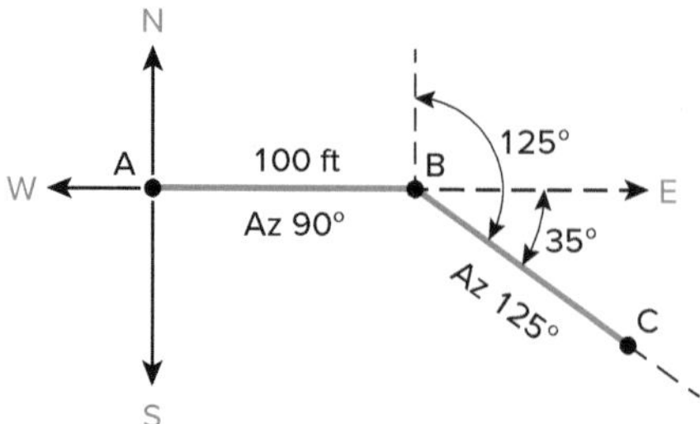

Example 6.2: Azimuths

Continuing on with Example 6.1, suppose your heading is now Az 125°. After walking exactly 100 ft, you reach point C, where you turn 90° to the left (counterclockwise). What is your new heading?

Solution

As shown in the figure, when heading from C to D, your heading is Az 125° − 90° = Az 35°. However, when heading from D to C, your new heading is Az 35° + 180° = Az 215°.

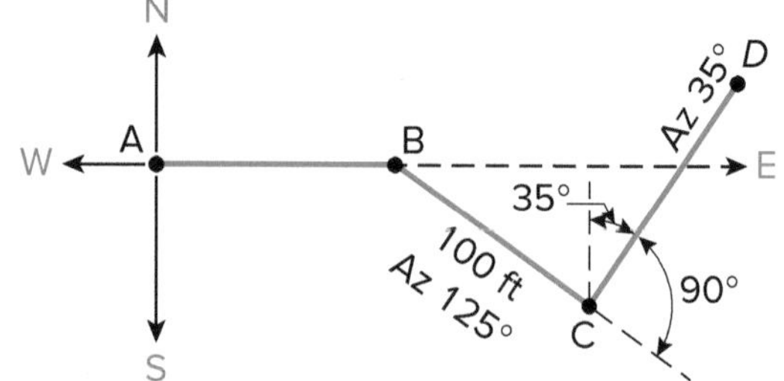

Note that by using the clockwise = positive or counterclockwise = negative (cw+/ccw−) convention, a turn to the right is always added to the previous azimuth, whereas a turn to the left is always subtracted. As discussed below, finding a new heading or the angle between two headings is not quite so simple when working with bearings.

6.1.1.3 Bearings

Bearings are generally more widely used than azimuths for purposes such as property legal descriptions and plats or highway right-of-way and construction plans. Examples of bearings are N25°30′E and S42°53′E. Note that the first letter (N or S) specifies if the line is heading north or south. Next, the **bearing angle** indicates the degree to which a line is turned, and the second letter (E or W) specifies if the angle is turned toward east or west. In so doing, this method divides two-dimensional space into four quadrants: NE, NW, SE, and SW. The four quadrants are divided by two perpendicular axes: the meridian (north-south) axis and the east-west axis.

> **TIP**
>
> Quickly find the deflection angle between two bearings in the same quadrant by simply finding the difference in the bearing angles.

Bearing angles are always equal to or between 0° and 90°. With bearings, some overlapping descriptions are possible. For example, due west can be described as N90°00′00″ W or S90°00′00″ W. Similarly, due north can be described as N0°00′00″W or N0°00′00″E. Due east and due south can also be described more than one way.

Example 6.3: Bearings

Suppose you're standing at point W and your heading is S42°W. You walk exactly 100 ft to point X, where you turn to the right to a new heading of S72°W and walk exactly 100 ft to point Y. How much did you turn to the right? In other words, what is the angle of deflection?

Example 6.3 *(continued)*

Solution

As shown in the figure, finding the angle of deflection is very sinple because the two headings are in the same SW quadrant. You simply take the difference in the bearing angles. The difference is a clockwise deflection of +30° (72° – 42°).

If the two bearings are in different quadrants, however, drawing a quick sketch is advisable.

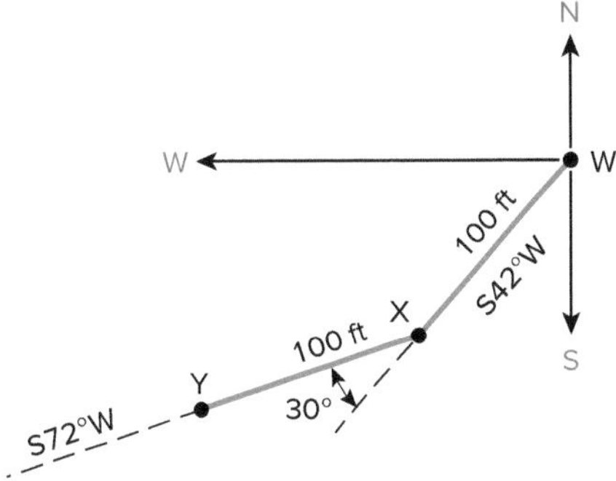

Example 6.4: Bearings

Continuing on with the above example, suppose you're now standing on point Y, where you turn to the right almost completely around onto a new heading of N50°E. Now, instead of heading into the SW quadrant, you are heading into the NE quadrant. How far did you turn to the right?

Solution

In this case, finding the angle of deflection is a little more involved, as shown in the figure at right. You must do more than simply take the difference in the bearing angles. You find that in turning from S72°W to N50°E, you've actually turned 180° – (72° – 50°) = 158°.

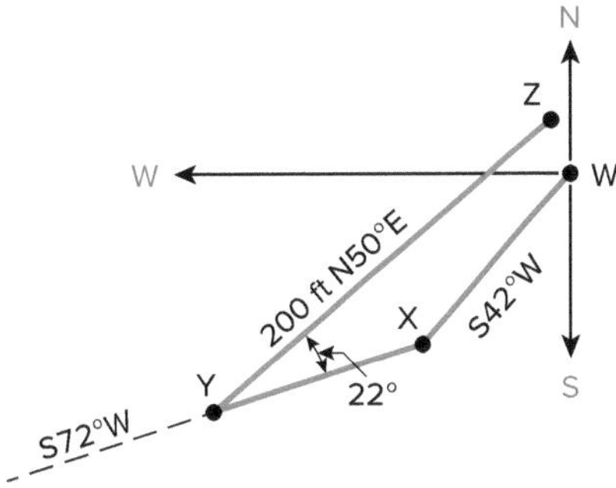

Alternative Solution

As shown in the figure below, if you were working in azimuths, your headings would have been WX: Az 222°, XY: Az 252°, and YZ: Az 50°.

To find the angle between lines XY and YZ, you would have done the following:

50° + (Az 360° – Az 252°)

50° + 108° = 158°

Visualization would have been easier and finding the solution somewhat faster.

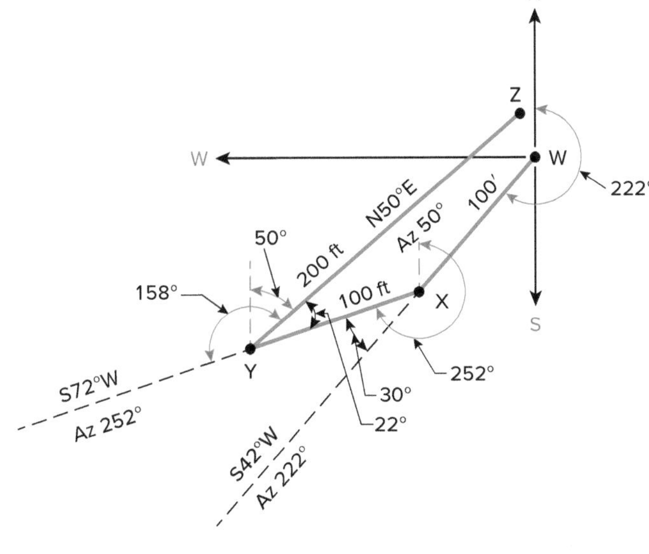

Geometrics

School of PE

6.1.1.4 Latitudes and Departures

The concept of latitudes and departures is a simple but very powerful technique for solving geometric problems and for balancing and determining the areas of closed traverses (Fig. 6.1). Latitudes and departures are widely used in surveying. They provide a simple, reliable, and accurate method of locating features, defining boundaries, and calculating the areas of closed traverses, such as a parcel of land.

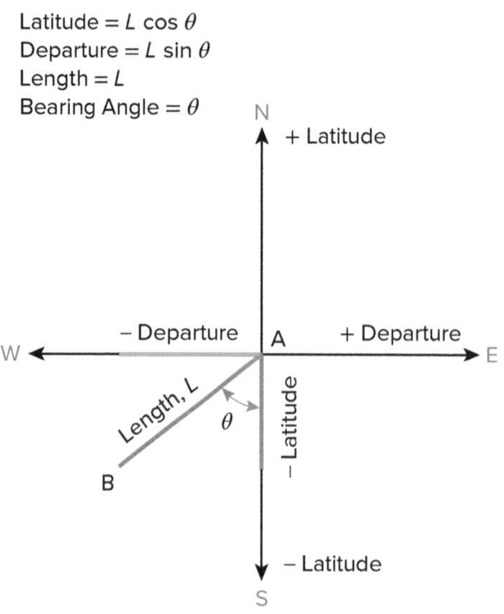

FIGURE 6.1 Example: Latitude and Departure of Line AB

Latitudes and departures are directly associated with straight lines, traverses, Cartesian coordinate systems, and trigonometric calculations. Every straight line has both length and direction, which with trigonometry can be used to subdivide a line into its components, latitude, and departure. Latitudes and departures are used extensively in coordinate geometry to determine the coordinates (northing and easting) of points.

The concept of latitudes and departures is represented by two intersecting perpendicular axes that subdivide a two-dimensional space into four parts. Like a Cartesian coordinate system, this figure indicates that the direction of a straight line may be determined by its components, which may be positive, negative, or zero.

The **vertical (latitude) axis** runs up and down (or north and south). By convention, heading upward (north) indicates positive latitude, whereas heading downward (south) indicates negative latitude. A line that runs due east-west would have zero latitude; its departure would be equal to its length.

The **horizontal (departure) axis** runs left to right (or west and east). Again, by convention, heading to the right (east) indicates positive departure, whereas heading to the left (west) indicates negative departure. A line that runs due north-south would have zero departure. Its latitude would be equal to its length and its departure would be zero.

Latitudes and departures are calculated using trigonometry. The key to using this technique is visualizing the right triangle formed by the line, its latitude, and departure. If we know the length and bearing (or azimuth) of a line, we can easily calculate its latitude and departure. The latitude of a line is associated with the side adjacent to the bearing angle and involves the cosine of the angle. The departure is associated with the side opposite the bearing angle and involves the sine of the angle.

6.1.1.5 Latitudes: Positive, Zero, and Negative

As shown in Figure 6.1, the latitude of a line is the north-south component of the line and is expressed as the change in the *y* direction, Δy. Depending on the direction of a line, its latitude may be positive, zero, or negative.

- Latitude is positive if the line is heading in a northerly direction. Lines heading northwest or northeast have positive latitudes. Examples: N25°15′W and N89° 15′E
- Latitude is zero if the line has no north/south rectangular component. Lines heading due west or due east have zero latitude. Examples: S90°00′W, N90° 00′W, S90°00′E, and N90°00′E (These are the only four bearings that have zero latitude.)
- Latitude is negative if the line is heading in a southerly direction. Lines heading southwest or southeast have negative latitudes. Examples: S85°30′W and S60°45′E

6.1.1.6 Departures: Positive, Zero, and Negative

Similarly, as shown in Figure 6.1, the departure of a line is the east-west component of the line and is expressed as the change in the *x* direction, Δx. Depending on the direction (bearing) of a line, a departure may be positive, zero, or negative.

- Departure is positive if the line is heading in an easterly direction. Lines heading northeast or southeast have positive departures. Examples: N25°15′E and S89°15′E
- Departure is zero if the line has no east/west component. Lines heading due north or due south have zero departure. Examples: N0°00′W, N0°00′E, S0°00′W, and S0°00′E (These are the only four bearings that have zero departure.)
- Departure is negative if the line is heading in a westerly direction. Lines heading southwest or northwest have negative departures. Examples: S85°30′W and N60°45′W

When working with a rectangular coordinate system, it is very important to follow the convention above and to use the appropriate signs for latitudes or departures. It is helpful to draw a quick sketch as well (see Fig. 6.2).

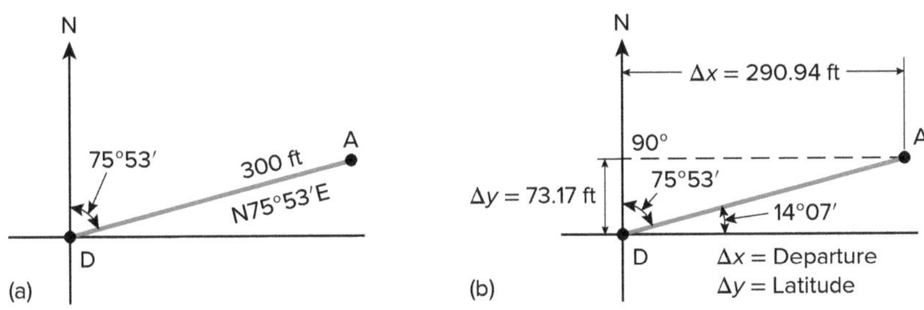

FIGURE 6.2 Example Calculation of Latitude and Departure of Line DA

6.1.1.7 Calculating Latitudes and Departures

Latitudes and departures are easily calculated using trigonometry. It is best to think of a straight line as the hypotenuse of a right triangle. Apply the sin and cos functions to the bearing angle (or azimuth), θ, and length of line, *L*, to find the latitude and departure.

$$\text{Latitude } (\Delta y) = \text{Length } (L) \times \cos \theta \qquad \textbf{Equation 6-1}$$

$$\text{Departure } (\Delta x) = \text{Length } (L) \times \sin \theta \qquad \textbf{Equation 6-2}$$

6.1.1.8 Use of Latitudes and Departures with Cartesian Coordinates

As shown in Example 6.5, if the coordinates of point A and the bearing and length of line AB are known, use latitudes and departures to determine the coordinates of point B.

Example 6.5: Latitudes and Departures with Cartesian Coordinates

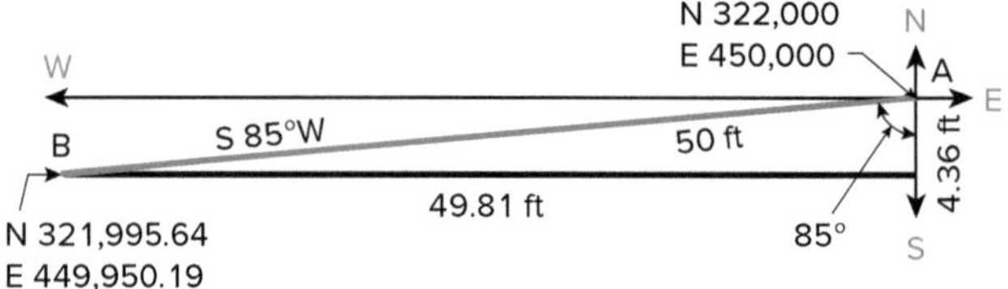

Length of AB = 50 ft

Bearing of line AB = S 85° 00′ W

Coordinates of point A = N 322,000, E 450,000

Find the coordinates of point B.

Solution

Latitude of line AB = $\Delta y = -50$ ft cos 85° = −4.36 ft

Departure of line AB = $\Delta x = -50$ ft sin 85° = −49.81 ft

Northing (N) of point B = N 322,000 –4.36 ft = N 321,995.64

Easting (E) of point B = E 450,000 –49.81 ft = E 449,950.19

Note:

- Drawing a quick sketch can help simplify the problem.
- The bearing angle is always measured form the north-south meridian.
- The Δy is the change in the northing, and Δx is the change in the easting.
- Proper use of the sign convention simplifies the math.

Example 6.6: Latitudes and Departures with Cartesian Coordinates

Line JK has a length of 1,000 ft and a bearing of N 18° E. By inspection, the northerly heading, N, indicates that line JK has a positive (+) latitude, and the easterly heading, E, indicates a positive (+) departure.

Latitude: +1,000 ft cos 18° = + 951.06 ft

Departure: +1,000 ft sin 18° = + 309.02 ft

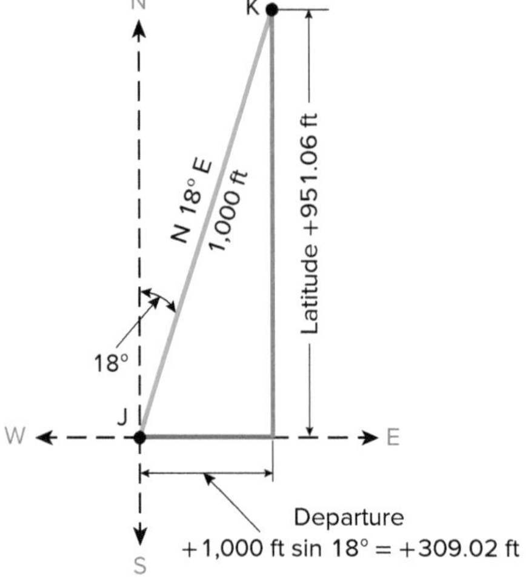

Geometrics

School of PE

Example 6.7: Latitudes and Departures with Cartesian Coordinates

Line PQ has a length of 500 ft and a bearing of S35°W. By inspection, the southerly heading, S, indicates that line PQ has a negative (−) latitude, and the westerly heading, W, indicates a negative (−) departure.

Latitude: −500 ft cos 35° = −409.58 ft

Departure: −500 ft sin 35° = −286.79 ft

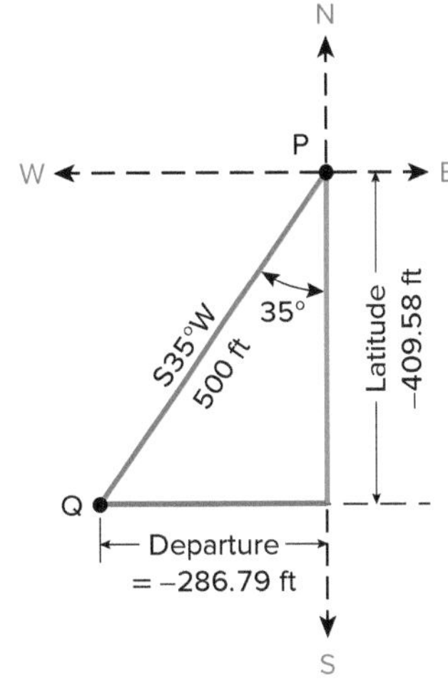

6.1.1.9 Inversing the Line

If the coordinates of the endpoints are known, one can work backward to establish the length and bearing (or azimuth) of the line connecting the endpoints. This technique is known as **inversing the line** and is often used when performing boundary surveys and establishing horizontal alignments.

$$\text{Latitude}_{GH} = N_H - N_G \qquad \textbf{Equation 6-3}$$

$$\text{Departure}_{GH} = E_H - E_G \qquad \textbf{Equation 6-4}$$

$$\text{Bearing Angle} = \tan^{-1}\left(\frac{\text{Departure}}{\text{Latitude}}\right) \qquad \textbf{Equation 6-5}$$

Example 6.8: Inversing the Line

The coordinates of property corners G and H are given below. Determine the bearing and length of the line GH.

POINT	NORTH	EAST
G	1,200	200
H	1,500	700

TIP

Inversing the line can be very useful. It can be used to determine the length and direction of a line connecting two points of known coordinates.

Solution

Draw a sketch like the one at the top of p. 248.

Determine the latitude and departure of line GH:

Latitude: 1,500 ft – 1,200 ft = 300 ft

Departure: 700 ft – 200 ft = 500 ft

Example 6.8 (continued)

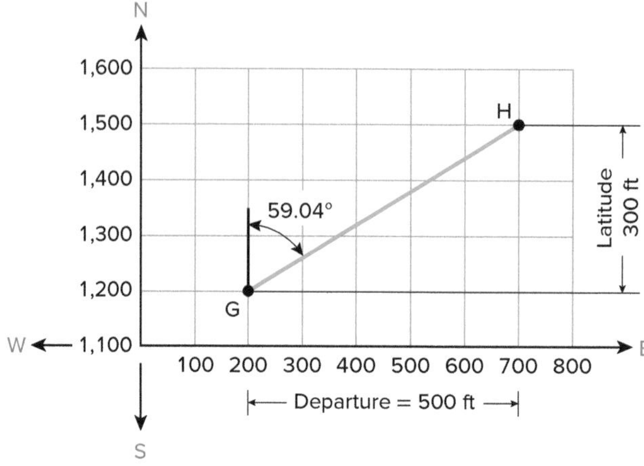

Length $\overline{GH}$ by Pythagorean theorem: $a^2 + b^2 = c^2$

$$c = \overline{GH} = \sqrt{(500\text{ ft})^2 + (300\text{ ft})^2} = 583.095\text{ ft}$$

$$\text{Bearing angle} = \tan^{-1}\left(\frac{\text{departure}}{\text{latitude}}\right) = \tan^{-1}\left(\frac{500\text{ ft}}{300\text{ ft}}\right) = 59.04°$$

Bearing of line $\overline{GH}$: N59°02′10.48″E

6.1.2 Locating Features Along a Horizontal Alignment

6.1.2.1 Center/Baseline

A horizontal alignment is represented graphically by a roadway center/baseline, which is a theoretical line established in a horizontal coordinate plane. A center/baseline consists of a continuous series of line segments and arcs that have length and direction, but no width. They are used to:

- establish control over the roadway design and construction.
- measure distances and locate points along the alignment.
- coordinate the various elements of geometric design, including right-of-way, drainage, utilities, and estimating earthwork volumes.

A horizontal alignment is the geometric projection of the roadway center/baseline onto a horizontal plane, which is defined by a rectangular coordinate system. This means that a horizontal alignment is defined in two dimensions only (northing and easting) and does not include elevation, which is the third dimension.

6.1.2.2 Stationing

The concept of stationing is fundamental to surveying, highway design, construction, and maintenance. As an integral property of a roadway center/baseline, stationing enables the center/baseline to function as a number line, providing a simple, convenient, and effective means of indicating where specific features are located along the center/baseline.

Stations are indicated by tick marks and integers that increase by one every 100 ft when heading in the direction of increasing stationing (and decrease by one every 100 ft when heading in the opposite direction). The direction of stationing is either increasing (stations ahead) or decreasing (stations back). Stationing also provides a convenient means of determining distances between points on a roadway.

By convention on horizontal curves, the **point of curve** (PC, the beginning of a curve) station is always to the back of the **point of tangent** (PT, the end of the curve) station. Also, by convention, stationing is typically established to increase from south to north and from west to east.

On projects prepared in US customary units, full stations are placed every 100 ft. The starting point of a center/baseline is often designated as station 0+00.00. The station of a point located 537.7 ft ahead of this starting point would be sta 5+37.70.

Example 6.9: Locating a Feature

Stations may be marked as sta 521+00, sta 522+00, sta 523+00, and so forth. If a feature is located on the baseline, 65.4 ft ahead of sta 522+00, its position is sta 522+65.4, offset 0.0 ft.

If a feature is located at this same station and 72.9 ft to the left of the baseline, its position is sta 522+65.4, offset 72.9 ft LT.

Note that stationing can be used in two ways. In the above example, stationing is used to indicate the exact location of a feature.

Example 6.10: Station as a Unit of Length

A second use of the term station is as a unit of length. In transportation, we often consider how long something is or how far apart two points are. Since the distance between full stations is 100 ft, one station is equal to 100 ft. Therefore, if a section of roadway is 2,500 ft long, we can say its length is 25 sta. If we are designing 3.5 miles of highway, we could say the project length is 184.8 sta (3.5 × 5,280 ft/100 ft).

A key point about roadway center/baselines and the stationing that runs along them is that they are purely two dimensional. They represent the roadway alignment in a theoretical horizontal plane in which all features are at the same elevation. In their physical reality, however, roadways are three dimensional. They curve to the left and right while ascending and descending as necessary to better fit the terrain, cross over rivers, and pass under railroads and other roadways. Therefore, because stationing is purely horizontal, it does not account for the slight increase in physical roadway length caused by changes in elevation. This is a key point that directly affects how we design vertical curves, which are discussed in Section 6.2.2.

TIP

A basic familiarity with stationing is needed for the exam because it often includes problems involving stationing. These problems may involve horizontal or vertical alignments, superelevation transitions, or earthwork calculations. Problems involving stationing on horizontal curves are common and deserve close attention.

6.1.2.3 Offsets

Offsets are lateral distances measured to the left or right of a center/baseline for the purpose of locating a feature, such as an inlet, headwall, or barrier terminal section. Offsets to the left are often listed as negative, and offsets to the right are listed as positive. When using stations and offsets, it is necessary to indicate the following:

- Designation of center/baseline
- Station—position along the center/baseline
- Offset—position perpendicular to the center/baseline
- Side—left or right of the center/baseline.

For the purposes of locating design and construction features, stations and offsets are much more convenient to use than coordinates (northings and eastings); however, stations and offsets are always relative to a specific center/baseline and can only be used after that center/baseline has been established.

Example 6.11: Offsets

Two inlets are to be installed on a straight section of roadway at sta 10+23, 20 ft LT and sta 10+38, 20 ft RT, and connected by a run of pipe. Determine the approximate length of the pipe.

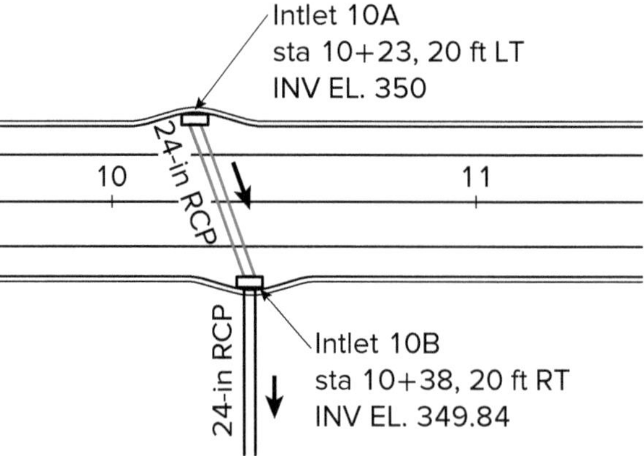

Solution

The longitudinal distance between inlets is the difference in stations:

sta 10+38 – sta 10+23 = 15 ft

The lateral distance between the inlets is the sum of the offsets:

20 ft + 20 ft = 40 ft.

The approximate pipe length is the straight-line distance between the inlets, which should be visualized as the hypotenuse of a right triangle created by the above longitudinal and lateral distances. Therefore, using the Pythagorean theorem, $c = \sqrt{a^2 + b^2}$, the approximate pipe length is:

$$\sqrt{(15\text{ ft})^2 + (40\text{ ft})^2} = 42.72\text{ ft} \cong 43\text{ linear feet}$$

6.1.3 Sight Distance

Sight distance has considerable influence on the geometric design of highways and streets because it directly affects safety and mobility. Sight distance is defined by AASHTO as the length of roadway ahead that is visible to the driver. A roadway design must provide enough sight distance for drivers to perceive dangerous or unexpected situations on the roadway ahead. The amount of sight distance needed depends on the design speed of the roadway and the driver's ability to react in time. Higher design speeds require more sight distance than lower design speeds, but drivers tend to process information and react to situations at the same rate regardless of travel speed. Therefore, roadways should be designed to provide motorists more sight distance, which equates to more time to process information at higher design speeds. Roadways that have restricted sight distance are often unable to realize their full traffic capacity because safety considerations require traffic to operate at reduced speeds.

As discussed later in this section, sight distance is a key geometric design criterion used to determine the minimum lateral clearance to sight obstructions on horizontal curves and the minimum lengths of vertical curves.

Sight distance may be constrained by various types of obstructions, including:

- The roadway pavement itself (on excessively short crest or sag vertical curves)
- Overhead structures, including bridges and sign structures

- Roadside sight obstruction (on horizontal curves)
  - Bridge abutments
  - Median barriers
  - Sign panels
  - Cut slopes
  - Vegetation
  - Parked vehicles
  - Buildings
  - Piles of plowed snow
  - Roadside barriers

6.1.3.1 Criteria for Measuring Sight Distance

The criteria for measuring sight distance depends on the following:

- **Height of driver's eye.** The standard height of the driver's eye above the pavement surface is 3.5 ft for passenger vehicles and 7.6 ft for large trucks.
- **Height of object.** The standard height of the object is 2.0 ft above the pavement surface for stopping sight distance (SSD) calculations, and 3.5 ft above the pavement surface for passing sight distance.
- **Sight obstructions.** Highway construction plans should be checked in both the vertical and horizontal plane. Sight distance obstructions may consist of the road surface on a crest vertical curve or roadside features such as longitudinal barriers, trees, or the backslope of a cut section.

The following three types of SSD problems have appeared frequently on the breadth portion of the PE Civil exam:

1. Stopping sight distance
2. Stopping sight distance on a vertical curve, which is the SSD associated with a minimum vertical curve length
3. Stopping sight distance on a horizontal curve, a special type of SSD, which is discussed in Section 6.1.5

6.1.3.2 Stopping Sight Distance

SSD is the length of roadway ahead that a driver must be able to see in order to have enough distance (and time) to perceive and react if an unexpected stationary object were located on the roadway within that length, and thus be able to bring the vehicle to a complete stop without hitting the object. This concept is applied in the design of horizontal and vertical curves.

SSD is directly related to the amount of perception-reaction time (PRT) that drivers are expected to need to perceive and react to an object in their path. Weather conditions, visibility, and driver attentiveness can affect PRT. Response time is generally shortest when there is only one response to be made to a single stimulus (for example, braking in response to the brake lights of a vehicle only a few feet ahead). In the case of choice reaction time, in which there is more than one stimulus or more than one possible response (for example, a toll plaza), driver reaction time increases as a function of the number of possible choices. Drivers may, for example, have to decide whether to steer or brake, or both, to avoid a pedestrian or object in their path. This is known as decision sight distance (DSD), which is unlikely to appear on the breadth portion of the PE Civil exam.

PRT is the amount of time required for a driver to perceive that a reaction is needed due to a road condition, to decide what maneuver is appropriate (in this case, stopping the vehicle), and to start the maneuver (taking the foot off of the accelerator and depressing the brake pedal). Based on the AASHTO *Green Book* [1], PRT used in design consists of two parts: 1.5 seconds for perception and decision making and 1.0 second for reacting, for a total of 2.5 seconds, which is generally considered adequate for all but the most complex driving situations. This PRT is based on the observed behavior of a large number of drivers and is conservatively attributed to the 85th percentile driver. Consequently, 85% of drivers are expected to react in less than 2.5 seconds.

Stopping sight distance is the sum of two distances:

- Brake reaction distance: the distance traversed by the vehicle from the instant the driver sees an object that requires stopping the vehicle to the instant the brakes are applied
- Braking distance: the distance needed to stop the vehicle from the instant the brake application begins

The SSD values used in design are conservative and represent a near worst-case situation, so that in most cases a below-average driver would be able to successfully bring a vehicle to a stop and avoid a collision. These values are based on a generous amount of PRT (2.5 s), and a fairly low rate of deceleration (11.2 ft/s^2).

Equation 6-6 is used to calculate the SSD required when a driver notices an unexpected object or event requiring a stop.

$$SSD = 1.47 \times V\,(\text{mph}) \times t_p + \frac{V^2(\text{mph})}{30\left(\frac{a}{32.2} \pm G\right)}$$ **Equation 6-6**

V = initial speed of vehicle (mph)
t_p = driver PRT (seconds), typically 2.5 seconds
a = vehicle rate of deceleration (ft/s^2), typically 11.2 ft/s^2
G = Roadway grade (%/100)
1.47 = conversion factor (mph to ft/s)

Example 6.12 demonstrates the use of the SSD equation.

Example 6.12: Stopping Sight Distance

What is the SSD of a vehicle on a level roadway traveling at a design speed of 40 mph with a deceleration rate of 4.0 ft/s^2, assuming the brake reaction time is $t_p = 2.5$ s?

A. 232 ft
B. 567 ft
C. 576 ft
D. 322 ft

Solution

SSD is calculated as follows:

$$SSD = 1.47Vt + \frac{V^2}{30\left(\frac{a}{32.2} \pm G\right)}$$

V = design speed = 40 mph
t_p = driver PRT (seconds), typically 2.5 s
a = deceleration rate = 4 ft/s^2
G = percent grade divided by 100 = 0

$$SSD = 1.47(40\text{ mph})(2.5\text{ s}) + \frac{(40\text{ mph})^2}{30\left(\frac{4.0\text{ ft/s}^2}{32.2} \pm 0\right)} \cong 576\text{ ft}$$

Answer: C

Geometrics

School of PE

The stopping distances for various speeds are shown in Table 6.1. These distances were calculated using the above SSD equation and the following assumptions: a brake reaction time of 2.5 s, a deceleration rate of 3.4 m/s^2 (11.2 ft/s^2), and a level grade (G = 0.0%).

TABLE 6.1 AASHTO *Green Book* Table 3-1
Stopping Sight Distance on Level Roadways

METRIC					U.S.CUSTOMARY				
			Stopping Sight Distance					Stopping Sight Distance	
Design Speed (km/h)	Brake Reaction Distance (m)	Brake Distance on Level (m)	Calculated (m)	Design (m)	Design Speed (mph)	Brake Reaction Distance (ft)	Brake Distance on Level (ft)	Calculated (ft)	Design (ft)
20	13.9	4.6	18.5	20	15	55.1	21.6	76.7	80
30	20.9	10.3	31.2	35	20	73.5	38.4	111.9	115
40	27.8	18.4	46.2	50	25	91.9	60.0	151.9	155
50	34.8	28.7	63.5	65	30	110.3	86.4	196.7	200
60	41.7	41.3	83.0	85	35	128.6	117.6	246.2	250
70	48.7	56.2	104.9	105	40	147.0	153.6	300.6	305
80	55.6	73.4	129.0	130	45	165.4	194.4	359.8	360
90	62.6	92.9	155.5	160	50	183.8	240.0	423.8	425
100	69.5	114.7	184.2	185	55	202.1	290.3	492.4	495
110	76.5	138.8	215.3	220	60	220.5	345.5	566.0	570
120	83.4	165.2	248.6	250	65	238.9	405.5	644.4	645
130	90.4	193.8	284.2	285	70	257.3	470.3	727.6	730
					75	275.6	539.9	815.5	820
					80	294.0	614.3	908.3	910

Note: Brake reactiion distance predicated on a time of 2.5 s; deceleration rate of 3.4 m/s^2 (11.2 ft/s^2) used to determine calculated sight distance.

From *A Policy on Geometric Design of Highways and Streets*, 2011, by the American Association of State Highway and Transportation Officials, Washington, DC. Used by permission.

6.1.4 Curvature of Horizontal Alignments

6.1.4.1 Horizontal Circular Curves

Geometrically, a horizontal alignment consists of a continuous series of straight lines (**tangents**) that are connected by circular curves to create a smooth and flowing path between two endpoints (Fig. 6.3). Circular curves are the most critical elements of a horizontal alignment because they must enable a smooth and safe change in direction at the design speed of the highway. Maintaining the proper relationship between design speed and curvature requires careful attention.

Circular curves are simply portions of a circle. Although computer-aided drafting and design (CADD) and roadway design software enable designers to lay out circular curves quickly and efficiently, for the purposes of the exam and general practice, engineers should still have an understanding of the elements of a circular curve and how curve data are calculated.

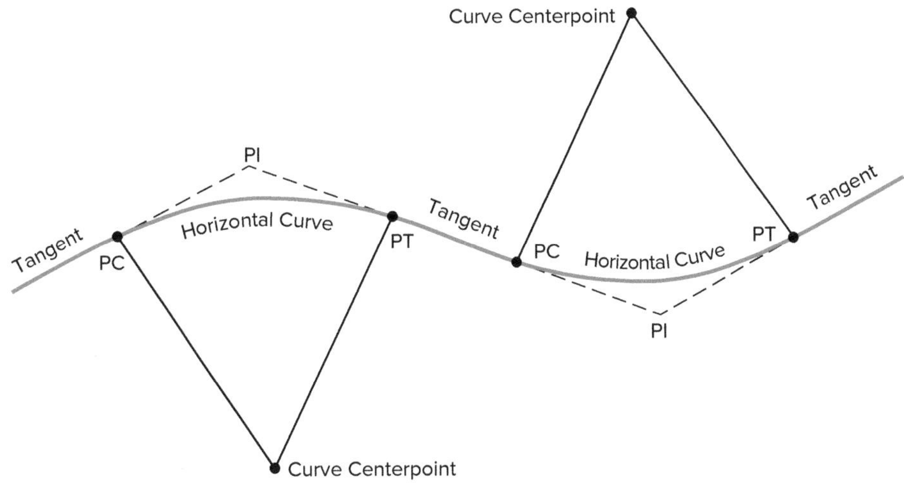

FIGURE 6.3 Example of an S-Curve

The two most important elements of horizontal curves are **radius** and **superelevation** rate. For safety and comfort, the radius of a curve must be long enough to meet the design speed criteria established for the highway. This section presents the geometric elements of a circular curve, including the terminology and equations used to calculate curve data, and several types of horizontal curves commonly used on horizontal alignments. As indicated by the equations, the mathematics of circular curves is largely based on trigonometry. This section also relates the minimum radius of a horizontal curve to design speed, superelevation, and side friction.

TIP

For the purpose of the exam, it is important to know the designations listed in Table 6.2 and how to calculate them.

6.1.4.1.1 Key Curve Points

All horizontal curves are described by several key alignment points that carry specific designations.

TABLE 6.2 Summary of Key Horizontal Alignment Points

ABBREVIATION	KEY POINT	DEFINITION
PI	Point of Intersection	Where the back and ahead tangents intersect
PC	Point of Curve	Where back tangent stationing ends and curve stationing starts
PT	Point of Tangent	Where curve stationing ends and ahead tangent stationing starts
PCC	Point of Compound Curvature	Where two curves with different radii curving in the same direction meet; the PT of the first curve coincides with the PC of the second curve
PRC	Point of Reverse Curvature	Where two curves in opposite directions meet; where the PT of the first curve coincides with the PC of the second curve
POT	Point on Tangent	Identifies a point located on a tangent
POC	Point on Curve	Identifies a point located on a curve

6.1.4.1.2 Geometric Elements of Horizontal Curves

A horizontal curve is an arc that connects two tangents (Fig. 6.4). The curve begins on the back tangent at the PC and ends on the ahead tangent at the PT.

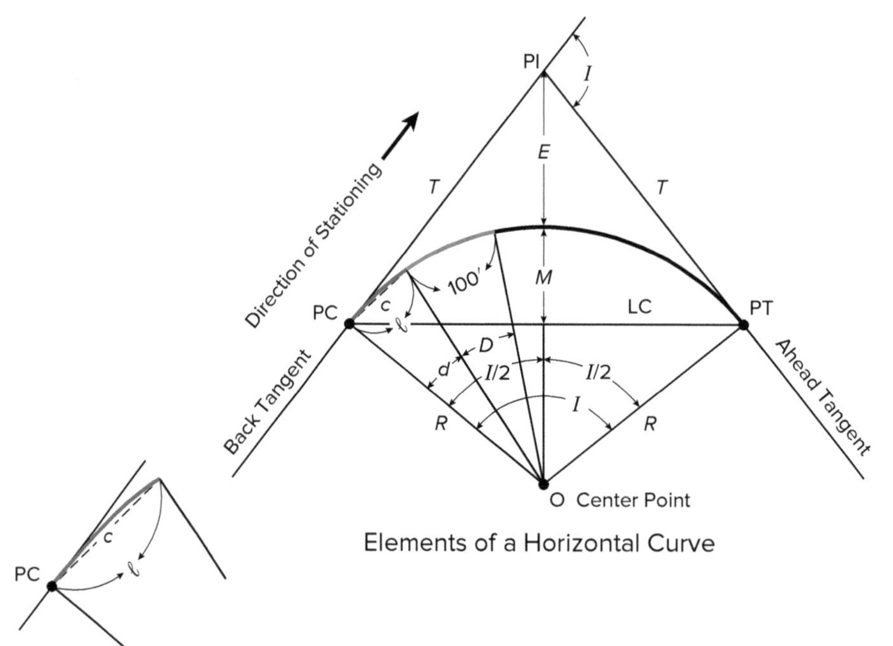

FIGURE 6.4 Elements of a Horizontal Curve

The stationing increases along the back tangent to the PC, and then continues in two paths: along the curve from the PC to the PT, and along the back tangent from the PC to the point of intersection (PI). Consequently, the exact same station may occur on both the curve between the PC and the PT and on the back tangent between the PC and the PI. Note that the PI is a surveyed point that is used to lay out the curve, but it is typically located some distance from the roadway pavement and is therefore rarely visible to motorists.

Equations 6-7 through 6-16 are based on trigonometry and geometry and are used to calculate circular curve data. Many of the circular curve problems on the exam can be solved by using these equations. A typical problem might involve finding the radius R associated with a given external E and middle ordinate M. Solving such a problem could involve using E and M to first solve for the angle I, and then using I to find R.

6.1.4.1.3 Equations for Circular Curve Calculations

$$R = \frac{5{,}729.58}{D}$$ **Equation 6-7**

$$R = \frac{LC}{2\sin\left(\frac{I}{2}\right)}$$ **Equation 6-8**

$$T = R\tan\left(\frac{I}{2}\right) = \frac{LC}{2\cos\left(\frac{I}{2}\right)}$$ **Equation 6-9**

$$L = RI\frac{\pi}{180} = \frac{I}{D}100$$ **Equation 6-10**

$$M = R\left[1 - \cos\left(\frac{I}{2}\right)\right]$$ **Equation 6-11**

$$\frac{R}{E+R} = \cos\left(\frac{I}{2}\right)$$ **Equation 6-12**

$$\frac{R-M}{R} = \cos\left(\frac{I}{2}\right)$$ **Equation 6-13**

$$c = 2R\sin\left(\frac{d}{2}\right)$$ **Equation 6-14**

$$l = Rd\left(\frac{\pi}{180}\right)$$ **Equation 6-15**

$$E = R\left[\frac{1}{\cos\left(\frac{I}{2}\right)} - 1\right]$$ **Equation 6-16**

$$\text{Deflection angle per 100 ft of arc length} = \left(\frac{D}{2}\right)$$

6.1.4.1.4 Stationing Along a Horizontal Curve

Proper stationing is critical to the layout of horizontal curves and is often the focus of exam questions. By convention, the PC station is always less than the PT station because stationing always increases along the curve from PC to PT. This is true regardless of whether the curve is turning to the left or to the right. Likewise, the PI station is always greater than the PC station because stationing also runs along the back tangent to the PI.

Although horizontal curves are typically shown curving to the right with stationing increasing from left to right (as in Figure 6.4), the exam may include problems in which they curve to the left and the stationing increases from right to left.

Also, by convention, the PI station is equal to the PC station plus the tangent length T, and the PT station is equal to the PC station plus the arc length L, as shown in Figure 6.5.

TIP

Do not underestimate the importance of these facts, because they have been and continue to be the basis of many exam questions.

Figure 6.5 illustrates the correct two-step procedure for determining the PT station given the PI station. Notice that although the distance between PI and PT is T, which is the same distance T between PI and PC, adding T to the PI station does not yield the correct PT station.

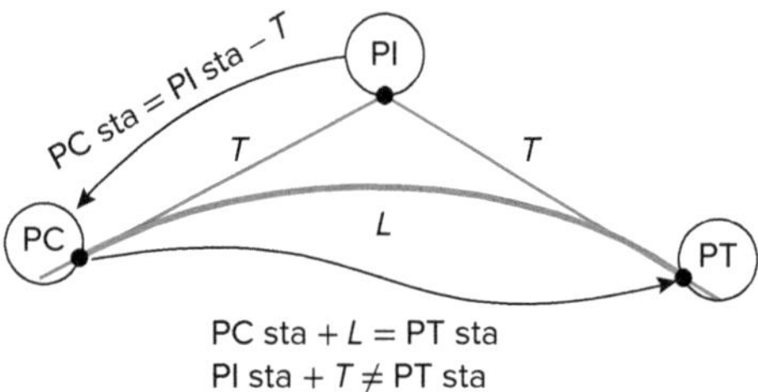

FIGURE 6.5 Stationing Key Points on a Horizontal Curve

There is one correct method of finding the PT station. The PT station is determined by first subtracting T from the PI station to find the PC station, and then adding L to the PC station to find the PT station.

Note that $2 \times T$ does not equal the arc length L.

Example 6.13: Stationing on Horizontal Curves

The PI of the curve is at sta 123+24.80, $T = 122.145$ ft, and $L = 242.164$ ft. Determine the stations of the PC and PT points.

Solution

PC sta = PI sta − T = 12,324.800 − 122.145 ft = 122 + 02.655

PT sta = PC sta + L = 12,202.655 + 242.164 ft = 124 + 44.819

6.1.4.1.5 Degree of Curvature, *D*

Curvature is the most important property of a circular curve. It is a measure of how much direction change occurs along the length of a curve (or how sharp or flat a curve is). Curvature is specified by either radius R or degree of curvature D. These properties are inversely related because curvature is the reciprocal of the radius. Consequently, a small radius indicates large curvature and a large radius indicates small curvature. There are two definitions of degree of curvature, D: arc and chord.

6.1.4.1.5.1 Arc Definition

The degree of curvature is the central angle subtended by a 100-ft arc length. The arc definition is routinely used in the layout of curves for highways and streets. Using Equation 6-17, the degree of curvature is easily found if the radius is known and vice versa.

$$D = 360° \frac{100 \text{ ft}}{2\pi R} = \frac{5{,}729.578}{R} \qquad \textbf{Equation 6-17}$$

6.1.4.1.5.2 Chord Definition

The degree of curvature is the central angle subtended by a 100-ft chord length, which has traditionally been used for railroad design. The reason for the choice of the chord length rather than the actual arc length is strictly practical: A chord length is more easily and directly measured than an arc length. A chord is a straight line and can therefore be measured by simply stretching a tape between endpoints.

Regardless of which definition is used, D can be converted to the corresponding R by using the appropriate equation. The equations for converting from the arc and chord definitions to the corresponding radius are presented Figure 6.6.

$$D = 360°\frac{100 \text{ ft}}{2\pi R} = \frac{5{,}729.578}{R} \qquad D = 2\left[\arcsin\left(\frac{50 \text{ ft}}{R}\right)\right]$$

FIGURE 6.6 Degree of Curvature - Two Definitions: Arc (a) and Chord (b)

> **TIP**
>
> For the exam, it is important to know the difference between the two definitions and to be able to recognize which definition was used in preparing the original design computations.

Field measurements of the curve with the tape must be made along the chord and not along the arc. When the arc basis is used, either a correction is applied for the difference between arc length and chord length or the chords are made shorter to reduce the error to a negligible amount.

Alignments are typically laid out in 100-ft stations. This continues through curves so that roadway length is always the length of a series of straight lines that can be directly measured. For curves with a degree of curvature of less than 5°, the difference between a 100-ft chord length and the actual arc length is inconsequential. For curves with a degree of curvature between 5° to 15°, 50-ft chords are typically used. For short radius curves with a degree of curvature between 15° to 30°, 25-ft chords are typically staked. For curves sharper than 30°, 10-ft chords are typically staked.

6.1.4.2 Types of Horizontal Curves

A simple curve consists of a single-radius circular arc that connects a back tangent to an ahead tangent. The key to solving simple curve problems is knowing the elements of a circular curve and the equations in Section 6.1.4.1.3.

> **TIP**
>
> The vast majority of the horizontal curve problems on the exam are of this type.

A **compound curve** consists of two or more consecutive curves of different radii curving in the same direction with no tangent sections in between. The end of one curve is the beginning of the next curve. The PT of the first curve coincides with the PC of the second curve at the point of compound curvature (PCC).

Compound curves are used extensively on interchange ramps where vehicles are changing speed as they exit or enter a high-speed roadway. They are also often used at intersections where truck-turning movements and roadway geometrics necessitate more than a simple single-radius curve. Compound curves are also frequently used on open highways in rugged terrain where fitting a roadway to a hillside or along a narrow river valley requires a curving alignment.

Compound curves are a series of two or more immediately adjacent, simple curves with deflections in the same direction. Designers frequently use compound curves for intersection curb radii, interchange ramps, and transitions into sharper curves. Compound curves are used to transition into and from a simple curve and to avoid some control or obstacle that cannot be relocated.

The components of a two-centered compound curve are shown in Figure 6.7.

6.1.4.2.1 Two-Centered Compound Curve Formulas

$I = I_1 + I_2$ **Equation 6-18**

$X = R_2 \times \sin I + (R_1 - R_2) \times \sin I_1$ **Equation 6-19**

$$Y = R_1 - R_2 \times \cos I - (R_1 - R_2) \times \cos I_1$$ **Equation 6-20**

$$T_L = \frac{R_2 - R_1 \times \cos I + (R_1 - R_2) \times \cos I_2}{\sin I}$$ **Equation 6-21**

$$T_S = \frac{R_1 - R_2 \times \cos I - (R_1 - R_2) \times \cos I_1}{\sin I}$$ **Equation 6-22**

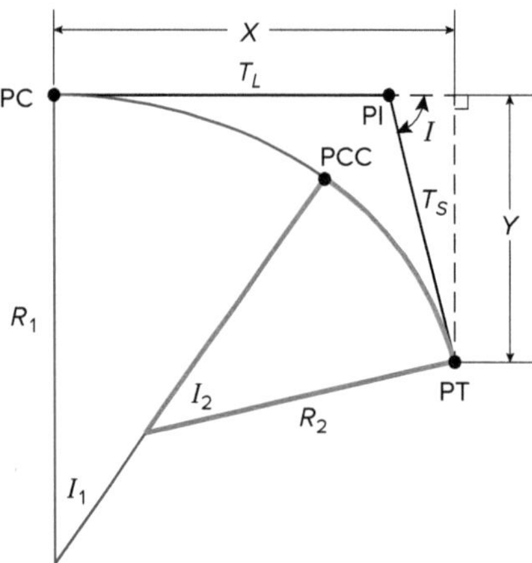

FIGURE 6.7 A Two-Centered Compound Curve

PI = point of intersection of back tangent and forward tangent

PC = point of curvature—point where back tangent meets circular curve

PT = point of tangency—point where circular curve meets forward tangent

PCC = point of compound curvature

T_L = long tangent of the compound curve

T_S = short tangent of the compound curve

I = total intersection angle of the compound curve

X = distance from PC to PT in the direction of the backward tangent

Y = perpendicular distance from the backward tangent to PT

I_1 = intersection angle of the flatter curve (decimal degrees)

I_2 = intersection angle of the sharper curve (decimal degrees)

R_1 = radius of the flatter curve

R_2 = radius of the sharper curve

6.1.4.2.2 Three-Centered Compound Curve

Compound curves with more than two curves are often used in rugged terrain where the horizontal alignment is determined by a river or stream or the steep slopes of a mountain (Fig. 6.8). They are also often used at high-volume intersections and interchanges for loop ramps. AASHTO recommends that the following ratios of larger to smaller radii be observed to avoid too abrupt a change in curvature on compound curves.

- On open highways, a maximum ratio of 1.5:1 should be used.
- At intersections where drivers accept more rapid changes in direction and speed, a maximum ratio of 2:1 may be used.
- Generally, a maximum ratio of 1.75:1 should be used.

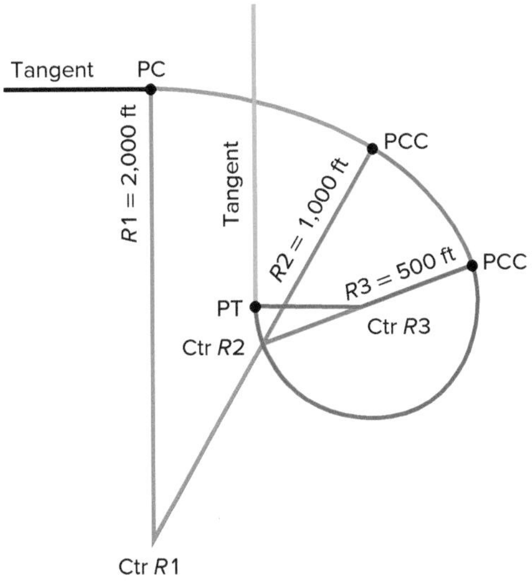

FIGURE 6.8 A Three-Centered Compound Curve

6.1.4.2.3 Reverse Curve

A reverse curve consists of two consecutive curves in opposite directions with no tangent length in between, as shown in Figure 6.9. The two curves meet at a point of reverse curvature (PRC) and typically involve an abrupt change in direction that drivers may not expect. This type of curve is therefore unacceptable, particularly on high-speed, open-road conditions. The lack of a tangent length makes it impossible to provide adequate superelevation transition between curves and can make it difficult for drivers unfamiliar with the curve to stay within their lanes. Reverse curves may be appropriate on low-speed local roads, on particularly rugged terrain, or in other low-speed situations, such as in a median where starting a left-turn lane.

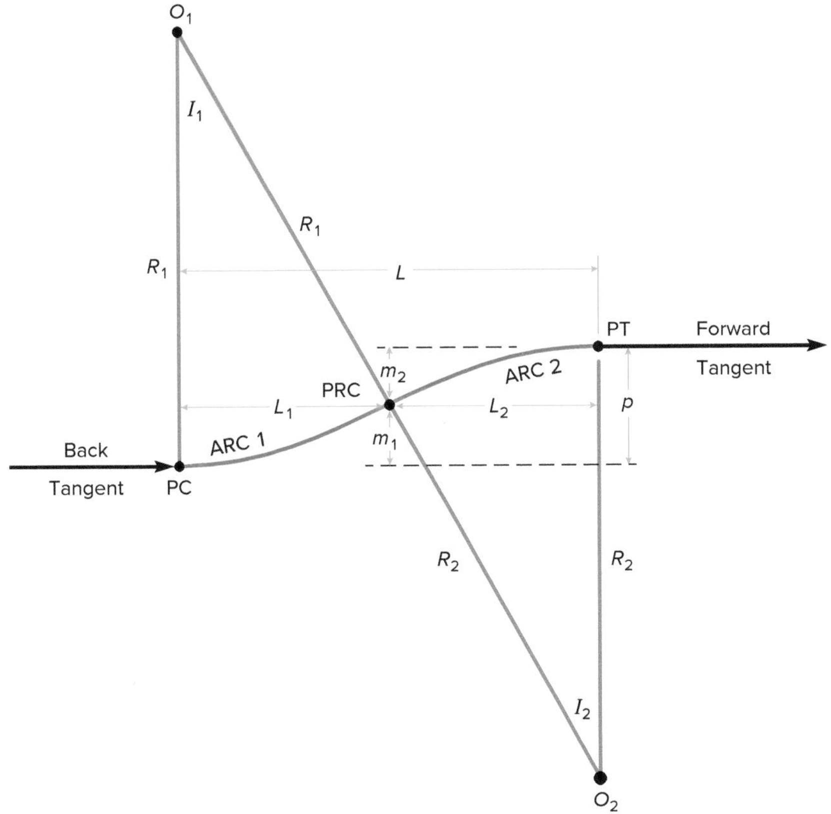

FIGURE 6.9 A Reverse Curve

6.1.4.2.4 Broken-Back Curve

A flat-back or broken-back curve consists of two curves in the same direction with a short length of tangent in between (Fig. 6.10). These types of curves should be avoided, but may be used where unusual topographical or right-of-way conditions control. Broken-back curves are undesirable because drivers do not expect the next curve to be in the same direction, and they are unpleasing in appearance. Where possible, broken-back curves should be replaced with one continuous single-radius curve.

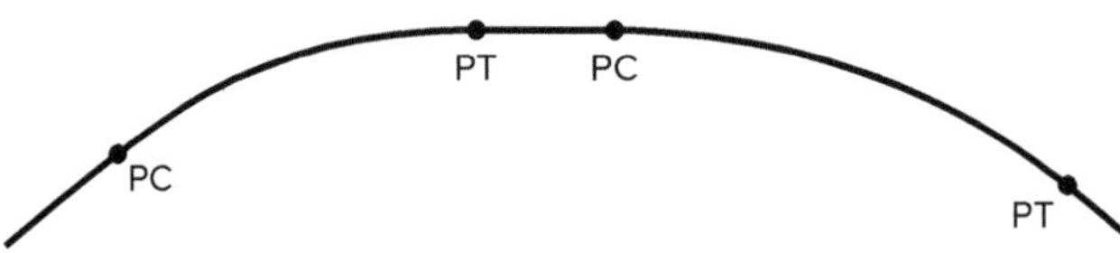

FIGURE 6.10 Broken-Back Curve

6.1.4.3 Methods of Curve Layout

6.1.4.3.1 Curve Layout by the Method of Tangent Offsets

Surveyors routinely stake out horizontal curves by driving wooden stakes into the ground at the PC, PT, and however many full stations are located in between.

In Figure 6.11, point P is a point on curve (POC) that is located using the method of curve layout by tangent offset. This method involves measuring a distance x from the PC (or PT) along the back (or ahead) tangent, and measuring an offset y for each POC to be staked.

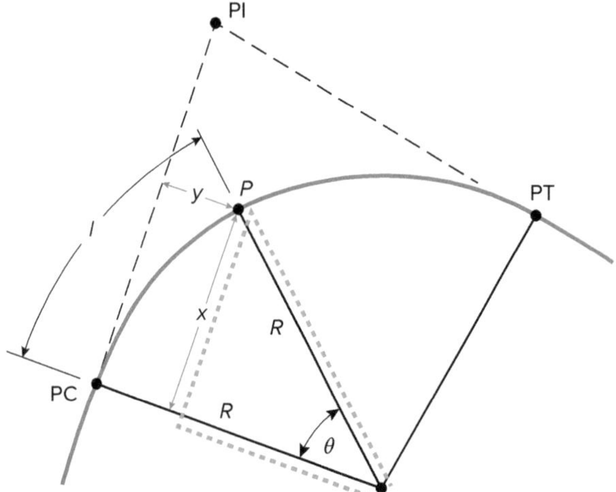

FIGURE 6.11 Curve Layout by Method of Tangent Offsets

Although the method of tangent offsets is relatively simple to execute in the field, it does require a little organization and planning and a few advance calculations. For each POC to be staked, the associated l, θ, x, and y values must first be calculated.

Note that l is the arc length from PC to the POC, and is equal to the POC station minus the PC station. The angle θ is calculated using the arc length l and the curve radius R. The parallel distance x is the side opposite angle θ in the green right triangle in Figure 6.11. The tangent offset y is the difference between R and the side adjacent to angle θ in the green right triangle. The following step-by-step procedure is recommended for curve layout using the method of tangent offsets.

Procedure:

1. Draw a rough sketch of the curve.
2. Calculate deflection angle I from the back and ahead tangent bearings.
3. Calculate the tangent length T.
4. Calculate the curve length L.
5. Calculate the PC station.
6. Calculate the PT station.
7. Determine the station of each POC to be located between the PC and PT.
8. Create a blank solutions table with a separate row for each POC to be located.
9. Fill in the row for each POC by calculating its corresponding l, θ, x, and y values.

Example 6.14: Curve Layout

Use the curve data in the table below and the method of tangent offsets to calculate the x, y, and θ values that correspond to the even stations on the curve shown in the figure below.

$I = 34°58' = 34.967°$

$$T = R\tan\frac{I}{2} = 500 \text{ ft} \tan\left(\frac{34.967°}{2}\right) = 157.49 \text{ ft}$$

$$L = RI\frac{\pi}{180.00°} = 500 \text{ ft} \times 34.967°\frac{\pi}{180.00°} = 305.14 \text{ ft}$$

CURVE DATA	
Radius	$R = 500.00$ ft
Back Tangent	S70°38′E
Ahead Tangent	S35°40′E
PI Station	500+00.00

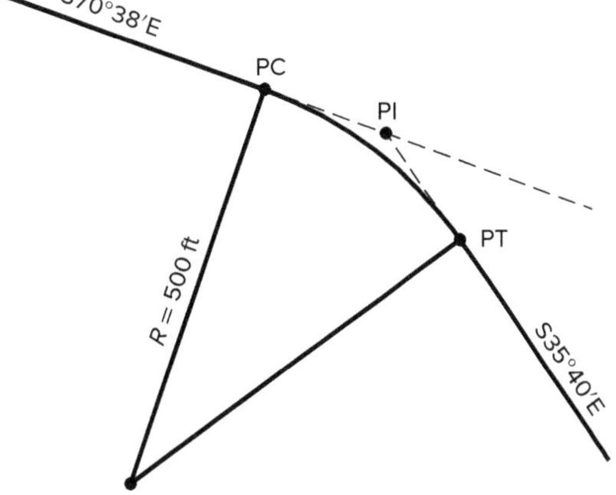

PI Station	=	500+00.00
T	=	−157.49
PC Station	=	498+42.51
L	=	305.14
PT Station	=	501+47.65

$$l = \theta R\frac{\pi}{180^0}$$

$$\theta = \arcsin\left(\frac{x}{R}\right) = \arccos\left(\frac{R-y}{R}\right) = \frac{l \times 180^0}{R\pi}$$

$$x = R(\sin\theta) = \sqrt{2Ry - y^2}$$

$$y = R(1 - \cos\theta) = R - \sqrt{R^2 - x^2}$$

Note that between the PC and PT, the full station POCs are at:

1. sta 499+00
2. sta 500+00
3. sta 501+00

Example 6.14 *(continued)*

Therefore, three rows of calculations are needed.

Fill in the rows for these two POCs by performing the calculations in the block at right once for each POC.

SOLUTIONS
Internal Angle, *I*: S70°38′E − S35°40′E = 34°58′ = 34.97°
Tangent Length, *T*: 157.49 ft
Length of Curve, *L*: 305.14 ft
PC Station: 498+42.51
PT Station: 501+47.65

POC Station	**Arc Length, *l***	θ	*x*	*y*
sta 499+00	57.49 ft	6.588°	57.36 ft	3.30 ft
sta 500+00	157.49 ft	18.047°	154.90 ft	24.60 ft
sta 501+00	257.49 ft	29.506°	246.26 ft	64.85 ft

6.1.4.4 Minimum Radius of a Horizontal Circular Curve

The minimum radius of a horizontal curve is based on the physics of a vehicle traveling at constant speed on a circular path (Fig. 6.12). Such a vehicle is subject to **centripetal acceleration**, which acts toward the center of the curve, and **centrifugal acceleration**, which causes the vehicle to drift to the outside of the curve.

$$a_n = \frac{V^2}{15R\text{ ft}}$$ **Equation 6-23**

V = vehicle speed (mph)
R = radius (ft)
a_n = normal acceleration (ft/s^2)

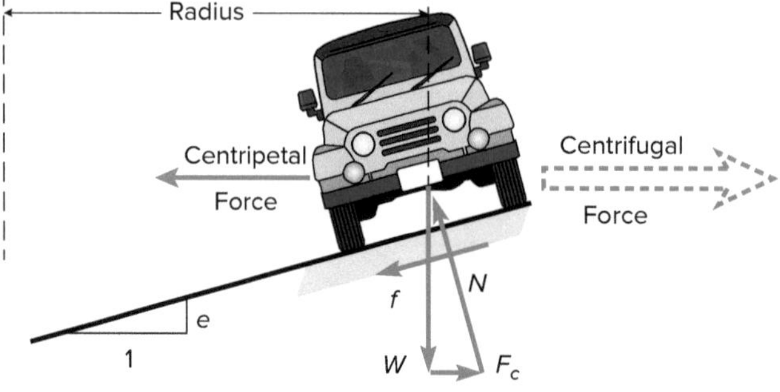

FIGURE 6.12 **Forces Acting on a Vehicle Traveling on a Superelevated Horizontal Curve**

As indicated by the equation above, the magnitude of the centripetal acceleration is determined by the vehicle speed squared and curve radius. Based on the above equation, centripetal/centrifugal acceleration is not linearly proportional to speed. On a constant radius curve, as vehicle speed increases, a_n increases by the square of the speed increase factor.

For example, a vehicle traveling on a constant radius curve at 30 mph is subject to a_n. If the vehicle speed increases by a factor of 2 to 60 mph, a_n increases by 4 (or 22) to $4a_n$. If the vehicle speed increases by a factor of 3 from 30 mph to 90 mph, a_n increases by a factor of 9 (or 32) to $9a_n$. If the vehicle speed increases by a factor of 4 from 30 mph to 120 mph, a_n increases by a factor of 16 (or 42) to $16a_n$, and so forth. When multiplied by the mass of a vehicle, centripetal acceleration becomes centripetal force.

6.1.4.4.1 Centripetal Force

Centripetal force is the force that allows an object to remain on a curved path. For example, if you tie a heavy object to a string and spin it around your head (like a ceiling fan), the tension in the string exerts a centripetal force on the object and keeps it on its circular path. Centripetal force is directed toward the center of rotation.

TIP

Note that the exam has included questions about centripetal force and centrifugal force in the past.

6.1.4.4.2 Centrifugal Force

Centrifugal force is considered an imaginary force because it exists due to the centripetal force. As you spin a heavy object at the end of a string around your head, you feel the object pulling away. That is the centrifugal force. Unlike centripetal force, centrifugal force is directed away from the center of rotation.

The minimum radius (R_{min}) is calculated from Equation 6-24:

$$0.01e + f = \frac{V^2}{15R} \quad \Rightarrow \quad R_{min} = \frac{V^2}{15(0.01e_{max} + f_{max})}$$ **Equation 6-24**

e_{max} = rate of roadway superelevation (%)

f_{max} = side friction factor

R_{min} = minimum radius (ft)

V = design speed (mph)

To balance the centripetal acceleration a_n, roadway designs must provide the appropriate amount of pavement banking or superelevation, which is the rotation of the pavement on the approach to and through a horizontal curve for the purpose of counteracting the centripetal acceleration a_n acting on a vehicle. Superelevation also helps drivers maintain vehicle control when traveling through a curve at a constant design speed.

Caution must be used not to specify too much superelevation, particularly in cold climates where excessive superelevation on an icy roadway can cause vehicles traveling at less than the design speed to slide off the pavement. Superelevation is typically expressed as a percentage and represents the cross slope (grade across the width of the pavement). The AASHTO *Green Book* lists maximum superelevation rates, e_{max}, that range from 0% to 12%.

Maximum superelevation rates are set by individual states based on several variables, such as climate, terrain, highway location (urban or rural), and frequency of very slow-moving vehicles. For example, northern states that experience ice and snow conditions may establish lower e_{max} rates than states that do not experience these conditions. Limiting maximum superelevation rates to lower values is intended to reduce the likelihood of vehicles sliding across the roadway when traveling at very low speeds during icy weather conditions. Typically, maximum **superelevation** rates of $e_{max} = 6.0\%$ are used in urban areas and $e_{max} = 8.0\%$ are used in rural areas. These limits are set based on climate conditions (frequency and amount of snow and ice) and the likelihood of very slow-moving traffic.

The minimum radius is therefore a limiting value for a given design speed and an important control in the determination of how much superelevation e_{max} and friction f is needed to maintain a constant speed V on curves of various radii. Note that using Equation 6-24 gives the same design speed V with a different maximum rate of superelevation, e_{max}, resulting in a different minimum radius R.

6.1.4.4.3 Side Friction

In addition to superelevation, a properly designed horizontal curve uses the side friction between pavement and tires to resist centrifugal force. As a vehicle travels around a horizontal curve, it is subject to centrifugal force and tends to drift to the outside of the curve. For the vehicle to remain in the center of the lane and continue traveling around the curve, enough side friction force must be produced to counter the centrifugal force. As shown in Equation 6-24 (the centripetal acceleration equation), side friction f is added to the superelevation e to balance the normal (centrifugal) force due to the ratio $\frac{V^2}{R}$.

Actual side friction is highly variable and is affected by many factors including the condition of tires and pavement, as well as moisture and temperature. The upper limit of side friction is the point of impending skid. The side friction factor is determined by the condition of the tires, the skid resistance of the roadway surface, and the speed of the vehicle.

Based on test track runs, the highest side friction factors that have been observed between wet concrete pavement and new tires range from 0.50 at 20 mph to 0.35 at 60 mph. Also, the highest side friction factor that has been observed between wet concrete pavement and smooth tires is 0.35 at 45 mph.

For design, the AASHTO *Green Book* recommends using the following conservative maximum friction factors:

	MAXIMUM FRICTION FACTORS														
Speed (mph)	10	15	20	25	30	35	40	45	50	55	60	65	70	75	80
Friction Factor	0.38	0.32	0.27	0.23	0.20	0.18	0.16	0.15	0.14	0.13	0.12	0.11	0.10	0.09	0.08

A properly designed horizontal curve allows drivers to maintain a uniform speed while traveling comfortably around the curve. If vehicle running speed is consistent with roadway design speed, a driver will not feel the vehicle tending to drift to the inside or outside of the curve and will not feel a need to over- or understeer.

A tendency to drift to the inside of the curve indicates that a vehicle is traveling too slowly for the superelevation provided. Conversely, a tendency to drift to the outside of the curve indicates that a vehicle is traveling too fast for the superelevation provided and that the driver should slow down.

6.1.4.5 Spiral (Transition) Curves

Transition curves, also referred to as spiral curves, are curves that provide a gradual transition from a straight path to a curved path at the beginning and end of horizontal curves (Fig. 6.13). Mathematically, a spiral may be described as a curve of continuously changing radius and degree of curvature, D. Throughout its length, the radius of a spiral varies from infinity at the tangent to the exact radius and D of the arc it is tying into.

Although mathematically much more complex than a simple circular curve, spirals have the following significant advantages:

1. A spiral provides drivers with a natural and easy-to-follow path along which centrifugal forces are applied gradually. This improves ride comfort, making it easier for drivers to control their vehicles and remain in the center of the travel lane.
2. The length of a spiral curve provides a convenient location to transition the pavement cross slope from normal crown to full superelevation.
3. If a roadway requires pavement widening around a curve, the spiral defines where the widening will occur.
4. The roadway appearance is enhanced by the application of spiral curves by smoothing out what appear to be kinks in a roadway.

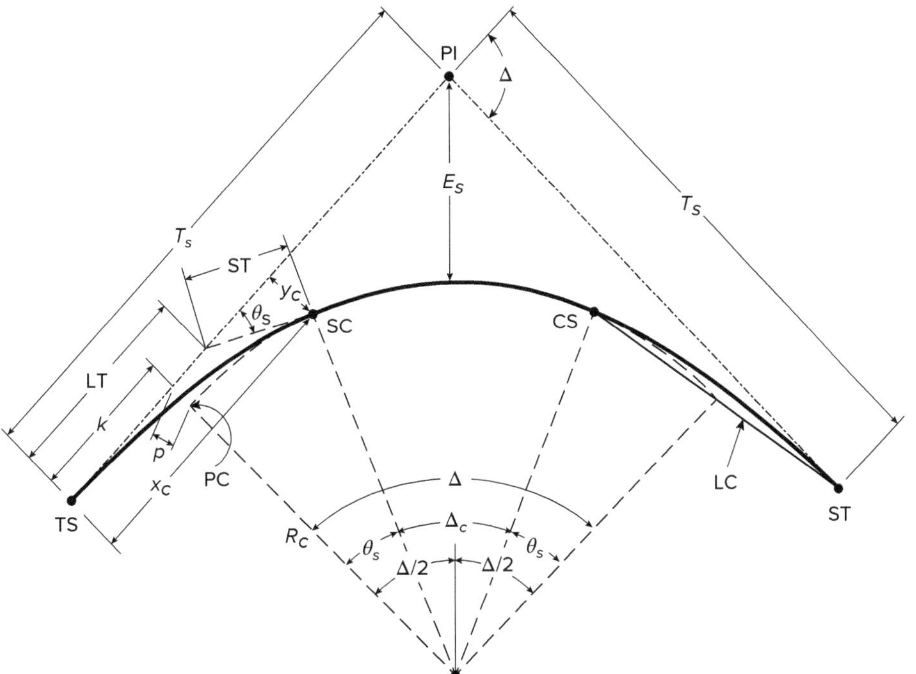

FIGURE 6.13 Elements of a Spiral Horizontal Curve

R_c = radius of circular curve

T_s = tangent distance

Δ = delta—external angle

θ_s = spiral angle

Δ_c = central angle between SC and CS

E_s = external distance

LC = long chord

LT = long tangent

ST = short tangent

x_c = tangent distance for SC

y_c = tangent offset of SC

k = simple curve—coordinate (abscissa)

p = simple curve—coordinate (ordinate)

TS = tangent to spiral point

SC = spiral to curve point

CS = curve to spiral point

6.1.4.5.1 Length of Spiral Curve

The minimum length of spiral (L_s) is determined by the roadway design speed, the radius of the circular curve, and the rate of increase of lateral acceleration.

$$L_s = \frac{3.15\ V^3}{RC}$$ **Equation 6-25**

L_s = length of spiral transition curve (ft)

V = design speed (mph)

R = radius (ft)

C = rate of increase of lateral acceleration (use 1 ft/s^3 unless otherwise stated)

The minimum length of superelevation runoff (L_r) is applicable to all superelevated curves and may be used for the minimum length of spiral (L_s). The AASHTO *Green Book* provides designers considerable flexibility in the determination of spiral curve lengths, and supplies formulas for calculating the minimum and maximum length of a spiral, as well as a table of desirable spiral lengths. The *Green Book* also acknowledges that a spiral should be at least as long as the minimum superelevation runoff (L_r) required for the design speed.

Example 6.15: Spiral Curve Length

A spiral transition segment is to be designed to connect a horizontal curve with a radius of 2,000 ft and a tangent for a highway with a design speed of 60 mph. Assuming a rate of increase of lateral acceleration of 2 ft/s^3, what is most nearly the minimum length of the transition segment?

A. 286 ft
B. 170 ft
C. 57 ft
D. 320 ft

Solution

The spiral transition length L_s can be calculated by:

$$L_s = \frac{3.15V^3}{RC}$$

R = horizontal curve radius = 2,000 ft

V = design speed = 60 mph

C = rate of increase of lateral acceleration = 2 ft/s^3

$$\text{Therefore, } L_s = \frac{3.15V^3}{RC} = \frac{3.15(60\text{ mph})^3}{(2{,}000\text{ ft})(2\text{ ft/s}^3)} = 170.1\text{ ft}$$

Answer: B

Example 6.16: Spiral Curve Length

A transition (spiral) curve has a spiral length of 250 ft and connects to a circular curve with a degree of curvature (D) of 3°. The rate of change in the degree of curvature D (in degrees per station) along the spiral is most nearly:

A. 1.20°
B. 1.35°
C. 1.50°
D. 2.25°

Solution

This exercise is asking for the rate of change in the degree of curvature D (degrees per station) along the spiral. Compute the rate of change of the degree of curvature per foot of spiral, known as K.

$$K = \frac{D}{L_s} = \frac{3°}{250\text{ ft}} = \frac{0.012°}{\text{ft}}$$

Convert the rate of change per foot to rate of change per station, K_s.

$$K_s = \left(\frac{100\text{ ft}}{\text{station}}\right)K$$

$$= \left(\frac{(100\text{ ft})}{\text{station}}\right)\left(\frac{0.012}{\text{ft}}\right) = 1.20°/\text{station}$$

Answer: A

6.1.5 Sight Distance on Horizontal Curves

The sight distance provided across the inside of a horizontal curve is an important design consideration (Fig. 6.14). If the SSD required for the design speed is not provided, drivers may be in danger of not being able to see an object in the road ahead until they are too close to stop and avoid hitting it.

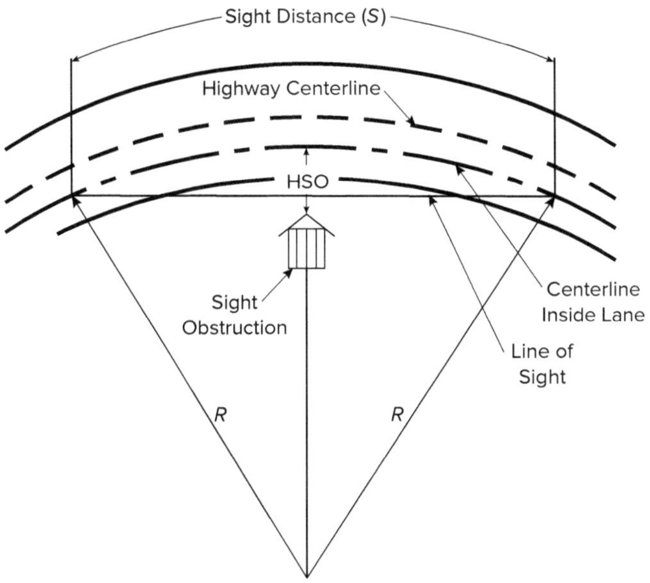

FIGURE 6.14 **Stopping Sight Distance on a Horizontal Curve**

From *A Policy on Geometric Design of Highways and Streets*, 2011, by the American Association of State Highway and Transportation Officials, Washington, DC. Used by permission.

Therefore, when laying out horizontal curves, roadside conditions should always be checked to ensure that the SSD required for the selected design speed is actually provided. The design parameter used to check the sight distance across a curve is the **horizontal sightline offset** (HSO).

As shown in Figure 6.14, the HSO is the lateral offset from the midpoint of the line of sight to the centerline of the inside lane. A sight obstruction (such as a sign, building, tree, parked vehicle, or even a snow bank) located within the HSO could block a driver's line of sight and reduce the available sight distance.

$$\text{HSO} = R \times \left(1 - \cos\frac{28.65 \times S}{R}\right) \qquad \textbf{Equation 6-26}$$

Equation 6-26 should be used to calculate the HSO:

S = SSD (ft)

R = radius of the centerline of inside lane (ft)

HSO = horizontal sightline offset (ft)

HSO is directly related to design speed, but design speed is not entered directly into Equation 6-26. Instead, the stopping sight distance S required for the specified design speed is entered.

HSO and the curve middle ordinate M are similar, but not interchangeable. The HSO varies with design speed and the associated stopping sight distance, S. M is based solely on the radius R and deflection angle I.

Stopping sight distance (S) should be determined by evaluating a roadway's vertical alignment. If the roadway is on a steep downgrade, the required S will be greater than on a level grade. Likewise, if the roadway is on a steep upgrade, the required S will be shorter than on a level grade. If no vertical alignment information is provided, it is safe to assume that the roadway is on a level grade and the S value for level terrain may be used.

Example 6.17: Sight Distance on a Horizontal Curve

Given: Design speed = 65 mph

Radius (center of inside lane) = 1,994 ft

Level grade

Determine: HSO

First, determine stopping sight distance, S.

For $V_d = 65$ mph, $S = 645$ ft

$$S = 1.47 \times V_{\text{mph}} \times t_p + \frac{V_{\text{mph}}^2}{30\left(\frac{a}{32.2} + G\right)} = 1.47\,\text{mph} \times 65\,\text{mph} \times 2.5\,\text{mph} + \frac{(65\ \text{mph})^2}{30\left(\frac{11.2}{32.2} + 0\right)}$$

$$= 644.4 \approx 645\ \text{ft}$$

Use the above equation to solve for HSO.

$$\text{HSO} = R\left(1 - \cos\frac{28.65 \times S}{R}\right) = 1{,}994\ \text{ft}\left(1 - \cos\frac{28.65 \times 645\ \text{ft}}{1{,}994\ \text{ft}}\right) = 26.03 \simeq 26\ \text{ft}$$

The following equation should be used when the actual distance from the centerline of the lane to obstruction (HSO) is known, and the question asks for the actual SSD.

$$S = \left(\frac{R}{28.65}\right)\left(\arccos\frac{R - \text{HSO}}{R}\right)$$ **Equation 6-27**

Example 6.18: Stopping Sight Distance on a Horizontal Curve

A six-lane divided highway with a design speed of 65 mph curves to the right on level grade and has the following typical section:

Grassed median: 14 ft wide

Median shoulder: 4 ft wide

Lanes: 12 ft wide, 3 lanes directional

Outside shoulders: 10 ft wide

Centerline degree of curvature, D: 3°45′

Part 1: If a single-face concrete barrier is to be installed along the right side of the highway without encroaching on the required horizontal sight distance, how far from the roadway centerline should the face of the barrier be located?

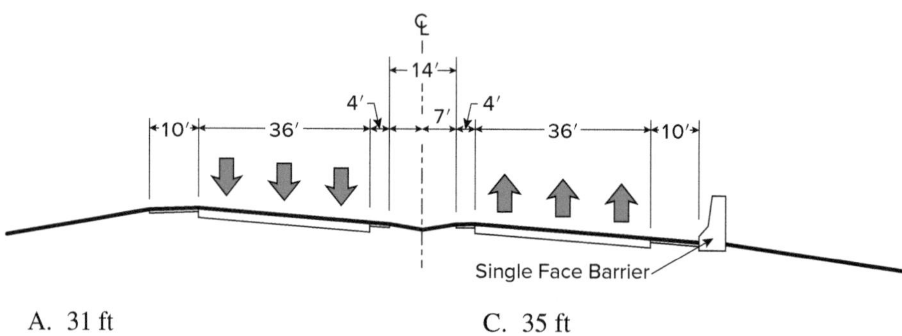

A. 31 ft

B. 33 ft

C. 35 ft

D. 39 ft

Example 6.18 *(continued)*

Solution

$V = 65$ mph

$S = 645$ ft

$D = 3°45'$, which is 3.75°; $R = 5{,}729.6$ ft/3.75° = 1,527.9 ft

The centerline of the inside lane is offset 41 ft (7 + 4 + 2(12) + 12/2) from the roadway centerline.

1,527.9 ft – 41 ft = 1,486.9 ft

$$\text{HSO} = R\left(1 - \cos\theta\left(\frac{28.65S}{R}\right)\right) = 1{,}486.9 \text{ ft} \times \left(1 - \cos\left(\frac{28.65 \times 645 \text{ ft}}{1{,}486.9 \text{ ft}}\right)\right)$$

$= 34.84 \approx 35$ ft

Answer: C

Part 2: Can a single-face concrete barrier be installed along the edge of the outside shoulder without encroaching on the horizontal sight distance?

Solution

No. The actual HSO would be only 16 ft = (6 + 10), which is less than the required HSO of 34.84 ft.

Geometrics

Example 6.19: Horizontal Sightline Offset

For a horizontal curve with a radius of 5,100 ft, the SSD is 1,900 ft. Determine the HSO within which no obstruction should exist.

A. 56 ft
B. 34 ft
C. 88 ft
D. 120 ft

Solution

The HSO is expressed by:

$$\text{HSO} = R\left[1 - \cos\left(\frac{28.65S}{R}\right)\right]$$

R = curve radius (ft)

S = stopping sight distance (ft)

In this problem, $R = 5{,}100$ ft and $S = 1{,}900$ ft.

Therefore, the HSO is:

$$\text{HSO} = (5{,}100 \text{ ft})\left[1 - \cos\left(\frac{28.65(1{,}900 \text{ ft})}{5{,}100 \text{ ft}}\right)\right] \cong 88 \text{ ft}$$

Answer: C

Example 6.20: Minimum Sight Distance

A corner of a tall building is located 50 ft from the centerline of a two-lane highway, which has 12-ft-wide lanes. The building is situated on the inside of a horizontal curve with a centerline radius of 600 ft. Assuming that there is no other sight restriction along the curve, determine the minimum sight distance along the curve.

A. 486 ft
B. 460 ft
C. 538 ft
D. 458 ft

Solution

$$\text{HSO} = 50 \text{ ft} - \frac{12 \text{ ft}}{2} = 44 \text{ ft}$$

The relationship between the HSO, sight distance, and centerline of inside lane radius is expressed by:

$$\text{HSO} = R\left[1 - \cos\left(\frac{28.65S}{R}\right)\right]$$

In this problem, $R = 594$ ft (600 ft – 6 ft). Therefore, $44 \text{ ft} = (594 \text{ ft})\left[1 - \cos\left(\frac{28.65S}{594 \text{ ft}}\right)\right]$.

$$\frac{28.65S}{594 \text{ ft}} = \arccos\left(1 - \frac{44 \text{ ft}}{594 \text{ ft}}\right) = 22.19°$$

$$S = \frac{22.19°}{28.65} 594 \text{ ft} \cong 460 \text{ ft}$$

Answer: B

6.2 VERTICAL ALIGNMENTS

Commonly Used Abbreviations

A	algebraic difference in grades
G_1	entering grade
G_2	exiting grade
Green Book	informal title of AASHTO publication, *A Policy on Geometric Design of Highways and Streets* [1]
h_1	height of driver's eye (3.5 ft, AASHTO standard)
h_2	height of object (2.0 ft, AASHTO standard)
HLSD	headlight sight distance
K	rate of change of vertical curvature; $K = L/A$
L	length of vertical curve (measured horizontally from PVC to PVT)
MO	middle ordinate
PVI	point of vertical intersection
PVC	point of vertical curve
PVT	point of vertical tangent
PVCC	point of vertical compound curvature
PVRC	point of vertical reverse curvature
r	rate of grade change
S	sight distance
x	horizontal distance measured from the PVC to a point of interest
x_t	horizontal distance from the PVC to a turning point

INTRODUCTION

The geometric design of highways and streets includes the development of vertical alignments (profiles) consisting of a series of intersecting grades (vertical tangents) connected by vertical curves. The goal of the vertical alignment design is to provide a smooth grade line that is safe, comfortable, pleasing in appearance, and adequate for drainage. The topography of the area where a proposed road is located also has a significant influence on the design of vertical alignments, which consist of two basic components: grades (vertical tangents) and vertical curves. Therefore, in its most basic form, vertical alignment design involves the selection of suitable grades and appropriate vertical curve lengths. While excessively steep grades may raise safety concerns and reduce capacity, all vertical curves must be designed to provide sufficient sight distance and are therefore typically the primary consideration in vertical alignment design.

6.2.1 Grades

Similar to straight lines (tangents) on horizontal alignments, grades are a fundamental component of vertical alignment design. As shown in Figure 6.15, grades may be positive (ascending), negative (descending), or zero (flat). They indicate numerically how steep or flat the slope of a line is. Slope is simply another word for grade, which is defined as the change in height (rise) divided by the corresponding horizontal distance (run). Grades are typically expressed in percentages.

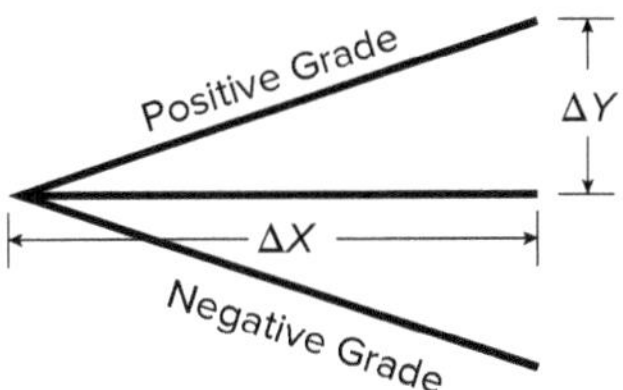

FIGURE 6.15 Grades Determine Change in Elevation

$$\text{Grade} = \frac{\Delta y}{\Delta x} \times 100$$ **Equation 6-28**

Example 6.21: Change in Elevation

Suppose that in traveling from point A to point B, you ascend a 4% grade while you move 1,000 ft horizontally. What is the change in elevation between points A and B?

Solution

1,000 ft = 10 stations

10 stations × 4% = 40 ft

Alternate solution

$$\frac{1{,}000 \text{ ft} \times 4\%}{100} = 1{,}000 \text{ ft} \times 0.04 = 40 \text{ ft}$$

Example 6.22: Grades

An engineer is designing a temporary access road for construction equipment to descend 23 ft to a river bed, and the maximum allowable grade is 10%. What is the required minimum horizontal length of the access road?

Solution

$$\text{horizontal distance} = \frac{\text{rise}}{\text{max grade}} = \frac{-23 \text{ ft}}{(10\%/100)} = \frac{-23 \text{ ft}}{0.10} = 230 \text{ ft}$$

Example 6.23: Grades

A roadway rises 26 ft in 400 ft. What is the percent grade of the roadway?

Solution

$$\frac{\text{rise}}{\text{run}} = \frac{26\text{ ft}}{400\text{ ft}} \times 100 = +6.5\%$$

Note that in this context, the 400-ft length is measured horizontally, rather than along the grade line itself. Grades must be compatible with the terrain, but also suitable for the types of vehicles that use the roadway with any frequency. Economic and environmental considerations dictate that roadway designs use steeper grades on mountainous terrain than on level or rolling terrain, but grades that are too steep can reduce the speed of heavy vehicles and reduce roadway capacity. In situations where long, steep upgrades are unavoidable, adding a climbing lane may be justified. The combination of long, steep downgrades and high truck percentages may pose a safety hazard that requires the construction of escape ramps.

Minimum grades depend on the drainage conditions of the highway. For example, 0% grades are generally not recommended, but may be used on uncurbed pavements if adequate cross slope is provided laterally to drain runoff. On curbed roadways, a minimum longitudinal grade (0.50%) should be provided to ensure adequate flow to drainage structures located along the curb line.

6.2.2 Vertical Curves

Vertical curves provide a smooth transition between grade lines and sufficient sight distance for the selected roadway design speed and traffic operations. Vertical curves enable drivers to travel safely along highways at a constant speed and avoid the need to slow down or stop whenever crossing over a bump or dip formed by the intersection of two grades.

As shown in Figure 6.16, there are two basic categories of vertical curves: crest and sag. A vertical curve is identified by its point of vertical intersection (PVI), which is where the entering grade line G_1 and the exiting grade line G_2 intersect. The PVI has a specific station and elevation, and the vertical curve lies either directly below the PVI (as in a crest vertical curve) or directly above the PVI (as in a sag vertical curve).

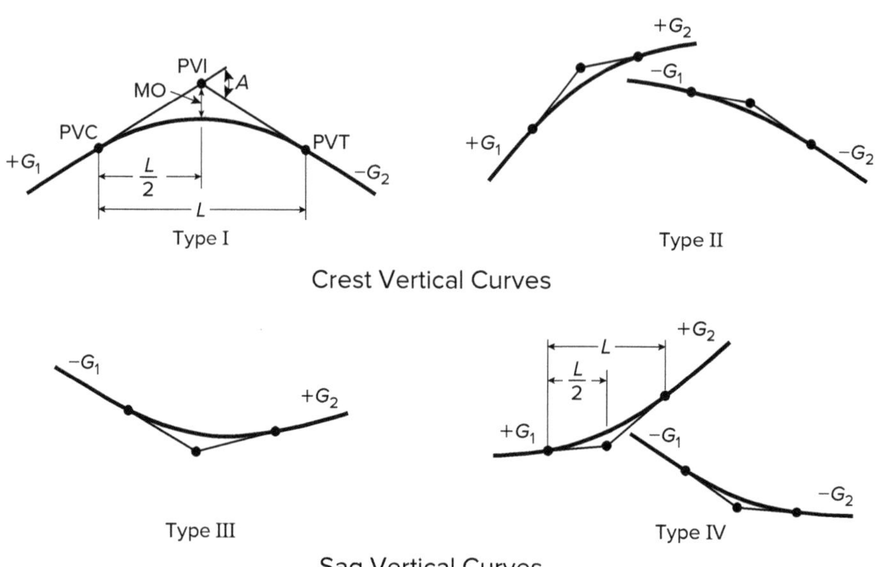

FIGURE 6.16 The Four Types of Vertical Curves

Depending on the signs of G_1 and G_2, these two basic categories are further subdivided into the four types shown.

Note that in Figure 6.16, type I and type III curves have turning points (high points or low points) where the slope of a line tangent to the curve changes sign, whereas type II and type IV vertical curves do not have turning points because their grades have the same sign throughout.

Sufficient curve length is needed for safety and comfort as well as aesthetics. Vertical curves should be long enough to provide the minimum required sight distance for stopping in all cases. Wherever practical, more than the minimum stopping sight distances should be provided. On two-lane highways, vertical curves are often designed to provide sufficient sight distance for passing. Curves that are too short for the roadway design speed or traffic operations do not provide sufficient sight distance and therefore may compromise safety, comfort, and roadway capacity.

The required length of a vertical curve is determined by several factors, including:

- Design speed
- Algebraic difference in grades
- Traffic operations and the associated sight distance

6.2.2.1 Vertical Curve Geometry and Equations

Unlike circular horizontal curves, which are based on trigonometry and geometry, vertical curves are parabolic and based on algebra. The geometric elements of a vertical curve and the equations used to calculate them are presented in Figure 6.17 and Section 6.2.2.2.

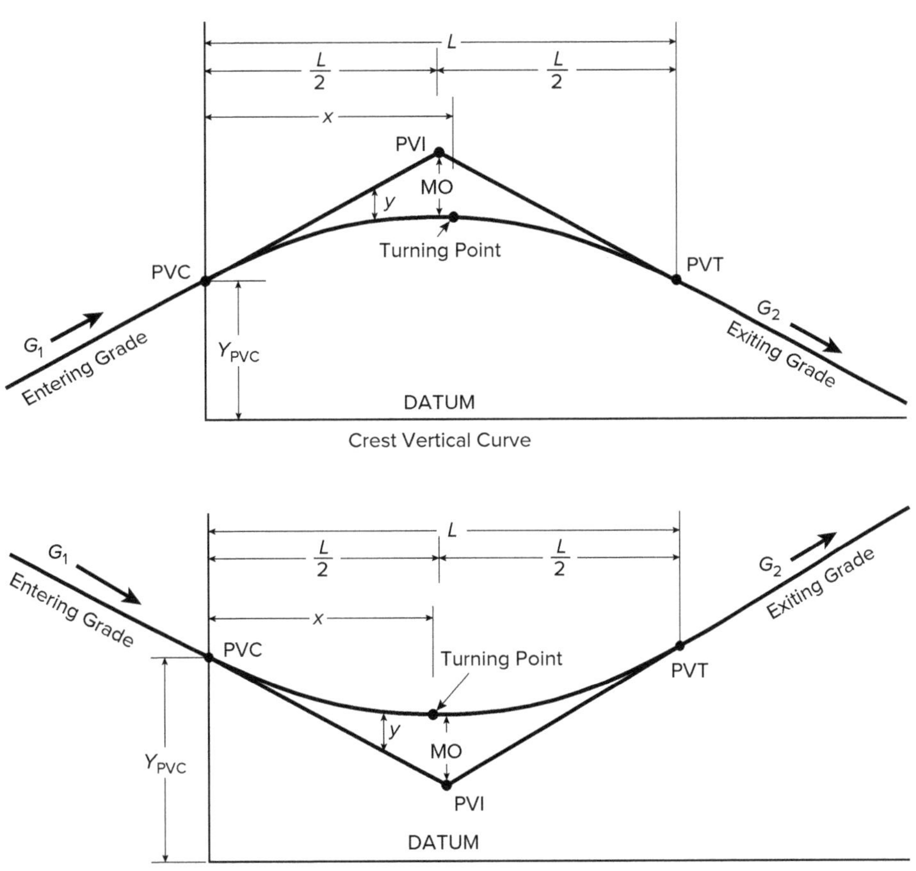

FIGURE 6.17 Elements of Crest and Sag Vertical Curves

L = length of curve

PVC = point of vertical curve

PVI = point of vertical intersection

PVT = point of vertical tangent

G_1 = entering grade

x = horizontal distance from PVC to turning point

G_2 = exiting grade

a = parabola constant

y = tangent offset

MO = tangent offset at PVI = $\frac{AL}{800}$

r = rate of change of grade

K = rate of vertical curvature

A = algebraic difference in grade

6.2.2.2 Vertical Curve Data

Vertical curves are defined completely by a concise set of curve data that includes:

- PVI station
- Elevation
- Grades (G_1 and G_2)
- Curve length (L)

This data provides a complete description of the curve so that all other associated stations and elevations can be derived from these data. Note that the PVI is located at the midpoint of a horizontal line between the **point of vertical curve** (PVC) and **point of vertical tangent** (PVT). Because of this property, a vertical curve is considered symmetrical. On a symmetrical curve, if the PVI station and curve length L are given, one can easily establish the stations of the PVC and PVT using Equations 6-29 and 6-30:

$$\text{PVC sta} = \text{PVI sta} - \frac{L}{2}$$ **Equation 6-29**

$$\text{PVT sta} = \text{PVI sta} + \frac{L}{2}$$ **Equation 6-30**

In certain unusual situations, because of critical clearance or other controls, the use of an asymmetrical vertical curve may be necessary. However, for the purpose of the exam, all vertical curves are expected to be symmetrical.

6.2.2.3 Calculating the Elevation of a Point on a Vertical Curve

The PE Civil exam is expected to include several problems requiring the calculation of the elevations of points on vertical curves. Equation 6-33 is particularly useful because it can be used to find the elevation of any point on a vertical curve. It is supported by Equations 6-31 through 6-40.

$$Y_{\text{PVC}} = \text{PVI}_{\text{ELEV}} - G_1 \times \frac{L}{2}$$ **Equation 6-31**

$$Y_{\text{PVC}} + G_1 x = Y_{\text{PVI}} + G_2\left(x - \frac{L}{2}\right)\text{(tangent elevation)}$$ **Equation 6-32**

$$Y_{\text{PVC}} + G_1 x + \left(\frac{G_2 - G_1}{2L}\right)x^2 \text{ (curve elevation)}$$ **Equation 6-33**

$$y = ax^2$$ **Equation 6-34**

$$A = |G_2 - G_1|$$ **Equation 6-35**

$$a = \frac{G_2 - G_1}{2L}$$ **Equation 6-36**

$$E = a(L/2)^2$$ **Equation 6-37**

$$r = \frac{G_2 - G_1}{L}$$ **Equation 6-38**

$$K = \frac{L}{A}$$ **Equation 6-39**

$$x_m = \text{horizontal distance to turning point} = \frac{-G_1}{2a} = \frac{G_1 L}{G_1 - G_2}$$ **Equation 6-40**

6.2.2.4 Mathematics of Vertical Curve Calculations

Vertical curve equations 6-31 through 6-40 are very different mathematically from horizontal circular curve equations. Whereas a horizontal circular curve is defined by its radius and intersection angle and requires the use of trigonometry and geometry, a vertical curve is defined by its curve length L and grades G_1 and G_2 and requires the use of algebra. Vertical curves are parabolic and defined by Equation 6-33, where the change in elevation y is a function of the horizontal distance x, which is always measured horizontally from the PVC station.

6.2.2.5 PVC Station and Elevation

The PVC is a very important reference point for vertical curve calculations. It serves as the origin of all horizontal measurements. All x distances are measured horizontally starting at the PVC station. The elevation of the PVC, Y_{PVC}, is the data point used to find the elevations of all other points on the curve. Y_{PVC} is easily determined using Equation 6-31 and the PVI elevation, Y_{PVI}.

6.2.2.6 Rate of Grade Change

Equation 6-38 defines the rate of grade change, r, as the algebraic difference in grade, A, divided by the length of the vertical curve L. The rate of grade change is a constant for all locations on a vertical curve. Note that the rate of grade change must be negative for type I curves and positive for type III curves.

6.2.2.7 Stationing of Vertical Curves

Unlike horizontal curves, the length of a vertical curve L is not measured along the curve itself. L is the projection of the vertical curve onto the horizontal plane. More precisely, L is the length of a straight, horizontal line between the PVC station and the PVT station. L does not include the slight increase in actual length due to grades and curvature. This simplification is necessary to ensure that the vertical alignment stationing exactly matches the horizontal alignment stationing. Therefore, the stationing of a vertical curve is not along the curve itself, but along the horizontal axis. The stationing increases from PVC to PVT, which is why vertical curves are always drawn with PVC to the left of the PVT, with the PVI in between.

Although in reality roadways are three-dimensional, traditional roadway design practice is to design the horizontal alignment (line) two dimensionally first, and then develop a complementary vertical alignment (profile), which is also two dimensional. The process is interactive so that the resulting design is fully coordinated and refined using cross sections cut at select locations. With this approach the stationing shown on the horizontal axis of a roadway profile exactly matches the horizontal alignment stationing, and the relatively small increase in roadway length due to grades and vertical curves is ignored. A single set of stationing for both horizontal and vertical alignments is essential to enable coordination and control of the two alignments in three dimensions.

Example 6.24 demonstrates the use of Equations 6-33 and 6-36.

Example 6.24: Stationing of Vertical Curves

A crest vertical curve connecting tangent grades of 3% and −2% has a length of 400 ft. If the elevation of the PVC is 50 ft, what is the elevation of the curve at a horizontal distance of 100 ft from PVC?

A. 48.65 ft
B. 51.23 ft
C. 52.38 ft
D. 53.58 ft

Solution

The parabola constant a is $a = \frac{G_2 - G_1}{2L}$, in which L = curve length (stations), G_2 = entering grade, and G_1= exiting grade.

In this problem, $L = 4$ stations, $G_2 = -2\%$, and $G_1 = 3\%$.

Therefore, $a = \frac{-2\% - 3\%}{2 \times 4} = -\frac{5\%}{8} = -0.625$

$$y = Y_{PVC} + G_1 x + ax^2 = 50 \text{ ft} + 3\%(1 \text{ station}) - \frac{5\%}{8 \text{ stations}}(1 \text{ station}) \cong 52.38 \text{ ft}$$

Answer: C

6.2.3 Minimum Length of Vertical Curve

6.2.3.1 The *K* Value Method

The reciprocal of r $\left(\text{or } \frac{A}{L}\right)$ is $\frac{L}{A}$. This ratio is the horizontal distance (in ft) needed to make a 1% change in grade and is therefore an important measure of curvature. The ratio *L*/*A* indicates the rate of vertical curvature and is generally known as *K*, which is useful in determining the horizontal distance from the PVC to a turning point (high or low point) on the curve where the longitudinal slope is zero. Turning points occur at a distance x_m from the PVC, which is equal to *K* times the approach gradient G_1. Locating turning points is necessary for establishing drainage areas and is therefore necessary for roadway drainage design. The term *K* is also useful as a design control in determining the minimum curve length needed to meet sight distance requirements. The AASHTO *Green Book* provides separate tables of *K* values for the following conditions:

- Stopping on a crest vertical curve (stopping sight distance criteria)
- Passing on a crest vertical curve (passing sight distance criteria)
- Stopping on a sag vertical curve (headlight sight distance criteria)

Minimum *K* values are determined by required sight distance under various conditions, including design speed, curve type (crest or sag), and traffic operations (stopping or passing). Therefore, the *K* value used to determine the length of a sag vertical curve needed for stopping sight distance and a 50-mph design speed cannot be used to determine the length of a crest vertical curve under the same operating conditions and design speed. Each curve type and operating condition has its own set of design controls.

6.2.3.2 Vertical Curve Length Based on *K* Value

The length of a vertical curve can be defined by the product of its rate of vertical curvature *K* and the algebraic difference of its grades *A*. This design relationship can be expressed mathematically as shown in Equation 6-41:

$$L = K \times |A| \quad \textbf{Equation 6-41}$$

The value K is a design control for vertical curves based on sight distance. The absolute value of A (the algebraic difference in grades) is needed because the K value and L are always positive. There are separate K values for each of the following sight distances:

- Stopping sight distance (crest only)
- Passing sight distance (crest only)
- Headlight sight distance (HLSD; sag only)

The vertical offset y from the vertical tangent G_1 varies as the square of the horizontal distance x from the PVC. The vertical offset y at any point along the curve is calculated as a proportion of the middle ordinate, MO, which is the vertical offset of the curve at the PVI.

6.2.3.3 Determining the Required Vertical Curve Length

6.2.3.3.1 Minimum Length of Vertical Curve

The minimum length of a crest vertical curve is based on the operative sight distance criteria (stopping or passing) and is generally long enough to satisfy safety, comfort, and appearance criteria. An exception may be at decision points, such as at ramp exit gores or traffic signals located beyond a crest vertical curve where more DSD is needed. The main design control for safe driving conditions on crest vertical curves is providing at least the minimum stopping sight distances for the design speed. The minimum SSD is based on the roadway design speed, and should be provided on all roadways (Fig. 6.18).

> **TIP**
>
> DSD is a topic that is unlikely to appear on the breadth portion of the PE exam.

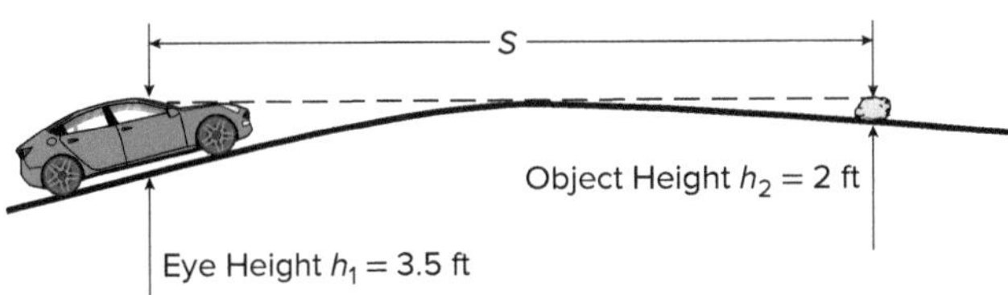

FIGURE 6.18 Stopping Sight Distance on a Crest Vertical Curve

The design controls for crest vertical curves based on stopping sight and the general formulas to determine minimum lengths of crest vertical curves are contained in the crest vertical curves section in the AASHTO *Green Book*, chapter 3.

6.2.3.4 Crest Vertical Curves

The length of a crest vertical curve is determined by several factors, including safety, drainage, riding comfort, and appearance. Curves that are too short and sharp are undesirable because they pose safety hazards by failing to provide enough sight distance. They also create the displeasing appearance of a sudden break in profile. Curves that are too long and flat do not have enough longitudinal slope for adequate roadside drainage. Riding comfort requires that the rate of change of a grade be kept within acceptable limits and that short roller-coaster type curves be avoided. Long curves are more pleasing in appearance than short curves.

The minimum length of a vertical curve is determined by design speed, the required SSD, and algebraic difference in grade, A. The AASHTO *Green Book* recommends a minimum length (in feet) for a crest curve that is three times the design speed (in mph). For example, on a freeway with a 75-mph design speed, the minimum length of a crest vertical curve, regardless of algebraic difference in grade, is 225 ft.

Crest vertical curves that are designed using minimum K values and that have small A values (algebraic differences) often result in very short curves or no curve at all. This is because with slight A values, the driver's sightline is unobstructed by the curve, and passes directly over it. Although no vertical curves or short curves are undesirable, they are often used as a practical solution in situations such as bridge replacements on low-speed or temporary roadways. Generous vertical curve lengths two to three times the minimum lengths required for SSD are generally recommended for improved safety and aesthetics.

For roadway safety, vertical curves must always be long enough to provide the sight distance required for traffic operations. All vertical curves must be long enough to meet SSD requirements. On two-lane highways where passing is permitted, vertical curves must be long enough to meet passing sight distance requirements, which do not apply to four-lane divided highways because the passing maneuver does not require the use of a lane of opposing traffic.

For aesthetics, the minimum length of a vertical curve (in feet) should be three times the design speed (in mph). Providing the minimum SSD is normally satisfactory for meeting all of the above requirements except passing sight distance.

For the purpose of the breadth portion of the PE Civil exam, this section is limited to the sight distance equations presented below. They include equations for SSD on crest and sag vertical curves, riding comfort criteria on sag vertical curves, and sight distance at overhead structures on sag vertical curves.

Minimum curve length based on SSD:

Crest vertical curve general equation:

for $S \le L$:

$$L = \frac{AS^2}{100\left(\sqrt{2h_1} + \sqrt{2h_2}\right)^2} \qquad \textbf{Equation 6-42}$$

for $S > L$:

$$L = 2S - \frac{200\left(\sqrt{h_1} + \sqrt{h_2}\right)^2}{A} \qquad \textbf{Equation 6-43}$$

L = length of vertical curve (ft)
S = stopping sight distance (ft)
A = algebraic difference in grades, $|G_2 - G_1|$ (%)
h_1 = height of eye above roadway surface (ft)
h_2 = height of object above roadway surface (ft)

When standard criteria are met—that is, the driver's eye height h_1 is 3.5 ft and the object height h_2 is 2.0 ft—the above general equations may be simplified and reduced as shown in Equations 6-44 and 6-45 below:

Crest vertical curve standard criteria: $h_1 = 3.5$ ft; $h_2 = 2.0$ ft

$$\text{for } S \le L, L = \frac{AS^2}{2{,}158} \Rightarrow S = \sqrt{\frac{2{,}158L}{A}} \qquad \textbf{Equation 6-44}$$

$$\text{for } S > L, L = 2S - \frac{2{,}158}{A} \Rightarrow S = \frac{1}{2}\left(L + \frac{2{,}158}{A}\right) \qquad \textbf{Equation 6-45}$$

Example 6.25: Crest Vertical Curves

A crest vertical curve will connect an entering grade of +3% to an exiting grade of −1%. Assuming that the minimum stopping sight distance is 600 ft, the height of a driver's eyes above the roadway surface is 3.5 ft, and the height of an object above the roadway surface is 2.0 ft, determine the horizontal distance from PVC to the maximum elevation on the curve.

Example 6.25 *(continued)*

A. 334 ft
B. 370 ft
C. 570 ft
D. 500 ft

Solution

The solution to this example involves five steps. The first three steps involve determining the crest vertical curve length. The last two steps involve finding the horizontal distance to the high point.

Step 1. Make an initial assumption about the stopping sight distance, S, relative to the crest vertical curve length, L. Note that there are only two possible assumptions: $S \leq L$ and $S > L$. Only one of these can be correct. For this example, select $S > L$.

Step 2. Test the selected assumption using the corresponding equation by calculating L. For $S > L$, use Equation 6-43.

$$L = 2S - \frac{200\left(\sqrt{h_1} + \sqrt{h_2}\right)^2}{A}$$

L = crest vertical curve length (in ft)

S = stopping sight distance = 600 ft

A = absolute value of algebraic difference in grades = $| -1\% - 3\% | = 4\%$

h_1 = height of driver's eyes above the roadway surface = 3.5 ft

h_2 = height of object above roadway surface = 2 ft

Therefore, the curve length is:

$$L = 2(600 \text{ ft}) - \frac{200\left(\sqrt{3.5 \text{ ft}} + \sqrt{2.0 \text{ ft}}\right)^2}{(4\%)} = 660.4 \text{ ft}$$

Notice that the answer, $L = 660.4$ ft, does not agree with the assumption $S > L$ because 600 ft is clearly not greater than 660.42 ft. Therefore, $S > L$ is not true, and the alternate assumption, $S \leq L$, must be true.

Step 3. Knowing that $S \leq L$ is the actual condition, use Equation 6-42 to calculate L.

$$L = \frac{AS^2}{100\left(\sqrt{2h_1} + \sqrt{2h_2}\right)^2}; \; L = \frac{(4\%)\,(600 \text{ ft})^2}{100\left(\sqrt{2(3.5 \text{ ft})} + \sqrt{2(2 \text{ ft})}\right)^2} = 667.2 \text{ ft}$$

Notice that this revised answer, $L = 667.2$ ft, agrees with the assumption $S \leq L$ because 600 ft is less than 660.42 ft.

Step 4. Knowing that the actual crest vertical curve length $L = 667.2$ ft, it can be useful to draw a sketch for visualization of the high point location.

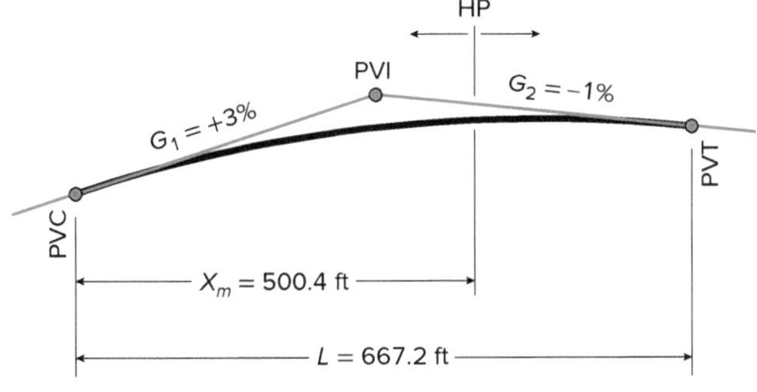

Example 6.25 (continued)

Step 5. Use Equation 6-40 to calculate the horizontal distance, x_m, from the PVC to the high point.

$$x_m = \frac{G_1 L}{G_1 - G_2} = \frac{(3\%)(667.2 \text{ ft})}{3\% - (-1\%)} \cong 500.4 \text{ ft}$$

Answer: D

6.2.3.5 Maximum Length of Vertical Curve

Generally, vertical curves should be designed to be as long as practical to maximize sight distance and for aesthetics, but excessive curve length can cause problems—specifically in areas with insufficient longitudinal slope for proper drainage. Many northern states require a minimum slope of +/− 0.50% because of typical winter conditions. Pavement surface drainage should be considered in the design of both crest and sag vertical curves because the minimum grade required for drainage is exactly the same for crest and sag vertical curves.

If a roadway is curbed and on a type III vertical curve, special attention should be given to roadway drainage, especially if the rate of vertical curvature is greater than 51 ($K >$ 51). For drainage, the critical areas are near the turning points of curves—specifically the high point of a crest curve or the low point of a sag curve. Designers should be particularly attentive to the pavement drainage near these points to confirm sufficient cross slope to drain the pavement surface adequately. This is particularly important if a turning point on the roadway profile coincides with, or is near, an adverse crown-removed section.

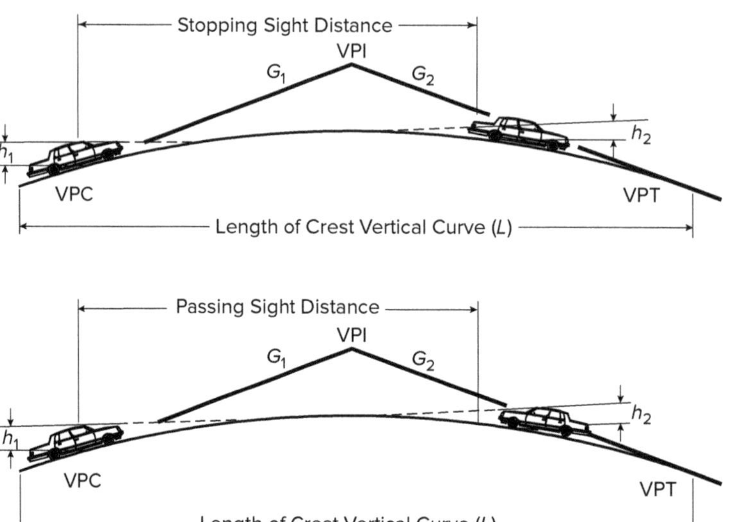

FIGURE 6.19

From *A Policy on Geometric Design*, 2011, by the American Association of State Highway and Transportation Officials, Washington, DC. Used by permission.

6.2.4 Sight Distance on Vertical Curves

6.2.4.1 Use of the *K* Value Method for the Selection of Crest Vertical Curve Length

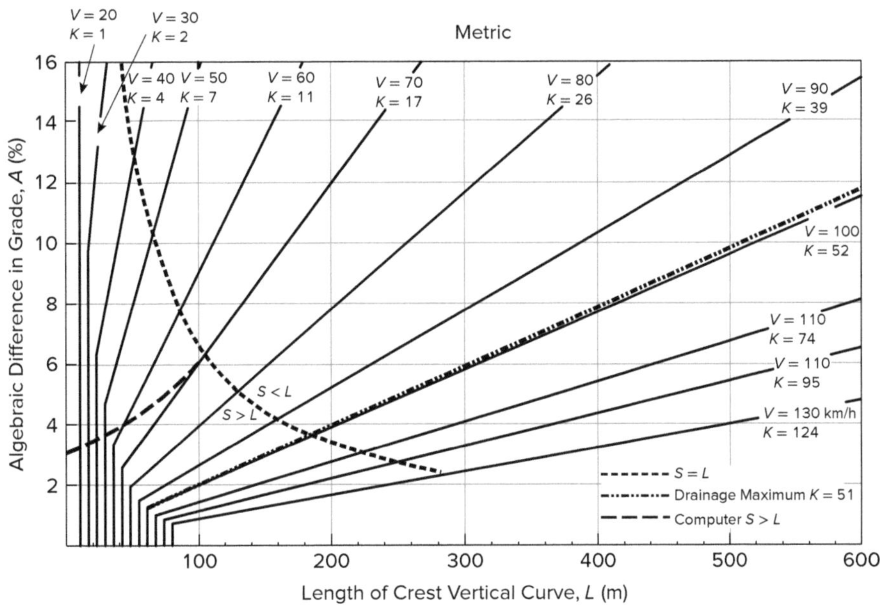

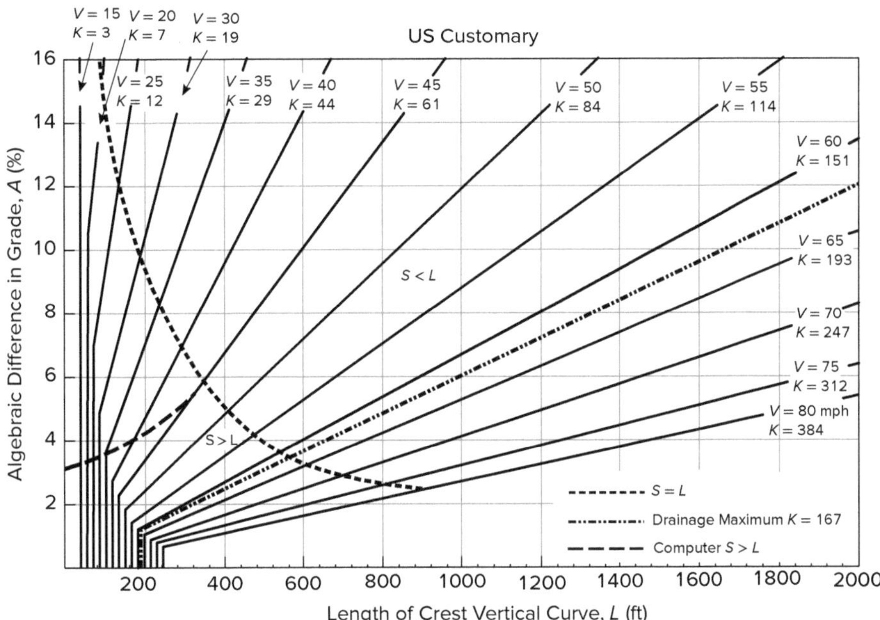

FIGURE 6.20

From *A Policy on Geometric Design*, 2011, by the American Association of State Highway and Transportation Officials, Washington, DC. Used by permission.

TABLE 6.3 AASHTO *Green Book* Table 3-34
Design Controls for Stopping Sight Distance for Crest Vertical Curves

US CUSTOMARY			
DESIGN SPEED (MPH)	STOPPING SIGHT DISTANCE (FT)	RATE OF VERTICAL CURVATURE, K^A	
		CALCULATED	DESIGN
15	80	3.0	3
20	115	6.1	7
25	155	11.1	12
30	200	18.5	19
35	250	29.0	29
40	305	43.1	44
45	360	60.1	61
50	425	83.7	84
55	495	113.5	114
60	570	150.6	151
65	645	192.8	193
70	730	246.9	247
75	820	311.6	312
80	910	383.7	384

[a]Rate of vertical curvature K is the length of curve per percent algebraic difference in intersecting grades (A). $K = L/A$

From *A Policy on Geometric Design*, 2011, by the American Association of State Highway and Transportation Officials, Washington, DC. Used by permission.

6.2.4.2 Sag Vertical Curve Design Considerations

6.2.4.2.1 Length of Sag Vertical Curves

At least four different criteria are recognized for the lengths of sag vertical curves (Fig. 6.21). They include:

1. Stopping sight distance based on headlight sight distance (HLSD)
2. Riding comfort
3. Drainage control
4. General appearance

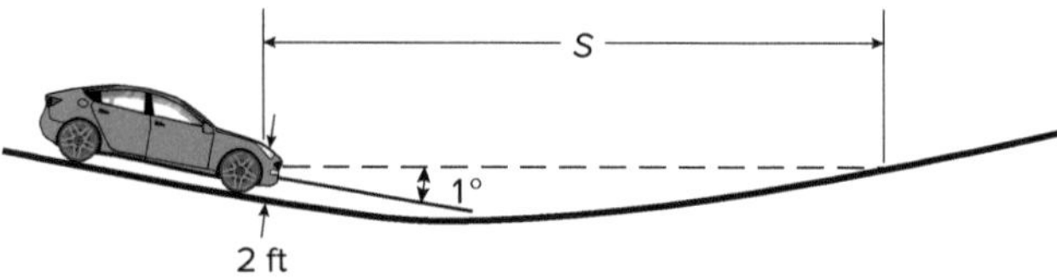

FIGURE 6.21 Headlight Sight Distance May Be Controlled by Sag Vertical Curve Length

Current design practice accepts that design based on HLSD is generally conservative and will normally provide a sag vertical curve as long as or longer than the other criteria and will therefore control.

6.2.4.2.2 Minimum Curve Length Based on Headlight Sight Distance

The nighttime sight distance in a sag vertical curve is based on the direction in which the vehicle headlight is pointed. Typically, the headlights will have an upward spread of about 1°. The roadway surface must come into view far enough ahead to allow a driver to stop a vehicle before reaching an object lying on the roadway. Therefore, the distance that can be seen by the driver is controlled by the headlight beam.

This means that the sag vertical curve must not interrupt the headlight beams at any point on the curve before the headlights provide a viewshed equal to or greater than the stopping sight distance. Sight distance for a sag vertical curve is also known as HLSD.

The following equations show the relationships between sight distance, length of curve, and algebraic difference of grades, using the headlight sight distance S as the distance between the vehicle and the point where the 1° upward angle of the light beam intersects the surface of the roadway:

for $S \leq L$, $L = \dfrac{AS^2}{400 + 3.5S}$ **Equation 6-46**

This equation can be reduced to $0 = AS^2 - 3.5LS - 400L$

for $S > L$, $L = 2S - \left(\dfrac{400 + 3.5S}{A}\right)$ **Equation 6-47**

This equation can be reduced to $S = \dfrac{LA + 400}{2A - 3.5}$

Example 6.26: Sag Vertical Curve

If an SSD of 500 ft is to be maintained on a sag vertical curve with tangent grades of −2% and 1%, what should the length of the curve be under standard headlight criteria?

A. 283 ft
B. 390 ft
C. 548 ft
D. 458 ft

Solution

Assuming the sight distance S is larger than the curve length L, L can be calculated by:

$$L = 2S - \left(\frac{400 + 3.5S}{A}\right)$$

L = curve length (ft)
A = absolute value of algebraic difference in grades (%)
S = stopping sight distance (ft)

In this problem, $S = 500$ ft and $A = (1\% - (-2\%)) = 3\%$.

Therefore, the curve length is $L = 2(500\text{ ft}) - \left(\dfrac{400 + 3.5(500\text{ ft})}{3}\right) \cong 283$ ft.

Since L is less than $S = 500$ ft, the original assumption is correct.

Answer: A

Example 6.27: Minimum Curve Length Based on Headlight Sight Distance

An equal-tangent sag vertical curve connects grades of −2.0% and +3.0%. The design sight distance on the curve is 645 ft. Determine the minimum curve length based on the stopping sight distance requirements following the standard headlight criteria.

A. 889.50 ft
B. 782.74 ft
C. 654.50 ft
D. 934.52 ft

Solution

The absolute value of the algebraic difference in grades (%) is:

$$A = |G_1 - G_2| = |-2 - (3)| = 5$$

Assume that the curve length L is greater than the sight distance S. Under the standard headlight criteria, the minimum curve length is:

$$L = \frac{AS^2}{400 + 3.5S}$$

$$= \frac{(5)(645\text{ ft})^2}{400 + 3.5 \times 645\text{ ft}}$$

$$\cong 782.74\text{ ft}$$

Example 6.27 *(continued)*

This is greater than S, so the assumption that $L > S$ is correct.

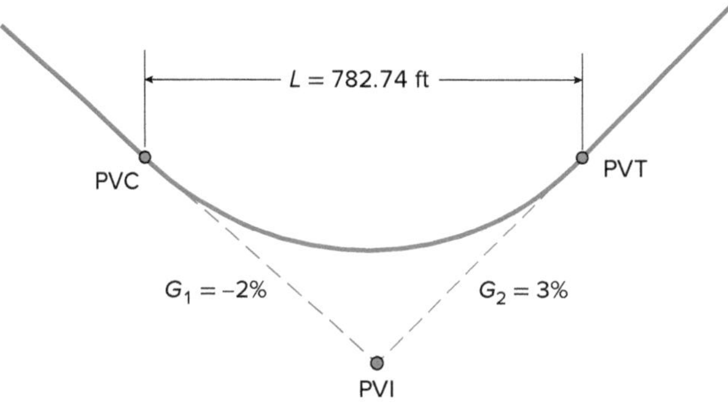

Answer: B

6.2.4.2.3 Sight Distance on a Sag Vertical Curve Under an Overhead Structure

The following equations are used to determine the actual SSD provided on a sag vertical curve under an overpass:

$$L = \frac{AS^2}{800\left(C - \frac{h_1 + h_2}{2}\right)} \quad \textbf{Equation 6-48}$$

$$L = 2S - \frac{800}{A}\left(C - \frac{h_1 + h_2}{2}\right) \quad \textbf{Equation 6-49}$$

Similar to other sight distance equations, these require an initial assumption of either $S \leq L$ or $S > L$. Also, these equations require that the driver's eye height h_1 and an object height h_2 be entered. What makes this situation unusual is that higher eye height results in more limited sight distance, and would therefore be a liability. In all other situations considered, a higher h_1 typically results in greater sight distance, which would be an advantage; however, in this case, a higher h_1 is a disadvantage. Therefore, in evaluating these situations, use the highest driver eye height expected to use this roadway with any frequency. That highest eye height would be 8 ft, which is the case for truck drivers.

6.2.4.2.4 Minimum Curve Length Based on Riding Comfort

Riding comfort is a design criterion because, on sag vertical curves, occupants experience a combination of centripetal and gravitational forces acting in the same downward direction, which, depending on speed, can cause discomfort. For riding comfort, the maximum centripetal acceleration should not exceed 1 ft/s^2.

Use the following equation to determine the minimum sag vertical curve length based on riding comfort:

$$L = \frac{AV^2}{46.5} \quad \textbf{Equation 6-50}$$

L = length of sag vertical curve (ft)
A = algebraic difference in grades, $|G_2 - G_1|(\%)$
V = design speed (mph)

Example 6.28: Minimum Curve Length Based on Riding Comfort

An equal-tangent sag vertical curve is to connect grades of −1.0% and +1.0%. The design speed on the curve is 80 mph. Determine the minimum curve length based on the riding comfort criteria.

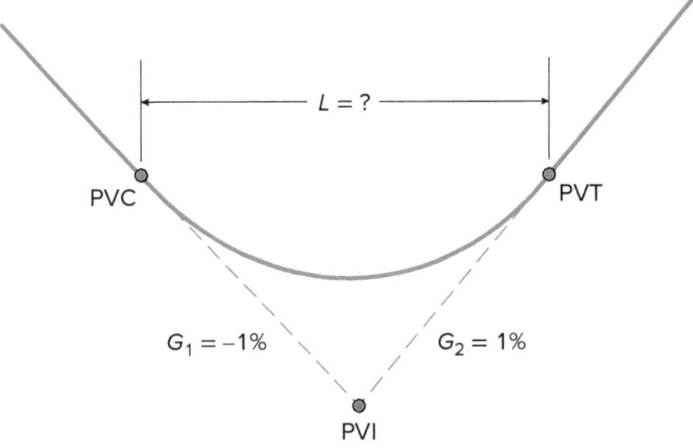

A. 27.50 ft
B. 64.74 ft
C. 275.27 ft
D. 161.62 ft

Solution

The absolute value of the algebraic difference in grades (%) is:

$$A = |G_1 - G_2| = |1 - (-1)| = 2$$

The minimum curve length L based on the riding comfort criteria is:

$$\begin{aligned} L &= \frac{AV^2}{46.5} \\ &= \frac{(2)(80)^2}{46.5} \\ &\cong 275.27 \text{ ft} \end{aligned}$$

Answer: C

6.2.4.2.5 Minimum Curve Length for Aesthetics

When entering a sag vertical curve, motorists are typically able to view the entire length of the vertical curve ahead of them. Therefore, providing sufficient curve length for aesthetics is a valid design consideration. The AASHTO *Green Book* mentions a rule of thumb for minimum curve length based on the following equation:

$$L = 100A$$

L = the minimum length of the sag vertical curve (ft)
A = the algebraic difference in grades

This approximation is a generalized control for small or intermediate values of A, and corresponds to a design speed of approximately 50 mph. On higher-speed roadways, longer curves are generally used to enhance appearance.

6.2.4.2.6 Minimum Length of Sag Vertical Curve for Drainage Criterion

Proper drainage on sag vertical curves must be provided, especially on curbed sections, because the curb contains the roadway runoff and may cause it to spread onto the roadway if enough catch basins (or inlets) are not provided.

6.3 TRAFFIC ANALYSIS

Commonly Used Abbreviations

AADT	average annual daily traffic
ADT	average daily traffic
D	density (vehicles per mile)
D_j	jam density
D_o	optimum density
E_T	passenger-car equivalent for trucks/buses and RVs in the traffic stream
FFS	free-flow speed
f_{HV}	heavy vehicle adjustment factor
HCM 6	*Highway Capacity Manual (6th ed.)* [2]
HV	heavy vehicle
i	particular time interval of speed observation
j	jam
LOS	level of service
o	critical or ideal
PCE	passenger car equivalents
pc/m/l	passenger car per mile per lane
PHF	peak hour factor
P_R	proportion of RVs in traffic stream
P_T	proportion of trucks/buses and RVs in the traffic stream
R	recreational vehicle
S	speed (mph)
SF	seasonal factor
S_s	space mean speed
S_t	time mean speed
SUT	single-unit truck
TT	tractor-trailers
V	flow (vehicles per hour)
veh/mi/ln	vehicles per mile per lane
Vm	maximum flow
v_p	demand flowrate under equivalent base conditions

INTRODUCTION

Traffic is a primary consideration in the planning and design of all highways and streets. Vehicle mix, flow, speed, density, and capacity are factors that contribute to traffic volume. Accurate estimates of the amount and type of traffic using a highway, as well as the geometric conditions (lane and shoulder widths), are needed to calculate the capacity.

This section reviews the basic principles of traffic analysis and explains how future traffic volumes are projected from current volumes and proposed land uses. The discussion of traffic terminology describes various time intervals used to describe traffic flow, and includes examples of how traffic volumes vary significantly throughout the day, week, and year, as well as by location and roadway classification. Also included is a discussion of the concept of design hourly volume (DHV) and how it represents a balance of roadway capacity and construction cost.

6.3.1 Volume

Volume is the number of vehicles that pass a point on a highway during a specified time interval. Several different time intervals are used to describe volume, including annual, daily, hourly, and 15-minute periods. Volume is measured in vehicles per unit of time. The most widely used units of time are day and hour.

6.3.1.1 Average Annual Daily Traffic

The average annual daily traffic (AADT) is used when data are collected over an entire year. AADT is calculated by dividing the number of vehicles counted by the number of days during which the counting took place. Partial days should be removed from the equation. The number of days in which the counting took place should be equal to or greater than 365. AADT is the national standard for measurement and reporting traffic data.

6.3.1.2 Average Daily Traffic

Average daily traffic (ADT) is used when the data collection period is at least one day and is calculated by dividing the number of vehicles counted by the number of whole days in the collection period.

6.3.1.3 Difference Between AADT and ADT

The difference between ADT and AADT is in the number of 24-hour traffic counts used to calculate the average volume. Whereas ADT is based on the volume over a collection period greater than one day and less than one year, AADT is based on the volume over at least one full year. Because of day-to-day and seasonal variations, a highway's ADT can vary considerably throughout the year. AADT eliminates these variations by providing a single number that reflects all traffic flows throughout a full year.

If you are a contractor preparing to spend three weeks in April rehabilitating a two-lane bridge deck, and your plan is to maintain two-way traffic in a single-lane pattern using portable traffic signals, which would be more useful: the AADT or the ADT of a typical week in April? Obviously, the ADT of a typical week in April would give you a much better estimate than the AADT of the actual number of vehicles your work would be affecting.

6.3.1.4 Variations in Traffic

Traffic volumes vary significantly with time and location. In making planning and design decisions concerning highway capacity and operations, it is necessary to consider when and where the peak volumes will occur, how long they will last, and how they will affect a location.

Table 6.4 shows typical traffic volume variations throughout an average day (hourly) on an urban arterial roadway. Notice that the single highest hourly percentage, 8.6%, occurs between 5:00 p.m. and 6:00 p.m. This is almost equal to the 9.0% of daily traffic registered over the eight-hour period between 10:00 p.m. to 6:00 a.m.

TABLE 6.4 Hourly Variation in Traffic Volume as a Percent of Average Daily Traffic

HOUR*	% OF ADT	HOUR*	% OF ADT
6 a.m.	5.7 %	1 p.m.	5.2 %
7 a.m.	8.2 %	2 p.m.	5.9 %
8 a.m.	7.0 %	3 p.m.	4.6 %
9 a.m.	5.2 %	4 p.m.	7.6 %
10 a.m.	4.6 %	**5 p.m.**	**8.6 %**
11 a.m.	5.2 %	6 p.m.	5.8 %
12 noon	5.4 %	7 p.m.	4.7 %
		8 p.m.	3.9 %
		9 p.m.	3.4 %
	10:00 p.m. to 6:00 a.m.		**9.0** %
*Starting Time			

Figure 6.22 shows hourly traffic variation on a typical weekday and on a typical weekend day. Notice that at this location, the heaviest weekday travel occurs between 18:00 and 18:30, and on the weekend, travel is consistently lower than on weekdays except during the early morning hours from 1:00 a.m. to 6:00 a.m.

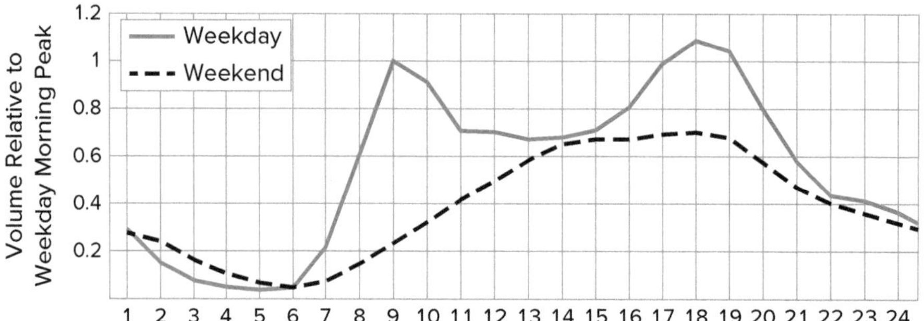

FIGURE 6.22 Hourly Traffic Variation—Weekday Hours and Weekend

6.3.1.5 Daily Variation

Table 6.5 shows typical traffic variations by day throughout an average week. Also notice that Friday has the highest ADT (110%) and Sunday has the lowest ADT (86%). Notice also that Tuesday's traffic (100%) is neither less than nor greater than the average. This implies that, in this case, Tuesday volumes are representative of the facility's average daily volume. At most locations, traffic studies are generally performed on Tuesdays, Wednesdays, and Thursdays.

TABLE 6.5 Daily Traffic Variations

DAY	ANNUAL DAILY TRAFFIC
Sunday	**86%**
Monday	103%
Tuesday	100%
Wednesday	99%
Thursday	103%
Friday	**110%**
Saturday	101%
Average	100%

Table 6.6 provides an even broader perspective. It shows traffic variation by month throughout an average year. Notice that the lowest average monthly traffic occurs in February (89%), and that the highest average monthly traffic (109%) occurs in June.

Notice also that October traffic (100%) is neither less than nor greater than the average. This implies that October volumes are representative of a typical facility's average monthly volume.

TABLE 6.6 Monthly Traffic Variations

MONTH	ANNUAL MONTHLY TRAFFIC
January	90%
February	**89%**
March	93%
April	103%
May	105%
June	**109%**
July	108%
August	107%
September	102%
October	**100%**
November	96%
December	98%
Average	100%

Variations in traffic volumes occur throughout the day, week, and year and are largely due to lifestyle-related travel patterns. These variations are more pronounced at various locations and on various types of routes. The following factors contribute to variations in traffic volumes and should be considered when planning roadway improvements:

- Daily (peak period commuter traffic volumes)
- Weekly (weekday volumes higher on commuter routes and lower on tourist routes)
- Seasonal (routes used to access recreational/tourist attractions)

A highway leading to a tourist attraction may experience an increase in traffic on weekends and, depending on climate, possibly also during the summer months. It will see a decrease during the week and during the winter months. A highway leading to a major sports facility, for example, may experience a significant increase in traffic at certain times of the year.

6.3.1.6 Seasonal Factor

The seasonal factor (SF) is a measure of how much the daily traffic volume varies (increases or decreases) during a certain season or period of the year compared with the average for the year. The equation for the SF is:

$$\text{SF} = \frac{\text{average daily traffic (vehicles)}}{\text{average annual daily traffic (vehicles)}} \qquad \textbf{Equation 6-51}$$

Example 6.29: Seasonal Factor

During the cold winter months of January and February, the ADT on a certain highway at a northern seashore tourist destination is 2,500 and 2,300 vehicles per day (vpd), respectively. During the hot summer months of July and August, the ADT on that same highway is 3,500 and 3,730 vpd, respectively. The AADT of the highway is 3,000 vpd. Find the SF for the indicated (a) winter months and (b) summer months.

Solution

(a) Seasonal factor for winter months:

$$\text{SF} = \frac{(2{,}500 + 2{,}300)/2}{3{,}000} = \frac{2{,}400}{3{,}000} = 0.80 = 80\%$$

(b) Seasonal factor for summer months:

$$\text{SF} = \frac{(3{,}500 + 3{,}730)/2}{3{,}000} = \frac{3{,}615}{3{,}000} = 1.205 = 121\%$$

6.3.1.7 Peak Hour Volume

The most important time interval for design purposes is typically the weekday peak hour, which is a single hour in a typical weekday when the heaviest traffic flow occurs. In most locations, two peaks are identified on weekdays because the highest traffic flows are normally in one direction in the morning and in the opposite direction in the evening. The morning peak usually occurs between 6:00 a.m. and 9:00 a.m., and the evening peak usually occurs between 4:00 p.m. and 6:00 p.m. In many areas, there is a midday peak between the hours of 11:00 a.m. and 1:00 p.m. due to increased lunchtime traffic.

Traffic data is most commonly collected in 15-minute intervals over a three- to four-hour period. The peak hour is determined by finding the highest four consecutive 15-minute intervals within the traffic count.

Examples of traffic data and identification of the peak hour:

TIME (15 MIN)	VEHICLES
6:30–6:45 a.m.	140
6:45–7:00 a.m.	135
7:00–7:15 a.m.	150
7:15–7:30 a.m.	140
7:30–7:45 a.m.	150
7:45–8:00 a.m.	160
8:00–8:15 a.m.	140
8:15–8:30 a.m.	140
8:30–8:45 a.m.	155
8:45–9:00 a.m.	100
9:00–9:15 a.m.	100
9:15–9:30 a.m.	110

TIME (1 HOUR)	HOURLY VOLUME
6:30–7:30 a.m.	565
6:45–7:45 a.m.	575
7:00–8:00 a.m.	600 (150+140+150+160)
7:15–8:15 a.m.	590
7:30–8:30 a.m.	590
8:30–9:30 a.m.	465

Over the course of one week, peak hour volumes (PHVs) vary. There are many reasons for this. Mondays and Fridays often have lower volume due to people taking long weekend vacations. Wednesday evening peak volumes may be distributed over a longer time period because of longer downtown shopping hours. Weekend traffic volumes are generally less than weekday traffic volumes and distributed over a longer time period because people's travel schedules are less uniform.

Because of the hourly, daily, monthly, and seasonal traffic fluctuations, PHV varies greatly. Designing a highway to accommodate the single highest PHV of the year would mean the highway would be overdesigned and underutilized most of the time, which would be very expensive and wasteful.

For example, the highest peak volume for the entire year often occurs on the day before Thanksgiving, which is typically an extremely heavy travel day. The peak volume on that day is usually several times the peak volume of an average day. To design a highway for the traffic conditions that occur on this one exceptionally busy day would, however, result in a highway with far more capacity than required under normal conditions. In addition, the highway would be prohibitively expensive to build. On the other hand, designing for the ADT volume would result in a highway that is underdesigned and overutilized about 50% of the time.

6.3.1.8 Design Hourly Volume

Determining the appropriate design volume for which a highway should be designed requires finding an acceptable compromise between its overuse and underuse. Designing for the PHV would clearly result in an overdesigned facility most of the time, but designing for the ADT would clearly result in a facility that is underdesigned more than half of the time. Thus, the appropriate design volume must be somewhere in between these two values.

The traffic volume for which the geometric and control elements of a facility should be designed is expressed in terms of the highest hourly volume (HV) in a typical year based on the relationship between the highest hourly volumes and ADT. DHV is typically the 30th highest hourly volume of the year (30 HV).

The appropriate design volume can be expressed in terms of the highest hourly volume that occurs on a roadway throughout a typical year. Referring to Figure 6.23, AASHTO provides the following explanation of DHV, as determined by the relationship between the highest hourly volumes and ADT on rural arterials: "The curves were prepared by arranging all of the hourly volumes for one year, expressed as a percentage of ADT, in a descending order of magnitude. The middle curve is the average for all locations studied and represents a highway with average fluctuation in traffic flow." [1] (This figure was produced from an analysis of traffic data covering a wide range of volumes and geographic conditions.)

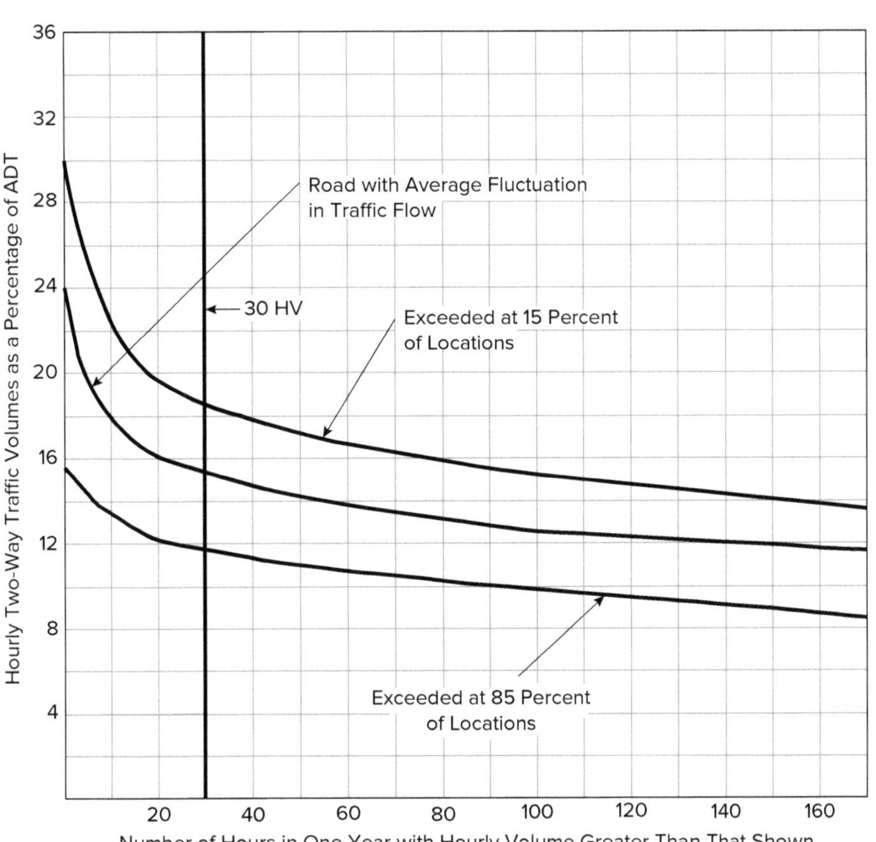

FIGURE 6.23

From *A Policy on Geometric Design*, 2011, by the American Association of State Highway and Transportation Officials, Washington, DC. Used by permission.

A review of these curves leads to the conclusion that the hourly traffic volume for which a roadway is designed should not be the single highest hourly volume of the year, or for that matter, the tenth highest hourly volume, or even the 29th highest hourly volume. To avoid providing excess capacity (overdesign), the 30th highest hourly volume (30 HV) should be used as the design control. As indicated in Figure 6.23, the 30 HV is where the slope of the descending curve changes from steep to nearly flat. The curve is much steeper to the left of the vertical line marked 30 HV, indicating much higher volumes for the inclusion of only a few more of the higher hourly volumes. The curve flattens to the right, indicating many hours in which volume is not much less than 30 HV.

On rural roads with average fluctuations in traffic flow, 30 HV corresponds to 15% of the ADT. Note that this is significantly higher than the 10% or 11% ADT typically observed in urban areas. This is because, in rural locations, weekday traffic volumes are typically very high during the morning and evening commuting periods and relatively low during the midday periods. In urban areas, there are typically many more people making trips throughout the entire day—not just during typical commuting periods. This tends to reduce the peak hour percentages in urban areas.

The *Highway Capacity Manual* [2] provides the following additional explanation:

> Customary practice in the United States is to base rural highway design on the 30th-highest hour of the year. There are few hours with higher volumes than this hour, while there are many hours with volumes not much lower. In urban areas, there is usually little difference between the 30th- and 200th-highest hours of the year, because of the recurring morning and afternoon commute patterns.

For many urban and rural highways, the *K* factor is typically between 0.09 and 0.10. For roadways with high peak period flows and relatively low off-peak flows, the *K* factor is often greater than 0.10. Alternatively, for highways with flows that are consistent and heavy throughout many hours of the day, the *K* factor is typically less than 0.09.

The design-hour factor (*K* factor) represents the percent of AADT occurring in the peak hour. This value is important in the design of roadways and capacity analysis studies.

6.3.2 Traffic Parameters

6.3.2.1 Design-Hour Factor (*K* Factor)

The *K* factor is the percent of AADT occurring in the peak travel hour of a day. The *K* factor is expressed as the ratio of the DHV to AADT:

$$K = \frac{\text{design hourly volume (vehicles)}}{\text{average annual daily traffic (vehicles)}} \qquad \textbf{Equation 6-52}$$

The *K* factor is used to determine DHV from AADT.

$$\text{DHV} = K \times \text{AADT}$$

The proportion of AADT occurring in the design hour is often referred to as *K*. It is expressed as a decimal and varies based on the hour selected for design and the characteristics of the subject route and its development environment. Where the *K* factor is based on the 30th highest hour of the year, several general characteristics can be noted:

1. *K* factors tend to decrease as AADT and density population increase.
2. Areas with greater residential and commercial development tend to have lower *K* factors than sparsely populated rural areas.
3. *K* factors are typically highest on low-volume rural highways.
4. *K* factors are typically lowest on high-volume urban highways.
5. The reduction rate for high *K* factors is faster than that for lower values.
6. *K* factors decrease as development density increases.
7. The highest *K* factors generally occur on recreational roads where a large number of motorists travel to an event that occurs for a limited time, such as a road to a ski resort or football stadium.

K factors are typically set by state transportation agencies and are based on observed volumes in different regions for similar facilities with similar demand characteristics. Table 6.7 provides suggested *K* values to be used for the various roadway functional classifications.

TABLE 6.7 ***K* Factor Typical Values**

FUNCTIONAL CLASSIFICATION	RURAL	URBAN
Interstate	0.11	0.09
Principal Arterial	0.10	0.10
Minor Arterial	0.11	0.11
Collector	0.11	0.11
Local	0.11	0.11

6.3.2.2 Peak Hour Factor

As discussed previously, traffic volumes can vary considerably throughout the day, week, and year. Likewise, traffic volumes can fluctuate considerably over the course of an hour, particularly during the peak hour. For example, a highway located near a commuter rail station can have a pronounced peak traffic flow just before and after a train arrives. If the fluctuation is large enough, traffic flow on the highway may not return to normal for several minutes after the surge.

The peak hour factor (PHF) is used to gauge the relative magnitude of the hourly fluctuation. A highway's PHF is calculated by dividing the PHV by four times the peak 15-minute traffic volume. The PHF typically ranges from 0.75 to 0.95 and is never greater than 1.0.

$$\text{PHF} = \frac{\text{actual hourly volume (vph)}}{\text{peak rate of flow (vph)}} = \frac{\text{peak hour volume}}{4 \times V_{15}}$$ **Equation 6-53**

V_{15} = maximum 15-minute volume within the hour

PHF Calculation

Use the peak hour data from page 290.

PHV = 600

$V_{15} = 160$

$$\text{PHF} = \frac{\text{PHV}}{4 \times V_{15}} = \frac{600}{4 \times 160} = 0.94$$

Example 6.30: Peak Hour Factor

The following traffic count data is given:

TIME INTERVAL	NUMBER OF VEHICLES
8:00–8:15 a.m.	1,400
8:15–8:30 a.m.	1,600
8:30–8:45 a.m.	2,200
8:45–9:00 a.m.	1,800

The PHF is closest to:

A. 0.795
B. 0.880
C. 0.650
D. 0.945

Solution

Use Equation 6-53 to determine the PHF.

$$\text{PHF} = \frac{V}{4 \times V_{15}} = \frac{1{,}400 + 1{,}600 + 2{,}200 + 1{,}800}{4 \times 2{,}200} = 0.795$$

Answer: A

Example 6.31: Peak Hour Volume and Peak Hour Factor

The traffic counts between 8 a.m. and 9 a.m. at a busy area in San Jose, CA are reported as follows:

TIME INTERVAL	LEFT TURN	RIGHT TURN	ST TRUCKS	ST CARS
8:00–8:15 a.m.	120	90	80	400
8:15–8:30 a.m.	70	100	90	450
8:30–8:45 a.m.	60	80	110	400
8:45–9:00 a.m.	50	70	60	400

Example 6.31 *(continued)*

A truck is equal to 2.5 passenger cars, a right turn is 1.5 passenger cars, and a left turn is equal to 2.0 passenger cars. Calculate the PHV and PHF.

A. 0.88
B. 0.79
C. 0.93
D. 0.83

Solution

The first step is to find the total traffic volume for each 15-minute period.

TIME INTERVAL	LEFT TURN	RIGHT TURN	ST TRUCKS	ST CARS	TOTAL
8:00–8:15 a.m.	120 × 2 = 240	90 × 1.5 = 135	80 × 2.5 = 200	400	975
8:15–8:30 a.m.	70 × 2 = 140	100 × 1.5 = 150	90 × 2.5 = 225	450	965
8:30–8:45 a.m.	60 × 2 = 120	80 × 1.5 = 120	110 × 2.5 = 275	400	915
8:45–9:00 a.m.	50 × 2 = 100	70 × 1.5 = 105	60 × 2.5 = 150	400	775

Refer to the *HCM (6th ed)* [2].

V_{15} = maximum 15-minute volume within the hour = 975

V = total one-hour volume = 975 + 965 + 915 + 775 = 3,630

$$\text{PHF} = \frac{V}{4 \times V_{15}} = \frac{3{,}630}{4 \times 975} = 0.93$$

Answer: C

6.3.2.3 Directional Split

Directional split is the ratio of the greater one-way direction hourly volume to the total (two-way) hourly volume on a highway, expressed as a percentage.

6.3.2.4 Directional Design Hourly Volume (DDHV)

In addition to knowing the total volume of traffic, the proportion of traffic traveling in either direction is also important. Most trips on a highway have a return trip that follows the same path, but in reverse. For example, most morning trips are home-to-work trips, and most evening trips are work-to-home trips. However, the amount of traffic flowing in each direction is not always equal. Therefore, the proportion of the peak hour traffic in the peak direction must be determined. This value is denoted as the directional distribution (D). The traffic volume heading in the peak direction during the DHV is determined by multiplying the DHV by D.

$$D = \frac{\text{peak direction volume}}{\text{total volume}} \qquad \textbf{Equation 6-54}$$

Directional Distribution Calculation

Peak direction volume = 1,025 vehicles per hour

Total volume = 2,000 vehicles per hour

$$D = \frac{\text{peak direction volume}}{\text{total volume}} = \frac{1{,}025}{2{,}000} = 0.51$$

This factor is used to calculate the peak hour traffic volume in the peak direction. The equation for this value (DDHV) is as follows in Equation 6-55:

$$\text{DDHV} = D \times K \times \text{AADT} \qquad \textbf{Equation 6-55}$$

The composition of traffic also needs to be considered in the design. Trucks, buses, and recreational vehicles have very different operating characteristics than passenger cars. These vehicles can have a significant effect on the capacity and operation of a roadway. In the case of a divided highway, the truck percentages for the DDHV are also needed.

Example 6.32: Directional Design Hourly Volume

The following traffic information is given:

AADT = 3,000 veh/day

K factor (proportion of AADT occurring in the peak hour) = 0.135

D factor (proportion of peak hour traffic in the peak direction) = 0.70

The DDHV is closest to:

A. 284
B. 386
C. 185
D. 234

Solution

Refer to the *HCM (6th ed.)* [2].

$\text{DDHV} = D \times K \times \text{AADT} = 0.70 \times 0.135 \times 3{,}000 = 284$ (veh/hr)

Answer: A

6.3.3 Traffic Volume Flow Theory

Greenshield's model relates speed, density, and flow. An understanding of these three interrelated parameters is needed to analyze traffic flow conditions. Bruce D. Greenshield, a civil engineer and renowned pioneer in the areas of traffic flow and highway capacity, created a model of uninterrupted traffic flow that predicts trends observed in actual traffic flows. While Greenshield's model does not match conditions exactly, it is relatively simple and accurate and is widely accepted in the analysis of traffic flow. Greenshield is responsible for developing the theory that, under uninterrupted traffic flow conditions, the relationship between speed and density is linear.

6.3.3.1 Traffic Flow Relationships

These parameters have well-established and predictable relationships. There is a direct and inverse relationship between density and speed: The lower the speed, the greater the density; the higher the speed, the lower the density. Mathematically, the relationship between speed and density is linear.

At certain times of the day (typically on weekdays from 12 a.m. to 5 a.m.), travel demand, and therefore density, are very low (because most people sleep at night), and the few vehicles that are on the roads can generally travel at free-flow speeds without interference from other vehicles. In terms of **level of service** (LOS), this is the best driving condition, LOS A.

Generally starting at around 5 a.m., however, more and more commuters (including other motorists, bicyclists, and pedestrians) are starting their workday home-to-work commute. Travel demand starts to increase steadily, so that vehicles can no longer travel at free-flow speeds because they must slow down to accommodate other road users. So, as more and more vehicles enter the traffic stream, the speed will continue to decrease as the density continues to increase. Ultimately (possibly around 7:45 a.m.), as density approaches jam density, the average running speed drops to zero. At jam density, vehicles are stopped, lined up bumper to bumper, and waiting for traffic to start moving.

This relationship is shown graphically in the speed versus density diagram in Figure 6.24 where free-flow speed, S_f, occurs when density is very close to zero. As density increases, speed S decreases and eventually reaches 0 mph at jam density D_j. The relationship between speed and density is linear and inverse. As density increases, speed decreases.

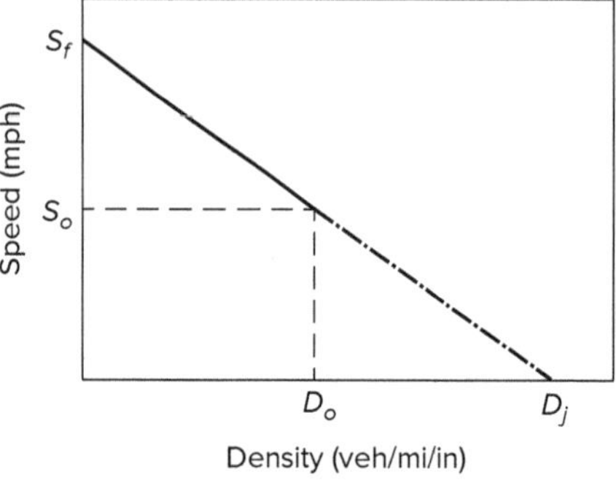

FIGURE 6.24 **Speed Versus Density**

From *A Policy on Geometric Design*, 2011, by the American Association of State Highway and Transportation Officials, Washington, DC. Used by permission.

The point of critical speed S_o represents a point of transition for stable flow to forced unstable flow. The point of critical density corresponds to the point of critical speed. Under ideal conditions, traffic flows would remain at or above the critical speed and unstable flow would be avoided.

Reduced speed and increased density may be caused by weather conditions, disabled vehicles, crashes, speed traps, and other types of interference. When interference becomes so great that the average speed drops below that necessary to maintain stable flow, congestion occurs. When the interference constricts flow and reduces the roadway capacity at a single location, the result is a bottleneck. To avoid this situation, care should be taken to design roadways with consistent capacity throughout.

Figure 6.25 indicates that the relationship between flow and density is more complicated than the linear relationship between speed and density. The flow versus density curve indicates a point of maximum flow. When density is low and speed is high, flow (the number of vehicles per hour) is low and stable. As the density increases, flow continues to increase, and speed decreases. However, at some point of critical density D_o, flow reaches a maximum value v_m and changes from stable to unstable flow, which is marked by decreasing flow and ever-increasing density.

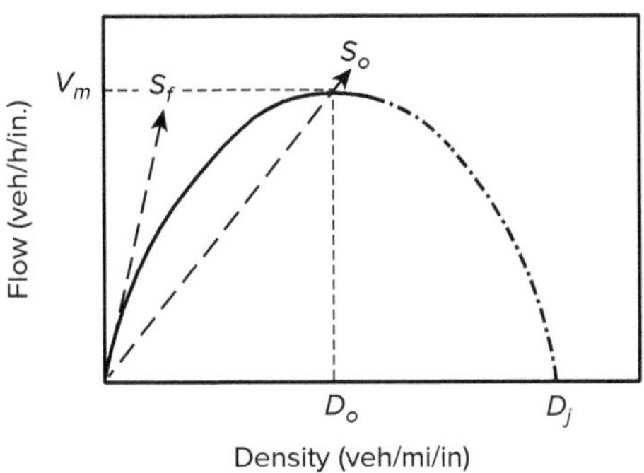

FIGURE 6.25 Flow vs. Density

From *A Policy on Geometric Design*, 2011, by the American Association of State Highway and Transportation Officials, Washington, DC. Used by permission.

As density continues to increase beyond D_o, flow continues to decrease, becomes more unstable, and eventually decreases to 0 vehicles per hour (vph)/lane, which marks the occurrence of jam density D_j.

Figures 6.24 and 6.25 show that the point of maximum flow occurs at points of critical speed S_o and critical density D_o.

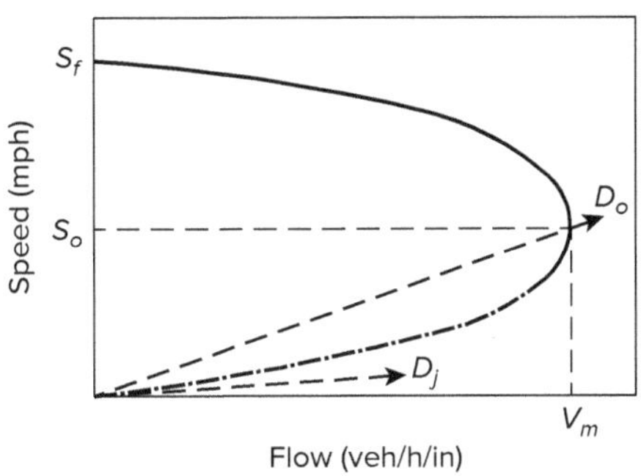

FIGURE 6.26 Speed vs. Flow

From *A Policy on Geometric Design*, 2011, by the American Association of State Highway and Transportation Officials, Washington, DC. Used by permission.

As shown at the top of the speed versus flow diagram (Fig. 6.26), as speed decreases from S_f to S_o, the flow increases from zero vph/lane to the maximum flow v_m, which is the capacity of the facility. This is also the point at which a facility operates at its critical speed S_o. Under ideal conditions, the capacity of a single freeway lane is 2,400 pc/hr/lane. Above the critical speed S_o, flow is stable. Below S_o, flow is forced or unstable and decreases as speed decreases. Consequently, unstable flow is an undesirable condition marked by decreasing flow and decreasing speed. In terms of roadway capacity, this condition corresponds to the lowest level of service, LOS F.

6.3.3.2 Basic Concepts

The above figures show graphically how the concepts of speed, density, and flow are interrelated. To determine actual numerical values of these parameters, we need to first define these terms more clearly, and then provide equations that can be used in calculations.

6.3.3.3 Traffic Flow Parameters

For the purpose of determining traffic flow and roadway capacity, there are two basic types of facilities:

1. Uninterrupted flow facilities: freeways, multilane highways, and two-lane highways
2. Interrupted flow facilities: signalized intersections and arterials

6.3.3.4 Flow

Flow rate is the number of vehicles passing a point in a given time period, usually expressed as an hourly flow rate. Flow or flow rate has gradually replaced the term volume over the years. Flow rate varies throughout the day. It is very high during morning and evening peak travel periods, and minimal in the early morning hours. Consequently, it is necessary to define not only flow rate, but also to specify when it was measured.

Example 6.33: Peak Rate of Flow

The following traffic count data is given:

TIME INTERVAL	NUMBER OF VEHICLES
8:00–8:15 a.m.	1,400
8:15–8:30 a.m.	1,600
8:30–8:45 a.m.	2,200
8:45–9:00 a.m.	1,800

The peak rate of flow (vph) is closest to:

A. 6,800 vph
B. 7,000 vph
C. 8,800 vph
D. 8,000 vph

Solution

Peak rate of flow (vph) = $4 \times 2{,}200 = 8{,}800$ vph

Answer: C

Traffic flow conditions on uninterrupted flow facilities can be characterized by the following interrelated parameters:

- **Flow rate, *v*.** The equivalent hourly rate at which vehicles pass a point on a roadway, computed as the number of vehicles passing the point divided by the time interval (usually less than 1 hr) in which they pass, and expressed as vehicles per hour.

$$v = \frac{\text{vehicles}}{\text{unit time}} = D \times S \qquad \textbf{Equation 6-56}$$

- **Running speed, *S*.** The distance a vehicle travels divided by running time (the portion of the travel time during which a vehicle is in motion); expressed in mph. *S* is directly related to density and is a key determinant of LOS.

$$\text{speed} = \frac{\text{distance}}{\text{time}};\ S = \frac{d}{t} \qquad \textbf{Equation 6-57}$$

- **Density, *D*.** The number of vehicles on a roadway segment averaged over space, usually expressed as passenger cars per mile per lane (pc/m/l). Density is directly related to speed and is a key determinant of LOS.

$$\text{density} = \frac{\text{flow}}{\text{speed}};\ D = \frac{v}{S} \qquad \textbf{Equation 6-58}$$

Example 6.34: Road Density

The flow rate of a segment of highway is 1,500 vph, and the average speed is 60 mph. What is the road density?

A. 20 veh/mi
B. 25 veh/mi
C. 30 veh/mi
D. 35 veh/mi

Solution
Refer to the *HCM (6th ed.)*[2].

$$D = \frac{v}{S} = \frac{1{,}500 \text{ vph}}{60 \text{ mph}} = 25 \text{ veh/mi}$$

Answer: B

6.3.3.5 Capacity
The capacity of a roadway is the maximum throughput measured in the number of vehicles per hour that can pass during a specified time period. Roadway capacity is measured in passenger cars per hour per lane (pc/hr/lane) for freeways and multilane highways, and in passenger cars per hour (pc/hr) for two-lane highways.

6.3.3.6 Capacity Analysis Methodologies
While density is a good indicator of the capacity of freeways and multilane highways where traffic flow is generally uninterrupted, it is not very useful in measuring the capacity of other facilities, such as signalized arterials, intersections, and bikeways, where delays due to stopping conditions and slow-moving vehicles are common. In analyzing traffic operations, it is important to use parameters that are appropriate for the operational characteristics of the specific facility.

6.3.4 Speed

Speed is one of the three fundamental characteristics of traffic flow. The term speed is often used in the study of traffic and transportation design. Speed is a scalar (rather than vector) quantity indicating a vehicle's rate of motion. Speed is based on the total distance traversed divided by the travel time, and it is expressed in miles per hour. Several types of speed are used in traffic flow analysis.

6.3.4.1 Spot Speed
Spot speed is the speed, in mph, of a vehicle as it passes a given location on a street or highway.

6.3.4.2 Running Speed
Running speed is the total distance traversed, divided by the total running time, expressed in mph.

6.3.4.3 Design Speed
Design speed is a key roadway design parameter that governs the selection of individual geometric design elements, including horizontal curve radius, vertical curve length, lateral clearance, sight distance, pavement superelevation, and so forth. AASHTO defines design speed as "a selected speed used to determine the various design features

of the roadway." The design speed of a roadway should always be greater than or equal to the expected running speed.

The design speed of a roadway should be selected based on topography, travel demand, adjacent land use (urban/suburban or rural), anticipated operating speed, and highway functional classification. Once selected, all of the pertinent highway features should be related to it to obtain a balanced design.

Design speed serves as a primary indicator of the safety and importance of a roadway. Freeways have much higher design speeds and far fewer crashes per vehicle miles driven than local roads. This is indicative of their higher design standards and greater economic importance. In an effort to achieve uniformity throughout the National Highway System, AASHTO has set minimum design speeds based on terrain, urban, or rural settings and highway functional classifications.

6.3.4.4 Time Mean Speed

The time mean speed S_t is the simple average (mean) of the spot speeds of vehicles passing a point on a roadway over a specified time interval. Speeds of individual moving vehicles are typically collected using a radar gun or laser device. Time mean speed S_t is given by Equation 6-59:

$$S_t = \frac{\sum_{i=1}^{n} S_i}{n}$$ **Equation 6-59**

S_i = speed of the ith vehicle

n = number of vehicles included in the measurement sample

Example 6.35: Time Mean Speed

This table depicts the travel times for different vehicles as they travel at a constant speed on a one-mile segment of a freeway. What is the time mean speed for these data?

VEHICLE ID	1	2	3	4
TRAVEL TIME (S)	50	55	54	48

A. 69.78 mph
B. 70.57 mph
C. 69.57 mph
D. 68.78 mph

Solution

Time mean speed is the average speed of all vehicles passing a point on a highway or lane over some specified time period. The speeds of the four vehicles in the 1-mile freeway segment are:

$$\text{Vehicle 1: } v_1 = \frac{1\text{ mile}}{50\text{ s}} = \frac{1\text{ mile}}{50\text{ s}} \times 3{,}600\frac{\text{s}}{\text{hr}} = 72\text{ mph}$$

$$\text{Vehicle 2: } v_2 = \frac{1\text{ mile}}{55\text{ s}} = \frac{1\text{ mile}}{55\text{ s}} \times 3{,}600\frac{\text{s}}{\text{hr}} \cong 65.45\text{ mph}$$

$$\text{Vehicle 3: } v_3 = \frac{1\text{ mile}}{54\text{ s}} = \frac{1\text{ mile}}{54\text{ s}} \times 3{,}600\frac{\text{s}}{\text{hr}} \cong 66.67\text{ mph}$$

$$\text{Vehicle 4: } v_4 = \frac{1\text{ mile}}{48\text{ s}} = \frac{1\text{ mile}}{48\text{ s}} \times 3{,}600\frac{\text{s}}{\text{hr}} = 75\text{ mph}$$

Example 6.35 *(continued)*

The time mean speed is:

$$v_t = \frac{\sum_{i=1}^{4} v_i}{4}$$

$$= \frac{72 \text{ mph} + 65.45 \text{ mph} + 66.67 \text{ mph} + 75 \text{ mph}}{4}$$

$$\cong 69.78 \text{ mph}$$

Answer: A

6.3.4.5 Space Mean Speed

Also known as the average travel speed, the space mean speed S_s is the total distance traveled by one or more vehicles divided by the total travel time required for all vehicles considered to travel that distance. S_s is defined by Equation 6-60:

$$S_s = \frac{nL}{\sum_{i=1}^{n} t_i} \qquad \textbf{Equation 6-60}$$

L = length of the segment

t_i = travel time of the ith vehicle to traverse the section (L)

Example 6.36: Space Mean Speed

The following travel times were measured for vehicles as they traversed a 1-mile segment of freeway at constant speeds. What is the space mean speed for these data?

VEHICLE ID	1	2	3	4
TRAVEL TIME (S)	50	55	54	48

A. 69.78 mph
B. 70.57 mph
C. 69.57 mph
D. 68.78 mph

Solution

Space mean speed is the average speed of all vehicles occupying a given section of highway or lane over some specified time period. The speeds of the four vehicles in the one-mile freeway segment are:

Vehicle 1: $v_1 = \frac{1 \text{ mile}}{50 \text{ s}} = \frac{1 \text{ mile}}{50 \text{ s}} \times 3{,}600 \frac{\text{s}}{\text{hr}} = 72 \text{ mph}$

Vehicle 2: $v_2 = \frac{1 \text{ mile}}{55 \text{ s}} = \frac{1 \text{ mile}}{55 \text{ s}} \times 3{,}600 \frac{\text{s}}{\text{hr}} \cong 65.45 \text{ mph}$

Vehicle 3: $v_3 = \frac{1 \text{ mile}}{54 \text{ s}} = \frac{1 \text{ mile}}{54 \text{ s}} \times 3{,}600 \frac{\text{s}}{\text{hr}} \cong 66.67 \text{ mph}$

Vehicle 4: $v_4 = \frac{1 \text{ mile}}{48 \text{ s}} = \frac{1 \text{ mile}}{48 \text{ s}} \times 3{,}600 \frac{\text{s}}{\text{hr}} = 75 \text{ mph}$

Example 6.36 *(continued)*

The space mean speed is:

$$v_s = \left(\frac{\sum_{i=1}^{4} \frac{1}{v_i}}{4} \right)^{-1}$$

$$= \left(\frac{\frac{1}{72 \text{ mph}} + \frac{1}{65.45 \text{ mph}} + \frac{1}{66.67 \text{ mph}} + \frac{1}{75 \text{ mph}}}{4} \right)^{-1}$$

$$\cong 69.57 \text{ mph}$$

Answer: C

6.3.4.6 Median Speed

Also known as the 50th percentile speed, median speed is as good of a central measure as the time mean speed, and it is easier to obtain. In a string of spot speed observations that have been ordered from lowest to highest, the median speed is the middle observation. If the total number of observations is an odd number, the median speed is the speed of the middle observation. If the total number of observations is an even number, there is no single middle observation, and the median speed is the average of the two observations on either side of the middle.

6.3.4.7 Modal Speed

Given a string of spot speed data, the modal speed is the value of the most frequently occurring observation.

Example 6.37: Median Speed, Modal Speed, and Time Mean Speed

The following spot speeds were observed on a section of a local road during a five-minute interval: 29, 31, 32, 30, 28, 32, 31, 29, 28, 30, 31, 33, 35, 25, 30, 30, 29, 31, 34, 29, 30

Determine:

(a) The median (50th percentile) speed
(b) The modal speed
(c) The time mean speed

Solution

Sort the string of data (observations) from lowest to highest:

25, 28, 28, 29, 29, 29, 29, 30, 30, 30, 30, 30, 31, 31, 31, 31, 32, 32, 33, 34, 35

(a) The median (50th percentile) speed. The data string of spot speeds includes a total of 21 observations. With an odd number of observations, the median speed is the value of the middle observation. In this example, the middle observation is the 11th one. Therefore, the median speed is 30 mph.

(b) The modal speed. The data string contains the following observation numbers:

One 25 mph observation

Two 28 mph observations

Four 29 mph observations

Example 6.37 *(continued)*

Five 30 mph observations

Four 31 mph observations

Two 32 mph observations

One 33 mph observation

One 34 mph observation

One 35 mph observation

The most frequently occurring value is 30 mph, which is therefore the modal speed.

(c) The time mean speed: The data string contains a total of 21 observations, and the sum of the observations is 637 mph. Therefore, the time mean speed is 637/21 = 30.33 mph.

6.3.4.8 Travel Time

Travel time is directly associated with space mean speed, S_S. It is the total time required to traverse a given distance, including all traffic stops and delays. The travel time used to determine S_s typically consists of two parts: running time and stopped time.

6.3.4.9 Running Time

Running time is the total time (seconds, minutes, or hours) required to traverse a given distance, excluding any stopped time.

6.3.4.10 Free-Flow Speed

As defined in the *Highway Capacity Manual* [2], **free-flow speed** (FFS or S_f) is the average speed of vehicles on a segment of highway under low-volume conditions, that is, when drivers are free to drive at their desired speed and are not constrained by the presence of other vehicles. FFS applies to uninterrupted flow facilities, such as freeways and multilane highways. Theoretically, FFS occurs only when both density and flow rate are zero. Practically, FFS is expected to occur on freeways at low flow rates, typically less than 1,000 passenger cars per hour per lane (pc/hr/ln). In this range, speed is insensitive to flow rate, and FFS can be measured directly in the field and from sensor data.

Example 6.38: Free-Flow Speed

A four-lane freeway (two lanes in each direction) has the following characteristics:

Lane width = 11 ft

Right-side lateral clearance = 2 ft

One cloverleaf interchange per mile

What is the FFS?

A. 70.5 mph
B. 60.8 mph
C. 80.4 mph
D. 40.6 mph

Example 6.38 *(continued)*

Solution

Solve using Exhibits 12-20 and 12-21 of the *HCM (6th ed.)*[2].

The FFS of the freeway is estimated as follows:

$FFS = 75.4 - f_{LW} - f_{LC} - 3.22TRD^{0.84}$

TRD = total ramp density

f_{LW} = adjustment to FFS for lane width

f_{LC} = adjustment to FFS for right-side lateral clearance

Determine f_{LW} and f_{LC}.

The adjustment for lane width is selected from Exhibit 12-20 (*HCM (6th ed.)* [2]). For 11-ft lanes, it is 1.9 mph. The adjustment for right-side lateral clearance is selected from Exhibit 12-21 (*HCM (6th ed.)* [2]). For a 2-ft right-side lateral clearance on a freeway with two lanes in one direction, it is 2.4 mph.

Determine total ramp density (TRD).

Given one cloverleaf interchange per mile, TRD is 4 ramps/mi (in each direction).

Compute FFS.

$FFS = 75.4 - 1.9 - 2.4 - 3.22 \times 4^{0.84} = 60.8$ mph

Answer: B

6.3.4.11 Running Speed

Measured in miles per hour, the running speed (*S*) is equal to the distance traveled divided by the running time (the portion of the travel time during which a vehicle is in motion). *S* is directly related to density and is a key determinant of the LOS. Average running speeds are used to determine the 85th percentile speed, which is commonly used for setting posted speed limits and evaluating the safety of an existing highway.

6.3.4.12 Stopped Time (or Delay)

Stopped time is the amount of time (seconds, minutes, or hours) that a vehicle is stopped for some reason (such as traffic signals, congestion, roadwork, a flat tire, or a pit stop).

6.3.4.13 Pace (10-mph Pace)

As shown in Figure 6.27, the 10-mph pace is presented graphically by a horizontal line of length 10 mph that connects the left and right sides of the speed distribution curve. It is the 10-mph speed range that contains the largest percentage of the vehicles on a speed distribution curve. The 10-mph pace is best suited to distributions of speeds on a rural highway where speeds tend to be higher. In urban areas, speeds are lower, and the 10-mph range usually covers too great a part of the curve.

6.3.4.14 85th Percentile Speed

The 85th percentile speed is the speed at which 85 percent of drivers will drive at or below in free-flowing conditions. The 85th percentile speed is often used to set the appropriate posted speed on a highway, which is typically the nearest 5 mph increment slightly above or below the calculated 85th percentile speed. The speed frequency distribution curve and cumulative speed distribution curve in Figure 6.27 show the 10-mph pace, 15th percentile speed, 50th percentile speed, and 85th percentile speed.

The three portions of a speed distribution curve of primary interest for traffic analysis are the low-speed group (below the 15th percentile), the central portion (between the 15th and 85th percentiles), and the high-speed group (above the 85th percentile).

The central portion contains the largest number of drivers who typically conform to traffic regulations. Maximum speed limits for speed zones are commonly set after determination of the 85th percentile speed and other factors such as tolerance in enforcement.

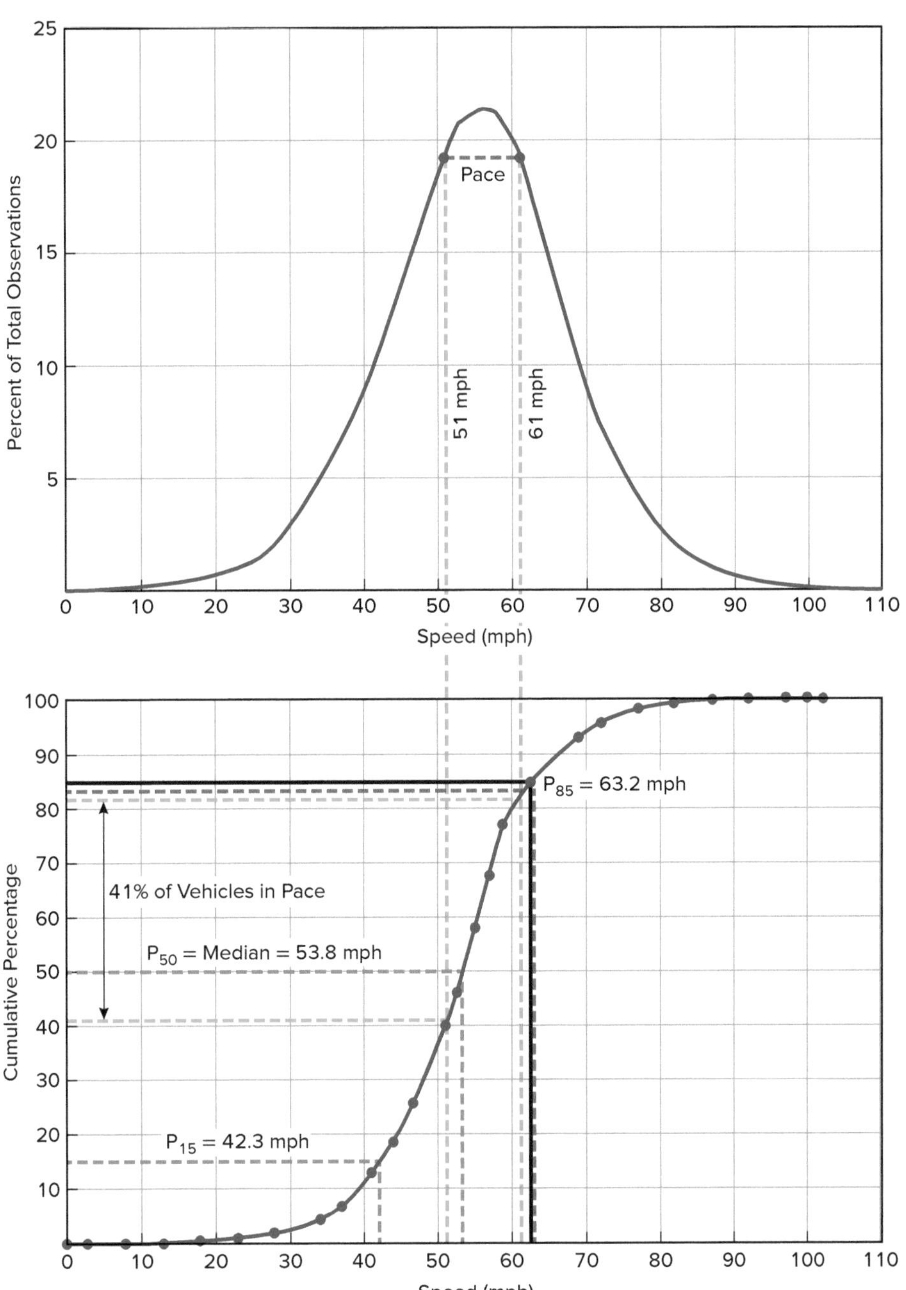

FIGURE 6.27 **Speed Distribution Curve (top) Cumulative Speed Distribution Curve (bottom)**

As shown in Figure 6.27, the cumulative speed distribution curve is much steeper between the 15th and 85th percentiles (representing 70% of the driving population) than on either side of this range. Below the 15th percentile and above the 85th percentile, the curve exhibits relatively flat slopes, indicating that observations of very low and very high speeds occur infrequently. From a traffic operations standpoint, vehicles traveling below the 15th percentile speed are viewed as a hindrance that has a disproportionately high impact on roadway capacity and safety. An example would be a heavily loaded truck traveling at a crawl up a steep grade.

On the opposite side of the cumulative speed distribution curve is the high-speed group, which again represents only 15% of the driving population but typically receives the most attention from speed enforcement. This group tends to resist voluntary compliance with posted speed limits and is generally more willing to risk being fined for speed violations than the rest of the driving population. From a traffic safety standpoint, drivers in this group are considered to be traveling at excessively high speeds and therefore pose a danger to themselves and other travelers.

The 85th percentile speed is easy to identify on a normally distributed speed frequency curve because it occurs approximately one standard deviation to the right of the mean (50th percentile) speed. This one standard deviation encompasses 34% of the total speed observations. Therefore, one standard deviation to the right of the mean accounts for 84% of the population, or approximately the 85% percentile speed (Fig. 6.28).

On the premise that the majority of drivers tend to travel at a reasonable speed without exposing themselves to unnecessary risks, the 85th percentile speed is widely considered representative of prudent driving behavior and therefore believed to be an appropriate upper speed limit.

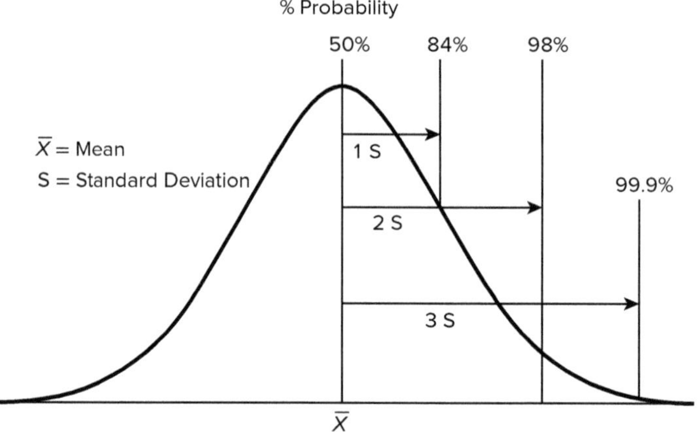

FIGURE 6.28 Normal Distribution Curve—Standard Deviations and Percentiles

Reprinted with permission from *The Superpave Mix Design Manual for New Construction and Overlays: SHRP A-407*, 1994 by the National Academy of Sciences, Courtesy of the National Academies Press, Washington, DC.

The 85th percentile speed is also used in setting the timing of a coordinated signal system. Drivers who travel at the system design speed will typically arrive at the next coordinated signal shortly after the signal changes from red to green, which enables them to continue through the signal without needing to stop and wait. Drivers who travel at speeds much higher than the system design speed tend to arrive at the next coordinated signal before it turns from red to green, which forces them to stop and wait at each signal, and thus repeatedly speeding up and braking to a stop.

6.3.4.14.1 Finding the 85th Percentile Speed

The procedure for determining the 85th percentile speed at a particular location involves the following steps:

1. Collect a large number of spot speed data points, n (say, $n = 500$ observations).
2. Calculate the 85th percentile observation. Multiply n by 0.85. (For example, $500 \times 0.85 = 425$.)
3. Put all n observations in order from lowest to highest speed.
4. Sort all n observations into 5-mph speed groups; for example, 15 mph $< n_1 \leq$ 20 mph, 20 mph $< n_2 \leq 25$ mph, and so forth.
5. Determine the assumed speed of each speed group.

6. Calculate the cumulative frequency of each speed group.
7. Identify the speed group containing the 85th percentile observation.
8. The 85th percentile speed is equal to the assumed speed of the speed group that contains the 85th percentile observation.

The procedure for calculating the 85th percentile speed is best demonstrated by Example 6.39.

Example 6.39: Speed Data Analysis

The following speed data was collected using a radar gun on an arterial street.

Estimate the following:

a) Time mean speed
b) 85th percentile speed
c) Modal speed
d) Pace

SPEED GROUP (mph)	FREQUENCY
$15 < S_1 \leq 20$	10
$20 < S_2 \leq 25$	17
$25 < S_3 \leq 30$	36
$30 < S_4 \leq 35$	66
$35 < S_5 \leq 40$	84
$40 < S_6 \leq 45$	70
$45 < S_7 \leq 50$	50
$50 < S_8 \leq 55$	21
$55 < S_9 \leq 60$	6

Solution

Expand the table above by adding the following three columns to the right side: Assumed Speed, Total Speed, and Cumulative Frequency. The average speed of each speed group is entered in the Assumed Speed column. Total Speed entry is the product of the entries in the Frequency and Assumed Speed columns. The Cumulative Frequency is the cumulative sum of the entries in the Frequency column.

SPEED GROUP	FREQUENCY	ASSUMED SPEED	TOTAL SPEED	CUMULATIVE FREQUENCY
$15 < S_1 \leq 20$	10	18	180	10
$20 < S_2 \leq 25$	17	23	391	27
$25 < S_3 \leq 30$	36	28	1,008	63
$30 < S_4 \leq 35$	66	33	2,178	129
$35 < S_5 \leq 40$	84	38	3,192	213
$40 < S_6 \leq 45$	70	43	3,010	283
$45 < S_7 \leq 50$	50	48	2,400	333
$50 < S_8 \leq 55$	21	53	1,113	354
$55 < S_9 \leq 60$	6	58	348	360
	360		13,820	

a) Time mean speed: $\bar{x} = \left(\frac{13{,}820}{360}\right) = 38.39$ mph

b) 85th percentile speed of vehicles:

Step 1: Total number of observations × 0.85 = 360 × 0.85 = 306

Step 2: The 306th observation is in the 45–50 mph speed group. Therefore the 85th percentile speed is 48 mph.

c) Modal speed or mode: The greatest frequency is 84 with an associated speed group 36–40 mph. Therefore, the mode is 38 mph.

d) Pace: The pace is the range that represents the largest number of observations for two consecutive speed groups. The two speed groups that fit this description are 36–40 mph and 41–45 mph. Therefore, the pace is between 36 and 45 mph.

6.3.5 Vehicle Separation by Time and Distance

6.3.5.1 Headway

Headway (h) is the time interval (minutes or seconds) between consecutive vehicles' arrivals at some point. Headway is determined by dividing the number of seconds in an hour by the flow (number of vehicles per hour).

$$h = \frac{3{,}600 \text{ s/hr}}{v \text{ vph}}$$ **Equation 6-61**

Saturation headway (or minimum headway) is the headway when the traffic volume reaches lane capacity, c.

$$c_{\text{vph}} = \frac{3{,}600 \text{ s/hr}}{h_{\min}}$$ **Equation 6-62**

Example 6.40: Headway

During the morning peak travel period (7 a.m. to 8 a.m.) the local transit agency runs a total of five buses. The first bus arrives at 7 a.m. The last bus arrives at 8 a.m. The bus headway (minutes) is most nearly which of the following?

A. 6 min
B. 12 min
C. 15 min
D. 18 min

Solution

Total time interval between bus arrivals: 60 minutes

$$h = \frac{\text{total time (min)}}{(n-1)\text{ intervals}} = \frac{60 \text{ min}}{(5-1)\text{ intervals}} = \frac{60 \text{ min}}{4 \text{ intervals}} = 15 \text{ min}$$

Answer: C

Example 6.41: Lane Capacity

A six-lane freeway has three lanes in each direction. If it is known that the saturation headway is 2.1 seconds per vehicle per lane (s/veh/lane), what is the capacity per lane of the freeway?

A. 10,286 veh/hr/lane
B. 1,714 veh/hr/lane
C. 5,143 veh/hr/lane
D. 3,560 veh/hr/lane

Solution

Headway (h) is the time interval between two consecutive vehicles traversing the same point. Saturation headway is the headway when the traffic volume reaches the freeway capacity. Since the saturation headway (h) is 2.1 s/veh/lane, the maximum number of vehicles that can pass through the freeway section in one hour is:

$$s = \frac{3{,}600 \text{ s/hr}}{h} = \frac{3{,}600 \text{ s/hr}}{2.1 \text{ s/veh/lane}} \cong 1{,}714 \text{ veh/hr/lane}$$

Answer: B

6.3.5.2 Spacing

Spacing is the distance (ft) between common points (for example, front or rear bumper) of successive vehicles. It is also the reciprocal of the density, or 1/density.

Example 6.42: Vehicle Spacing

On a certain two-lane highway, traffic is heading eastbound at 70 mph. The demand flow rate v_p is 280 vph.

Part 1: The average spacing between eastbound vehicles (in ft) is most nearly which of the following?

A. 1,280 ft
B. 1,320 ft
C. 1,400 ft
D. 1,420 ft

Solution

speed, $S = 70$ mph

demand flow rate, $v_p = 280$ vph

$$\text{density, } D = \frac{v_p}{S} = \frac{280 \text{ vph}}{70 \text{ mph}} = \frac{4 \text{ vehicles}}{\text{mile}}$$

$$\text{Therefore, spacing} = \frac{1}{D} = \frac{\text{mile}}{4 \text{ vehicles}} = 0.25 \text{ mile between vehicles}$$

$$\text{Average spacing} = 5{,}280 \frac{\text{ft}}{\text{mi}} \times 0.25 \text{ mile} = 1{,}320 \text{ ft}$$

Note: The flow is in vehicles per hour (vph), not vehicles per hour per lane (vphpl), and density is in vehicles per minute (vpm), not vehicles per minute per lane (vpmpl), because this problem is asking about only one lane, the eastbound lane.

Answer: B

Part 2: The headway of eastbound vehicles (in seconds) is most nearly which of the following?

A. 12.9 s
B. 13.2 s
C. 13.6 s
D. 13.9 s

Solution

$$h = \frac{3{,}600 \text{ s/hr}}{v \text{ vph}} = \frac{3{,}600 \text{ s/hr}}{280 \text{ vph}} = 12.9 \text{ s}$$

Answer: A

Example 6.43: Highway Capacity

A highway segment has two lanes of different widths in the southbound direction. Under prevailing conditions, the minimum headway is 1.8 seconds on lane 1 and 2.0 seconds on lane 2. What is the capacity of this highway segment in the southbound direction?

A. 1,600 vph
B. 3,800 vph
C. 3,600 vph
D. 3,900 vph

Example 6.43 *(continued)*

Solution

The southbound capacity of the highway segment is the sum of the capacities of the two lanes.

The capacity of lane 1 is:

$$c_1 = \frac{3{,}600}{h_1} = \frac{3{,}600 \text{ s/hr}}{1.8 \text{ s}} = 2{,}000 \text{ vph}$$

The capacity of lane 2 is:

$$c_2 = \frac{3{,}600}{h_2} = \frac{3{,}600 \text{ s/hr}}{2.0 \text{ s}} = 1{,}800 \text{ vph}$$

The southbound capacity is:

$$c_1 + c_2 = (2{,}000 \text{ vph}) + (1{,}800 \text{ vph}) = 3{,}800 \text{ vph}$$

Answer: B

Example 6.44: Highway Capacity

A freeway has three lanes in one direction. If it is known that the saturation headways are 2.2 s/veh, 2.3 s/veh, and 2.4 s/veh on the three lanes, respectively, what is the capacity of the freeway in that direction?

A. 1,636 vph
B. 4,500 vph
C. 4,701 vph
D. 1,567 vph

Solution

Headway is the time interval between two consecutive vehicles traversing the same point. Saturation headway is the headway when the traffic volume reaches the lane capacity. Since the saturation headway h is known for the three lanes, the saturation flow s on each lane is:

$$\text{Lane 1: } s_1 = \frac{3{,}600}{h} = \frac{3{,}600 \text{ s/hr}}{2.2 \text{ s/veh}} = 1{,}636.4 \text{ vph}$$

$$\text{Lane 2: } s_2 = \frac{3{,}600}{h} = \frac{3{,}600 \text{ s/hr}}{2.3 \text{ s/veh}} = 1{,}565.2 \text{ vph}$$

$$\text{Lane 3: } s_3 = \frac{3{,}600}{h} = \frac{3{,}600 \text{ s/hr}}{2.4 \text{ s/veh}} = 1{,}500 \text{ vph}$$

Since there are three lanes in each direction, the freeway capacity in that direction is the sum of three lane capacities:

$$S = s_1 + s_2 + s_3 = 1{,}636.4 \text{ vph} + 1{,}565.2 \text{ vph} + 1{,}500 \text{ vph} \cong 4{,}701 \text{ vph}$$

Answer: C

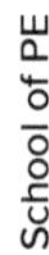

6.3.6 Growth in Travel Demand

Travel demand is estimated to grow at a constant rate per year. Transportation agencies monitor the growth in travel demand constantly to ensure that adequate funding is available to add new capacity where and when needed and to ensure that programming for new capacity planning, design, and construction is started on time.

Growth in travel demand is similar to the growth in value of an economic asset, and can be estimated using the principles of engineering economic analysis. Like the value of a financial asset, travel demand is expected to increase at a constant rate per year. Current travel demand can be thought of as the present worth of an asset, and future travel demand as the future worth of that asset. The amount of future travel demand is calculated based on current demand (present worth), the number of years n between present and future years (term), and the annual rate of growth in travel demand (interest rate).

Example 6.45: Growth in Average Daily Traffic

The current ADT volume on a two-lane road is 6,000 vpd. It is assumed travel demand is increasing at 2.5% per year. What is the projected future ADT based on $n = 5$ years?

A. 6,075 vpd
B. 6,150 vpd
C. 6,789 vpd
D. 6,958 vpd

Solution

$F = P(1+i)^n = 6{,}000(1+0.025)^5 = 6{,}000(1.1314) = 6{,}788.46$, round to 6,789 vpd

Answer: C

Example 6.46: Growth in Average Daily Traffic

The current ADT volume on a two-lane road is 30,000 vpd. Travel demand is expected to grow 3.2% per year for the next 12 years. What is the projected future ADT based on $n = 12$ years?

A. 31,172 vpd
B. 35,760 vpd
C. 36,241 vpd
D. 43,780 vpd

Solution

$F = P \times (F/P \text{ factor})$

$= P(1+i)^n = 30{,}000(1+0.032)^{12}$

$= 30{,}000(1.4593) = 43{,}780.19$, round to 43,780 vpd

Answer: D

Example 6.47: Growth in Travel Demand

In how many months will the four-lane basic freeway segment described below reach capacity?

The facts:

- The existing volume is 3,700 vph in one direction.
- There are two lanes in each direction.
- FFS = 55 mph
- Heavy vehicles factor = 0.932
- PHF = 0.90
- Traffic growth rate = 1.9% per year

A. 10
B. 11
C. 12
D. 13

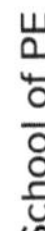

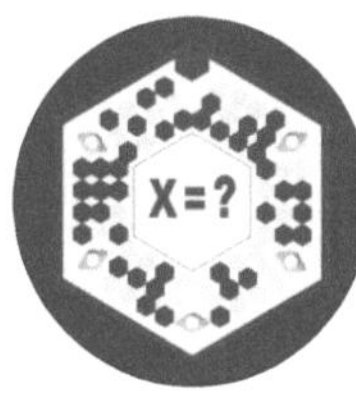

Example 6.47 *(continued)*

Solution

Step 1. Find the saturation flow SF at its capacity:

$SF = MSF \times N \times f_{HV}.$

MSF at LOS E = max capacity

MSF = 2,250 pc/hr/ln (per *HCM (6th ed.)*[2], Exhibit 12-37)

$SF = 2{,}250 \times 2 \times 0.932 = 4{,}194$ vph

Step 2. Find service volume (*SV*) at LOS E.

$SV = SF \times PHF = 4{,}194 \times 0.90 = 3{,}774.6$, round to 3,775 vph

Note that SV = 3,775 vph is the maximum number of vehicles that this basic freeway segment can carry under the given conditions. Therefore, with an existing volume of 3,700 vph, this freeway segment is getting close to capacity. It currently has only 75 vph of excess capacity.

Step 3. Determine how long it will take for this freeway segment to reach capacity.

Current traffic = 3,700 vph

Capacity for the given conditions = 3,775 vph

$3{,}700\,(1.019)^n = 3{,}775$

$(1.019)^n = 3{,}775/3{,}700 \rightarrow (1.019)^n = 1.0203$

$n = \ln(1.0203)/\ln(1.019) = 0.0201/0.0188 = 1.07$ years

Based on the given conditions, this freeway will reach its capacity in approximately 13 months.

Answer: D

REFERENCES

1. American Association of State Highway and Transportation Officials (AASHTO). 2011. *A Policy on Geometric Design of Highways and Streets (6th ed.).* Washington, DC: AASHTO.
2. Transportation Research Board of the National Academy of Sciences. 2016. *Highway Capacity Manual (6th ed.).* Washington, DC: TRB.

CHAPTER

Materials

7

Amr M. Sallam, PhD, PE

Daniel A. Howell, PhD, PE

Muhammad Elgammal, PE, PMP

Amir Mousa, MSCE, PE

CONTENTS

CONTENTS (*continued*)

EXAM GUIDE

VII. Materials

A. Soil classification and boring log interpretation
B. Soil properties (e.g., strength, permeability, compressibility, phase relationships)
C. Concrete (e.g., nonreinforced, reinforced)
D. Structural steel
E. Material test methods and specification conformance
F. Compaction

Approximate Number of Questions on Exam: 6

NCEES Principles & Practice of Engineering Examination, Civil Breadth Exam Specifications

COMMONLY USED ABBREVIATIONS

A	area
CPT	cone penetration test
GI	group index
i	hydraulic gradient
k	coefficient of permeability
L	length
LL	liquid limit
l	distance water travels in direction of flow
P	axial compressive force
PI	plasticity index
PL	plastic limit
Q	flow rate
SPT	standard penetration test
USCS	Unified Soil Classification System
V	volume
v	average discharge of velocity
W	weight
w	moisture content

COMMONLY USED SYMBOLS

$(\gamma_d)_{max}$	maximum dry unit weight
$(\gamma_d)_{min}$	minimum dry unit weight
γ_d	in situ dry weight
τ'	soil shear strength
σ'	effective stress normal to shearing plane
ϕ'	effective internal angle of friction
c'	effective cohesion, or apparent cohesion
Δh	head difference

7.1 SOIL CLASSIFICATION

The two main soil classification systems are the **AASHTO** classification system and the **Unified Soil Classification System (USCS)**. The AASHTO is mainly utilized to classify soils for transportation-related geotechnical applications, such as the suitability of soils as pavement subgrade, handling, spreading, working out, and compaction in-field. The AASHTO is a US national system and is not used internationally except for projects that

strictly follow US codes. The USCS is the internationally recognized soil classification system. It is used for the whole spectrum of soil mechanics and geotechnical engineering.

For simplicity, soils can be divided into two main categories:

1. Cohesionless soils, coarse-grained soils, or granular soils (sands and gravels)
2. Cohesive soils or fine-grained soils (silts and clays)

7.1.1 AASHTO Soil Classification System

The current AASHTO system originated in the late 1920s when the US Bureau of Public Roads (now the Federal Highway Administration) developed a system to identify the use of soils for local and secondary road construction. Currently, the AASHTO system can be used to determine the relative quality of soils for use in embankments, subgrades, subbases, and bases. In the AASHTO system, soils are divided into granular and silty-clayey soils based on the percentage passing sieve No. 200. Soils are classified into groups from A-1 to A-7 with organics classified as A-8. A **group index** (GI) number is added to compare soils within the same group with better soils associated with a lower GI.

7.1.1.1 Particle Size Definitions

Gravel: 75 mm (3 in) to 2.0 mm (sieve No. 10)
Sand: 2.0 mm (sieve No. 10) to 0.075 mm (sieve No. 200)
Silt and clay: less than 0.075 mm (passing sieve No. 200)

Group index (GI):

$$GI = (F_{200} - 35)(0.2 + 0.005(LL - 40)) + 0.01(F_{200} - 15)(PI - 10) \qquad \textbf{Equation 7-1}$$

$$PGI = 0.01(F_{200} - 15)(PI - 10)(\text{for soil groups A-2-6 and A-2-7}) \qquad \textbf{Equation 7-2}$$

TABLE 7.1 AASHTO Classification Table

AASHTO soil classification system based on AASHTO M 145 (or ASTM D 3282)											
GENERAL CLASSIFICATION	**GRANULAR MATERIALS (35% or less of total sample passing No. 200 sieve (0.075 mm)**							**SILT-CLAY MATERIALS (More than 35% of total sample passing No. 200 sieve (0.075 mm)**			
GROUP CLASSIFICATION	A-1		A-3	A-2				A-4	A-5	A-6	A-7
	A-1-a	A-1-b		A-2-4	A-2-5	A-2-6	A-2-7				A-7-5, A-7-6
Sieve analysis, percent passing:											
No. 10 (2 mm)	50 max										
No. 40 (0.425 mm)	30 max	50 max	51 min								
No. 200 (0.075 mm)	15 max	25 max	10 max	35 max	35 max	35 max	35 max	36 min	36 min	36 min	36 min
Characteristics of fraction passing No. 40 (0.425 mm)											
Liquid limit				40 max	41 min	40 max	41 min	40 max	41 min	40 max	41 min
Plasticity index	6 max		NP	10 max	10 max	11 min	11 min	10 max	10 max	11 min	11 min*
Usual significant constituent materials	Stone fragments, gravel and sand		Fine sand	Silty or clayey gravel and sand				Silty soils		Clayey soils	
Group Index**	0		0	0		4 max		8 max	12 max	16 max	20 max

Classification procedure: *With required test data available, proceed from left to right on chart; correct group will be found by process of elimination. The first group from left into which the test data will fit is the correct classification.*
**Plasticity Index of A-7-5 subgroup is equal to or less than LL minus 30. Plasticity Index of A-7-6 subgroup is greater than LL minus 30.*
***See group index formula. Group index should be shown in parentheses after group symbol as: A-2-6(3), A-4(5), A-7-5(17), etc.*

From *AASHTO M-145 Standard Specification for Classification of Soils and Soil-Aggregate Mixtures for Highway Construction Purposes*, 1995, by the American Association of State Highway and Transportation Officials, Washington, DC. Used by permission.

7.1.1.2 AASHTO Soil Classification Procedures

1. Identify the percent passing the No. 10, No. 40, and No. 200 sieves. Identify the **liquid limit** (LL) and **plasticity index** (PI) of the fraction passing the No. 40 sieve.
2. Using the AASHTO classification table (Table 7.1), move from left to right using the process of elimination until the first group that is consistent with the laboratory data is identified.

3. Shortcut: If the percent passing the No. 200 sieve is less than or equal to 35% and the PI > 6, use Figure 7.1, utilizing the LL and PI to get the full classification of the soil.
4. Shortcut: If the percent passing the No. 200 sieve is greater than 35%, then use Figure 7.1, utilizing the LL and PI to get the full classification of the soil.
5. Calculate the GI or partial group index (PGI) using Equations 7-1 and 7-2 (shown above Table 7.1). If the equation yields a negative number, then report it as zero. Always round the GI to the nearest whole number.
6. Shortcut: Groups A-1, A-3, A-2-4, and A-2-5: GI = 0. Groups A-2-6 and A-2-7: use the PGI equation (Equation 7-2).

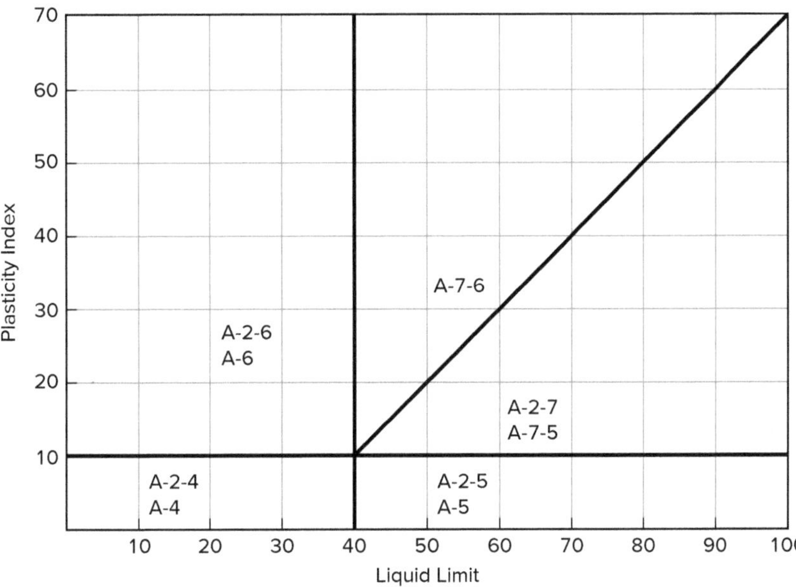

FIGURE 7.1 AASHTO Graphical Classification Chart

Source: ASTM D3282-15, *Standard Practice for Classification of Soils and Soil-Aggregate Mixtures for Highway Construction Purposes*, ASTM International, 2015.

Example 7.1: AASHTO Soil Classification

Determine the AASHTO Classification of soils A, B, and C.

SIEVE NO.	% PASSING SOIL A	% PASSING SOIL B	% PASSING SOIL C
No. 10	82	74	100
No. 40	46	52	90
No. 200	23	30	87
Liquid Limit	NP	25	71
Plastic Limit	NP	14	40

Refer to Sections 7.2.3 and 7.2.4 for descriptions of sieve analysis and Atterberg limits, respectively.

Solution

Sample A:

No. 200 = 23 < 35%: cohesionless soil

Both LL and PI are zero: nonplastic soil

PI < 6: use Table 7.1 (AASHTO Classification Table)

Start at the left and try to match the given percent passing within the table criteria.

First match: soil type A-1-b

GI for A-1-b = 0

Example 7.1 *(continued)*

(Note: Using Equation 7.1 will yield a negative GI, which should be reported as zero.)

Classification: A-1-b (0)

Sample B:

No. 200 = 87 < 35%: cohesionless soil

LL = 25, PI = 25 − 14 = 11 > 6: use Fig. 7.1 (AASHTO Classification Table)

Plot LL = 25 and PI = 11 on Figure 7.1.

First match: soil type A-2-6

$PGI = 0.01(F_{200} - 15)(PI - 10) = 0.01(30 - 15)(11 - 10) = 0.15$ (reported as 0) (from Equation 7-2)

Classification: A-2-6 (0)

Sample C:

No. 200 = 30 > 35%: cohesive soil

LL = 71, PI = 71 − 40 = 31: use Fig. 7.1 (AASHTO Classification Table)

Plot LL = 71 and PI = 31 on Figure 7.1.

First match: A-7-5

$GI = (F_{200} - 35)(0.2 + 0.005(LL - 40)) + 0.01(F_{200} - 15)(PI - 10) = 33.58$ (reported as 34) (from Equation 7-1)

Classification: A-7-5 (34)

7.1.2 Unified Soil Classification System

The USCS was originally developed by A. Casagrande in 1948. In this system, cohesionless soils are classified based mainly on their grain size, whereas cohesive soils are classified based on their plasticity. Therefore, only the grain size distribution in the form of sieve analysis and plasticity in the form of Atterberg limits can completely classify the soil in the USCS. Cohesionless soils such as sands and gravels are classified—based on the percentage of fines (smallest particles in the soil)—into clean, borderline, and with fines. Cohesive soils such as silts and clays are classified—based on Atterberg limits (LL and PI)—into low and high plasticity silts and clays.

7.1.2.1 Particle Size Definitions

Gravel: 75 mm (3 in) to 4.75 mm (No. 4)
Sand: 4.75 mm (No. 4) to 0.075 mm (No. 200)
Silt and clay: less than 0.075 mm (No. 200)

7.1.2.2 Group Symbols

The USCS classification of a soil consists of two symbols: The first represents the main soil, and the second represents a description of the main soil or the minor soil included within the main soil.

First letter:	G	Gravel
	S	Sand
	M	Silt
	C	Clay
	O	Organic

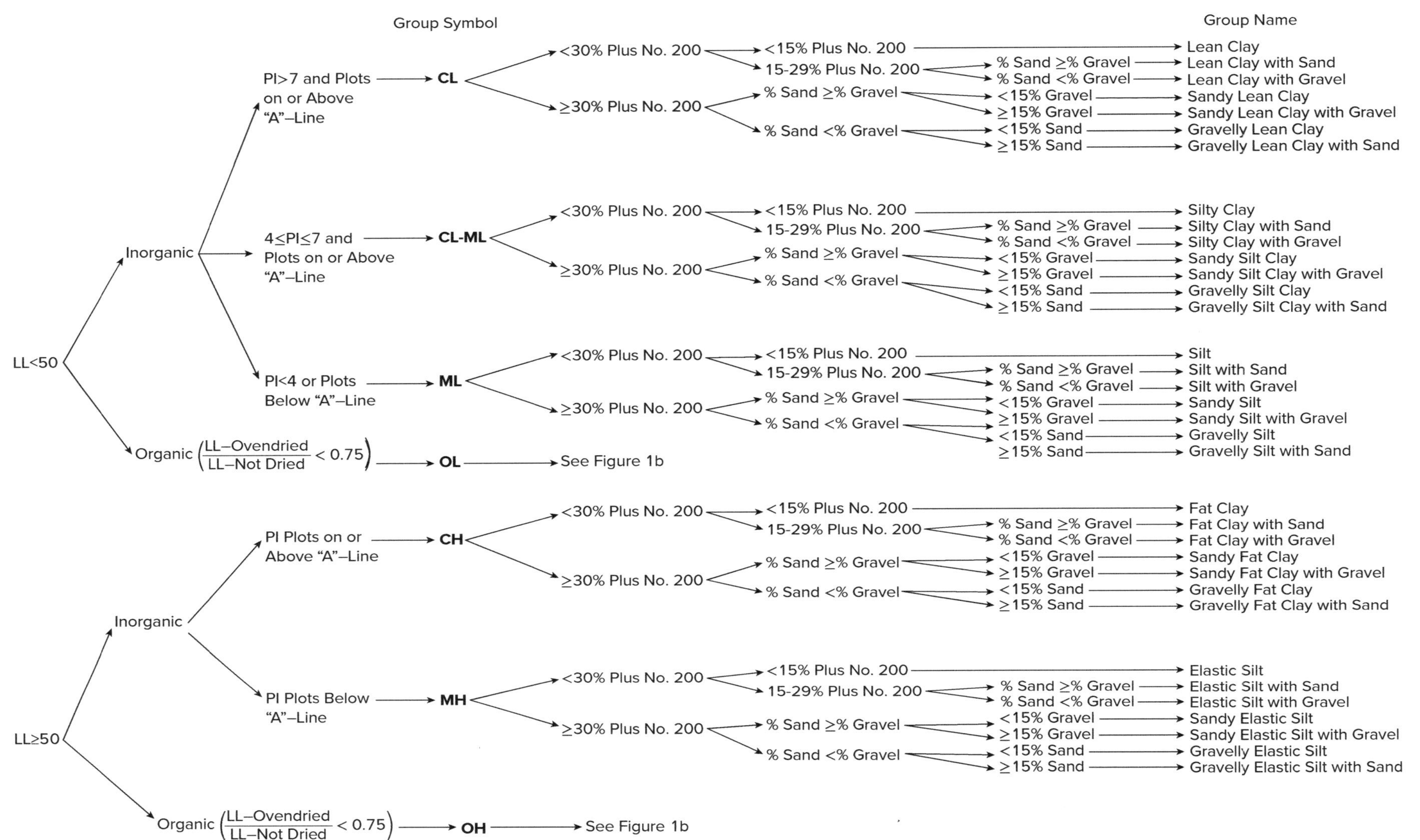

FIGURE 7.2 **USCS Inorganic Fine-Grained Classification Flowchart (% Passing No. 200 Sieve ≥ 50%)**

Source: ASTM D2487-17, *Standard Practice for Classification of Soils for Engineering Purposes (Unified Soil Classification System)*, ASTM International, 2017.

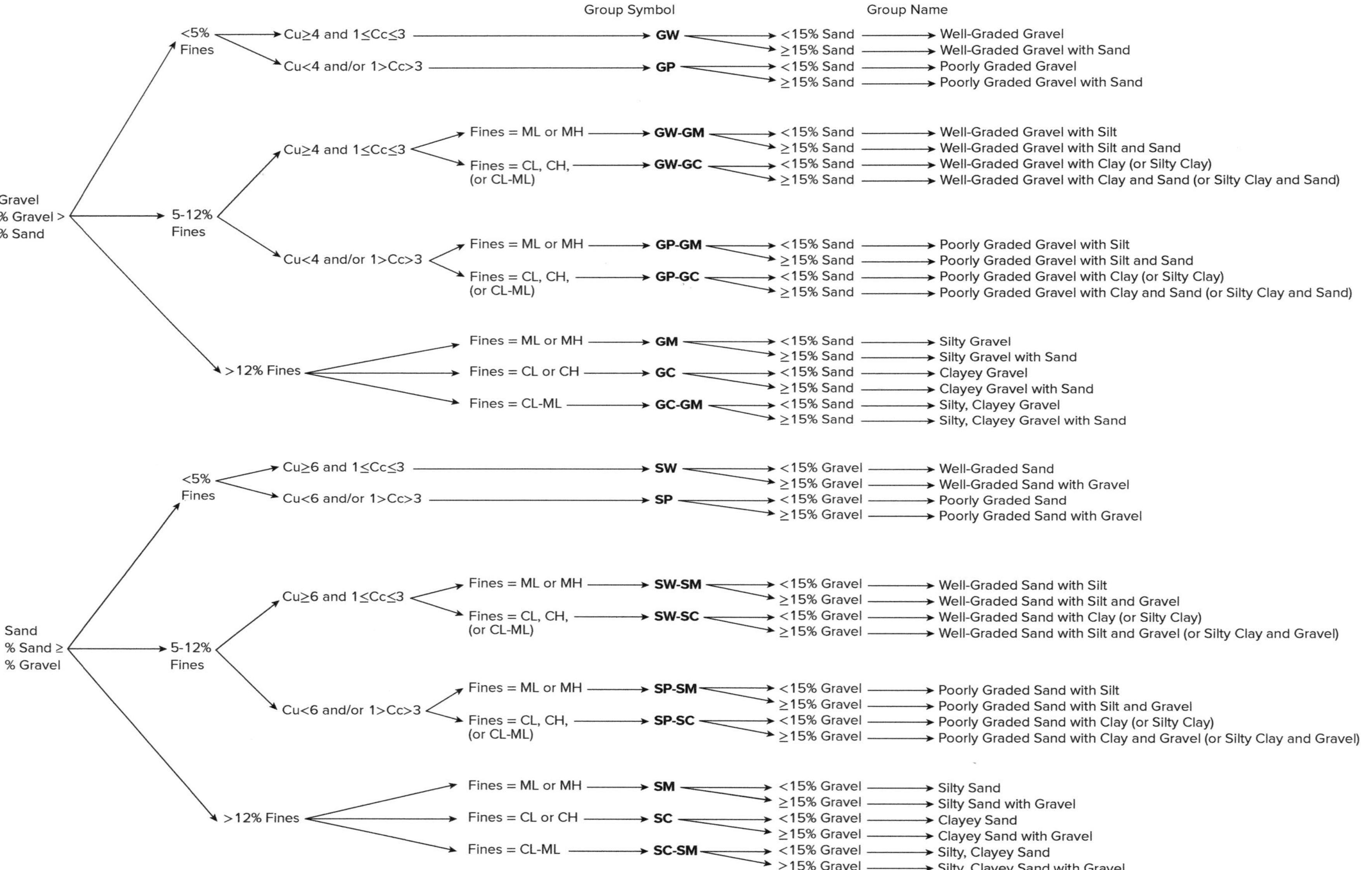

FIGURE 7.3 **USCS Coarse-Grained Classification Flowchart (% Passing No. 200 Sieve < 50%)**

Source: ASTM D2487-17, *Standard Practice for Classification of Soils for Engineering Purposes (Unified Soil Classification System)*, ASTM International, 2017.

Second letter: For granular soils: G or S

P	Poorly graded
W	Well graded
M	Silty
C	Clayey

For fine-grained soils: M, C, or O

L	Low plasticity
H	High plasticity or elastic

7.1.2.3 USCS Soil Classification Procedures

1. Identify the percent gravel, percent sand, and percent fines using No. 4 and No. 200 sieves as follows:

   Percent gravel in the sample = 100 – passing No. 4

   Percent sand in the sample = passing No. 4 – passing No. 200
2. If the percent passing the No. 200 sieve is greater than or equal to 50%, then the sample is fine-grained.
3. Determine the LL and PI. Plot the results on the Casagrande Plasticity Chart (Fig. 7.4), and read the USCS classification directly from the chart. You might also use Figure 7.2 to determine the group symbol. Note that nonplastic soil (PI < 4) is classified as silt (usually low-plasticity silt [ML]).
4. If the percent passing the No. 200 sieve is less than 50%, then the soil is coarse grained.
5. For soils with less than 5% fines, determine the uniformity coefficient C_u and coefficient of gradation C_z (see Section 7.2.3). Use Figure 7.3 to determine the group symbol.
6. For soils with greater than 12% fines, determine the LL and PI of the fraction passing the No. 40 sieve. Plot the results on the Casagrande Plasticity Chart (Fig. 7.4). Use Figure 7.4 to determine the group symbol. Note that if the fines plot in the CL-ML area, the group symbol will either be GC-GM or SC-SM.
7. If the soil has 5% to 12% fines, the soil will have a dual symbol.
8. Determine the C_u and C_z to determine the first symbol (GP/GW or SP/SW). Determine the LL and PI of the fraction passing the No. 40 sieve. Plot the results on the Casagrande Plasticity Chart (Fig. 7.4) to determine the second symbol (GM/GC or SM/SC). Use Figure 7.3 to determine the group symbol.

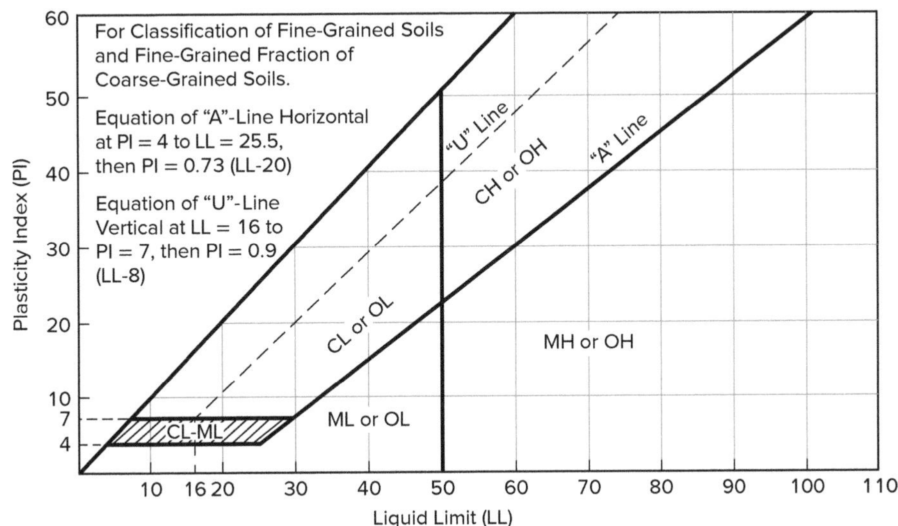

FIGURE 7.4 **The Casagrande Plasticity Chart for USCS (ASTM D2487)**

Source: ASTM D2487-17, *Standard Practice for Classification of Soils for Engineering Purposes (Unified Soil Classification System)*, ASTM International, 2017.

Example 7.2: USCS Soil Classification

Determine the USCS classification of soils A, B, and C.

SIEVE NO.	% PASSING SOIL A	% PASSING SOIL B	% PASSING SOIL C
No. 4	100	98	77
No. 40	90	46	42
No. 200	87	23	10
LL	75	20	22
PI	20	6	5
C_u			26
C_c			0.2

Solution

Sample A:

No. 200 = 87 > 50%: cohesive (fine-grained) soil

Use Fig. 7.2. Plot LL = 75 and PI = 20.

Classification: MH

Sample B:

No. 200 = 23 < 50%: cohesionless (coarse-grained) soil (sand or gravel)

$$\left.\begin{array}{l}\%\ \text{gravel} = 100 - 98 = 2\% \\ \%\ \text{sand} = 98 - 23 = 75\%\end{array}\right\}\ \text{It is sand.}$$

Use Figure 7.3. Follow the flowchart to > 12% fines. To get the type of fines, use the plasticity chart (Fig. 7.2). Plot LL = 20 and PI = 6. The intersection will be in the ML-CL zone.

Classification: SC-SM

Sample C:

No. 200 = 10 < 50%: cohesionless soil (sand or gravel)

$$\left.\begin{array}{l}\%\ \text{gravel} = 100 - 77 = 23\% \\ \%\ \text{sand} = 77 - 10 = 67\%\end{array}\right\}\ \text{It is sand.}$$

To get the type of fines, use the plasticity chart (Fig. 7.4). Plot LL = 22 and PI = 5. The intersection will be in the ML-CL zone. Make a choice between silts and clays. Always choose the worst-case scenario (clays).

Classification: SP-SC

7.1.3 Boring Log Graphical Representation

The boring log is typically included in the appendix portion of geotechnical reports. A **boring log** is a graphical representation of the data collected from a field exploration, field testing, visual and manual classification, and laboratory testing, all summarized in a simple graphical sheet. An example boring log is shown in Figure 7.5.

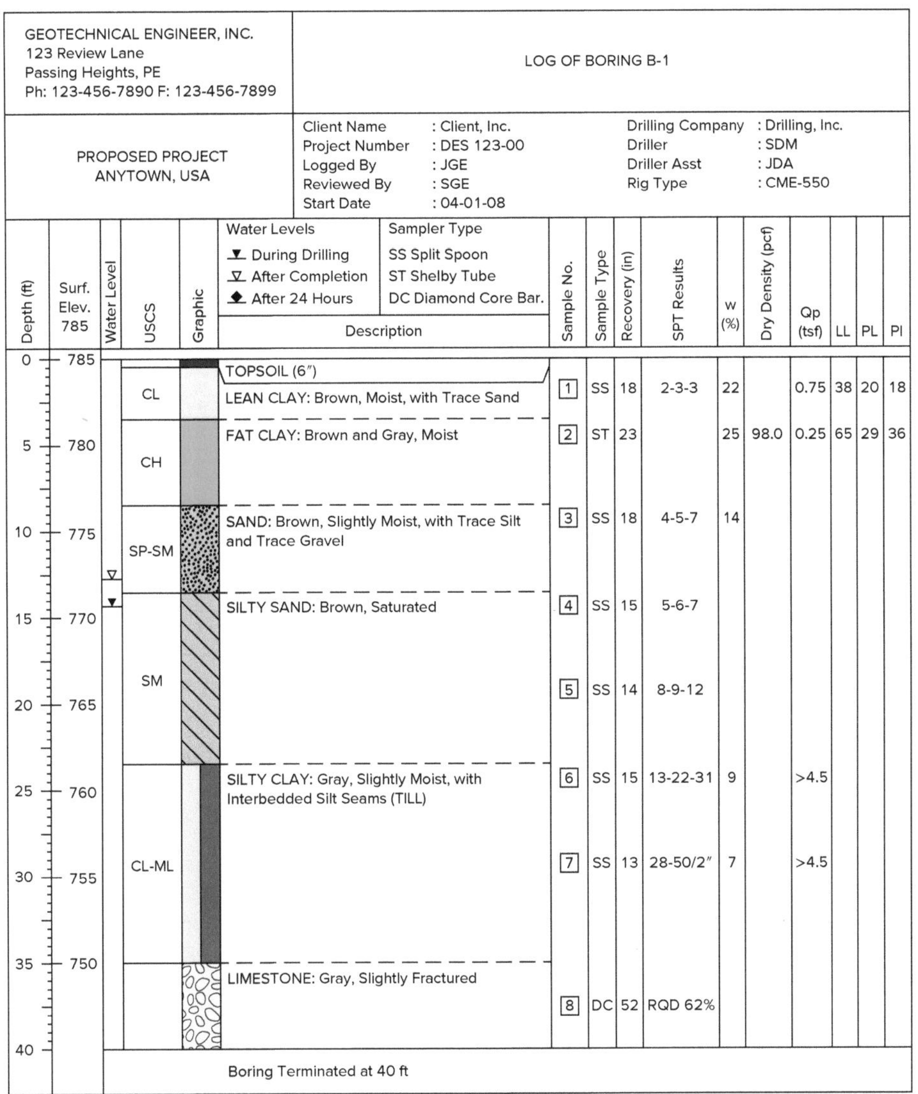

FIGURE 7.5 Sample Boring Log

7.1.4 Field Explorations and Testing

Field exploration is carried out utilizing soil borings, test pits, soil soundings, and geophysical methods. The number of borings and the depth of the exploration are selected based on the proposed structure and loading conditions. Common field (in situ) testing includes the **standard penetration test** (SPT) and the **cone penetration test** (CPT). Other methods presented in Table 7.2 (p. 325) are more advanced techniques that target measuring specific soil properties usually needed for advanced geotechnical calculation and analysis methods.

The SPT is the most common field test and is widely used in the US. This test uses a split-spoon sampler with an inner diameter that is commonly 1.375 in. The spoon is driven into the ground with a drop hammer. The hammer weighs 140 lb, and the drop is 30 in. The test involves driving the hammer three consecutive increments of 6 in into the undisturbed soil. The field N value is taken as the sum of the number of blows for the second increment (6 to 12 in) and third increment (12 to 18 in), representing the number of blows per foot.

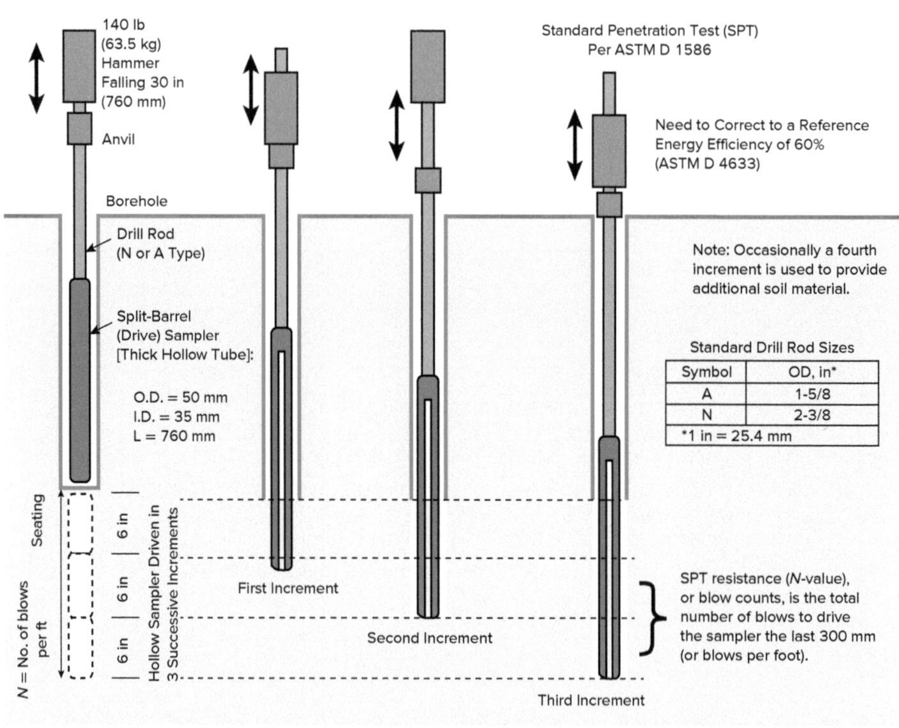

FIGURE 7.6 **Sequence of Driving Split-Barrel Sampler During the Standard Penetration Test**

Source: *Soils and Foundations Reference Manual (FHWA-NHI-06-088),* 2006 [1].

The measured field N value is generally inconsistent due to equipment and operator variability. Therefore, an energy correction is applied to the measured field N value (N_m) to achieve a standardized N value (N_{60}).

$$N_{60} = N_m \frac{E}{60}$$ **Equation 7-3**

Additionally, because the SPT is sensitive to overburden pressure, some applications require that the N value also be corrected for the influence of overburden pressure (generally only in cohesionless soils).

$$N_{\text{corr}} = N_{60}\sqrt{\frac{p_a}{\sigma'_v}}$$ **Equation 7-4**

N_m = field-measured SPT N value (blows per foot)

N_{60} = N value corrected for driving energy

N_{corr} = N value corrected for overburden pressure

E = field-measured driving energy (hammer efficiency)

p_a = atmospheric pressure (14.7 psi, which is about 2,000 psf)

σ'_v = effective overburden pressure

TABLE 7.2 Summary of Common In Situ Tests for Soils

TEST	SUITABLE FOR	NOT SUITABLE FOR	PROPERTIES THAT CAN BE ESTIMATED
SPT	All soils finer than gravel: may require retainer in granular soils	Gravel	Stratigraphy, undrained shear strength in clay, relative density and friction angle in sand
CPT	Sand, silt, clay, and peat	Gravel; soils with gravel or cobbles	Continuous evaluation of stratigraphy, strength of sand, undrained shear strength of clay, relative density; with piezocone also estimate in situ stress, pore pressure
FVT	Soft to medium clay, some silt and peat	Sand and gravel	Undrained shear strength
PMT	Clays, silt; soft rock, dense sand, non-sensitive clay, gravel, and till	Soft, sensitive clays, loose silts and sands	Strength, Ko, OCR, in situ stress, lateral compressibility, elastic modulus (E) and shear modulus (G)
DMT	Sand, silt, clay, and peat	Gravel	Soil type, stratigraphy, Ko, OCR; undrained shear strength, compressibility at small strain, and elastic modulus (E) in clay, friction angle in sand

SPT = Standard penetration test (ASTM D1586)
CPT = Cone penetration test (ASTM D3441)
FVT = Field vane test (ASTM D2573)
PMT = Pressuremeter test (ASTM D4719)
DMT = Flat plate dilatometer test

Source: *Soils and Foundations Reference Manual* (FHWA NHI-06-088) [1].

Example 7.3: SPT *N* Value

If an SPT yields the following data, what is the field N value?

First increment (0 to 6 in.): 3 blows

Second increment (6 to 12 in.): 7 blows

Third increment (12 to 18 in.): 6 blows

Fourth increment (18 to 24 in.): 9 blows

Solution

The N value is the sum of the blows for the second and third increments:

N value = 7 + 6 = 13 blows

The SPT N values are correlated to most soil design properties. Tables 7.3 and 7.4 are examples of such correlations. Many empirical correlations are also available, which correlate the N value to the strength and compressibility characteristics of both sandy and clayey soils.

TABLE 7.3 Strength and Consistency of Fine-Grained Soils

N_{60}	CONSISTENCY	UNCONFINED COMPRESSIVE STRENGTH, q_u, ksf (kPa)	RESULTS OF MANUAL MANIPULATION
<2	Very soft	<0.5 (<25)	Specimen (height = twice the diameter) sags under its own weight; extrudes between fingers when squeezed.
2–4	Soft	0.5–1 (25–50)	Specimen can be pinched in two between the thumb and forefinger; remolded by light finger pressure.
4–8	Medium stiff	1–2 (50–100)	Specimen can be imprinted easily with fingers; remolded by strong finger pressure.
8–15	Stiff	2–4 (100–200)	Specimen can be imprinted with considerable pressure from fingers or indented by thumbnail.
15–30	Very stiff	4–8 (200–400)	Specimen can barely be imprinted by pressure from fingers or indented by thumbnail.
>30	Hard	>8 >400	Specimen cannot be imprinted by fingers or difficult to indent by thumbnail.
Note that N_{60}-values should not be used to determine the design strength of fine-grained soils.			

Source: *Soils and Foundations Reference Manual (FHWA NHI-06-088)* [1].

TABLE 7.4 Strength of Cohesionless Soils Based on SPT *N* value

N VALUE (BLOWS/FT OR 305 MM)	RELATIVE DENSITY	APPROXIMATE $\bar{\phi}_{TC}$ (DEGREES)	
		(A)	(B)
0–4	Very loose	<28	<30
4–10	Loose	28–30	30–35
10–30	Medium	30–36	35–40
30–50	Dense	36–41	40–45
>50	Very dense	>41	>45

Source: *Manual on Estimating Soil Properties for Foundation Design (Project 1493-6)*, Electric Power Research Institute, 1990. Used with permission [2].

7.1.5 Undisturbed Sampling

Field explorations of the subsoil conditions include drilling holes and collecting soil samples, which might be disturbed or undisturbed.

The objective of undisturbed sampling is to keep the in-place structure of the soil intact during sampling and transportation. Undisturbed samples are required for in situ unit weight, permeability, compressive strength, shear strength, and consolidation tests. They are generally only collected for softer cohesive soils. The most commonly used method in the US to obtain undisturbed samples is to push a thin-walled (Shelby) tube sampler (Fig. 7.7) into the undisturbed soil and extract the tube with the sample inside. The diameter and wall thickness can vary, but generally should have an area ratio of less than 10%. The area ratio is a function of the **sampler outer diameter** (OD) and **inner diameter** (ID), and is shown in Equation 7-5:

$$\text{Area ratio (\%)} = \frac{(OD)^2 - (ID)^2}{(ID)^2} \times 100\% \qquad \textbf{Equation 7-5}$$

Other undisturbed sampling methods include thin-walled piston samplers and block sampling.

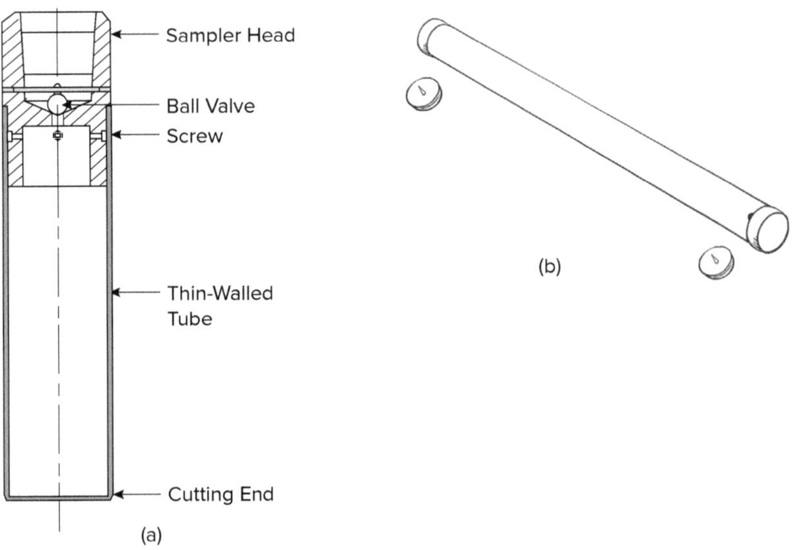

FIGURE 7.7 Schematics of Thin-Walled (Shelby) Tube and Tube with End Caps

Source: *Soils and Foundations Reference Manual (FHWA-NHI-06-088)*, 2006 [1].

7.1.6 Disturbed Sampling

Disturbed sampling does not require that the in-place structure of the soil remain in place. Split-barrel (split-spoon) samplers are commonly used to obtain disturbed samples.

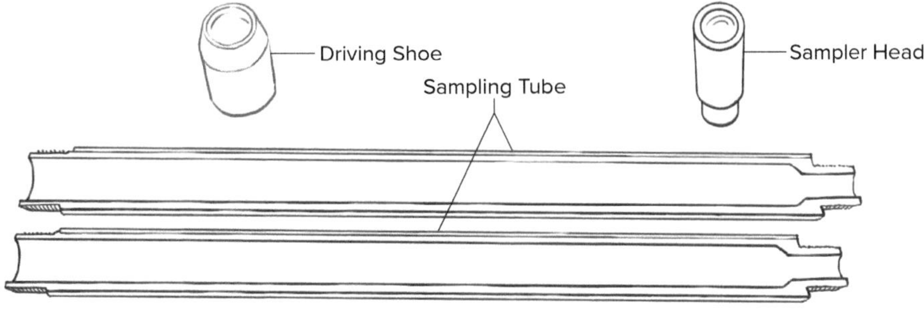

FIGURE 7.8 Split-Barrel Samples (for the SPT)

7.1.7 Rock Coring and Quality

When encountered, rock is sampled utilizing rock coring. A 5-ft core equipped with hard cutting teeth is advanced into the rock by rotating and pushing down. Water is used to cool down the cutting teeth. The recovered rock sample typically includes many pieces.

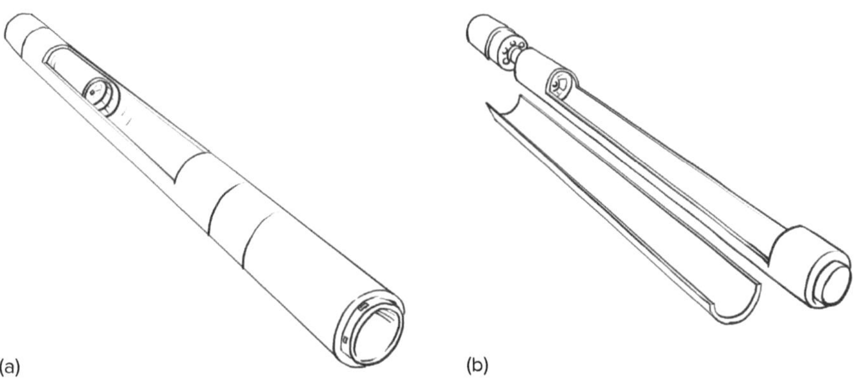

FIGURE 7.9 Double Tube Core Barrel: (a) Outer Barrel Assembly, (b) Inner Barrel Assembly

The recovery ratio is defined as the total rock recovered divided by the length of the rock core. The **rock quality designation** (RQD) is defined as the summation of the recovered rock pieces larger than 4 in divided by the total length of the core (typically 5 ft).

$$\text{RQD} = \frac{\Sigma \text{ lengths of intact pieces of core} > 100 \text{ mm}}{\text{length of core advance}} \times 100\% \qquad \textbf{Equation 7-6}$$

The RQD is used to classify the rock quality according to Figure 7.10.

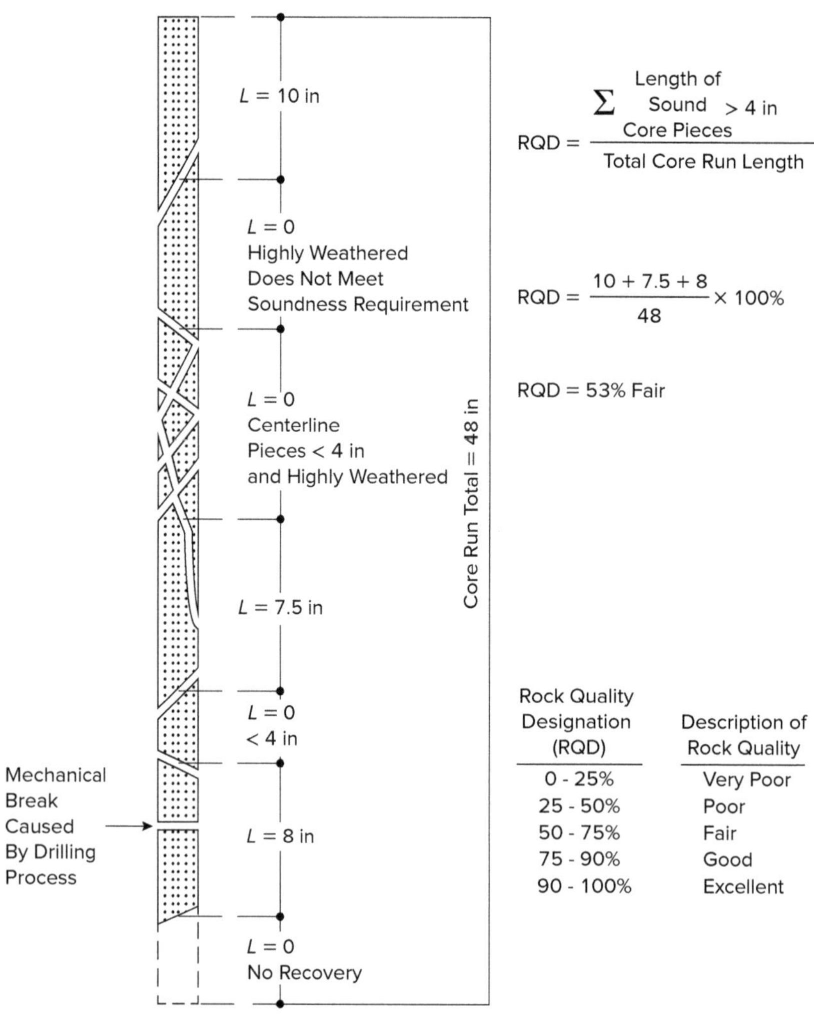

FIGURE 7.10 **Modified Rock Recovery as an Index of Rock Mass Quality**

Source: *Soils and Foundations Reference Manual (FHWA-NHI-06-088)*, 2006 [1].

Example 7.4: Rock Coring and Quality

Use the boring log in Figure 7.5 (p. 323) to answer the following questions:

Part 1: Which description would be most appropriate for the fat clay layer encountered between the depths of 3.5 and 8.5 ft?

A. Very loose
B. Very soft
C. Loose
D. Soft

Example 7.4 *(continued)*

Solution
Using Table 7.3 (p. 326), the clay the question refers to has an unconfined strength of 0.25 tons per square foot (tsf) and a soft consistency.

Answer: D

Part 2: The best estimate of the shear strength parameter ϕ of the sand layer encountered at EL 776.5 to EL 771.5 is:

A. 28
B. 35
C. 42
D. 49

Solution
For the sand layer, *N* value = 5 + 7 = 12. From Table 7.4 (p. 326), the best estimate for the shear strength parameter is between 30° to 35°.

Answer: B

Part 3: The estimated unconfined compressive strength (tsf) of the lean clay encountered at a depth of 1.0 to 3.5 ft is most nearly:

A. 0.5
B. 0.75
C. 1.0
D. 1.25

Solution
The unconfined compressive strength is given in the table to be 0.75 tsf.

Answer: B

Part 4: Which RQD term best describes the limestone encountered at a depth of 35 ft?

A. Poor
B. Fair
C. Good
D. Excellent

Solution
Utilizing Figure 7.10 (p. 328), for RQD of 62%, the rock quality is fair.

Answer: B

7.2 SOIL PROPERTIES

Soils encountered in the field generally consist of solids and voids. The solids constitute most of the soil volume. The voids may be filled with water or air or both. The soil is considered fully saturated when all voids are filled with water; however, for dry soils, all voids are filled with air. In the field, the soil structure is random; solids and voids are interconnected and mixed. Such random structure is very complicated for engineers to utilize in soil mechanics calculations and geotechnical applications.

7.2.1 Phase Diagram and Relationships

The soil phase diagram was introduced to idealize the random soil structure and simplify soil mechanics definitions and calculations. Imagine a block of soil that has a total weight of W_t and a total volume of V_t, as shown in Figure 7.11. If all of the solids in the soil block are crushed and compacted within the soil block, their weight will be W_s and their volume will be V_s. Again, combine all of the water to have a weight of W_w and a volume of V_w. The air volume is V_a and its weight is neglected. If the volume of water and air is combined, the total volume of voids (V_v) will be defined. The full phase diagram is now defined and can be used to provide basic soil definitions.

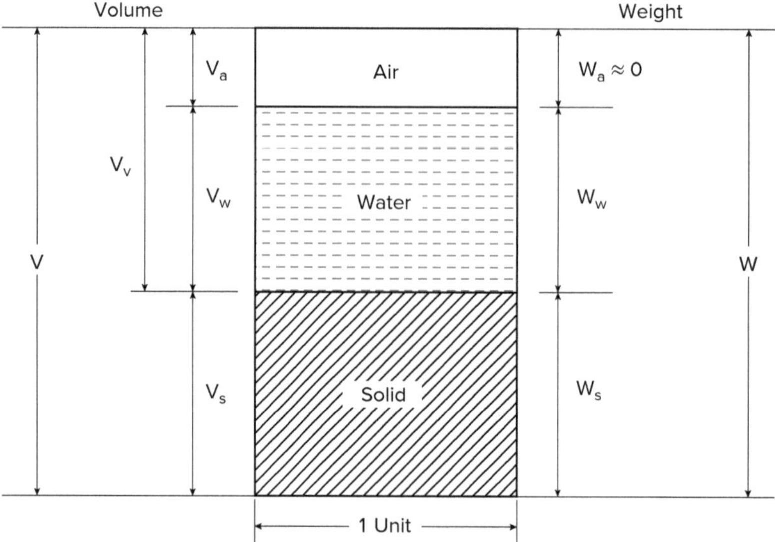

FIGURE 7.11 Soil Phase Diagram

Source: *Soils and Foundations Reference Manual (FHWA-NHI-06-088)*, 2006 [1].

Moisture (or water) content is the percentage of water to solids by weight. For most soils the moisture content is less than one or 100%, except for organic soils and very soft clays, which might have very high moisture content. These soils are sometimes called problematic soils.

$$w = \frac{\text{weight of water}}{\text{weight of solids}} = \frac{W_w}{W_s} \times 100\% \qquad \textbf{Equation 7-7}$$

Degree of saturation defines how much water is in the soil voids. For fully saturated soils, the degree of saturation is one or 100%, while it is zero for dry soils.

$$S = \frac{\text{volume of water}}{\text{volume of voids}} = \frac{V_w}{V_v} \times 100\% \qquad \textbf{Equation 7-8}$$

Total unit weight is defined as the total weight divided by the total volume. It is also called bulk, wet, or moist unit weight.

$$\gamma_t = \frac{\text{total weight}}{\text{total volume}} = \frac{W_t}{V_t} \qquad \textbf{Equation 7-9}$$

Dry unit weight is the unit weight of the soil when it is in a dry state, in which the total weight is equal to the solids weight (no water).

$$\gamma_d = \frac{\text{weight of solids}}{\text{total volume}} = \frac{W_s}{V_t} \qquad \textbf{Equation 7-10}$$

Solids unit weight is the unit weight of the solids only and should not be confused with the dry unit weight.

$$\gamma_s = \frac{\text{weight of solids}}{\text{volume of solids}} = \frac{W_S}{V_s} \qquad \textbf{Equation 7-11}$$

Void ratio is one of the most important soil parameters, and it is defined as the ratio between voids and solids by volume. For most soils, the void ratio is less than one or 100%, except for organic soils and very soft clays, which might have very high void ratios.

$$e = \frac{\text{volume of voids}}{\text{volume of solids}} = \frac{V_v}{V_s} \qquad \textbf{Equation 7-12}$$

Porosity is a very similar concept to the void ratio; however, it correlates the volume of voids to the total volume. Porosity must be less than one.

$$n = \frac{\text{volume of voids}}{\text{total volume}} = \frac{V_v}{V_t} \qquad \textbf{Equation 7-13}$$

Specific gravity is the unit weight of solids normalized by the unit weight of water (62.4 pcf). For most soils, the specific gravity ranges from 2.6 to 2.74.

$$G_s = \frac{\text{unit weight of solids}}{\text{unit weight of water}} = \frac{\gamma_{\text{solids}}}{\gamma_w} \qquad \textbf{Equation 7-14}$$

Some of the basic soil parameters can easily be measured in the laboratory; however, other properties are extremely difficult and time consuming to measure. Therefore, phase relationships were developed to estimate those parameters. Table 7.5 presents the most popular phase relationships correlating soil weights, volumes, and unit weights.

It should be noted that these relationships can also be written in terms of mass and/or density by substituting weight (W) with mass (M) and unit weight (γ) with density (ρ).

$$G_s = \frac{W_s}{V_s\gamma_w} \quad \text{or} \quad G_s = \frac{M_s}{V_s\rho_w} \qquad \textbf{Equation 7-15}$$

Where the relationship of weight to mass is as follows:

$$W = Mg, \quad \text{where } g = 32.2 \text{ ft/s}^2 \qquad \textbf{Equation 7-16}$$

TABLE 7.5 Phase Relationships

WEIGHT							
W	$W_s + W_w$	$W_s(1+w)$	$V\gamma$				
W_s	$W - W_w$	$\frac{W}{1+w}$	$\frac{W_w G_s}{Se}$	$V\gamma_d$	$V_s G_s \gamma_w$	$\frac{VG_s\gamma_w}{1+e}$	$VG_s\gamma_w(1-n)$
W_w	$W - W_s$	wW_s	$\frac{W_s Se}{G_s}$	$V_v S\gamma_w$	$V_w\gamma_w$		

VOLUME							
V	$V_s + V_w + V_a$	$\frac{W}{\gamma}$	$\frac{W_s}{\gamma_d}$	$V_s(1+e)$	$\frac{V_s}{1-n}$	$\frac{V_v(1+e)}{e}$	$\frac{V_v}{n}$
V_s	$V - V_w - V_a$	$\frac{W_s}{G_s\gamma_w}$	$\frac{V}{1+e}$	$V(1-n)$	$\frac{V_v}{e}$	$\frac{V_v(1-n)}{n}$	
V_w	$V - V_s - V_a$	$\frac{W_w}{\gamma_w}$	$\frac{eSV}{1+e}$	nSV	eSV_s	$\frac{nSV_s}{1-n}$	SI_v
V_v	$V - V_s$	$V - \frac{W_s}{G_s\gamma_w}$	$\frac{eV}{1+e}$	nV	eV_s	$\frac{nV_s}{1-n}$	$\frac{V_w}{S}$
V_a	$V - V_s - V_w$	$\frac{eV(1-S)}{1+e}$	$nV(1-S)$	$eV_s(1-S)$	$V_v(1-S)$	$\frac{nV_s(1-S)}{1-n}$	

UNIT WEIGHT					
γ	$\frac{W}{V}$	$\gamma_d(1+w)$	$\frac{\gamma_w(G_s+eS)}{1+e}$	$\frac{\gamma_w(1+w)}{\frac{w}{S}+\frac{1}{G_s}}$	$\frac{G_s\gamma_w}{1+e}(1+w)$
γ_d	$\frac{W_s}{V}$	$\frac{\gamma}{1+w}$	$\frac{G_s\gamma_w}{1+e}$	$G_s\gamma_w(1-n)$	$\frac{G_s\gamma_w}{1+\frac{wG_s}{S}}$
γ_{sat} ($S = 100\%$)	$\frac{W}{V}$	$\frac{W_s + V_v\gamma_w}{V}$	$\gamma_d + \gamma_w\left(\frac{e}{1+e}\right)$	$\frac{\gamma_w(1+w)}{w+\frac{1}{G_s}}$	$\frac{\gamma_w(G_s+e)}{1+e}$
$\gamma_b = \gamma'$ (buoyant)	$\gamma_{sat} - \gamma_w$	$\gamma_d - \gamma_w(1-n)$	$\gamma_w\left(\frac{G_s+e}{1+e} - 1\right)$		
γ_w	$\frac{W_w}{V_w}$	$\frac{W_s}{V_s G_s}$	$\frac{\gamma_d(1+e)}{G_s}$	$\frac{\gamma_d}{G_s(1-n)}$	
γ_{solid}	$\frac{W_s}{V_s}$	$\frac{W_w}{V_w G_s}$	$\gamma_w G_s$		

WEIGHT-VOLUME RELATIONSHIPS						
w [× 100%]	$\frac{W_w}{W_s}$	$\frac{W}{W_s} - 1$	$\frac{\gamma}{\gamma_d} - 1$	$S\left(\frac{\gamma_w}{\gamma_d} - \frac{1}{G_s}\right)$	$\frac{eS}{G_s}$	
S [× 100%]	$\frac{V_w}{V_v}$	$\frac{W_w}{\gamma_w V_v}$	$\frac{wG_s}{e}$	$\frac{w}{\frac{\gamma_w}{\gamma_d} - \frac{1}{G_s}}$		
e	$\frac{V_v}{V_s}$	$\frac{V}{V_s} - 1$	$\frac{W_w G_s}{W_s S}$	$\frac{G_s\gamma_w}{\gamma_d} - 1$	$\frac{n}{1-n}$	$\frac{wG_s}{S}$
n	$\frac{V_v}{V}$	$1 - \frac{V_s}{V}$	$1 - \frac{W_s}{VG_s\gamma_w}$	$1 - \frac{\gamma_d}{G_s\gamma_w}$	$\frac{e}{e+1}$	
G_s	$\frac{\gamma_{solid}}{\gamma_w}$	$\frac{W_s}{V_s\gamma_w}$	$\frac{\gamma_d(1+e)}{\gamma_w}$	$\frac{eS}{w}$		

Example 7.5: Phase Relationships

A soil has a volume of 0.3 ft^3 and weighs 36 lb. The specific gravity (G_s) of the soil sample is 2.67 and the moisture content (w) is 18%. Determine the moist unit weight (γ), dry unit weight (γ_d), void ratio (e), porosity (n), and degree of saturation (S).

Solution

Use Table 7.5 to solve directly for the required parameters:

$$\gamma = \frac{W}{V} = \frac{36 \text{ lb}}{0.3 \text{ ft}^3} = 120 \text{ lb/ft}^3$$

$$\gamma_d = \frac{\gamma}{1+w} = \frac{120}{(1+0.18)} = 101.7 \text{ lb/ft}^3$$

$$e = \frac{G_s\gamma_w}{\gamma_d} - 1 = \frac{(2.67 \times 62.4)}{101.7 \text{ lb/ft}^3 - 1} = 0.64$$

$$n = \frac{e}{e+1} = \frac{0.64}{(1+0.64)} = 0.39$$

$$s = \frac{wG_s}{e} = \frac{0.18 \times 2.67}{0.64} = 0.75$$

Example 7.6: Phase Relationships

A soil sample has a dry unit weight of 85 lb/ft^3 and a porosity of 0.4. What is the specific gravity of the solids?

Solution

Use Table 7.5 to solve directly for the required parameters:

$$e = \frac{n}{1-n} = \frac{0.4}{1-0.4} = 0.667$$

$$G_S = \frac{\gamma_d(1+e)}{\gamma_w} = \frac{95(1+0.667)}{62.4} = 2.54$$

Sample dry unit weight = $\gamma_d = 85$ lb/ft^3

Specific gravity of water = $\gamma_w = 62.4$ lb/ft^3

$$\text{Specific gravity of solids, } G_S = \frac{85 \text{ lb/ft}^3 (1+0.667)}{62.4 \text{ lb}}$$

$$= 2.27$$

Example 7.7: Phase Relationships

The wet unit weight of a soil sample is 110 lb/ft^3 and the moisture content is 18%. The specific gravity of the solids is 2.7. Determine the dry unit weight, void ratio, and degree of saturation.

Solution

Use Table 7.5 to solve directly for the required parameters:

Wet weight of soil sample, $y = 110$ lb/ft^3

Moisture content, $w = 18\% = 0.18$

Specific gravity of solids, $G_s = 2.7$

Specific gravity of water, $y_w = 62.4$ lb/ft^3

Example 7.7 *(continued)*

Dry unit weight, $\gamma_d = \frac{\gamma}{1+w} = \frac{110 \text{ lb/ft}^3}{1+0.18} = 93.22 \text{ lb/ft}^3$

Void ratio, $e = \left[\frac{G_S \gamma_w}{\gamma_d} - 1\right] = \left[\frac{2.7(62.4 \text{ lb/ft}^3)}{93.22 \text{ lb/ft}^3} - 1\right] = 0.81$

Degree of saturation, $S = \frac{wG_S}{e} = \frac{0.18(2.7)}{0.81} = 0.60$

7.2.2 Relative Density

Relative density is a special weight-volume relationship used for sands and gravels only. Relative density measures how dense the in-field soil is compared with its loosest and densest conditions, as measured in the laboratory. It can be expressed in terms of dry unit weight or void ratio.

$$D_r = \left(\frac{\gamma_d - (\gamma_d)_{\min}}{(\gamma_d)_{\max} - (\gamma_d)_{\min}}\right)\left(\frac{(\gamma_d)_{\max}}{\gamma_d}\right) \times 100\% \qquad \textbf{Equation 7-17}$$

$$\rightarrow \quad \gamma_d = \frac{(\gamma_d)_{\min}}{1 - \left(\frac{D_r}{(\gamma_d)_{\max}}\right)((\gamma_d)_{\max} - (\gamma_d)_{\min})} \qquad \textbf{Equation 7-18}$$

or

$$D_r = \frac{e_{\max} - e}{e_{\max} - e_{\min}} \times 100\ \% \quad \rightarrow \quad e = e_{\max} - D_r(e_{\max} - e_{\min}) \qquad \textbf{Equation 7-19}$$

D_r = relative density

$e_{\max}$ = maximum void ratio

$e_{\min}$ = minimum void ratio

e = in situ void ratio

$(\gamma_d)_{\max}$ = maximum dry unit weight

$(\gamma_d)_{\min}$ = minimum dry unit weight

γ_d = in situ dry unit weight

Table 7.6 presents correlations of relative density to soil compactness.

TABLE 7.6 Evaluation of Apparent Density of Coarse-Grained Soils

N_{60}	APPARENT DENSITY	RELATIVE DENSITY, %
0–4	Very Loose	0–20
>4–10	Loose	20–40
>10–30	Medium Dense	40–70
>30–50	Dense	70–85
>50	Very Dense	85–100
The above guidance may be misleading in gravelly soils.		

Source: *Soils and Foundations Reference Manual (FHWA-NHI-06-088)*, 2006 [1].

Example 7.8: Relative Density—Dry Unit Weight

For a given sand sample, laboratory tests show $e_{\max} = 0.41$, $e_{\min} = 0.25$, and $G_s = 2.72$. An in situ test indicates that the relative density of the sand is approximately 65%. Estimate the in situ dry unit weight of the sand.

Example 7.8 *(continued)*

Solution

Use Table 7.6 to pick an equation for dry unit weight that matches information that is known or can be solved for.

Given: e_{max}, e_{min}, D_r.

Solve for $e_{in\ situ}$.

$$e = e_{max} - D_r(e_{max} - e_{min})$$
$$e = 0.41 - 0.65(0.41 - 0.25) = 0.31$$

Given: G_s

Determine dry unit weight from: $\gamma_d = \dfrac{G_s\gamma_w}{1+e}$.

$$\gamma_d = \frac{G_s\gamma_w}{1+e} = \frac{2.72\left(62.4\ \text{lb/ft}^3\right)}{1+0.31} = 130\ \text{lb/ft}^3$$

7.2.3 Grain Size Distribution

It is important for geotechnical engineers to understand soil particle sizes and how they are distributed in a soil field sample. Two boundary or extreme soil grain size distribution conditions may exist:

1. **Well-graded soil**, in which particle sizes range from sands and gravels to silts and clays with normal distribution
2. **Uniform soil**, also called poorly graded soil, in which most of the soil particle sizes are limited to a very narrow range (uniform soil), or the soil sizes lack a specific range of particle sizes (**gap-graded soil**)

To separate the particle sizes in a soil sample by weight, two methods are available based on grain sizes:

1. **Sieve analysis** is used to obtain the grain size distribution of coarse-grained soils (sands and gravels) larger than 0.075 mm (retained above No. 200 sieve).
2. **Hydrometer analysis** is used to obtain the grain size distribution of fine-grained soils (silts and clays) smaller than 0.075 mm (passing No. 200 sieve).

Various sieves are shown in Figure 7.12.

FIGURE 7.12 Sieve Analysis

Source: *Soils and Foundations Reference Manual (FHWA-NHI-06-088)*, 2006 [1].

Table 7.7 summarizes the sizes of US standard sieves used in grain size analysis. Figure 7.13 shows some typical grain size distribution curves, sizes of the US standard sieves, and the typical sizes of different soil types. The grain size distribution curve is a plot of the percent of finer, passing, or smaller versus the particle diameter (plotted on a logarithmic scale). The shape of the curve could be smooth, mildly sloped, and continuous (a well-graded soil); steeply sloped and limited in a small range of particle sizes (a uniform soil); or containing a broken or flat portion (a gap-graded soil). Although an engineer can visually differentiate between these types of soils (assuming the scale of the plot is kept constant), there are some tools that use mathematical methods to define the slope and continuity of a curve.

The coefficient of uniformity (C_u) defines the slope of the curve with higher values corresponding to a flat curve with a mild slope, which is a feature of well-graded soils.

$$C_u = \frac{D_{60}}{D_{10}}$$ **Equation 7-20**

The coefficient of curvature/gradation (C_c or C_z) defines the continuity of a curve and captures gaps, if any.

$$C_c \text{ (or } C_z) = \frac{(D_{30})^2}{D_{60} D_{10}}$$ **Equation 7-21**

The mean particle size (D_{50}) defines the average particle diameter, which is the diameter corresponding to 50% passing (from the curve).

The effective particle size (D_{10}) is the diameter that corresponds to 10% passing (from the curve), which is significant in problems related to permeability, soil filter design, and seepage through soils.

TABLE 7.7 US Standard Sieves

Sieve Size	Sieve Opening (mm)
4 in	100
3 in	75
2 in	50
1 ½ in	37.5
1 in	25
¾ in	19
½ in	12.5
⅜ in	9.5
No. 4	**4.75**
No. 8	2.36
No. 10	**2.00**
No. 16	1.18
No. 20	0.850
No. 30	0.600
No. 40	**0.425**
No. 50	0.300
No. 60	0.250
No. 70	0.212
No. 100	0.150
No. 140	0.106
No. 200	**0.075**

Source: ASTM E11-17, *Standard Specification for Woven Wire Test Sieve Cloth and Test Sieves*, ASTM International, 2017.

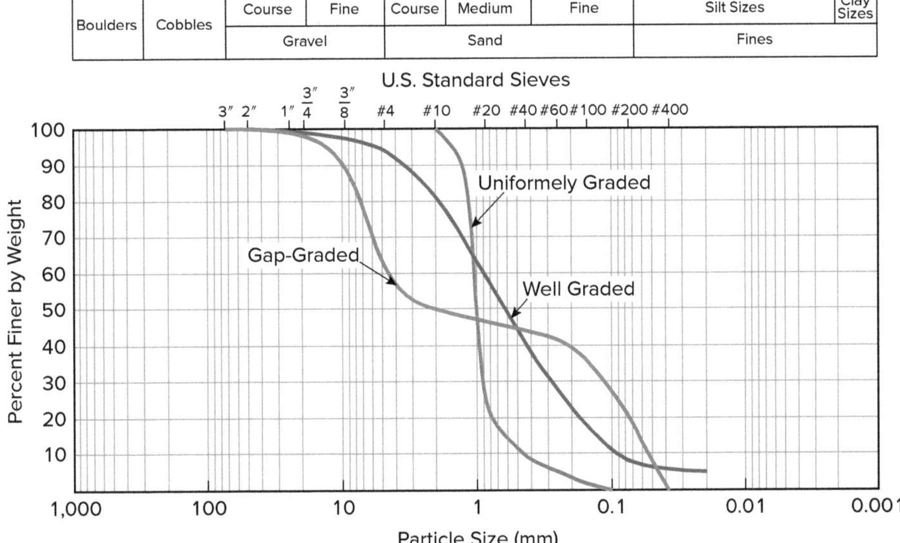

FIGURE 7.13 Typical Grain Size Distribution Curves

Example 7.9: Grain Size Distribution

Determine the coefficient of uniformity and the coefficient of curvature of gap-graded and well-graded soils, as shown in Figure 7.13.

Solution

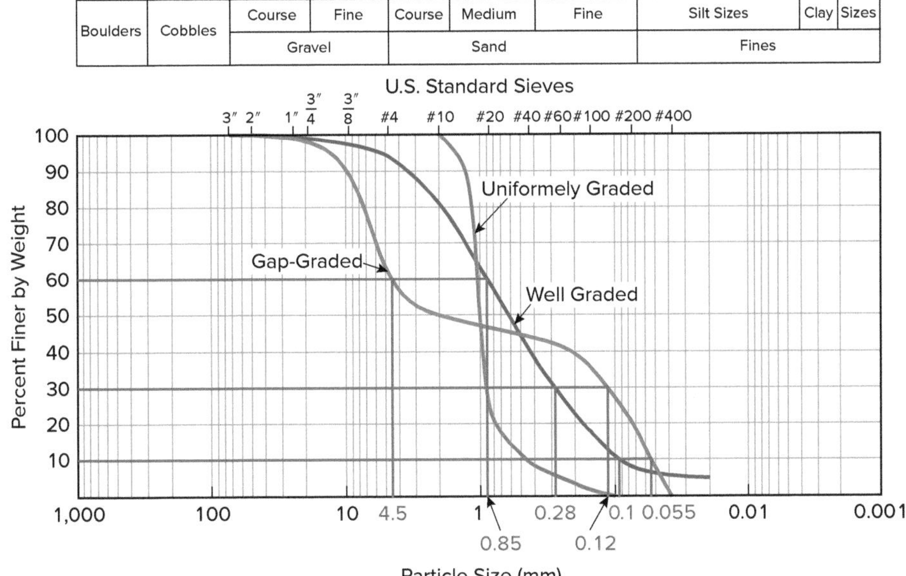

For gap-graded soil:

$D_{60} \approx 4.50$ mm, $D_{30} \approx 0.12$ mm, $D_{10} \approx 0.055$ mm

$$C_u = \frac{D_{60}}{D_{10}} = \frac{4.50\text{ mm}}{0.055\text{ mm}} = 81.8$$

$$C_c = \frac{{D_{30}}^2}{D_{60}D_{10}} = \frac{(0.12\text{ mm})^2}{(4.50\text{ mm})(0.055\text{ mm})} = 0.58$$

For well-graded soil:

$D_{60} \approx 0.85$ mm, $D_{30} \approx 0.28$ mm, $D_{10} \approx 0.1$ mm

Example 7.9 *(continued)*

$$C_u = \frac{D_{60}}{D_{10}} = \frac{0.85 \text{ mm}}{0.1 \text{ mm}} = 8.5$$

$$C_c = \frac{{D_{30}}^2}{D_{60}D_{10}} = \frac{(0.28 \text{ mm})^2}{(0.85 \text{ mm})(0.1 \text{ mm})} = 0.92$$

Example 7.10: Grain Size Distribution

Based on the sieve analysis results below, determine the mean grain size, effective grain size, coefficient of uniformity, and coefficient of gradation.

SIEVE SIZE	PERCENT FINER
No. 4	100
No. 10	72
No. 20	60
No. 30	50
No. 40	44
No. 60	30
No. 200	10

Solution

Mean grain size: $D_{50} = 0.60$ (from Table 7.7, size of Sieve No. 30)

Effective grain size: $D_{10} = 0.075$ (from Table 7.7, size of Sieve No. 200)

Coefficient of uniformity: $C_u = \frac{D_{60}}{D_{10}} = 11.33$ (Get D_{60} and D_{10} from Table 7.7, size of Sieve No. 20 and Sieve No. 200, respectively.)

Coefficient of curvature: $C_c(\text{or } C_z) = \frac{(D_{30})^2}{D_{60}D_{10}} = 0.98$ (Obtain D_{30}, D_{60}, and D_{10} from Table 7.7, for sieve sizes No. 60, No. 20, and No. 200, respectively.)

7.2.4 Atterberg Limits

Atterberg limits were developed to assess the plasticity of fine-grained soils. Moisture content has a significant effect on the shear strength of fine-grained soils. Increasing the moisture content of fine-grained soils decreases their shear strength. Atterberg defined three specific types of moisture content at which fine-grained soils have specific shear strengths: shrinkage limit, **plastic limit** (PL), and LL. Figure 7.14 presents the states of fine-grained soils as they relate to moisture content. Moving from the bottom up, the moisture content increases, and the corresponding shear strength decreases.

Liquid State (Liquid Behavior)	
	Liquid Limit (*LL*)
Plastic State (Add Water - Remold)	*PI* = *LL*−*PL*
	Plastic Limit (*PL*)
Semi-Solid State (Add Water - Remold)	
Solid State (Add Water - Remold)	

FIGURE 7.14 Consistency of Fine-Grained Soils (Atterberg Limits)

The LL is the moisture content at which the soil sample passes from a plastic state to a liquid state.

Materials

School of PE

The PL is the moisture content at which the soil sample passes from a semisolid to a plastic state.

The plasticity index (PI) is the difference in moisture content from the threshold of the plastic-to-liquid state to the threshold of the semisolid-to-plastic state. The plasticity index is a main indicator of the amount of real clay particles in the sample. The higher the clay particle content, the higher the PI and the higher chance of problems such as shrink-swell potential and heave. The PI is also the main parameter needed to classify fine-grained soils.

$$PI = LL - PL$$ **Equation 7-22**

7.2.5 Strength

Materials resist applied stress by their strength. Applied stresses might include compression stresses, tension stresses, shear stresses, and so forth.

Given that soil is a semi-infinite medium that extends from the surface deep into the ground, the only logical mode of failure is shear failure, which is caused by a portion of the soil separating from the semi-infinite medium. Shear stresses develop in soil due to applied loads that might be compression loads or moments. When the applied loads or stresses are high enough to create shear stresses in the soil that exceed the soil shear strength, shear failure occurs, which is a catastrophic brittle failure.

Soil resists shear in two main ways: friction between the particles (in cohesionless soils) and cohesion between the particles (cohesive soils). Soil shear strength is expressed by the Mohr-Coulomb failure criterion as shown in Equation 7-23:

$$\tau' = c' + \sigma' \tan'\phi'$$ **Equation 7-23**

τ' = soil shear strength

σ' = effective stress normal to shearing plane

ϕ' = effective internal angle of friction (also called angle of shearing resistance)

c' = effective cohesion, or apparent cohesion

The angle of shearing resistance and cohesion are often called the shear strength parameters of soils. They can be estimated using:

- Correlations to field-exploration and field-testing methods, such as the SPT, CPT, DMT, PMT, vane shear test (VST), and other field methods (see Table 7.2).
- Laboratory tests such as the direct shear test (all soil types, drainage is not controlled); the triaxial test (all soil types, the test can represent existing field conditions and accurately follow loading conditions, and drainage is fully controlled); and the unconfined compression test (only for clay soils and very similar to concrete cylinder breaks, undrained condition only).

7.2.6 Permeability

7.2.6.1 Darcy's Law and Bernoulli's Equation

Darcy's law provides the basic formula that describes the discharge of water from any cross section and is defined in Equation 7-24:

$$Q = vA = kiA$$ **Equation 7-24**

Q = flow quantity

v = average discharge velocity (for flow in soils, $v = ki$)

A = cross-sectional area perpendicular to the flow direction

k = coefficient of permeability

i = hydraulic gradient defined as: $i = \dfrac{\Delta h}{l}$

Δh = loss in total head between upstream and downstream

l = distance water travels in direction of flow

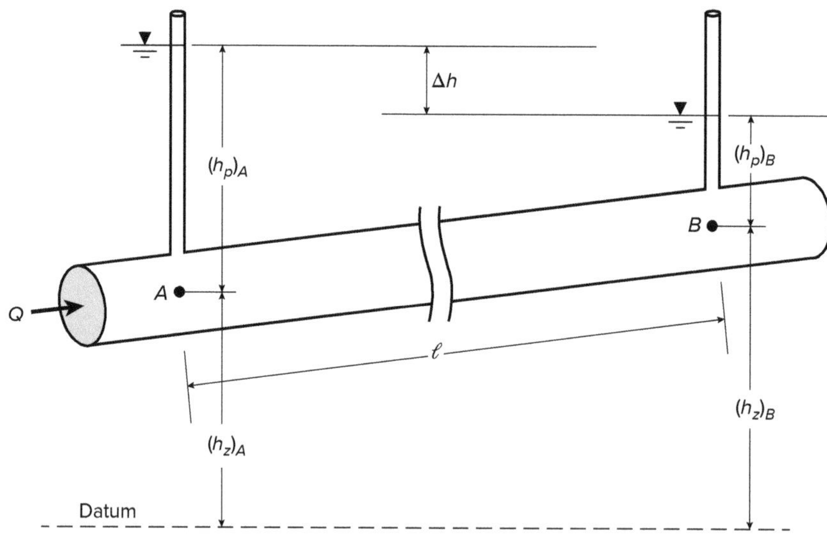

FIGURE 7.15 Hydraulic Gradient as Head Loss Between Two Points

In soil and rock applications, total head is defined in Equation 7-25 (neglecting velocity head because the flow through soils is slow):

$$h = h_z + h_p \qquad \textbf{Equation 7-25}$$

h = total head, neglecting velocity head

h_z = elevation head, which is the difference between the datum and a given point

h_p = pressure head, which is the difference between a given point and the free water level shown in the piezometer

7.2.6.2 One-Dimensional Flow of Water Through Soils

The flow of water through soils involves losses of the driving head due to friction with soil particles, as well as other factors. A water particle entering the soil at the upstream face with a total head h_{in} will leave the soil at the downstream face at a total head of h_{out}. The difference between h_{in} and h_{out} is the total head loss Δh, which is typically assumed to be linear through the soil length; hence, the total head may be calculated at any location in the soil.

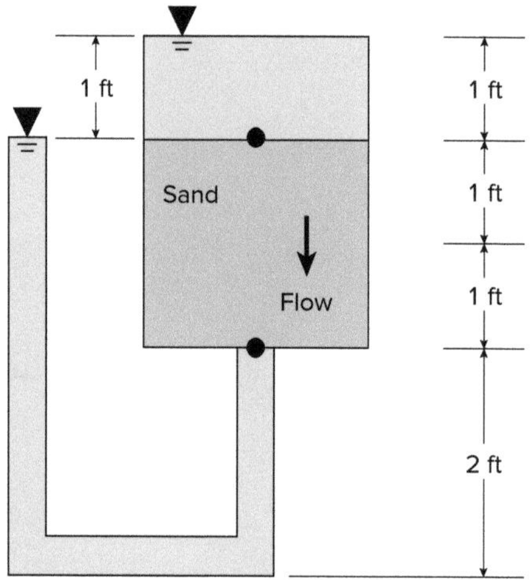

FIGURE 7.16 Downward Flow

Materials

School of PE

Figure 7.16 represents a one-dimensional downward flow through sand. The hydraulic gradient in the sand column is calculated as follows:

$$i = \frac{\Delta h}{l} = \frac{5 \text{ ft} - 4 \text{ ft}}{2 \text{ ft}} = 0.5$$

Assuming a datum at the bottom of the tube, the total head at the entrance (top) of the sand column is:

$$h_{\text{ENT}} = h_z + h_p = 4 \text{ ft} + 1 \text{ ft} = 5 \text{ ft}$$

The total head at the exit (bottom) of the sand column is:

$$h_{\text{EXT}} = h_z + h_p = 2 \text{ ft} + 2 \text{ ft} = 4 \text{ ft}$$

The total head and pressure head at the center (midpoint) of the sand column are:

$$h_{\text{MID}} = h_{\text{ENT}} - i\Delta l = 5 \text{ ft} - 0.5(1 \text{ ft}) = 4.5 \text{ ft}$$
$$h_p = h_{\text{MID}} - h_z = 4.5 \text{ ft} - 3 \text{ ft} = 1.5 \text{ ft}$$

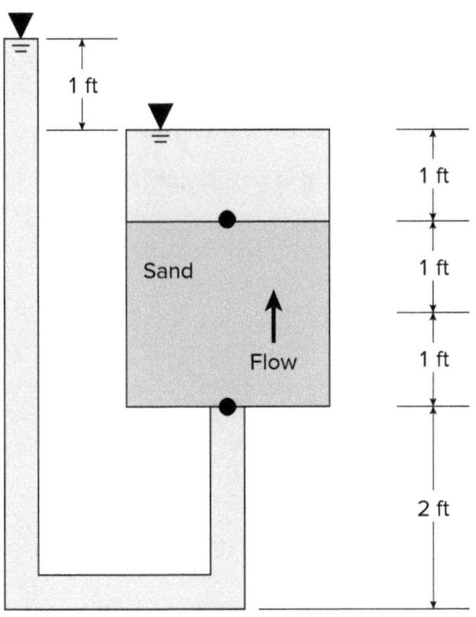

FIGURE 7.17 Upward Flow

Figure 7.17 represents a one-dimensional upward flow through sand. The hydraulic gradient in the sand column is calculated as follows:

$$i = \frac{\Delta h}{l} = \frac{6 \text{ ft} - 5 \text{ ft}}{2 \text{ ft}} = 0.5 \text{ ft}$$

Assuming a datum at the bottom of the tube, the total head at the entrance (top) of the sand column is:

$$h_{\text{ENT}} = h_z + h_p = 2 \text{ ft} + 4 \text{ ft} = 6 \text{ ft}$$

The total head at the exit (bottom) of the sand column is:

$$h_{\text{EXT}} = h_z + h_p = 4 \text{ ft} + 1 \text{ ft} = 5 \text{ ft}$$

The total head and pressure head at the center (midpoint) of the sand column are:

$$h_{\text{MID}} = h_{\text{ENT}} - i\Delta l = 6 \text{ ft} - 0.5(1 \text{ ft}) = 5.5 \text{ ft}$$
$$h_p = h_{\text{MID}} - h_z = 5.5 \text{ ft} - 3 \text{ ft} = 2.5 \text{ ft}$$

7.2.6.3 Seepage Velocity

Seepage velocity is the rate of movement of an element of water through soil. It is faster than Darcy's superficial velocity because the flow only occurs through the voids in the soil rather than the whole cross section.

$$v_s = \frac{ki}{n_e} = \frac{v}{n_e}$$ **Equation 7-26**

v_s = seepage velocity

v = average discharge velocity

k = hydraulic conductivity (coefficient of permeability)

i = hydraulic gradient

n_e = effective porosity (for sandy soils $n_e \approx n$)

7.2.6.4 Coefficient of Permeability Laboratory Tests

Based on the soil permeability level, two laboratory tests are typically used to estimate the coefficient of permeability of soils:

1. For high-permeability soils (with permeability greater than 10^{-3} cm/sec), such as sands and gravels, the constant head permeability test is typically used.
2. For low-permeability soils (with permeability less than 10^{-3} cm/sec), such as silts and clays, the falling or variable head permeability test is typically used.

Field permeability tests, also known as pumping tests, may also be used to estimate the coefficient of permeability for in-place soils. Field-pumping tests are extremely expensive, but the results are much more accurate and dependable compared with laboratory tests.

Constant head permeability test:

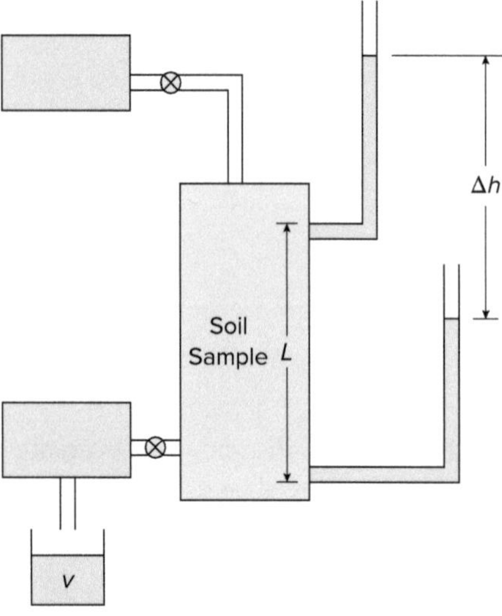

FIGURE 7.18 Constant Head Sketch

The coefficient of permeability via constant head can be estimated:

$$k = \frac{VL}{\Delta hAt}$$ **Equation 7-27**

V = volume of water

L = length of specimen between piezometers

A = cross-sectional area of specimen

t = duration of water collection

Δh = head difference

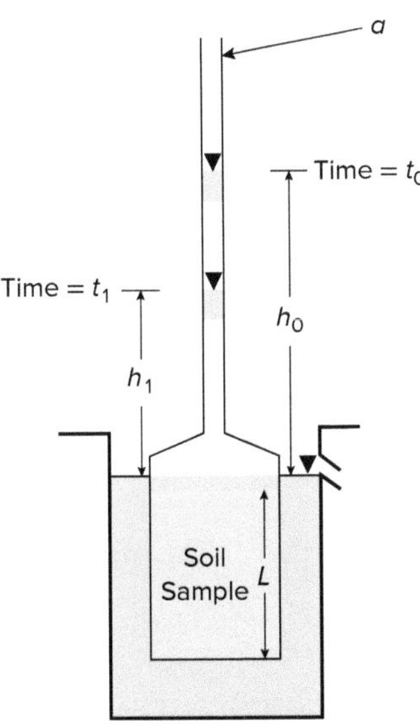

FIGURE 7.19 Falling Head Permeability Test

The coefficient of permeability via falling head can be estimated as shown in Equation 7-28:

$$k = \left(2.303\frac{aL}{At}\log_{10}\frac{h_0}{h_1}\right) = \left(\frac{aL}{At}\ln\frac{h_0}{h_1}\right) \quad \textbf{Equation 7-28}$$

h_0 = head at the start of the test (t_0)

h_1 = head at the end of the test (t_1)

L = length of specimen

A = cross-sectional area of specimen

a = cross-sectional area of standpipe

$t = (t_1 - t_0)$: duration of time when water level falls from h_0 to h_1

7.2.6.5 Coefficient of Permeability from Empirical Relationships

There are many available empirical relationships used to estimate the coefficient of permeability based on some basic soil index parameters such as void ratio, specific gravity, porosity, unit weight, and effective grain size. The following relationship may be used for natural, uniform sands and gravels and for nonplastic silty sands:

$$k(\text{cm/s}) = 2.4622\left[(D_{10}\ \text{mm})^2\frac{e^3}{1+e}\right]^{0.7825} \quad \textbf{Equation 7-29}$$

D_{10} = effective particle size from the grain size distribution curve

e = void ratio

7.2.6.6 Flow Nets

Seepage or flow through soil is typically two dimensional and sometimes three dimensional, which cannot be described by Darcy's law. Rather, Laplace's equation, which represents energy loss through a resistive medium similar to the flow through soil, is utilized to describe the flow through soil. A **flow net** is a simple two-dimensional graphical solution of Laplace's equation for flow through soils. A flow net is a combination of flow lines and equipotential lines. A **flow line** is a line along which a water particle travels from upstream to downstream. There is no flow along equipotential lines, which are 90° normal to flow lines. The total head along an equipotential line is equal at all points.

Rules of flow net construction:

1. Flow lines cannot cross other flow lines.
2. Equipotential lines cannot cross other equipotential lines.
3. Equipotential lines intersect the flow lines at right angles (perpendicular to each other).
4. Flow net must be constructed such that each element, bounded by two flow lines and two equipotential lines, is a curvilinear square (sides may be curved, but a circle must be inscribed within it that touches all four of its sides).

If all of the above conditions are satisfied, the flow net may be used to estimate the total flow rate, as shown in Equation 7-30:

$$Q = k\Delta h \frac{N_f}{N_d} L$$ **Equation 7-30**

Q = total flow rate (volume per time)

N_f = number of flow channels in a flow net (not number of flow lines)

N_d = number of potential drops

Δh = head difference between upstream and downstream

k = coefficient of permeability

L = length of structure (perpendicular or into the page)

Example 7.11: Flow Nets

For the flow net below, determine the head loss at points A, B, and C. The structure is approximately 100 ft long. Determine the flow rate in ft^3/min through the permeable layer. Assume hydraulic conductivity $k = 1.64 \times 10^{-4}$ ft/min.

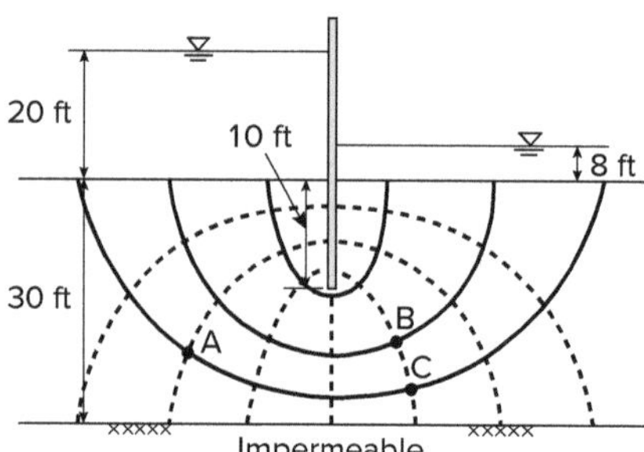

Solution

A flow channel is bounded between two flow lines. The first flow line is the sheet pile wall. The last flow line is the impermeable layer. There are three additional flow lines in between the first and last flow lines for a total of five flow lines, creating four flow channels. Therefore, in this example, $N_f = 4$. Moving from one equipotentional line to the next signifies a drop in head. The first equipotentional line is the upstream channel bed and the last is the downstream bed. From upstream to downstream, there are eight drops; therefore, $N_d = 8$.

$N_f = 4$ and $N_d = 8$

Total head loss = 20 ft – 8 ft = 12 ft

Head loss per drop $= \dfrac{12 \text{ ft}}{8 \text{ ft}} = 1.5$ ft

Head loss at point A = 2 ft × 1.5 ft = 3 ft

Example 7.11 *(continued)*

Head loss at point B = 5 ft × 1.5 ft = 7.5 ft

Head loss at point C = 5 ft × 1.5 ft = 7.5 ft (same as A; both are located on the same equipotential line)

Calculate seepage:

$$Q = \frac{k\,\Delta h\,N_f}{N_d\,L} = \frac{1.6 \times 10^{-4} \times 12 \text{ ft} \times 4}{8 \times 100 \text{ ft}} = 0.0984 \text{ ft}^3/\text{min}$$

7.2.7 Compressibility

Soil compressibility is one of the main concepts that defines soil mechanics as a subdiscipline of civil engineering. Total settlement of soil consists of elastic settlement, consolidation settlement, and secondary settlement.

7.2.7.1 Elastic or Primary Settlement

Elastic or primary settlement is the main type of settlement in cohesionless soil. It occurs relatively fast and typically by the time the structure is topped out. The soil modulus of elasticity (E) is needed to estimate the soil elastic settlement. The modulus of elasticity is the ratio between stress and strain for elastic material; however, for soils it is dependent on the strain level. It can be estimated from empirical correlations to field tests such as SPT, CPT, DMT, and PMT (see Table 7.2). It can also be estimated from laboratory tests such as direct shear, consolidation, or triaxial tests.

7.2.7.2 Consolidation Settlement

Consolidation settlement is the main type of settlement in clays. It occurs over long periods of time that might extend to months or years due to slow transfer of the excess pressure from the pore water to the clay particles. Clays are either normally consolidated (NC) clay, in which the present effective overburden pressure (σ_0') is the maximum pressure the soil has been subjected to in the recent past, or overconsolidated (OC) clay, in which the present effective overburden pressure is less than what the soil has seen in the recent past. The past maximum effective overburden pressure is called the preconsolidation pressure σ_c'.

For normally consolidated (NC) soils: $\sigma_0' \approx \sigma_c'$

For overconsolidated soils (OC): $\sigma_0' < \sigma_c'$

σ_0' = initial (present) effective overburden pressure

σ_c' = preconsolidation pressure

The overconsolidation ratio (OCR) is defined as follows: $\text{OCR} = \frac{\sigma_c'}{\sigma_0'}$ **Equation 7-31**

Compression indices are used to estimate consolidation settlement based on the condition of the clays being NC or OC. For NC clays, the compression index (C_c) is needed. For OC clays, the recompression index (C_r) is needed. However, for problems where heave or swelling is estimated, the swelling index (C_s) is needed.

Figure 7.20 shows that the compression index is defined as the slope of the virgin compression curve (the part of the curve between B and C); the recompression index is defined as the slope of the recompression curve (the part of the curve between A and B); and the swelling index is defined as the slope of the unloading curve (the part of the curve between C and D).

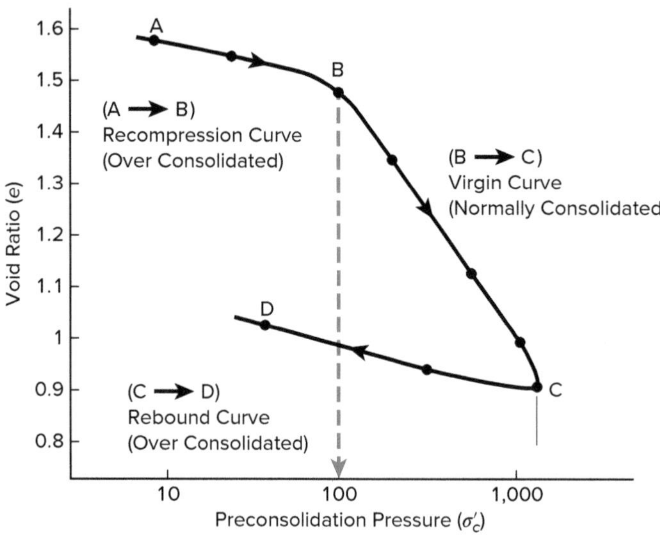

FIGURE 7.20 Typical One-Dimensional Consolidation Test

If consolidation tests were not performed, the compression and recompression indices could be estimated from the following empirical relationships:

For all clays: $C_c = 1.15(e_0 - 0.35)$ and C_r = 5% to 10% of C_c

e_0 = initial void ratio

For undisturbed clay of low to moderate sensitivity: $C_c = 0.009$ (LL – 10)

To estimate the rate of consolidation, the coefficient of consolidation C_v, which constrains the material properties that govern the consolidation process, is needed. It is defined as follows:

$$C_v = \frac{k}{\rho_w g} m_v \qquad \textbf{Equation 7-32}$$

C_v = coefficient of consolidation (ft^2/day or length square per time)

k = coefficient of permeability of water

ρ_w = density of water

g = acceleration of gravity 32.2 (ft/s^2)

m_v = coefficient of compressibility (ft^2/lb or inverse of stress units)

7.3 CONCRETE MATERIALS

7.3.1 Unreinforced Concrete

The use of plain unreinforced concrete is outlined in ACI 318-14 [3] (chapter 14), which is limited to:

1. Members continuously supported by soil or by other structural members capable of providing continuous vertical support
2. Members for which arch action provides compression under all conditions of loading
3. Walls
4. Pedestals

Note that structural plain concrete is not permitted in column design and pile caps.

7.3.2 Reinforced Concrete Mix Design

Reinforced concrete mix design encompasses several topics, but focuses mainly on the various types of portland cement that can be added to a mix, as well as the various possible admixtures. There are five types of portland cement that can be added to a concrete mix.

TIP

See Appendix A.8 for a chart summarizing the types of portland cement.

Type I relates to normal portland cement for general use where the heat of hydration is not an issue.

Type II is a modified portland cement that provides a moderate sulfate resistance and is used for mass concrete applications such as a gravity dam or mat foundation for a large building.

A sulfate attack on concrete causes the concrete paste around the coarse aggregate to break down and form what is known as **ettringite**. This can be seen in the field from leaching through the finished concrete surface, which is an indication of the chemical process that has occurred at the microscopic level. In the worst case, the concrete can break off, exposing the reinforcing steel to the elements and resulting in possible section loss. A sulfate attack on concrete is an issue with permeable mixes. Providing a different portland cement mixture and the use of a pozzolanic admixture such as fly ash can reduce the sulfate ion used in the concrete mix.

Type III is known as high-early concrete, which achieves concrete strength (f'_c) in several days compared with the more typical 28 days required for other types of concrete. Type III is commonly used in civil applications for roadways and bridges where a major corridor must remain open to traffic, and shutting down lanes for weeks at a time is unacceptable. However, the downside of high-early concrete is that the mix cures at a rapid rate, which can result in extensive cracking in the finished concrete.

Type IV portland cement results in a lower heat of hydration and is ideal for larger mass concrete applications. Gravity dams are a good example, where the dominant feature of the concrete is not necessarily the strength at 28 days, but rather the unit weight of 150 lb/ft^3.

Type V provides a higher sulfate resistance compared with Type II. It is used where high-alkaline soils are anticipated.

7.3.3 Extreme Temperature Mix Design

Extreme temperature placement of concrete is an area of concern for engineers. In an extremely hot or arid environment, the risk is the possible evaporation of the water in the mix prior to hardening. If water evaporates, then the water-to-cement ratio will be altered, and the 28-day strength will change as well. To combat this, there are several methods to ensure that the water stays in the liquid phase. The initial use of ice in the mix is permitted, as long as the ice used is a very finely shaved consistency, such that the temperature is reduced, but it will quickly change phases back to a liquid form. If larger ice chunks are used, the change of phases will be more difficult and would result in pockets of ice in the mix, which would be undesirable. Another method to combat extreme heat is to provide a sprinkler system with burlap and plastic tarps over the concrete as it cures. In bridge deck casting, most state departments of transportation require a sprinkler system to ensure saturation of the deck and to prevent any possible evaporation of the water in the mix. Casting at night can also reduce the rise in the initial temperature of the mix, compared with temperature change while working in the heat of the day.

Extreme cold-temperature concrete casting results in the water turning to ice, which is also undesirable. To combat this, the coarse aggregate could be heated prior to casting. Alternatively, the formwork can be insulated, thermal blankets can be used, or the project site can be enclosed with portable heaters. In each scenario, the objective is the same: to help increase the temperature of the concrete as it cures and to deter the outside environment from encroaching on the work site.

7.3.4 Concrete Admixtures

TIP

Refer to Appendix A.7 for a quick-reference admixtures chart.

Concrete strength (f'_c) is directly related to the water-to-cement ratio. For a concrete mix with a low slump value, and thus a very stiff mix, the typical response is to add water to the mix for improved flowability or workability. However, this will change the water-to-cement ratio and reduce the strength of the concrete and its long-term durability. As an alternative, certain **admixtures** can be provided in the mix design that will improve workability without changing the water-to-cement ratio.

There are admixtures that relate to specific niche markets within civil engineering, including anti-wash-out admixtures for pouring concrete under water and other admixtures that are chosen for their bonding or grouting properties.

Retarder admixtures are useful in high-temperature applications, where the mix sets up faster than usual. The retarder reduces the initial setup of the concrete and permits longer workability for finishing. Many retarder admixtures also provide water-reducing properties. Some examples of retarder materials include calcium sulfate, gypsum, sugar, starch, and salts of acids.

Air-entrainment admixtures provide additional air bubbles throughout the mix as the concrete cures and provide increased durability when subjected to freeze-thaw temperature cycles. The workability of the mix is improved. Air entrainment reduces segregation and bleeding of the mix. Segregation occurs when the larger and smaller particles separate based on weight, and bleeding refers to water forming on the surface of freshly poured concrete.

Accelerator admixtures provide concrete strength f'_c in an increased time frame. They are useful for roadway work on heavily traveled corridors, often requiring only a few hours to reach initial strength. The admixtures increase the rate of hydration, but may result in cracking of the concrete as it begins to dry out. Calcium chloride is a commonly used accelerator, but it may not be used in prestressed applications or with aluminum or galvanized metals. Silica fume and calcium formate are other examples of accelerators that are commonly used.

Water-reducing admixtures require less water for the concrete and increase the workability of the mix. They are categorized into three different types: plasticizers, midrange plasticizers, and superplasticizers, which reduce the water demand by about 10%, 15%, and 30%, respectively. The concrete mix also exhibits additional strength with decreased permeability and shrinkage. This results in a stronger mix that can lead to reduced cracking.

Corrosion-inhibiting admixtures are used in locations exposed to water with high salinity or in areas with high chloride concentrations. Industrial applications where abrasive chemicals are present may benefit from corrosion-inhibiting admixtures that will protect the reinforcing steel in the structure. Sodium nitrate, sodium nitrite, and sodium benzoate are common types of corrosion-inhibiting admixtures.

Pozzolanic admixtures are composed of very fine microscopic particles, compared with typical cement particles. They fill in the smaller gaps in the concrete material matrix, resulting in a mix that is less permeable and thus less susceptible to chemical infiltration. Also, this concrete mix is much more water resistant, reduces the heat of hydration during curing, and results in less shrinkage. There are natural and synthetic pozzolanic materials. Naturally occurring examples include clay, pumicite, and shale, whereas synthetic examples include silica fume and fly ash. Silica fume—a byproduct of the electric furnace process—is composed of noncrystalline silicon dioxide (SiO_2), and fly ash is a byproduct of burning coal in electric power plants.

7.3.5 Concrete Testing

Many materials are used for the construction of engineering elements. To appropriately assess material conformance to project specifications, certain material test methods are utilized. Concrete is considered one of the primary building materials in construction because of its durability, longevity, and accessibility. Its composition can vary due to scope, location, and placement requirements.

The most common concrete test completed at a jobsite is the slump test, which provides a measure of how easily the concrete flows but is not a measure of strength. The test is performed with a thin-walled cylindrical cone that is anchored to a rigid baseplate, which is typically metallic. Concrete is added in lifts to account for one-third of the total, and between each lift, the concrete is rodded 25 times just within the lift height under investigation. This provides a level of compaction in the concrete and is mandated by an ASTM International standard. Once the concrete has reached the top of the cone, the baseplate is released, and the steel cone is removed from the concrete within three to five seconds. Too fast or two slow of a lift may alter the results of the test. Keep in mind that this test is more for workability of the mix, that is, to get it into the formwork and to vibrate it into smaller areas in order to achieve a finish that is free of imperfections. Once the steel cone is removed, the concrete will spread out laterally and move down vertically. The difference in height from the original location to the final one is the slump value. A range of three to five inches provides good workability with the required design compressive strength.

7.3.5.1 Compression Testing

Concrete cylinders are used to determine compression strength at 28 days, f'_c, with either a 6-in-diameter × 12-in-tall cylinder or a 4-in-diameter × 8-in-tall cylinder. ACI 318-14 designates a strength test consisting of a minimum of two 6-in-diameter cylinders or three 4-in-diameter cylinders. The required number of strength tests is based on the class of concrete for a project. For example, in bridge design, there are different classes for substructure concrete in the abutment compared with superstructure concrete in the deck or driving surface. As such, ACI 318-14 requires samples for each class of concrete placed each day: they cannot be taken less than once a day, nor less than once for each 150 yd^3, nor less than once for each 5,000 ft^2 of surface area for slabs or walls. There are a few additional requirements as outlined in ACI 318-14, such as if the total volume poured results in less than five strength tests per class of concrete, tests shall be made from at least five randomly selected batches or from each batch if fewer than five are used.

Example 7.12: Compressive Strength of Concrete

A 4-in-diameter × 8-in-tall mold was used to cast a cylinder. After 28 days, the specimen failed at an axial compressive force of 63,500 lbf. Determine the compressive strength of the concrete specimen.

Solution

To determine the compressive strength of concrete, divide the axial compressive force that resulted in the failure (fracture) of the specimen by the cross-sectional area given in the relationship in Equation 7-33.

$$f'_c = \frac{P}{A} \qquad \textbf{Equation 7-33}$$

Example 7.12 *(continued)*

$$f'_c = \frac{\text{axial compressive force}}{\text{cross-sectional area of cylinder}} = \frac{63{,}500 \text{ lbf}}{(\pi \times 4^2)/4}$$

$$f'_c = 5{,}055.73 \text{ psi}$$

7.3.5.2 Tension Testing

Tension tests for concrete are used with the larger 6-in-diameter cylinders placed horizontally, as shown in Figure 7.21. Axial compression is applied, and a vertical crack will propagate through the length of the specimen. While the tension capacity of concrete is very small (approximately 10% of f'_c), it is utilized in several equations in ACI 318-14.

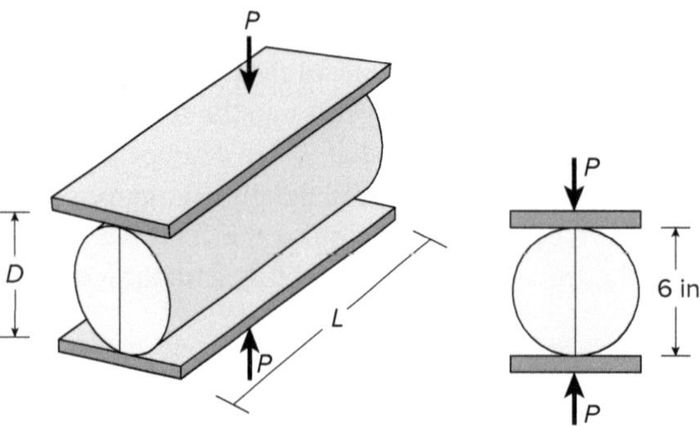

FIGURE 7.21 Tension Testing of Concrete Cylinders

Example 7.13: Tensile Strength of Concrete

A 6-in-diameter × 12-in-long specimen has a tensile strength of 510 psi. Use Equation 7-34 to determine the load that would cause the cylinder to split.

Solution

$$f_{ct} = \frac{2P}{\pi DL}$$ **Equation 7-34**

$$510 \text{ psi} = \frac{2 \times P}{\pi \times 6 \text{ in} \times 12 \text{ in}}$$

Solve for P.

$$P = 57{,}679.64 \text{ lbf}$$

7.3.5.3 Modulus of Rupture Testing

The modulus of rupture is an important property of concrete in the progression of stress in a concrete beam after curing. The modulus of rupture indicates the stress associated with concrete cracking at the extreme fiber of an element in tension. This is important for loading a concrete beam prior to cracking, such that the stresses through the cross section are the same for the concrete and steel in tension. The tension force is not taken by the steel alone until the concrete reaches the modulus of rupture. The value for the modulus of rupture (or concrete tensile strength) is fairly small—often a few hundred psi. It is shown in Equation 7-35, with the note that the value of f'_c must be in psi.

$$f_r = 7.5 \times \sqrt{f'_c}$$ **Equation 7-35**

The test for modulus of rupture is conducted on a 6- × 6-in square cross section measuring 18 in long. The specimen is subjected to four-point bending, such that the center third of the beam results in a constant moment, as shown in Figure 7.22.

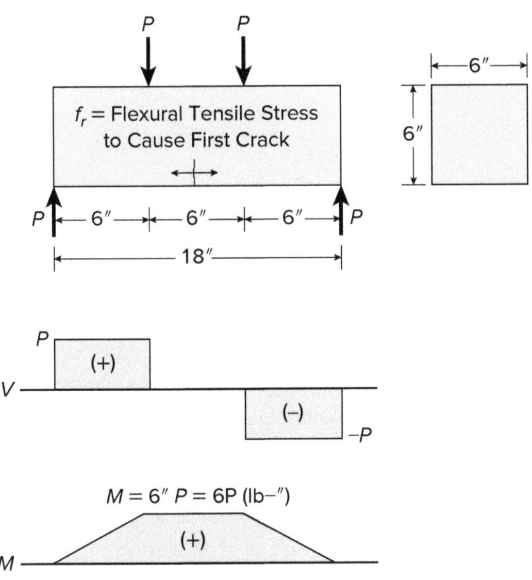

FIGURE 7.22 Modulus of Rupture Testing Specimen and Moment Diagram

7.3.6 Water-to-Cement Ratio

As mentioned previously, workability is essential to the proper placement of concrete, but increasing the amount of water to achieve more workability compromises (reduces) concrete strength. A lower ratio (more cement than water) implies that the mix can have a higher compressive strength. Antithetically, more water or a higher **water-to-cement ratio** will have a lower compressive strength in comparison. A water-to-cement ratio can be presented by a weight ratio: divide all of the water content (free water and water content associated with aggregates) by all of the weight of cementitious material (cement plus any pozzolans, fly ash, blast furnace slag, and so on).

$$w/c = \frac{\text{water content (lb)}}{\text{cement (lb)} + \text{pozzolan (lb)}}$$ **Equation 7-36**

The water-to-cement ratio can also be presented as gallons of water per sack of cement (for instance, 5 gal/sack).

Example 7.14: Water-to-Cement Ratio

For a water-to-cement ratio of 0.45, how much cement mix in pounds is permitted if the total amount of water is 27 gallons? (There are 8.33 lb/gal of water.)

A. 500 lb
B. 200 lb
C. 417 lb
D. 60 lb

Example 7.14 (continued)

Solution

$$\text{Water-to-cement ratio} = \frac{\text{weight of water}}{\text{weight of mix}}$$

Weight of water = 27 gal × 8.33 lb/gal = 224.91 lb

$$\text{Water-to-cement ratio} = \frac{224.91 \text{ lb (water)}}{\times \text{ lb (mix)}} = 0.45$$

$$\text{Cement mix} = \frac{224.91 \text{ lb (water)}}{0.45} = 500 \text{ lb}$$

Answer: A

7.3.7 Concrete Mix Design

A standard concrete mix design is generally composed of four ingredients: cement, fine aggregates, coarse aggregates, and water. Before beginning a mix design, all components of the concrete batch are measured either through weight or volume. Using a method of proportioning, indeterminate volumes or weights of specific ingredients can be determined. The most conventional example is this: a 1:2:3 weight ratio implies one part cement to two parts fine aggregates to three parts coarse aggregates. For example, in a 1-yd^3 yield, given the amount of cement and a proportion (1:2:3), the remaining ingredients—fine and coarse aggregates and free water—can be determined.

7.4 STRUCTURAL STEEL MATERIAL

The carbon content of mild and structural steel is a very small part of the overall makeup. Strength and hardness increase when more carbon is introduced, but the ductility decreases and the steel becomes more brittle. Also, the weldability of the material decreases with increasing amounts of carbon. The highest carbon steel produced commercially is reserved for metal-cutting tools and for some automotive applications.

7.4.1 Rolled Structural Steel

The steel shapes outlined in the American Institute of Steel Construction (AISC) *Steel Manual* [4] are referred to as rolled steel shapes. They start out as a solid mass of steel that is then forced through a series of rollers to obtain the final shape. There is a physical limit to the depth at which a rolled shape can be formed. The limit from the *Steel Manual* is a 44-in-deep beam. Beyond that limit, an I-beam called a built-up plate girder section must be formed from individual plates for the web and flange, which are then welded along the length of the beam to form the I shape.

7.4.2 Plate Steel

As noted above, a steel plate girder utilizes individual steel plates for the top and bottom flange as well as the web section. The important item to recall for a steel plate girder is that local buckling or compactness can be an issue. Plate girders are also used for curved steel bridges, which are somewhat unique in construction.

7.4.3 Grades of Structural Steel

Table 7.8 illustrates the various grades of structural steel that are currently available. The industry standard for many decades was 36-ksi yield strength steel. Recently, 50-ksi yield strength steel has been more commercially available, and this is the current standard. The tensile strength of the steel (F_u) is used in several instances in the *Steel Manual*, including connection design and tensile member capacity.

TABLE 7.8 Structural Steel Grades

ASTM Designation	Grade	F_y (ksi) Yield	F_u (ksi) Tensile
A36	–	36	58–80
A572	42	42	50
	60	50	65
	60	60	75
	65	65	80
A588	50	50	70
A852	Plate steel	70	90–110
A514	Plate steel	90	90–130
A514	Plate steel	100	100–130
A992	Rolled shapes	50	65

Based on the above noted ASTM standards (www.astm.org).

TABLE 7.9 ASTM Standard Reinforcing Bars

Bar Size, no.	Nominal Diameter, in.	Nominal Area, in.^{2}	Nominal Weight, lb/ft
3	0.375	0.11	0.376
4	0.500	0.20	0.668
5	0.625	0.31	1.043
6	0.750	0.44	1.502
7	0.875	0.60	2.044
8	1.000	0.79	2.670
9	1.128	1.00	3.400
10	1.270	1.27	4.303
11	1.410	1.56	5.313
14	1.693	2.25	7.65
18	2.257	4.00	13.60

From *ACI 318-14: Building Code Requirements for Structural Concrete and Commentary*, 2014. Reprinted by permission of American Concrete Institute.

7.4.4 Structural Steel Testing

7.4.4.1 Uniaxial Tension Testing

The yield strength (F_y) and ultimate strength (F_u) of a steel specimen are found by utilizing a uniaxial tension test of a given cross-sectional area. The applied load on the sample is converted to a stress similar to the compression testing for concrete samples as P/A, where P is the axial tensile load and A is the cross-sectional area of the steel sample, or coupon.

7.4.4.2 Toughness Testing

The main test for strain energy absorption is referred to as the Charpy V-notch impact test. It consists of a 10-mm × 10-mm steel sample with a 2-mm-deep V-shaped notch on one face subjected to a high rate of strain from a moving mass. The test indicates how brittle or ductile a grade of steel is relative to a high strain event (see Table 7.8, for steel grades).

7.5 COMPACTION

Compaction is the densification of soil by reducing air in soil voids. Compaction is measured in dry unit weight (dry density).

7.5.1 Purposes of Compaction

- Reduce subsequent settlement under working loads
- Increase the shear strength of the soil
- Reduce the void ratio, which results in reducing seepage through soils

- Prevent buildup of water pressure: Deep soil compaction of loose saturated sands prevents the buildup of large pore water pressures that cause soil to liquefy during earthquakes.

7.5.2 Factors Affecting Compaction

- Water content of the soil. Compacting dry soil is very difficult due to friction between the particles. The water in very wet soils absorbs most of the compaction energy, making it hard to compact. An optimum moisture content is needed for the most efficient compaction.
- The type of soil being compacted: gravel, sand, silt, or clay
- The amount of compaction energy (lift thickness, number of roller passes, weight of roller, and compaction technique)
- Type of compaction equipment (smooth drum, sheepsfoot, pneumatic tire, vibrating or static, and so forth). Figure 7.23 provides a summary of available compaction equipment along with the corresponding soil type.
- Speed of application

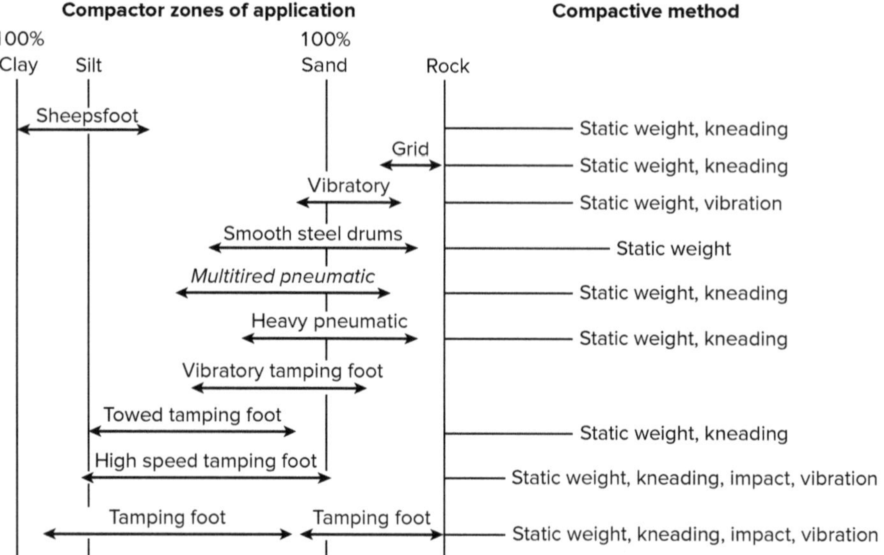

FIGURE 7.23 Compactors Recommended for Various Types of Soil and Rock

Source: *Part 645 Construction Inspection: National Engineering Handbook (210–VI–NEH, Amend. 7),* Natural Resources Conservation Service (USDA), 2015.

7.5.3 Compaction Testing

7.5.3.1 Proctor Laboratory Tests

Two tests are available to determine soil compaction properties: the standard Proctor test (ASTM D698) and the modified Proctor test (ASTM D1557). Proctor tests are performed on soil samples of the material to be compacted on site. Bag samples are brought to the laboratory and divided into four or five smaller samples, which are mixed with different amounts of water to produce samples with different moisture contents. The small samples are packed in layers in a standard mold using a standard hammer drop from a standard height. Figure 7.24 shows the resulting dry density of the samples with varying moisture contents. From Figure 7.24, the maximum dry density and optimum moisture contents can be estimated.

Note that the Proctor curve cannot plot above the no voids line, which is a plot of dry unit weight (γ_d) versus moisture content (w), at 100% saturation (S = 100%). At 100% saturation, the volume of air is zero ($V_a = 0$); therefore, there can be no further reduction in air voids by compaction.

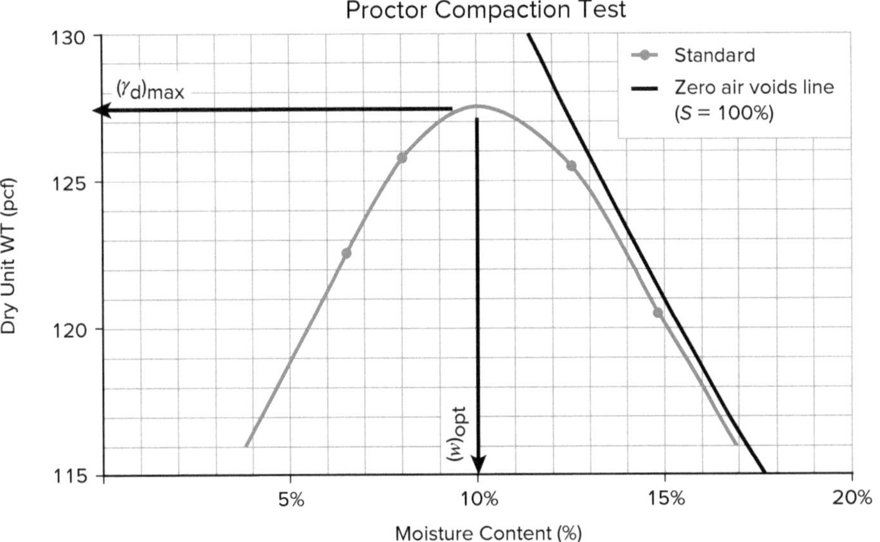

FIGURE 7.24 A Typical Compaction Test Proctor Curve

The main difference between the standard Proctor and the modified Proctor tests is the compaction energy applied during the laboratory tests:

$$\text{Theoretical Energy} = \frac{(\text{Wt})(\text{Drop})(\text{Blows})(\text{Layers})}{\text{Volume}} \qquad \textbf{Equation 7-37}$$

$$\text{Standard Proctor} = \frac{(5.5\text{ lb})(1\text{ ft})(25)(3)}{1/30\text{ ft}^3} = 12{,}375\text{ lb/ft}^2$$

$$\text{Modified Proctor} = \frac{(10\text{ lb})(1.5\text{ ft})(25)(5)}{(1/30\text{ ft}^3)} = 56{,}250\text{ lb/ft}^2$$

7.5.3.2 Field Density Tests

After the results of the Proctor tests are presented by the testing firm, the earth contractor starts field compaction according to the optimum moisture content and the target maximum dry density. After compacting a layer of fill, in-place field density and moisture content tests must be performed to assess the field compaction and move on to the next layer. Methods used to estimate the in-place field density and moisture content include the **rubber ballon method** (ASTM D2167), the **sand cone method** (ASTM D1556), the **nuclear density test** (ASTM D6938), and the **speedy moisture test** (ASTM D4944).

7.5.3.3 Sand Cone Method (ASTM D1556)

The sand cone method test is performed as shown in Figure 7.25. A 12- × 12-in plate with a 6-in-diameter hole is placed on top of the compacted soil. The plastic jar is filled with a standard sand (uniform sand with known density) and has a cone with calibrated volume. The soil is dug inside the hole to about 6 in deep, and the excavated soil is stored and weighed. The jar is set on top of the plate and the uniform sand is allowed to flow to replace the excavated soil volume. The jar is then weighed after replacement.

Known information:

Unit weight of dry uniform sand in jar (γ_{sand})

Volume of cone (V_{cone}) or weight of sand to fill cone (W_{cone})

Measured information:

Weight of jar with sand before test (W_o)

Weight of jar with sand after test (W_f)

Total weight of soil excavated from hole (W_{hole})

Moisture content of the soil excavated from the hole (w)

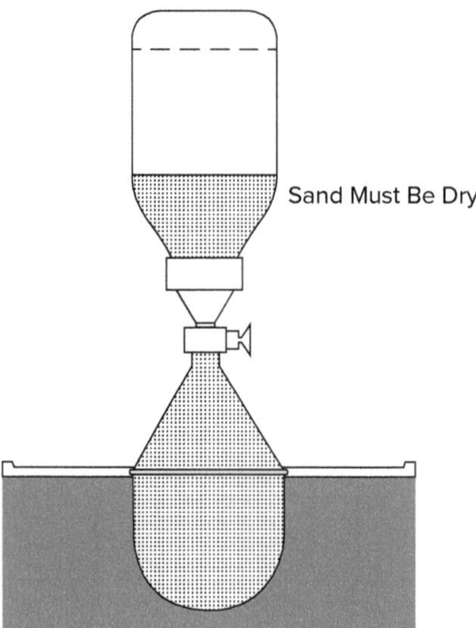

FIGURE 7.25 Sand Cone Test Method

Determine the dry unit weight of the excavated soil.

Solve for the volume of soil excavated from the hole:

$$V_{hole} = \frac{(W_o - W_f) - W_{cone}}{\gamma_{sand}} = \frac{(W_o - W_f)}{\gamma_{sand}} - V_{cone} \quad \textbf{Equation 7-38}$$

Solve for the dry unit weight of soil excavated from the hole:

$$\gamma_d = \frac{\gamma}{1 + w} = \frac{W_{hole}}{V_{hole}(1 + w)} \quad \textbf{Equation 7-39}$$

7.5.3.4 Nuclear Density Test (ASTM D6938)

The nuclear density test is a rapid, nondestructive testing method for in-place measurements of wet and dry density and water content of soil and soil aggregate through direct or backscatter methods (Fig. 7.26). This method is used for quality control and acceptance testing of compacted soil and soil-aggregate mixtures (that is, bituminous asphalt). This method is inappropriate for testing virgin subgrade.

This method assumes that the material is homogeneous, and therefore anomalies, such as irregular aggregate size, irregular voids, or deleterious materials (such as construction debris, heavy metals, organic soils, and so forth), can result in false readings.

Density assumes that the gamma ray (photon) will lose energy and rebound in a different direction as it interacts with the orbital electron. This is called Compton scattering.

Water content assumes that the hydrogen ions present in the soil and soil aggregate are in the form of water (H_2O).

Effect of frost: Moisture content will be exaggerated, and the dry density reading will be lower.

Effect of lime treatment: Lime treatment adds many hydroxyl ions to soil. If the nuclear density gauge is not properly recalibrated, it will interpret these as hydrogen atoms (that is, water), which will result in a falsely high water content reading and a falsely low dry density reading.

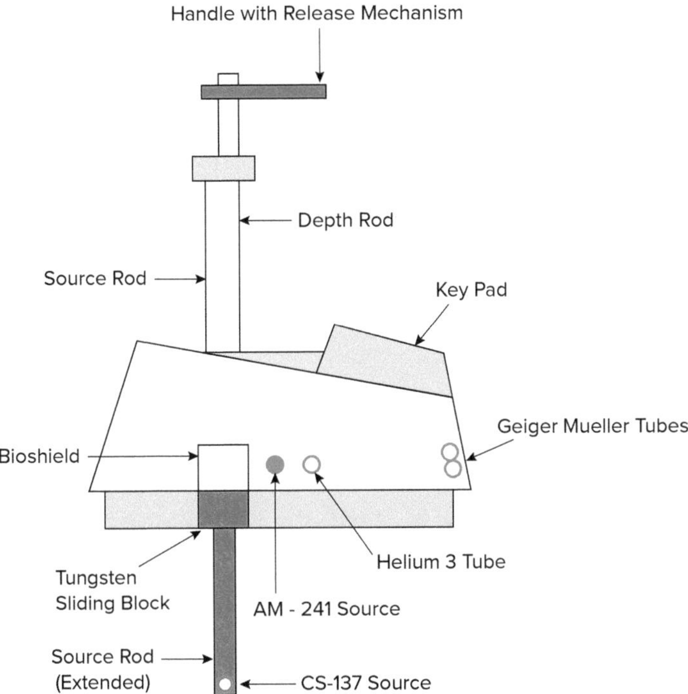

FIGURE 7.26 Typical Schematic of a Nuclear Density Gauge

Photo courtesy of the American Portable Nuclear Gauge Association.

7.5.4 Relative (Degree of) Compaction

Relative compaction is the ratio of the field dry unit weight to the maximum laboratory dry unit weight. Relative compaction is also known as the compaction specification that is typically included in the earthwork section of the geotechnical report.

$$RC = \frac{\text{field dry unit weight}}{\text{max dry unit weight}} = \frac{(\gamma_d)_{\text{field}}}{(\gamma_d)_{\max}} \times 100\% \qquad \textbf{Equation 7-40}$$

Example 7.15: Relative Compaction

A field sand cone test is performed on a layer of compacted fill. The results of the sand cone test are shown below. A laboratory Proctor test determined that the fill soil has a maximum dry unit weight of approximately 120 pcf. Determine the relative compaction of the fill.

Unit weight of dry uniform of sand, $\gamma_{\text{sand}} = 95\ \text{lb/ft}^3$

Weight of sand to fill cone, $W_{\text{cone}} = 1.10\ \text{lb}$

Weight of jar + cone + sand (before test), $W_o = 16.5\ \text{lb}$

Weight of jar + cone + sand (after test), $W_f = 10.5\ \text{lb}$

Weight of moist soil excavated from hole, $W_{\text{hole}} = 6.60\ \text{lb}$

Moisture content of soil from hole, $w = 10\%$

Solution

Calculate the volume of soil excavated from the hole:

$$V_{\text{hole}} = \frac{(W_o - W_f) - W_{\text{cone}}}{\gamma_{\text{sand}}} = \frac{16.5\ \text{lb} - 10.5\ \text{lb} - 1.10\ \text{lb}}{95\ \text{lb/ft}^3} = 0.0516\ \text{ft}^3$$

Materials

Example 7.15 *(continued)*

Calculate the dry unit weight of soil excavated from the hole:

$$\gamma_d = \frac{W_{\text{hole}}}{V_{\text{hole}}(1 + w)} = \frac{6.60 \text{ lb}}{(0.0516 \text{ ft}^3)(1 + 0.10)} = 116.3 \text{ lb/ft}^3$$

Calculate the relative compaction:

$$RC = \frac{(\gamma_d)_{\text{field}}}{(\gamma_d)_{\text{max}}} \times 100\% = \frac{116.3 \text{ lb/ft}^3}{120 \text{ lb/ft}^3} \times 100\% = 97\%$$

7.5.5 California Bearing Ratio (ASTM D1883)

The California Bearing Ratio (CBR) test is used only for flexible pavement design (bituminous). It is an empirical measure of mechanical subgrade strength compared with that of high-quality crushed rock (CBR = 100). The harder the surface, the higher the CBR rating.

The test is performed by measuring the pressure required to penetrate a soil sample with a plunger of standard area (3 in^2). In the laboratory, the sample is compacted into the CBR mold and is usually immersed in a water bath prior to testing. Load is applied to the plunger, and the applied stress and penetration for a series of loads are recorded and plotted.

The CBR rating is calculated for the stress at 0.1 and 0.2 in of penetration by the formulas below, where 1,000 psi and 1,500 psi are the standard stresses for 0.1 and 0.2 in of penetration, respectively:

$$\text{CBR} = \frac{\text{actual stress}}{\text{standard stress}} \times 100$$ **Equation 7-41**

$$\text{CBR}_{0.1} = \frac{\sigma_{0.1}}{1{,}000 \text{ psi}} \times 100$$

$$\text{CBR}_{0.2} = \frac{\sigma_{0.2}}{1{,}500 \text{ psi}} \times 100$$

The CBR rating is typically taken as the $\text{CBR}_{0.1}$ (in percent and rounded to the nearest whole number), where the $\text{CBR}_{0.2}$ is less than the $\text{CBR}_{0.1}$. Note that if $\text{CBR}_{0.2}$ is greater than $\text{CBR}_{0.1}$, the sample must be retested. If the retest results are consistent, then the CBR rating is taken as the $\text{CBR}_{0.2}$ (in percent and rounded to the nearest whole number).

Materials

7.5.6 Subgrade Reaction Modulus (*K*)

Rigid pavement slabs are designed based on the modulus of subgrade reaction. The subgrade modulus is determined using the plate bearing load test, which is performed in situ on compacted soil using a standard 12-in or 30-in diameter steel plate. The plate is loaded ten times. The corrected load verses the corrected deflection is graphed for the tenth repetition. The subgrade modulus is taken as the slope of the line of stress versus deflection.

7.5.7 Resilient Modulus

The resilient modulus (M_R) is a subgrade material stiffness test that is primarily used for the design of flexible pavements. A material's resilient modulus is an estimate of its modulus of elasticity (E), where the resilient modulus is stress divided by strain for rapidly repetitive applied loads (like those experienced by pavements).

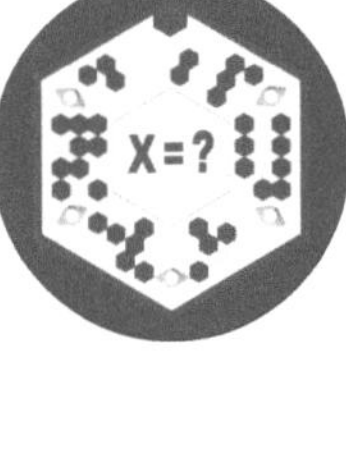

The M_R can be correlated using the widely used empirical relationship with the CBR value (applicable only for fine-grained soils with CBR $\leq$ 10):

$M_R(\text{psi}) = (1{,}500)\text{CBR}$ **Equation 7-42**

The AASHTO *Design Guide* [5] uses the following correlation:

$M_R(\text{psi}) = (2{,}555)\text{CBR}^{0.64}$ **Equation 7-43**

7.5.8 Expansive Soils

The expansive property of soil is related to the swell potential, which is a function of the clay minerals present in a given soil. Swelling occurs when water particles infiltrate in and between the clay minerals. Swell potential depends on the clay mineralogy; for example, the potential for swell-shrink for montmorillonite is more than for illite, which is more than for kaolinite. Foundations bearing on expansive soil will be subjected to heave, uplift of the foundation, and differential settlement problems.

Ground improvement techniques utilized to decrease heave include the following:

- Chemical soil stabilization (use of lime and cement)
- Installation of moisture barrier (prevention of water flow)
- Compaction of expansive soil (lower permeability)
- Removal of expansive soil under foundation

REFERENCES

1. Samtani, N. C.; Nowatzki, E. A. 2006. *Soils and Foundations Reference Manual (FHWA-NHI-06-088).* Washington, DC: US Dept. of Transportation, Federal Highway Administration.
2. Kulhawy, F. H.; Mayne, P. W. 1990. EL-6800 (Research Project 1493-6): *Manual on Estimating Soil Properties for Foundation Design*. Palo Alto, CA: EPRI.
3. American Concrete Institute (ACI). 2014. ACI 318-14: *Building Code Requirements for Structural Concrete and Commentary*. Farmington Hills, MI: ACI.
4. American Institute of Steel Construction (AISC). 2011. *Steel Construction Manual (14th ed.)*. Chicago, IL: AISC.
5. American Association of State Highway and Transportation Officials (AASHTO). 2011. *Roadside Design Guide (4th ed.)*. Washington, DC: AASHTO.

CHAPTER

Site Development

8

Timothy R. Sturges, PE

George Stankiewicz, PE

CONTENTS

CONTENTS (*continued*)

EXAM GUIDE

VIII. Site Development

A. Excavation and embankment (e.g., cut and fill)
B. Construction site layout and control
C. Temporary and permanent soil erosion and sediment control (e.g., construction erosion control and permits, sediment transport, channel/outlet protection)
D. Impact of construction on adjacent facilities
E. Safety (e.g., construction, roadside, work zone)

Approximate Number of Questions on Exam: 5

NCEES Principles & Practice of Engineering Examination, Civil Breadth Exam Specifications

COMMONLY USED ABBREVIATIONS

BV	bank volume
CFR	Code of Federal Regulations
CGP	construction general permit
CV	compacted volume
EDA	earth disturbed area
LDF	load factor
LV	loose volume
OSHA	Occupational Safety and Health Administration
PEL	permissible exposure limit
RC	relative compaction
SHF	shrinkage factor
SWF	swell factor
SWPPP	stormwater pollution prevention plan

8.1 EXCAVATION AND EMBANKMENT

Excavation and embankment are commonly used terms to describe earth-moving operations. **Excavation** refers to the act of removing earth (cut), and **embankment** refers to the act of placing earth (fill). These two operations are crucial to earth moving and site development. When given a particular elevation as the final product, the contractor has to determine how much material must be removed or excavated and how much material must be placed, within the project specifications. In this section, the relationships between excavation and embankment will be explored in detail, as well as the considerations necessary to properly and safely perform the work in the field.

8.1.1 Soil Properties

8.1.1.1 Density

Soil density is composed of three different elements: soil particles or solids, water, and air. As soil increases in air and water content, its density decreases. As air and water content in soil decreases, its density increases. This introduces the concept of **compaction**. As soil is compacted, and the air and water are removed, the soil's density increases.

Moisture content is a vital property of soil with relation to density and compaction. Moisture or water allows the soil particles to break apart and slide into allowable voids, thus removing air voids from the soil mass and increasing density. Too much moisture,

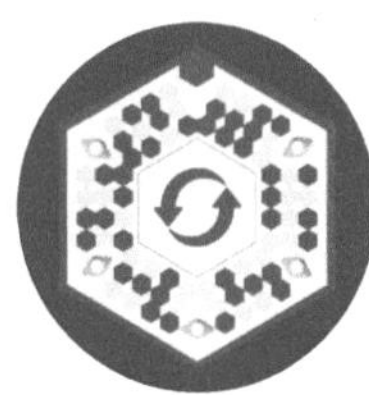

however, and the soil mass cannot be fully compacted, as it begins to turn into mud. At optimum moisture content, the soil will be able to be compacted to the maximum possible density for the compactive effort. This maximum compaction and optimum moisture percentage is determined using a laboratory Proctor test.

8.1.1.2 Swell/Shrinkage

Swell: percentage increase in soil volume when soil is excavated; measured in relation to bank volume or in situ volume

Swell factor (SWF): used for stockpiling and hauling, not in final earthwork balancing; the ratio of loose density to bank density

Shrinkage: percentage decrease in soil volume when soil is compacted; measured in relation to bank volume or in situ volume

Shrinkage factor (SHF): used for converting bank density to compacted density

Load factor (LDF): used to show the relationship between bank density and loose density; inverse of the swell factor

Loose volume (LV): volume of soil once it has been excavated, typically used for a temporary state. Excavation increases air voids and breaks up soil, leaving it in an uncompacted state; used for transportation in trucks, stockpiling, or any other temporary, uncompacted state

Bank volume (BV): volume of soil in an in situ, or untouched, state; soil taken from a borrow pit or that has been allowed to settle or compact over a long period of time

Compacted volume (CV): volume of soil once the final compactive effort has been applied; soil state when it has been placed and compacted to specifications (Tables 8.1 and 8.2)

TABLE 8.1 Swell Relationship Equations

SWELL DENSITY	SWELL VOLUME
$\text{Swell (\%)} = \left(\left(\frac{\text{Bank Density}}{\text{Loose Density}} - 1\right) \times 100\right)$	$V_{loose} = \left(\frac{100\% + \%\ \text{Swell}}{100\%}\right) \times V_{bank} = \frac{V_{bank}}{\text{Load Factor}}$
$\text{Load Factor} = \frac{\text{Loose Density}}{\text{Bank Density}}$	$\text{Load Factor} = (1 + \text{Decimal Swell})^{-1}$ or, $\text{Load Factor} = 1 \div (1 + \text{Decimal Swell})$
	Bank Volume = Loose Volume × Load Factor

TABLE 8.2 Shrinkage Relationship Equations

SHRINKAGE DENSITY	SHRINKAGE VOLUME
$\text{Shrinkage (\%)} = \left(\left(1 - \frac{\text{Bank Density}}{\text{Compacted Density}}\right) \times 100\right)$	$V_{compacted} = \left(\frac{100\% - \%\ \text{Shrinkage}}{100\%}\right) V_{bank}$
Shrinkage Factor = 1 – Shrinkage (% Decimal)	Compacted Volume = Bank Volume × Shrinkage Factor

The preceding can be applied to an example of an excavation operation involving clay with a bank density of 3,300 lb/yd^3 and a loose density of 2,700 lb/yd^3. If one ton of the soil is excavated from the initial borrow area, this mass would occupy 0.61 bank cubic yards (BV) in the ground.

$$\frac{2{,}000\text{ lb}}{3{,}300\text{ lb}} = 0.61\text{ BV}$$

When this same soil is hauled to be stockpiled, it occupies 0.74 loose cubic yards (LV).

$$\frac{2{,}000\text{ lb}}{2{,}700\text{ lb}} = 0.74\text{ LV}$$

The resultant decimal swell factor (SWF) is 0.22.

$$\text{SWF} = \frac{\text{BV}}{\text{LV}} - 1 \qquad \textbf{Equation 8-1}$$

$$= \frac{3{,}300 \text{ lb}}{2{,}700 \text{ lb}} - 1 = 0.22 \text{ SWF}$$

If this same clay is used to construct an embankment that, after compaction, occupies 0.55 yd^3, the shrink factor would be 0.90.

$$\text{SHF} = \frac{\text{CV}}{\text{BV}} \qquad \textbf{Equation 8-2}$$

$$= \frac{0.55 \text{ yd}^3}{0.61 \text{ yd}^3} = 0.90 \text{ SHF}$$

Using the above calculations to plan the earthwork operation, the contractor needs to haul 122 yd^3 for every 100 yd^3 that he removes from the borrow pit. Once this material is placed and compacted, the soil will only occupy 90 yd^3 in its final place.

Example 8.1: Swell/Shrinkage

During an excavation operation, the contractor came across a location within the borrow pit where the in situ soil conditions changed. Rather than excavating 1,000 yd^3 of clay, the contractor excavated 1,000 yd^3 of weathered shale and silt having an in situ density of 3,400 lb/yd^3 and a loose density of 2,700 lb/yd^3. One ton (2,000 lb) of this material will occupy, most nearly, how much volume in the in situ condition and how much volume in the stockpiled condition?

A. 0.72 yd^3 in situ, 0.75 yd^3 stockpiled
B. 0.59 yd^3 in situ, 0.74 yd^3 stockpiled
C. 0.59 yd^3 in situ, 0.75 yd^3 stockpiled
D. 0.62 yd^3 in situ, 0.91 yd^3 stockpiled

Solution

2,000 lb of the weathered shale-silt mix soil will occupy 0.59 yd^3 in the in situ condition (or bank), and 0.74 yd^3 in the stockpiled condition (loose).

$$\text{BV} = \frac{2{,}000 \text{ lb}}{3{,}400 \text{ lb/yd}^3} = 0.59 \text{ yd}^3$$

$$\text{LV} = \frac{2{,}000 \text{ lb}}{2{,}700 \text{ lb/yd}^3} = 0.74 \text{ yd}^3$$

Answer: B

Example 8.2: Swell/Shrinkage

20,000 yd^3 of in situ soil from a borrow pit is trucked to the jobsite. The soil has 22% swell and shrinkage of 15%. The final volume of the compacted soil is most nearly:

A. 24,900 yd^3
B. 24,400 yd^3
C. 18,700 yd^3
D. 17,000 yd^3

Example 8.2 *(continued)*

Solution

Shrinkage is measured with respect to the bank condition.

$$\text{Volume of compacted soil} = \left(\frac{100\% - 15\%}{100\%}\right)(20{,}000\ \text{yd}^3) = 17{,}000\ \text{yd}^3$$

Answer: D

Example 8.3: Swell/Shrinkage

A contractor was awarded a contract to excavate and haul 500,000 yd^3 of clay to cap a construction and demolition debris (C&DD) landfill. Job specifications require a final compaction of 95% of maximum dry density. The on-site geotechnical engineer has determined that the clay has a bulking factor of 22%. The contractor has elected to use dump trucks with a capacity of 26 yd^3. The time to load the trucks, travel to the site, unload, and travel back to the borrow site has been determined to be 45 minutes. The contract features an incentive payment of $15,000/day for each day completed prior to 80 calendar days, with a maximum incentive of $150,000. The number of dump trucks working 8 hours per day, 5 days per week that must be utilized to achieve the maximum incentive payment is most nearly:

A. 39 trucks
B. 35 trucks
C. 40 trucks
D. 43 trucks

Solution

1. Apply a bulking factor (swell) of 22% to the total volume.
2. 500,000 yd^3 × 1.22 = 610,000 yd^3 (volume to be trucked off site)
3. $150,000/$15,000 = 10 calendar days (working days to achieve maximum incentive)
   80 calendar days – 10 calendar days = 70 calendar days, or 10 workweeks
4. 5 days/week × 8 hr/day × 10 weeks = 400 work hours (working hours in 10 workweeks)
5. $\dfrac{610{,}000\ \text{yd}^3}{400\ \text{hr}} = 1{,}525\ \text{yd}^3/\text{hr}$ (required haulage per hour)
6. $\left[26\ \text{yd}^3/\text{truck}/\dfrac{45\text{–min/cycle}}{60\ \text{min/hr}}\right] = 39.0\ \text{yd}^3/\text{truck-hr}$ (total possible haul per truck per hour)
7. $\dfrac{1{,}525\ \text{yd}^3/\text{hr}}{39\ \text{yd}^3/\text{truck-hr}} = 39.1$ trucks (total trucks needed)
8. Remember that to receive the maximum incentive payment, the contract requires completion prior to 70 days. Since a fraction of a truck cannot be used, the answer must be rounded up or down. If 39 trucks are used, the total hauled in the 70-day period would only be 608,400 yd^3. This would result in the contractor failing to meet the maximum incentive requirements. The answer is 40 trucks.

Answer: C

8.1.1.3 Angle of Repose

The **angle of repose** (angle of internal friction) is the maximum slope angle at which a granular material, such as loose rock, soil, or sand, will stand and remain stable. Stockpiled soil will stand stable with a conical shape at this maximum angle. For soils, this can be estimated to be 30 degrees. When determining stockpile heights and area requirements, this angle must be checked to determine if the stockpile can be safely built within those parameters.

Example 8.4: Angle of Repose

An earthwork contractor must install a cofferdam and excavate for a bridge pier footer. This excavated material must be stockpiled on site for future backfill and embankment use. The inside dimensions of the cofferdam are 50 ft in length, 50 ft in width, and 15 ft in depth. The on-site stockpile area that is available limits the proposed stockpile to a 40-ft radius. The soil has a swell of 22% and an angle of repose at 30 degrees.

The stabilized height of the stockpile is most nearly:

A. 28 ft
B. 27 ft
C. 24 ft
D. 23 ft

Solution

1. Calculate the cubic volume of the excavation and add the swell to the soil volume:

   $50 \text{ ft} \times 50 \text{ ft} \times 15 \text{ ft} = 37{,}500 \text{ ft}^3 \times 1.22 \text{ (22\% swell)} = 45{,}750 \text{ ft}^3$

2. The stockpile, to remain stable, will need to be placed in a conical shape. Calculate the theoretical maximum stable height of the stockpile based on the angle of repose:

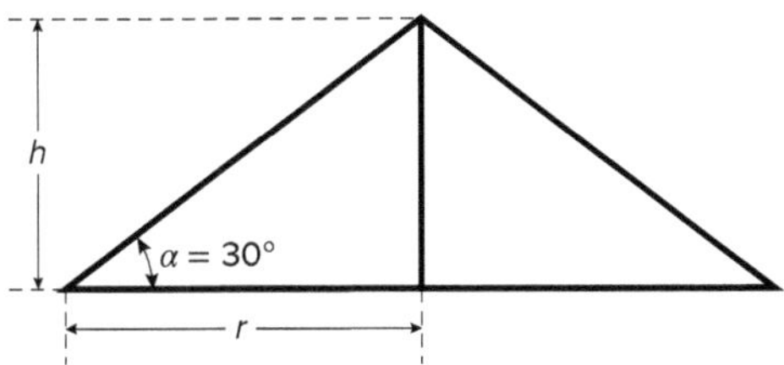

$$r = \frac{h}{\tan \alpha°}$$

$$40 \text{ ft} = \frac{h}{\tan 30°}$$

$$h = 23.1 \text{ ft}$$

3. Determine the required height of the stockpile to hold the excavated soil:

$$V = \frac{\pi r^2 h}{3};\ 45{,}750 \text{ ft}^3 = \frac{\pi (40 \text{ ft})^2 h}{3};\ h = 27.3 \text{ ft}$$

4. Although the stockpile will need to be 27.31 ft high to hold the required soil, the question asks for the stabilized height. Since the height of the stockpile will be higher than the calculated maximum stable height, 23.1 ft must be used. The answer is 23 ft.

Answer: D

TIP

If the calculated value for the stockpile height were below the theoretical maximum, that height would be stable and the calculated value would be the correct answer.

TIP

It should be noted that rounding up the answer would not be appropriate in this case because it would exceed the stabilized height; however, if the calculated required height for the stockpile were less than the theoretical max, typical rounding would be logical.

8.1.2 Excavation and Embankment Equipment

There are five main types of equipment to compact soil. This equipment presses or vibrates the soil to fill available air voids and squeeze out water. These five types are used for different purposes.

8.1.2.1 Smooth-Drum Vibratory Rollers

Smooth-drum vibratory rollers are typically used to finish the compaction of sandy or clayey soils, or compact granular soils or granular aggregate (Fig. 8.1). The vibration is achieved by using off-center weights. This piece of equipment can also be used with the vibration turned off, to provide compactive effort over shallow utility lines or shallow culverts that cannot be vibrated.

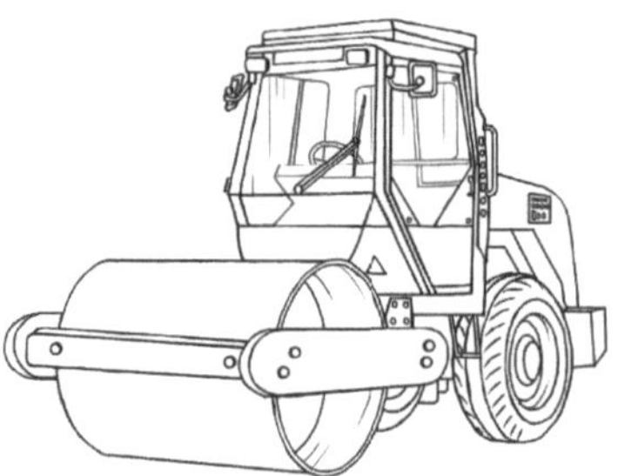

FIGURE 8.1 Smooth-Drum Vibratory Roller

8.1.2.2 Pneumatic Rubber-Tire Rollers

Pneumatic rubber-tire rollers use a combination of pressure and a kneading action to compact sandy and clayey soils (Fig. 8.2). While not the most common option used, they can provide excellent compactive effort.

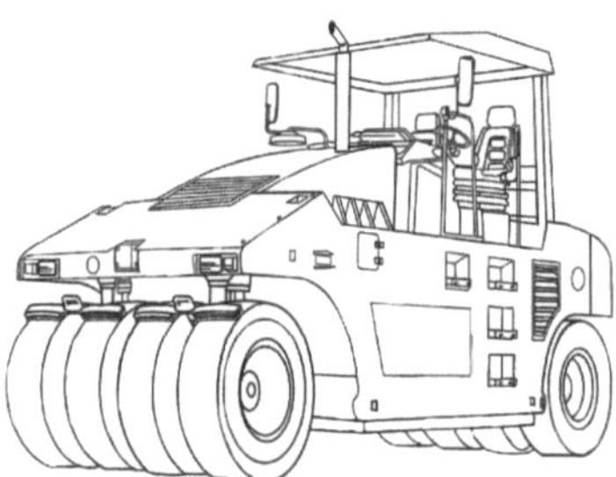

FIGURE 8.2 Pneumatic Rubber-Tire Roller

8.1.2.3 Sheepsfoot Rollers

Sheepsfoot rollers are the best option for compacting clayey soils as part of a roller train (Fig. 8.3). The roller drums have a series of projections that continuously knead and push the soil. The use of these rollers allows greater lift thicknesses to be properly compacted. After the first few passes with the sheepsfoot roller, a smooth-drum vibratory roller is commonly used to seal the surface, as the sheepsfoot will leave small depressions.

FIGURE 8.3 Sheepsfoot Roller

8.1.2.4 Hand Compaction Efforts

Hand compaction is the least effective method for imparting compactive effort to soils; however, it is appropriate to use when trying to provide the best compaction possible in small areas and against adjacent facilities, such as building foundation walls. Hand compaction equipment includes plate tampers, hand rollers, and jumping jacks (Fig. 8.4). There are also commercially available trench rollers that feature sheepsfoot or smooth-drummed options for narrow or small areas.

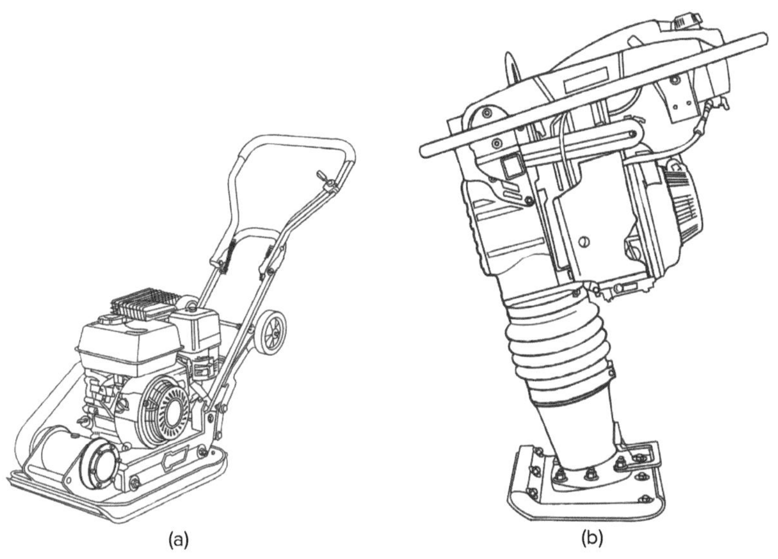

FIGURE 8.4 Plate Tamper, Jumping Jack

8.1.2.5 Hoe-Packs

Hoe-packs are vibratory plate attachments for excavators (Fig. 8.5). These plates connect to the hydraulic system of the excavator to provide compactive effort in small, inaccessible areas.

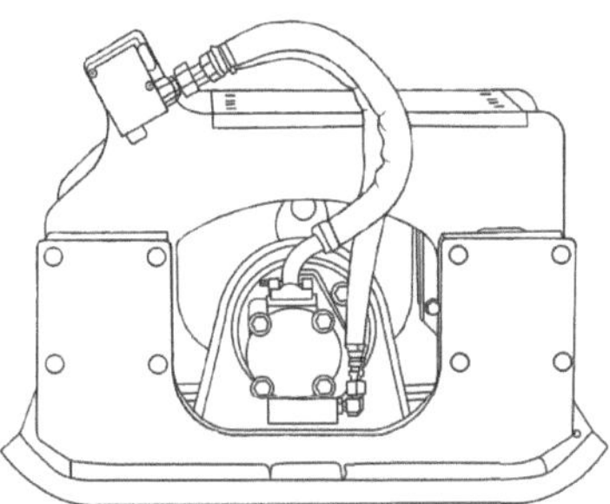

FIGURE 8.5 Hoe-Pack

8.1.3 Specifications

8.1.3.1 Relative Compaction

Relative compaction is the relationship between the initial density and the maximum density; it is the percentage ratio of these states of compaction. When determining the additional compactive effort necessary, by determining the relative compaction, it can be shown how much additional material will be necessary to meet the required density. The relative compaction percentage is the in situ dry density percentage divided by the maximum laboratory dry density, as determined by a Proctor test.

$$RC = \left(\frac{\text{In Situ Dry Density}}{\text{Laboratory Maximum Dry Density}}\right) \times 100\%$$ **Equation 8-3**

$$\text{Dry Unit Density} = \frac{\text{Total Unit Weight In Situ}}{(1 + \text{Water Content})}$$ **Equation 8-4**

Example 8.5: Relative Compaction

Project specifications require the compaction of a soil to be 95% of maximum dry density (modified Proctor). Construction of a highway embankment requires 5,000 yd^3 of fill. The on-site geotechnical engineer has determined that the borrow soil has an in situ density of 100.2 lb/ft^3 with 4% water content and a laboratory maximum dry density of 122.5 lb/ft^3. The total volume of soil that must be excavated from the borrow site is most nearly:

A. 5,000 yd^3
B. 10,000 yd^3
C. 4,750 yd^3
D. 6,043 yd^3

Solution

1. Determine the in situ dry density for comparison to the maximum laboratory dry density:

$$\text{Dry Unit Density} = \frac{\text{Total Unit Weight In Situ}}{(1 + \text{Water Content})}$$

$$= \frac{100.2\ \text{lb/ft}^3}{(1 + .04)} = 96.3\ \text{lb/ft}^3$$

2. Determine the relative compaction of the in situ material:

$$RC = \left(\frac{\text{In Situ Dry Density}}{\text{Laboratory Maximum Dry Density}}\right) \times 100\%$$

$$= \frac{(96.3\ \text{lb/ft}^3)}{(122.5\ \text{lb/ft}^3)} \times 100\% = 78.6\%$$

Based on the fact that the relative compaction is less than the required compaction percentage, additional material will be necessary to meet the project specifications.

3. Determine the amount of material required to complete the fill at the project-specified compaction percentage:

$$\text{Amt of Fill} = \frac{(\text{Fill Required})(\text{Required Compaction \%})}{\text{Relative Compaction}}$$ **Equation 8-5**

$$= \frac{(5{,}000\ \text{yd}^3)(0.95)}{(0.786)} = 6{,}043\ \text{yd}^3$$

Project specifications often are reported in terms of percent of maximum density. For example, if the maximum density, as determined via modified Proctor, is 122.5 lb/ft^3, a 95% compaction specification would require a final in-place compacted dry density of 116.4 lb/ft^3. The in-place compaction percent is typically derived with the use of a nuclear gauge.

Answer: D

8.1.4 Quantity Estimation

There are several ways that earthwork quantities are estimated. Although three-dimensional models are becoming more popular, most owners still develop plans that compute estimated earthwork quantities based on cross sections cut at regular intervals along the project centerline. The most common method for horizontal roadwork or other long distances is the **average end area method**. **Grid estimation calculations** are performed on smaller, more spread-out areas, typically for commercial and residential building sites.

8.1.4.1 Average End Area Method

The end area method takes the cross-sectional area of cuts and fills at specific stations and averages them along the distance to the next cross section to determine an estimated amount of earthwork necessary to meet the plans. When there are cuts and fills that come to zero, or near zero, at a specific station, a more accurate way of estimating this quantity is to assume a pyramid, that is, that at some point along that area, the quantity will come to zero (or near zero), and not necessarily directly at the station.

When cuts and fills are not zero at both cross sections:

$$V = \frac{L(A_1 + A_2)}{2}$$ **Equation 8-6**

When cuts or fills come to zero at one of the cross sections:

$$V = \frac{L(A_{\text{base}})}{3}$$ **Equation 8-7**

Example 8.6: Average End Area Method

Using the information given in the figure below, the volume of the embankment in cubic yards is most nearly:

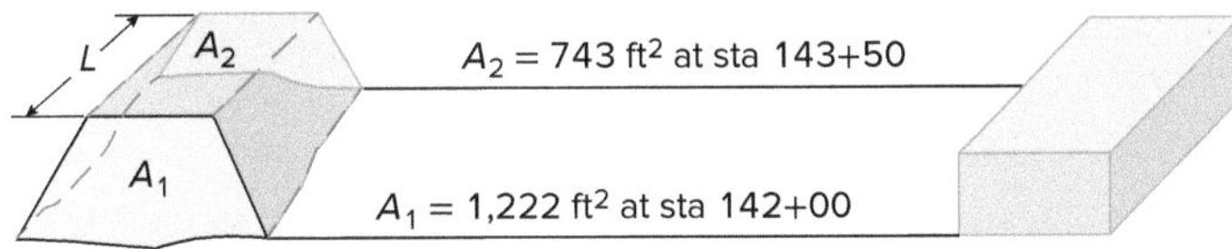

A. 7,500 yd^3
B. 5,500 yd^3
C. 10,000 yd^3
D. 2,850 yd^3

Solution

1. Use the equation for end area (Equation 8-6) since neither of the areas is zero, or close to zero. Remember that a station is 100 ft.
2. $L = 14{,}350\text{ ft} - 14{,}200\text{ ft} = 150\text{ ft}$
3. $V = \dfrac{150\text{ ft}\,(1{,}222\text{ ft}^2 + 743\text{ ft}^2)}{2} = 147{,}375\text{ ft}^3$
4. $= \dfrac{147{,}375\text{ ft}^3}{27\text{ ft}^3/\text{yd}^3} = 5{,}458\text{ yd}^3$

Answer: B

School of PE

Site Development

Example 8.7: Average End Area Method

For the cross-section areas listed in the table, answer the following questions.

Station	End Area	
	Cut (ft^2)	Fill (ft^2)
220+00	55	0
221+00	155	0
222+00	234	0
222+50	47	0
223+00	432	0
223+50	98	42
224+00	0	134
225+00	44	431
225+75	0	214
226+50	23	250
227+00	145	123
228+00	22	0

1. Is the project's earthwork balanced?
   A. Yes
   B. No
2. Does it produce waste or require borrow?
   A. It produces waste.
   B. It requires borrow.
3. In response to question 2 above, the volume of waste or borrow in cubic yards is most nearly:
   A. 400 yd^3
   B. 330 yd^3
   C. 300 yd^3
   D. 275 yd^3

Solution

1. Set up a table to easily track the calculated values and determine the total cut and fill volumes. Since the units of the answers are in yd^3, track the quantities as such.
2. Calculate the volumes. Examples of the calculations are located below the table.

Station	Distance (ft)	Cut (ft^2)	Fill (ft^2)	Cut Vol (yd^3)	Fill Vol (yd^3)
220+00	100 *(a)*	55	0	389 *(b)*	0
221+00		155	0		
222+00	100	234	0	720	0
222+50	50	47	0	260	0
223+00	50	432	0	444	0
223+50	50	98	42	491	26
224+00	50	0	134	60 *(c)*	163
225+00	100	44	431	54	1,046
225+75	75	0	214	41	896
226+50	75	23	250	21	644
227+00	50	145	123	156	345
228+00	100	22	0	309	152
			Total	2,945	3,273

Example 8.7 *(continued)*

Example table calculations:

(a) $L = (22{,}100 \text{ ft} - 22{,}000 \text{ ft}) = 100 \text{ ft}$

(b) $$V = \frac{L(A_1 + A_2)}{2} = \frac{100 \text{ ft}(55 \text{ ft}^2 + 155 \text{ ft}^2)}{2} = \frac{10{,}500 \text{ ft}^3}{27 \text{ ft}^3/\text{yd}^3} = 389 \text{ yd}^3$$

(c) $$V = \frac{L(A_{\text{base}})}{3} = \frac{50 \text{ ft}(98 \text{ ft}^2)}{3} = \frac{1{,}633.3 \text{ ft}^3}{27 \text{ ft}^3/\text{yd}^3} = 60 \text{ yd}^3$$

Remember that using Equation 8-7 (volume of a pyramid) is appropriate since the area is zero (224+00). If the average end area were used (Equation 8-6), the answer would be 91 yd^3.

Answers:

1. **B.** Earthwork is not balanced because the difference between cuts and fills is not zero.
2. **B.** Since the fill amount is greater than the cut amount, borrow is required.
3. **B.** Total Fill – Total Cut = Total Borrow; 3,273 yd^3 – 2,945 yd^3 = 328 yd^3

8.1.4.2 Grid Estimations

When estimating cuts and fills at a building site, where regular cross sections are impractical, quantities of excavation and embankment are often estimated with a grid survey. This survey is typically done at regular intervals, but can also start and stop at important features, like a wetland that cannot be disturbed, a stream, or the property line itself. By determining the elevations at the corner points of the grid, average elevation can then be used to estimate the amount of excavation or embankment that is necessary to get that grid to a certain specified elevation.

Example 8.8: Grid Estimations

The site design requires an elevation of 102 ft. Using the given grid survey, the required earthwork volume necessary to reach the specified elevation is most nearly:

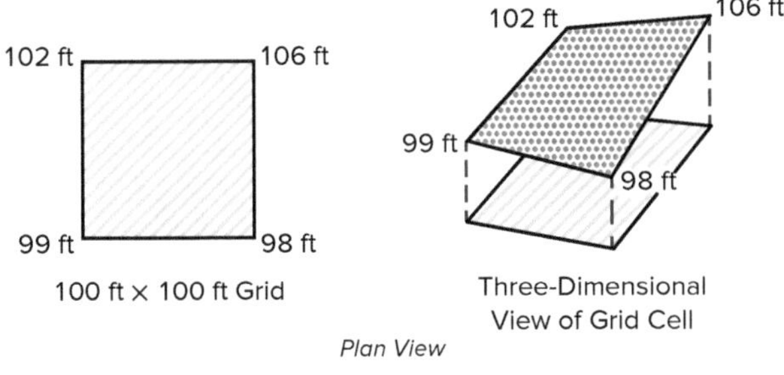

Plan View

A. 7,260 ft^3

B. 250 yd^3

C. 280 yd^3

D. 8,300 ft^3

Solution

1. Compute the total area of the site:

   A = (Length × Width) = (100 ft)(100 ft) = 10,000 ft^2

2. Compute the average elevation change:

$$\text{Avg Elevation of Site} = \frac{(\text{Elevation of Points})}{(\text{\# of Points})} \quad \textbf{Equation 8-8}$$

$$= \frac{(102 \text{ ft} + 106 \text{ ft} + 99 \text{ ft} + 98 \text{ ft})}{4} = 101.25 \text{ ft}$$

$$= 102.00 \text{ ft} - 101.25 \text{ ft} = 0.75 \text{ ft}$$

Example 8.8 *(continued)*

3. Compute the estimated volume of earthwork:

   $0.75 \text{ ft} \times 10{,}000 \text{ ft}^2 = 7{,}500 \text{ ft}^3 = 278 \text{ yd}^3$

4. Remember that since the answers are in multiple units, the solution must be compared for the "most nearly" value. In this case, the most nearly value is 280 yd^3.

Answer: C

8.2 CONSTRUCTION SITE LAYOUT AND CONTROL

Site layout and control are the primary methods of determining the exact geophysical location of the construction site and the elements that are proposed to be built. This is performed using survey techniques to project a known position onto an unknown position, to determine the location in a three-dimensional space (x, y, z coordinates).

8.2.1 Survey Equipment

Most survey work is currently done using a combination of survey instruments. These include, but are not limited to, a robotic total station, GPS rover, laser levels, and transits. Depending on the necessary accuracy or precision and the dimension that needs to be located, a combination of these instruments may be employed.

Robotic total stations have the ability to automatically locate a target and record the angle, distance, and elevation of the target, with respect to the total station (Fig. 8.6). This makes measuring and staking much faster, and it can be performed with only one surveyor.

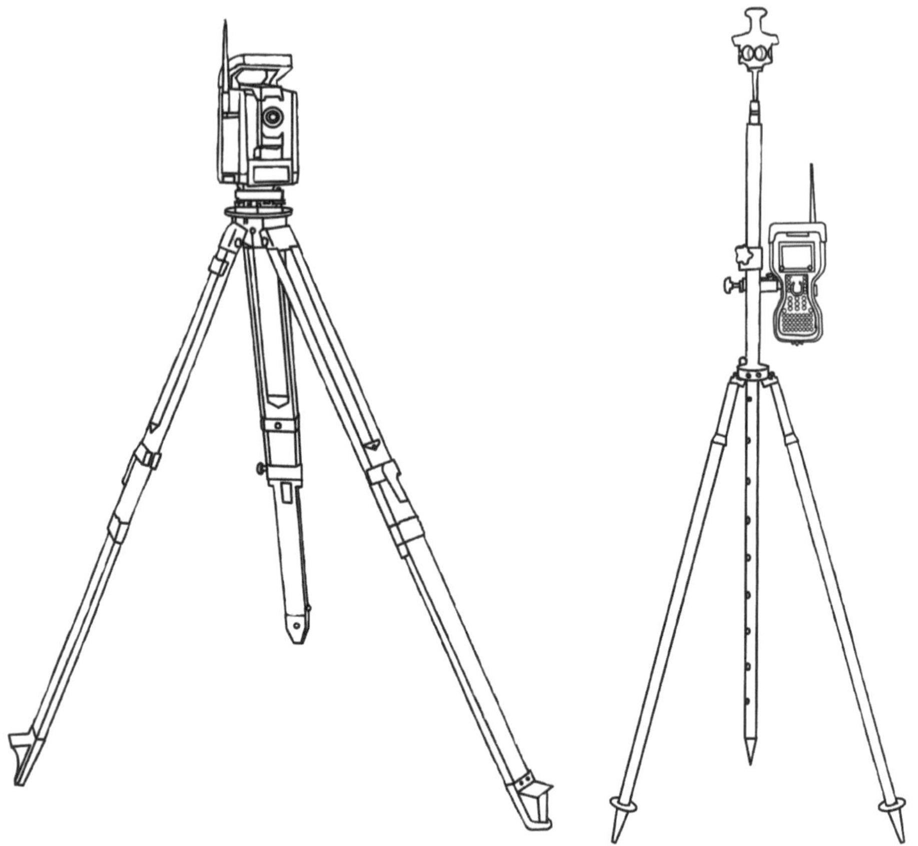

FIGURE 8.6 Robotic Total Station

GPS rovers use GPS satellites to determine the *x*, *y*, and *z* coordinates of a location (Fig. 8.7). They are also capable of determining latitude and longitude, as well as determining northing and easting when projecting into a state plane. The rover also can use a three-dimensional model developed by the survey team to project a current location to a centerline and determine station, offset, and elevation with respect to the surface model.

One notable drawback of using GPS rovers is their accuracy limitation. The rover's accuracy is dependent on the number of satellites available, satellite geometry, and whether or not the rover is connected to a base station. The **base station** is a known surveyed point that reads the information given by the GPS satellites and issues a correction to the GPS rover. All flaws considered, GPS rovers can get within 0.05 ft in the horizontal plane and 0.1 ft in elevation with the proper corrections from the base station.

Laser levels are transits that send out a laser pulse in a 360-degree direction (Fig. 8.8). The level laser pulse allows the user to adjust the prism to get a rod reading, which gives a relative elevation to the laser pulse.

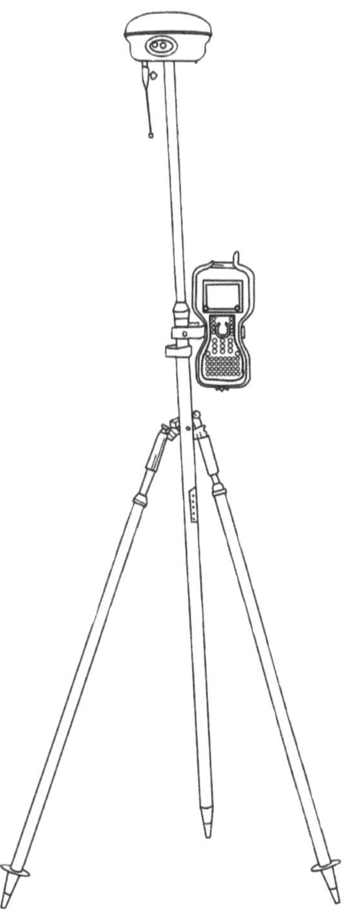

FIGURE 8.7 GPS Rover

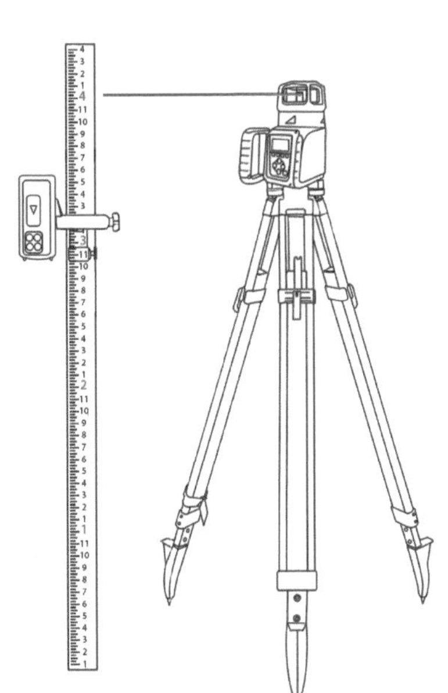

FIGURE 8.8 Laser Level

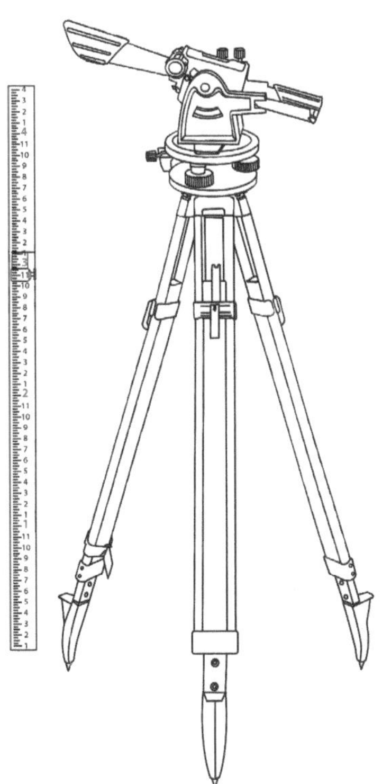

FIGURE 8.9 Transit Level

A transit, simply referred to as a level, is a precision level that allows the user to manually focus in on a rod and determine a relative elevation with respect to the transit (Fig. 8.9).

8.2.2 Survey Markers

Benchmarks are relatively permanent points of known elevation that can be used to triangulate around a site (Fig. 8.10). Benchmarks do not provide locations within the horizontal plane.

FIGURE 8.10 Benchmark

Source: Chris Curtis/Shutterstock.com.

Survey monuments are permanent pins that provide *x*, *y*, and *z* coordinates for a known location (Fig. 8.11). These pins are set in concrete in a protected box and are commonly located at the intersections of roadways or along points of interest like right-of-way (R/W) limits.

Control points, also commonly referred to as pins or nails, are either relatively permanent or temporary, depending on the application. These control points can be set by the survey crew during design to assist the contractor in laying out the design features, or they can be set by the contractor to aid in referencing for future layout work.

8.2.3 Stationing

Stationing is the positioning of any point along a line. Stations usually reference a centerline and are expressed in multiples of 100 ft, plus the remainder. For example, a station given as sta 953+50 is 95,350 ft past sta 0+00. Offsets are the parallel distance from that centerline, expressed in terms of feet left (LT) or right (RT) of the line (direction is referenced from the upstation direction).

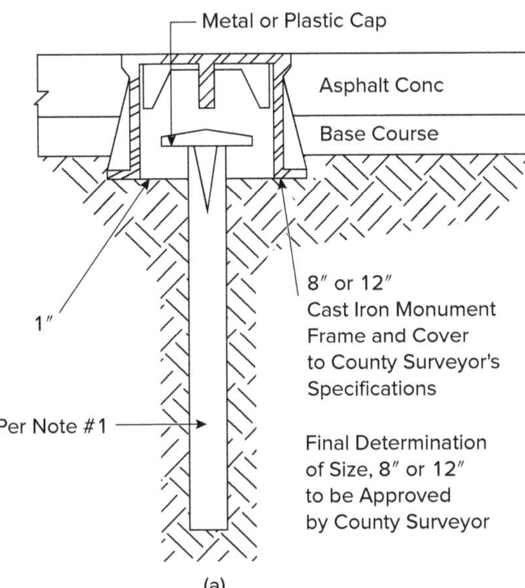

(a)

Source: Washington County, Oregon Surveyor's Office.

(b)

FIGURE 8.11 Survey Monument (a), Survey Monument Permanent Pin (b)

Photo Courtesy of Tim Sturges.

Example 8.9: Stationing

A 72-in inside diameter culvert pipe is installed along a roadway. The inlet location is determined to be at sta 42+56, 56 ft RT, and the outlet location is determined to be at sta 45+54, 23 ft LT. The length of the culvert is most nearly:

A. 79 ft

B. 308 ft

C. 298 ft

D. 423 ft

Example 8.9 *(continued)*

Solution

1. Sketch the problem.

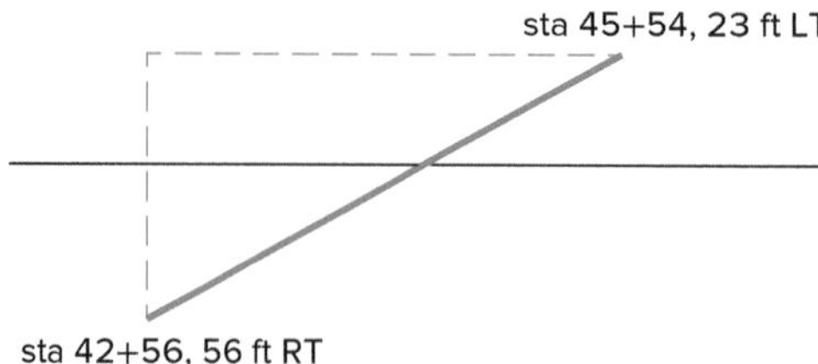

2. Determine legs of the triangle (distances). Remember that the RT and LT designations are in reference to the distance from the centerline.

   $4{,}554 \text{ ft} - 4{,}256 \text{ ft} = 298 \text{ ft}; 56 \text{ ft} + 23 \text{ ft} = 79 \text{ ft}$

3. Use the Pythagorean theorem to determine the length of the hypotenuse.

   $\text{Length} = \sqrt{298 \text{ ft}^2 + 79 \text{ ft}^2} = 308 \text{ ft}$

Answer: B

8.2.4 Staking

Survey stakes are used to delineate construction features for the workers so that a surveyor does not need to be present at all times. Stakes can give information such as R/W line, cut depths, foundation centerline locations, and so forth.

There are several main types of stakes:

- **Alignment stakes** delineate the location and alignment of a line, such as a centerline or R/W line. These stakes are typically provided at whole, half, and quarter stations, as necessary.
- **Offset stakes** delineate a line or feature, but to facilitate construction, they are located at a given offset from the center of that line or feature. This distance is typically written directly on the stake and offset with the centerline. Common uses for offset stakes are to mark the centerline of a foundation or an elevation that must be excavated to.
- **Grade stakes** delineate a specific elevation at the top of the stake.
- **Slope stakes** indicate where cuts and fills begin and end. These stakes are often used in conjunction with offset stakes.

8.2.5 Differential Leveling

Differential leveling is the process used for determining the elevation difference between two points. By using a level, the elevation of an instrument can be determined by measuring the difference on a rod from a known point and subtracting the height of the instrument. Once the elevation at the instrument is known, any visible point can be determined in a 360-degree radius.

Backsight is the elevation difference to a known point, and foresight is the elevation difference to an unknown point. Even if the unknown point is out of reach, successive backsight and foresight measurements can allow the elevation to be determined. The nomenclature for differential leveling is that a "+" is an elevation higher than the instrument and a "–" is an elevation lower than the instrument used to perform the measurement.

Example 8.10: Differential Leveling

Based on the information provided below, the elevation at point TP2 is most nearly:

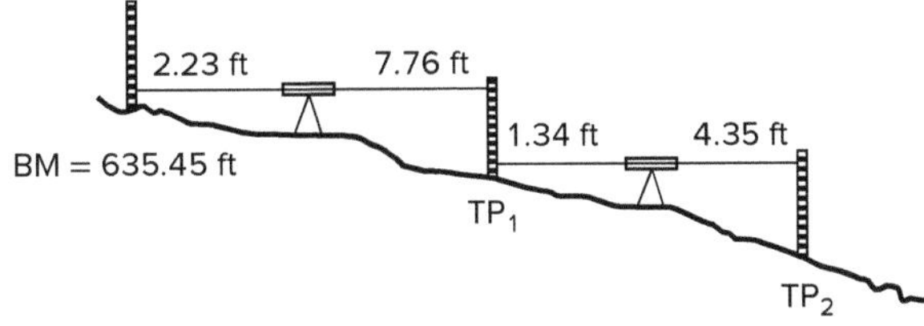

A. 643.99 ft
B. 626.91 ft
C. 623.34 ft
D. 642.46 ft

Solution

1. In the example, where the elevation determination is being projected along the level line, the height of the instrument is not necessary to take into account. The difference of the sums of the backsights and foresights can be used to determine the correct elevation. Pay close attention to the direction in which the ground is changing in relation to the height of the instrument to ensure that the correct sign is used.

$$\text{Backsights} = +2.23\text{ ft} + 1.34\text{ ft} = +3.57\text{ ft}$$
$$\text{Foresights} = (-7.76\text{ ft}) + (-4.35\text{ ft}) = -12.11\text{ ft}$$
$$+3.57\text{ ft} - 12.11\text{ ft} = -8.54\text{ ft}$$

2. To calculate the elevation at TP2, subtract the difference in elevation from the BM elevation:

$$635.45\text{ ft} - 8.54\text{ ft} = 626.91\text{ ft}$$

Answer: B

8.2.6 Trigonometric Leveling

Trigonometric leveling uses the concepts of trigonometry to determine elevation differences. This is done by using differential leveling and lowering or raising the instrument to a specific angle. When a clear line of sight is possible, this can eliminate the need to reset the instrument several times to determine this same elevation by using differential leveling.

The angle of the reading can be determined using both degree-minute-second (DMS) or decimal degree. Conversion from DMS to decimal degree is performed by the following equation:

$$\text{deg}^\circ\text{min}'\text{sec}'' = \text{deg}^\circ + \left(\frac{\text{min}' + \dfrac{\text{sec}''}{60}}{60}\right) \qquad \textbf{Equation 8-9}$$

Example 8.11: Trigonometric Leveling

The surveyor is laying out the elevation of a proposed storm-sewer system. The rod reading that must be obtained to ensure that the invert of the pipe is at the plan elevation is most nearly:

A. 7.51 ft
B. 635.56 ft
C. 628.05 ft
D. 4.63 ft

Solution

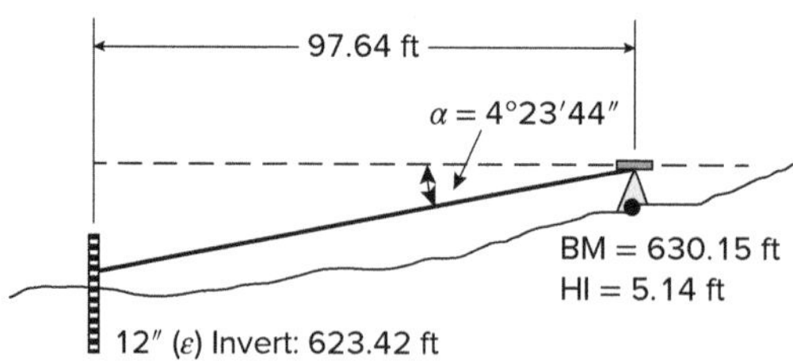

1. Determine the unknown elevation differences on the drawing. From observation, it can be determined that the two unknowns are the elevation difference from the invert point to the rod reading and the vertical distance from the instrument level to the rod reading.
2. Convert to decimal degrees:

$$4°23'44'' = 4° + \left(\frac{23 + \frac{44}{60}}{60}\right) = 4° + \left(\frac{23.73}{60}\right) = 4.40°$$

3. Determine the vertical distance from rod reading to level of instrument:

$$\tan\alpha = \frac{\text{opposite}}{\text{adjacent}};\ \tan(4.40°) = \frac{\text{elev}}{97.64\text{ ft}};\ \text{elev} = 7.51\text{ ft}$$

4. Following graphically, project the known elevation (BM) to the instrument level line and back down to the rod height, then subtract the invert elevation to determine the height of the rod reading necessary to ensure that the invert is installed at the plan elevation.

   Rod reading elevation = 630.15 ft + 5.41 ft − 7.51 ft = 628.05 ft
   Required rod height = 628.05 ft − 623.42 ft = 4.63 ft

Answer: D

8.3 TEMPORARY AND PERMANENT SOIL EROSION AND SEDIMENT CONTROL

Temporary and permanent soil erosion and sediment control are necessary considerations during site development. The **Environmental Protection Agency (EPA)**, under authority from the Clean Water Act, is the regulatory agency charged with minimizing and permitting such discharges into US waters. This discharge is regulated under the **National Pollutant Discharge Elimination System (NPDES)** permitting process. To simplify the permitting process, most construction sites are permitted under a **construction general permit** (CGP).

The federal EPA sets most regulations for this, but most states delegate the state environmental agency, with its own version of the CGP, as the permitting authority. These state CGPs often include local restrictions tailored to the unique needs of the state or specific watershed regions while meeting minimum requirements for the federal regulations. A construction site is required to obtain coverage under the applicable CGP when construction activities will disturb one or more acres, or for sites smaller than one acre if they are part of a larger plan of development, such as a subdivision. One of the permit requirements is a **stormwater pollution prevention plan** (SWPPP), which is site specific and details necessary sediment and erosion controls to meet the CGP requirements.

For government or larger projects, the owner is typically required to submit a **notice of intent** (NOI) to the regulatory agency (US EPA or state environmental agency). The prime contractor is then required to register as a co-permittee. For smaller and private projects, the contractor must submit the NOI to obtain coverage. Once the project is complete and final stabilization has occurred, the permittee then submits a **notice of termination** (NOT) to notify the regulatory agency that the project is complete and that the facility number is no longer in operation.

8.3.1 Temporary Controls

Erosion and sediment controls are used during construction in an attempt to keep sediment in place (erosion controls) and to capture any sediment that is moved by stormwater before it is allowed to leave the site (sediment controls). Earth disturbed area (EDA) is the area of exposed earth from a construction activity that is prone to increased erosion as a result of its stabilizing feature (vegetation, cover, rocks, and the like) being removed.

The SWPPP is an engineered plan by a certified professional, who typically holds a Certified Professional in Erosion & Sediment Control (CPESC) license and a professional engineer's license and who evaluates contributing site conditions, evaluates planned construction and phasing, and designs the appropriate controls to ensure that erosion and the resulting sediment are properly controlled. The SWPPP designer or designee will perform required continued inspections (as defined in the CGP) and make changes to the SWPPP as necessary. In addition, during the inspections, maintenance needs and new or no-longer-needed controls are documented. It is in any contractor's best interest to pay close attention to erosion and sediment controls, not just from a regulatory standpoint (public relations, notices of violation, and so forth) but also from a fiscal one. Once erosion begins, the site will eventually need to be repaired to final grade. Continued erosion will create larger and more expensive rework.

8.3.1.1 Erosion Controls

Erosion is the process of displacing, or moving, soil particles from one place to another via wind or water. While wind erosion is a concern, it is much harder to plan for this occurrence, and controls are typically limited to watering dry, exposed soil areas, or using chemical stabilizing agents.

Erosion from water carries a much higher risk and is the typical focus of any effective SWPPP. An SWPPP should rely on erosion controls as the primary means of preventing stormwater pollution. While many proprietary products for erosion control exist, generic categories of controls and their primary functions are often used first.

Some common forms of erosion:

- **Raindrop erosion** happens when exposed earth is subjected to rain. When the rain hits the exposed earth, it splashes the soil and dislodges it. Once the initial soil bond is broken, it begins to move with the water.
- **Sheet erosion** is a very shallow, widespread flow from the movement of water. This is typically found along embankment areas and large sites where the slope is constant and not steep.

- **Rill erosion** occurs when the moving sheet flow begins to congregate in low areas and develops small, shallow channels (rills) of concentrated flow.
- **Gully erosion** occurs when deep channels begin to form and large amounts of soil have begun to move. This typically occurs from unchecked or uncontrolled rill erosion as water continues to concentrate along the same path.
- **Streambank or channel erosion** is caused by the continued flow of established, or newly constructed, planned areas of water flow. Some examples are relocated streams, newly constructed culverts or storm sewers, and ditch lines.
- **Trucking** and equipment tracking can also move soil. Tires or tracks disturb the soil surface and can carry soil off site if not properly planned for.

Some common forms of erosion controls:

- **Diversion**, or removing or diverting flow from going across the EDA, should be the first consideration in any cost-effective SWPPP. If the water does not pass over the EDA, the water typically does not need to be treated for the removal of possible sediments, since it never passed over exposed earth.
- **Project sequencing** can make a large impact on erosion. By delaying clearing and grubbing operations until absolutely necessary, the original stabilization measures (vegetation, grass, trees, and the like) hold the soil particles together and in place.
- The application of straw or hay cover is used to dissipate raindrop erosion and keep soil in its original place. It can also be an effective control when dealing with sheet erosion.
- **Slope drains**, typically used in conjunction with dikes, can be effective for raindrop, rill, and gully erosion. These are constructed at the top of a sloped area and use dikes to channel water to a pipe and down to an area of planned water conveyance, thus eliminating the flow over an area of exposed surface. Slope drains can also be particularly effective controls when attempting to establish vegetative growth.
- Temporary or permanent **vegetative cover** (70% minimum coverage) dissipates raindrop erosion, and the root structure holds the soil particles together under sheet flow, reducing the risk for rill erosion.
- **Rolled erosion-control products** are typically used in conjunction with seeding to dissipate raindrop erosion and keep rill erosion from getting worse. Certain products can also be used in ditches.
- **Rock ditch checks** can be used in ditches and channels to slow the velocity of concentrated flows, thus reducing the shear stress in the channel.
- Rock construction entrances and exits can be used in areas where trucks or equipment are entering and exiting work areas. These rocks provide a stable surface for soil to fall off the equipment and trucks prior to them leaving the site.

8.3.1.2 Sediment Controls

Sediment controls attempt to capture any sediment before it leaves the construction site. Once erosion has occurred and sediment is moving across the site through sheet flow, shallow concentrated flow, or in-channel flow, proper best management practice (BMP) controls must be installed to minimize the incidence of these soil particles leaving with the water. Identifying the type and amount of sediment flow helps with determining the proper controls.

Some common forms of sediment controls follow.

- **Perimeter sheet flow controls** are designed to hold, or pond, water and release it slowly. By stopping and ponding the water, this control gives the sediment a chance to settle out of the water prior to being discharged. Some examples are filter fabric fence, compost filter socks, and hay bales (less common due to their ineffectiveness).

- **Sediment traps** are used to intercept flows and hold water, giving the sediment a chance to settle out. Sediment traps need to be monitored and emptied once the designed sedimentation zone fills up.
- **Inlet protections** are necessary around areas where water enters a storm-sewer system or a culvert. These are typically constructed using filter fabric fencing around the catch basin or inlet, in the same manner as a perimeter fence, to pond water and allow for settlement.
- **Ditch checks** can be modified and installed using filter fabric fence to allow for ponding and settlement to occur in low-flow ditch lines. The fencing must extend up the sides of the ditch, but considerations need to be made for potential flooding with respect to the elevations of critical elements in the vicinity.

8.3.2 Permanent Controls

Permanent controls are as important as temporary ones. Construction or temporary erosion is closely monitored through the SWPPP process; however, once the site has been completed and the contractor leaves, water continues to flow across the site. Although proper vegetative establishment minimizes the erosion potential of a completed site, areas of concentrated flow such as ditches, streams, culvert and storm-sewer outlets, and slope obstructions need to be protected in the long term.

Examples of permanent controls follow:

- Final seeding or sod should be used as appropriate to eliminate or reduce raindrop and sheet erosion from moving soil from its original location.
- Rock protection can be placed in areas of high flow to dissipate the velocity of the water and hold the soil particles.
- Turf reinforcement mats can be used in channels to reinforce the vegetation and resist the water from moving the soil that holds it in place.
- Concrete caps (concrete channel protection) can be placed in areas of high shear stress to keep water from hitting the soil and causing scour. These are typically found along areas of high slope, high flow, culvert and storm-sewer outlets, and so forth.
- Retention or detention basins are used to collect stormwater from a site and store or slowly release the water into the receiving stream. Given that development reduces the area where water can enter the ground, these controls are necessary to avoid overloading the receiving stream and causing excessive bank erosion and flooding.

8.4 IMPACT OF CONSTRUCTION ON ADJACENT FACILITIES

8.4.1 Excavation Bracing

8.4.1.1 OSHA Soil Classification

OSHA dictates construction operations to ensure that worker safety is lawfully paramount to all contractor operations. Although there are many OSHA regulations that apply to construction, this section focuses on the provisions of the Code of Federal Regulations (CFR) 1926.652 and 1926 Subpart P, Appendices A and B [1].

In 29 CFR 1926 Subpart P, Appendix A, OSHA defines four main types of soils—stable rock, Type A, Type B, and Type C soils—in order of their stability from cave-ins when excavated. Depending on the classification of the soils encountered in the field and the available space for layback, different methods of bracing to safely perform the excavation may be necessary.

Type A soil is defined as a soil with an unconfined compressive strength of 1.5 ton/ft^2 (144 kPa) or greater. Examples of cohesive soils are clay, silty clay, sandy clay, clay loam, and, in some cases, silty clay loam and sandy clay loam. No soil is considered Type A if:

(i) The soil is fissured; or

(ii) The soil is subject to vibration from heavy traffic, pile driving, or similar effects; or

(iii) The soil has been previously disturbed; or

(iv) The soil is part of a sloped, layered system where the layers dip into the excavation on a slope of four horizontal to one vertical (4H:1V slope) or greater; or

(v) The material is subject to other factors that would require it to be classified as a less stable material.

Type B soil is defined as:

(i) A cohesive soil with an unconfined compressive strength greater than 0.5 ton/ft^2 (48 kPa), but less than 1.5 ton/ft^2 (144 kPa); or

(ii) Granular cohesionless soils including: angular gravel (similar to crushed rock), silt, silt loam, sandy loam, and, in some cases, silty clay loam and sandy clay loam.

(iii) Previously disturbed soils except those that would otherwise be classified as Type C soils; or

(iv) Soil that meets the unconfined compressive strength or cementation requirements for Type A, but is fissured or subject to vibration; or

(v) Dry rock that is not stable; or

(vi) Material that is part of a sloped, layered system where the layers dip into the excavation on a slope less steep than four horizontal to one vertical (4:1 slope), but only if the material would otherwise be classified as Type B.

Type C soil is defined as:

(i) Cohesive soil with an unconfined compressive strength of 0.5 ton/ft^2 (48 kPa) or less; or

(ii) Granular soils including gravel, sand, and loamy sand; or

(iii) Submerged soil or soil from which water is freely seeping; or

(iv) Submerged rock that is not stable, or

(v) Material in a sloped, layered system where the layers dip into the excavation or a slope of four horizontal to one vertical (4H:1V) or steeper.

Adapted from *OSHA Safety and Health Regulations for Construction [1].*

Depending on the type of soil encountered, the soil may be properly braced against cave-in via cutback of the slope.

8.4.1.2 Cutback Sloping

Cutback sloping specifications from OSHA 29 CFR 1926 Subpart P, Appendix B are shown in Table 8.3.

TABLE 8.3 Maximum Allowable Slopes

SOIL OR ROCK TYPE	MAXIMUM ALLOWABLE SLOPES (H:V)(1) FOR EXCAVATIONS LESS THAN 20 FEET DEEP (3)
STABLE ROCK	VERTICAL (90°)
TYPE A (2)	3/4:1 (53°)
TYPE B	1:1 (45°)
TYPE C	1½:1 (34°)

Source: Occupational Safety and Health Administration (OSHA). *Safety and Health Regulations for Construction (Standards - 29 CFR, Part 1926).*

Footnote (1): Numbers shown in parentheses next to maximum allowable slopes are angles expressed in degrees from the horizontal. Angles have been rounded.

Footnote (2): A short-term maximum allowable slope of 1/2H:1V (63°) is allowed in excavations in Type A soil that are 12 feet (3.67 m) or less in depth. Short-term maximum allowable slopes for excavations greater than 12 feet (3.67 m) in depth shall be 3/4H:1V (53°).

Footnote (3): Sloping or benching for excavations greater than 20 feet deep shall be designed by a registered professional engineer.

Layered soils must also be considered (Fig. 8.12). When excavating, different soil classifications are often found at different depths. As a general rule, the stability of the bottom layer must be taken into account to determine how the excavation can be safely cut back. For example, if a Type C soil is found on top of a Type A soil, the Type C must be cut back at a 1½:1 slope, whereas the Type A may be cut back at a ¾:1 slope. Conversely, if a Type A is found on top of a Type C, the entire excavation must be treated as a Type C and cut back at a 1½:1 slope (Table 8.4).

TABLE 8.4 Sloping Requirements for Layered Slopes

LAYERED SOIL TYPE	TYPE A LAYER	TYPE B LAYER	TYPE C LAYER
B over A	¾:1	1:1	
C over A	¾:1		1½:1
C over B		1:1	1½:1
A over B	1:1	1:1	
A over C	1½:1		1½:1
B over C		1½:1	1½:1

Source: Occupational Safety and Health Administration (OSHA). *Safety and Health Regulations for Construction (Standards - 29 CFR, Part 1926).*

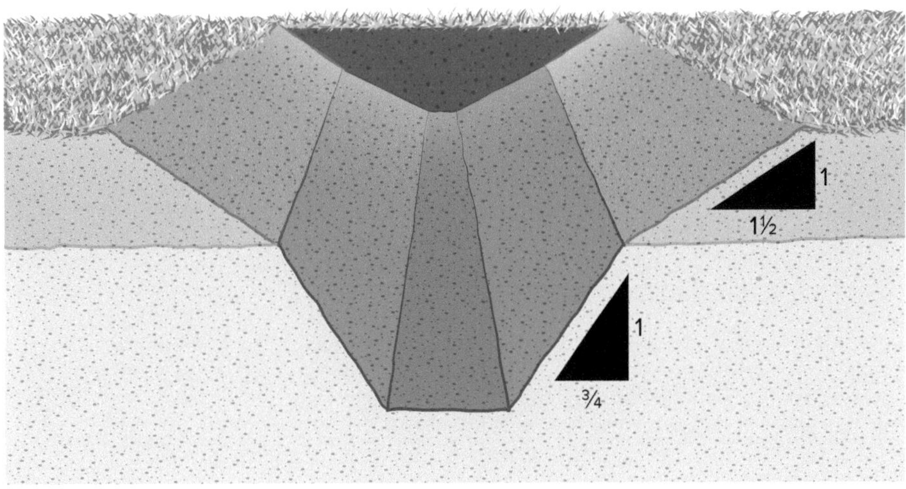

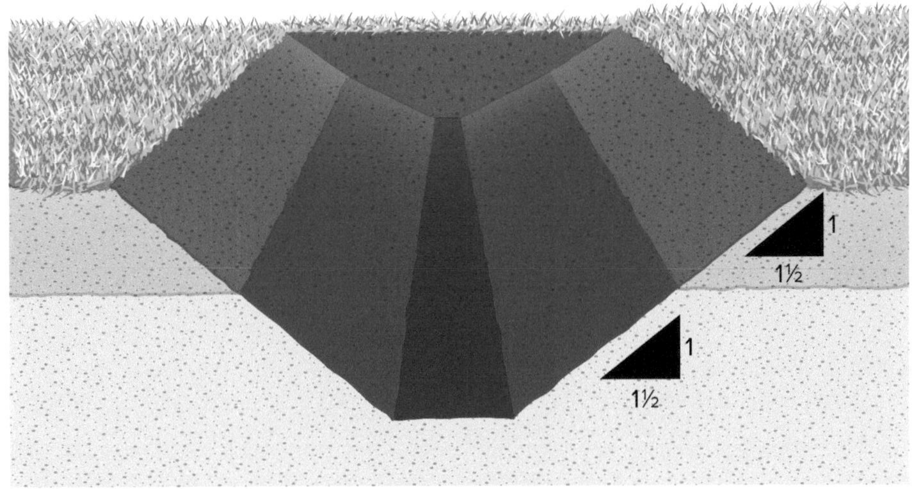

FIGURE 8.12 OSHA Cutback Slope Ratios for Soil Types

Example 8.12: Cutback Sloping

A contractor plans to cut back a slope to brace for an excavation with a spread footing measuring 32 ft wide. The carpenters require a 3-ft stand-off distance from the edge of the footer for formwork and access. The total depth of the excavation is 10 ft from the existing ground. On one side of the footer, the soil borings indicate Type B soils for the top 6 ft and Type C soils below 6 ft. On the other side, the borings show Type A soils. The total width of the excavation at ground level is most nearly:

A. 60 ft
B. 61 ft
C. 58 ft
D. 59 ft

Solution

1. Sketch the problem:

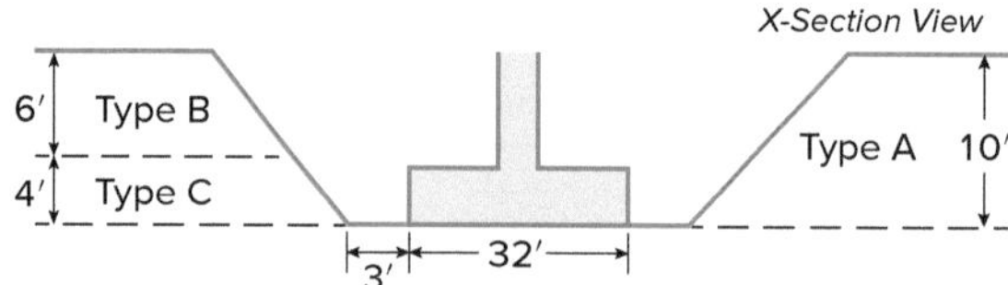

2. Working left to right, determine the distances necessary for the excavation.

Cutback for Type B soil:

1.5:1 = 1.5(6 ft) = 9 ft

Cutback for Type C soil:

1.5:1 = 1.5(4 ft) = 6 ft

Stand-off distance for workers = 3 ft

Width of footer = 32 ft

Stand-off distance for workers = 3 ft

Cutback for Type A soil:

0.75:1 = 0.75(10 ft) = 7.5 ft

Add the total widths:

9 ft + 6 ft + 3 ft + 32 ft + 3 ft + 7.5 ft = 60.5 ft (Round up.)

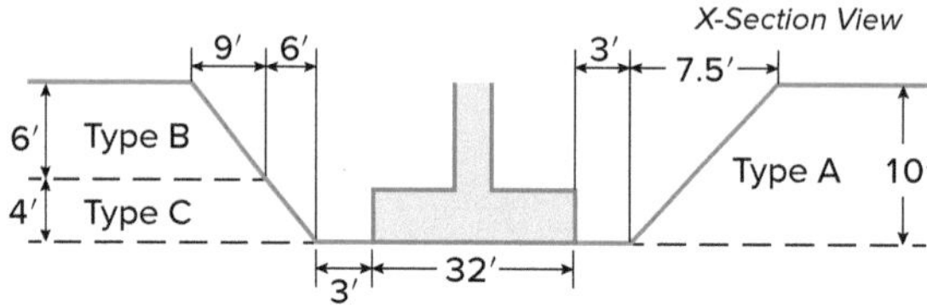

TIP

Any soil type over a lower soil type requires the cutback for the lower soil type.

TIP

Rounding down will not provide the width necessary to meet OSHA requirements.

Answer: B

8.4.1.3 Trench Boxes

The trench box is described in Figure 8.13.

OSHA design specifications apply only to trenches that do not exceed 20 feet. The soil type in which the excavation is made must be determined in order to use the OSHA data. The specifications do not apply in every situation experienced in the field; the data was developed to apply to the most common trench situations.

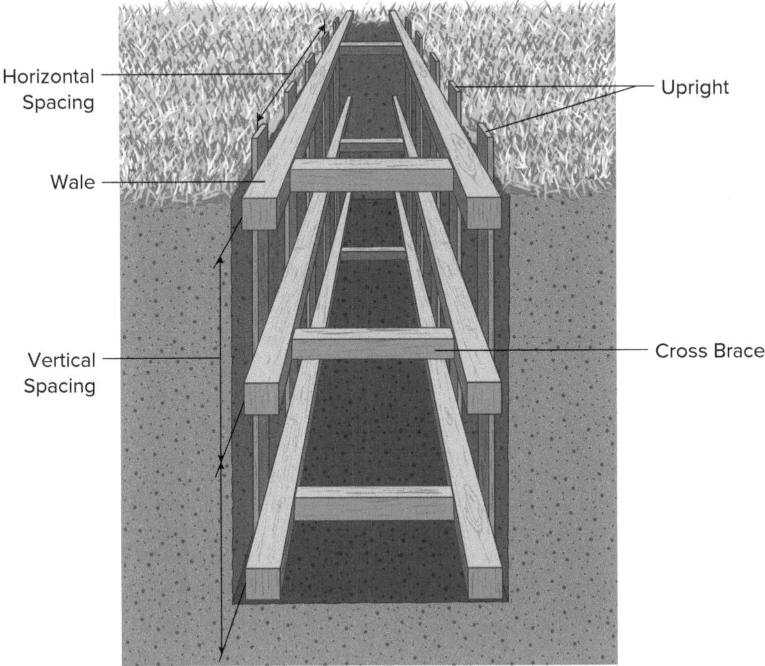

FIGURE 8.13 Trench Box

For example, Figure 8.13 illustrates timber shoring in a trench approximately 13 ft deep and 5 ft wide in Type B soil. The OSHA specifications for timber trench shoring–minimum timber requirements describe the construction as follows: space 6 × 6 crossbraces at 6 ft horizontally and 5 ft vertically; 8- × 8-ft wales at 5 ft vertically; and 2 × 6 uprights every two feet horizontally.

8.5 SAFETY

8.5.1 Construction

OSHA 29 CFR sections 1910 [2] (general industry) and 1926 (construction industry) [1] specify regulations meant to ensure worker safety. While there are many regulations that pertain to the construction industry, some particular points are important to review for the exam.

> **TIP**
>
> The exam will likely ask a question or two in which the solution can be found within the codes, possibly requiring minor calculations. It is imperative that the index is used to help quickly find the solutions.

Pertinent Sections of OSHA Code:

- Subpart E: Personal Protective and Life-Saving Equipment
- Subpart I: Tools, Hand and Power
- Subpart L: Scaffolds
- Subpart M: Fall Protection
- Subpart N: Cranes
- Subpart P: Excavations
- Subpart Q: Concrete and Masonry Construction
- Subpart R: Steel Erection
- Subpart S: Underground Construction, Caissons, Cofferdams
- Subpart U: Explosives
- Subpart X: Ladders

8.5.1.1 Fall Protection

OSHA regulates fall protection under 29 CFR 1926.501 (Subpart M). This standard requires that any construction employers provide fall protection for any workers that are exposed to a fall distance of 6 ft or greater. This fall protection is typically provided by use of a horizontal lifeline and a harness system. The lifelines need to be designed and installed under the supervision of a qualified person (as further defined in the OSHA regulations) and maintain a factor of safety of at least two.

8.5.1.2 Personal Protective Equipment

Personal protective equipment (PPE) is necessary to keep workers safe from the inherent dangers of performing tasks in the construction industry. This section is regulated under 29 CFR 1926 Subpart E (1926.95 – 1926.107). The official responsibility is on the employer to ensure the use of PPE by workers when exposed to hazards. Examples in this subpart include eye and face protection, foot protection, head protection, ear protection, respiratory protection, and electrical protection.

8.5.1.3 Crane Safety

With respect to OSHA and cranes, it is imperative that all manufacturer specifications, instructions, inspection schedules, and maintenance schedules are followed. When sizing a crane, the manufacturer's tables and charts must be followed with no exception.

While the operation of the crane itself is the main concern in a lifting operation, one thing that is overlooked is the clearance to electrical lines, which, if not taken into account, can cause injury or death.

Table A in 29 CFR 1926.1408 (Subpart CC) lists the required standoff distances to electrical lines, depending on the line voltage (shown here in Table 8.5).

TABLE 8.5 Minimum Clearance Distances

VOLTAGE (NOMINAL, *kV*, ALTERNATING CURRENT)	MINIMUM CLEARANCE DISTANCE (FEET)
Up to 50	10
Over 50 to 200	15
Over 200 to 350	20
Over 350 to 500	25
Over 500 to 750	35
Over 750 to 1,000	45
Over 1,000	(As established by the utility owner/operator or registered professional engineer who is a qualified person with respect to electrical power transmission and distribution).

Note: The value that follows to is up to and includes that value. For example, over 50 to 200 means up to and including 200 kV.

Source: Occupational Safety and Health Administration (OSHA). *Safety and Health Regulations for Construction (Standards - 29 CFR, Part 1926).*

8.5.1.4 Noise Exposure

Exposure to noise is a frequent concern in the construction industry. As a result, OSHA has regulated the amount of noise exposure that a worker can experience in an 8-hour time-weighted average and permissible exposure limits (PEL). These are dictated under 29 CFR 1926.52 and 29 CFR 1910.95.

Example 8.13: Noise Exposure

A construction employee receives the following exposure to noise on a construction site during an 8-hour period:

0.5 hours at 110 dBA

1.5 hours at 97 dBA

Example 8.13 *(continued)*

3 hours at 95 dBA

The percent of the OSHA permissible exposure limit (PEL) for this exposure level is most nearly:

A. 176%
B. 100%
C. 75%
D. 225%

Solution

1. In the OSHA 29 CFR 1926 reference manual, look up the applicable standard. If it is not given, use the index.
2. The OSHA PELs at these sound levels are:

   Sound level permissible duration:

   95 dBA: 4 hours
   97 dBA: 3 hours
   110 dBA: 0.5 hours

   The calculation is as follows:

$$F = \left(\frac{C_1}{T_1} + \frac{C_2}{T_2} + \cdots \frac{C_i}{T_i}\right)$$

   F = exposure level

   C = time of exposure at noise level

   T = allowed time of exposure

3. Calculate the combined effective exposure to determine if the value of F is over PEL, or unity (1.0).

$$F = \left(\frac{0.5}{0.5} + \frac{1.5}{3} + \frac{3}{4}\right) = 2.25 = 225\%\ \text{PEL}$$

Answer: D

8.5.1.5 Reportable Injuries

The revised OSHA 29 CFR Part 1904 [3], entitled *Recording and Reporting Occupational Injuries and Illnesses*, requires the reporting of the following:

1. OSHA Form 300, Log of Work-Related Injuries and Illnesses
2. OSHA Form 300A, Summary of Work-Related Injuries and Illnesses
3. OSHA Form 301, Injury and Illness Incident Report

OSHA 300 log data uses an incidence rate that takes the number of injuries and illnesses per 100 full-time workers. The OSHA 300 log instructs the approximation of total hours worked per 100 full-time employees using 40 hours per week for 50 weeks per year (or, 200,000 hr/year). Comparing the actual number of hours against the 100 full-time workers' data allows OSHA to compare a company's safety record to other companies in the same sector, and to determine which sectors pose higher risks to employees. OSHA also uses this data to determine scheduled safety inspections for companies that routinely exceed the average.

Example 8.14: Incident Rate

A construction company had 24 recordable injuries, with one of them resulting in 10 days of lost time and one resulting in 60 days of lost time. The annual total number of hours worked was 347,320. What is the calculated incident rate associated with these statistics?

A. 7.43
B. 0.023
C. 13.82
D. 3.45

Solution

$$\text{Incident Rate (IR)} = \frac{(\text{number of recordable incidents} \times 200{,}000)}{(\text{total number of hours worked})} \qquad \textbf{Equation 8-10}$$

$$\text{IR} = \frac{24 \times 200{,}000}{347{,}320 \text{ hr}}$$

$$\text{IR} = 13.82$$

Answer: C

8.5.2 Roadside

According to the Federal Highway Administration (FHWA), the *Manual on Uniform Traffic Control Devices for Streets and Highways* (*MUTCD*) [4] defines the standards used by road managers nationwide to install and maintain traffic control devices on all public streets, highways, bikeways, and private roads open to public travel. The *MUTCD* is published under 23 CFR, Part 655, Subpart F. The intent of having uniform traffic control devices (signs, signals, pavement markings, and the like) is to ensure that anyone driving within the US will be able to navigate traffic, regardless of the state.

8.5.2.1 Clear Zone

According to the FHWA, "a clear zone is an unobstructed, traversable roadside area that allows a driver to stop safely, or regain control of a vehicle that has left the roadway." Although the *AASHTO Roadside Design Guide* [5] (also known as the *Green Book* due to its color) provides guidance for the minimum clear zone widths, it is vital to examine site-specific roadway characteristics to determine if additional width must be provided. Remembering that the point of providing this unobstructed width is to provide errant vehicles the opportunity to safely stop or correct course, considerations should include, but are not limited to, travel speed, number of fixed objects, traffic volumes (average daily traffic), curvature, elevation differences (slope), and severity risk. For example, although the minimum clear zone for a 60 mph urban interstate placed on fill (6H:1V slope) is recommended to be 30-to-34 feet, it would be logical to provide as much distance as possible or provide a barrier if a multi-use path is located at 35 feet, horizontally. The safety risk if an errant vehicle could not recover is exponentially greater, and requires the use of good engineering judgment.

Although the above considerations would most likely be part of a set of engineering plans, for the general purpose of construction, the storage of materials and parking of construction vehicles and equipment should be outside of the clear zone. Where this is not possible or practical, a proper work zone should be utilized to warn drivers of impending hazards. Depending on the actual hazard, positive protection (portable concrete barriers, work zone attenuators) rather than passive protection (barrels, delineators) may be necessary.

8.5.2.2 Work Zone

Work zones should be set whenever: roadside work must occur within the clear zone, contractor ingress and egress from the roadway is necessary, or site constriction requires the storage of materials or equipment in or near the clear zone. Work zone standardization is important not only for motorist and contractor safety, but also for liability purposes in the event of a crash. Appropriate advance signing, buffer lengths, and taper lengths are all based on the legal signed speed of the adjacent traffic and can be found in the *MUTCD* tables in chapter 6 (Figs. 8.14 and 8.15).

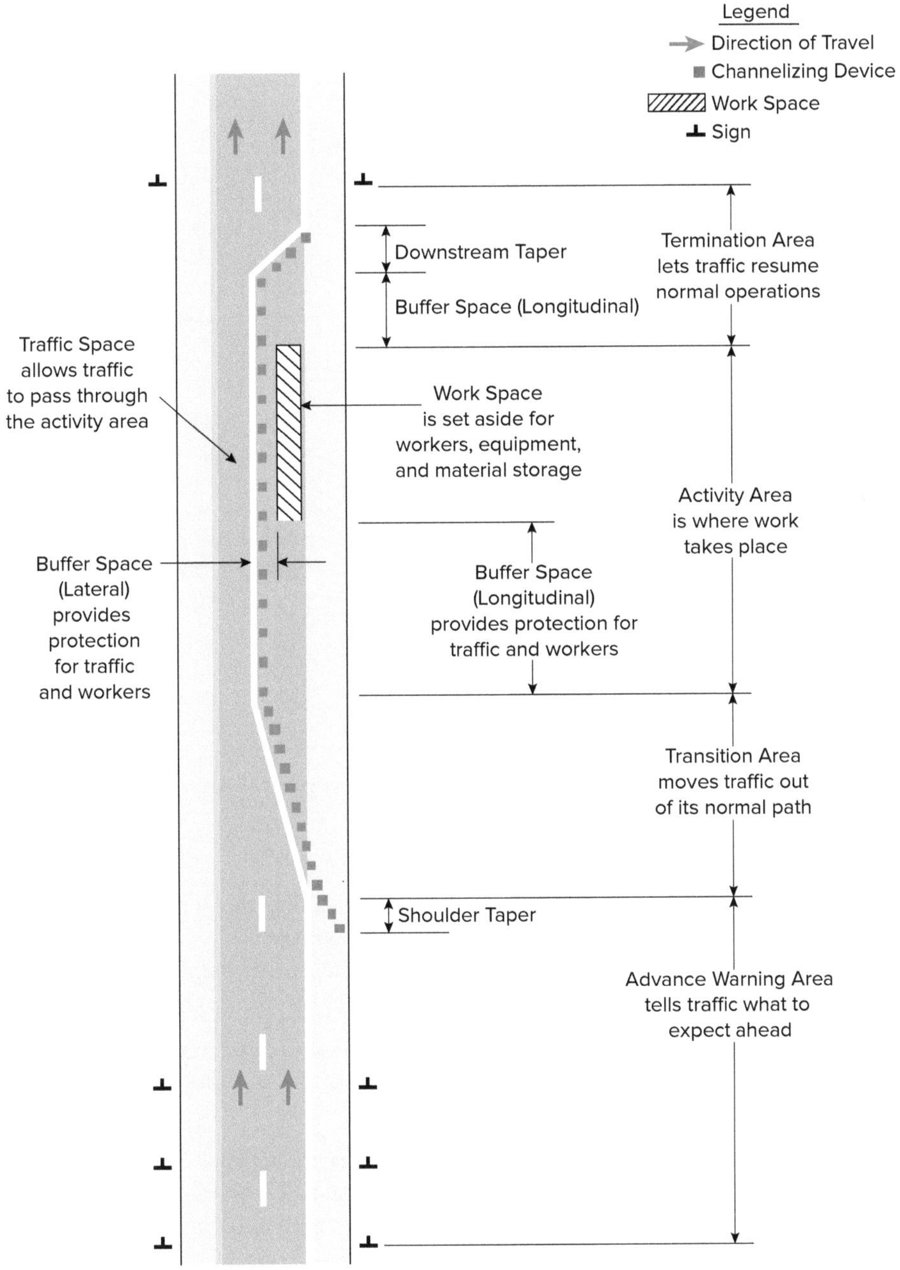

FIGURE 8.14 Component Parts of a Temporary Traffic Control Zone

Source: US Department of Transportation, Federal Highway Administration. *Manual on Uniform Traffic Control Devices for Streets and Highways (MUTCD)*, 2009.

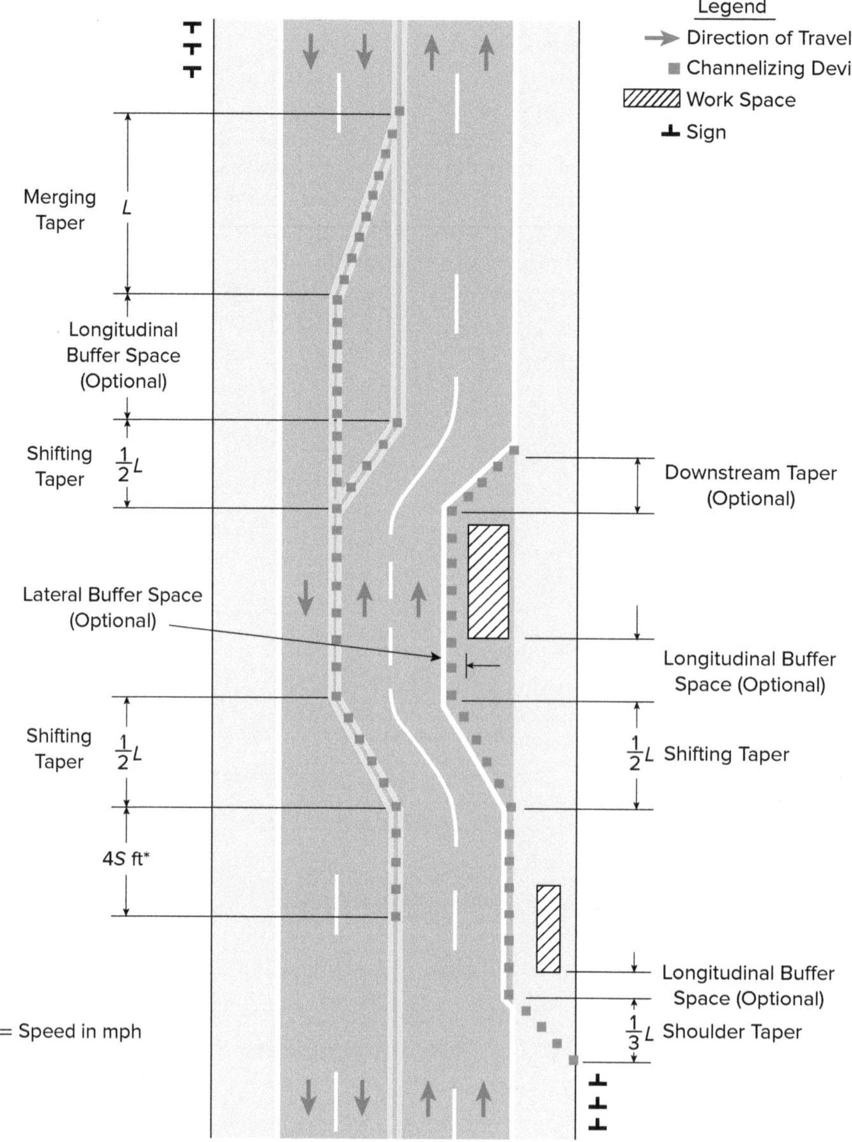

FIGURE 8.15 Types of Tapers and Buffer Spaces

Source: US Department of Transportation, Federal Highway Administration. *Manual on Uniform Traffic Control Devices for Streets and Highways (MUTCD)*, 2009.

8.5.2.2.1 Taper Length Criteria for Work Zones

The typical method of moving traffic longitudinally along the traveled way is to introduce a taper. Using channelizing devices (barrels, pavement markings, and so forth), traffic is transitioned from one location to another. Taper lengths are important as they are engineered to safely move traffic at a specific speed.

Merging tapers move traffic from a lane that is ending and direct traffic into another lane where active traffic is flowing. Some examples of this are drop lanes, entrance ramps, and temporary lane closures.

Shifting tapers move traffic longitudinally to another location. One example of this is a work zone shift to allow for construction to occur. If a full taper distance with applicable buffer space is not available, this shift could also be accomplished via reverse curves.

Shoulder tapers are used on high-speed facilities where construction vehicles are in use or when traffic could accidently mistake the area as open to use (an example is when no edge lines are present). Because vehicles are not meant to be traveling on these areas, shorter taper lengths can be used, mostly as a visual indicator of the activity.

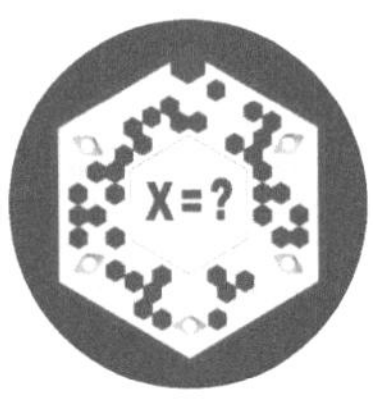

Downstream tapers are mostly used in work zone termination areas where a lane is being opened up to use or when a new lane is being introduced.

One-lane, two-way tapers are used on two-lane roadways where one of the lanes is closed. Traffic is allowed, with proper traffic control devices, to advance in both directions, alternating directions, one at a time.

Table 8.6 shows the taper length criteria used in work zone traffic control, and Table 8.7 provides formulas for determining taper lengths.

TABLE 8.6 Taper Length Criteria for Temporary Traffic Control Zones

TYPE OF TAPER	TAPER LENGTH
Merging Taper	At least L
Shifting Taper	At least 0.5 L
Shoulder Taper	At least 0.33 L
One-Lane, Two-Way Traffic Taper	50 ft min, 100 ft max
Downstream Taper	50 ft min, 100 ft max

Source: US Department of Transportation, Federal Highway Administration. *Manual on Uniform Traffic Control Devices for Streets and Highways (MUTCD)*, 2009.

TABLE 8.7 Formulas for Determining Taper Length

SPEED(S)	TAPER LENGTH (L) IN FEET
40 mph or less	$L = \frac{WS^2}{60}$
45 mph or more	$L = WS$

Where:
L = Taper Length (ft)
W = Width of Offset (ft)
S = Posted Speed Limit, or Off-Peak 85th percentile speed prior to work starting, or the anticipated operating speed (mph)

Source: US Department of Transportation, Federal Highway Administration. *Manual on Uniform Traffic Control Devices for Streets and Highways (MUTCD)*, 2009.

Example 8.15: Taper Length

According to the *MUTCD*, for a roadway that has a posted speed limit of 70 mph, what is the minimum shifting taper length needed in order to laterally move traffic 12 ft?

A. 840 ft
B. 980 ft
C. 420 ft
D. 1,020 ft

Solution

Refer to Table 8.7. Since the speed is over 45 mph, use this formula:

$$L = WS$$

$$L = 70\text{ mph}(12\text{ ft}) = 840\text{ ft}$$

The example asks for the minimum shifting taper length, so the taper length criteria in Table 8.6 apply.

Shifting taper = at least 0.5 L

$$\min L = 0.5\,L$$

$$840\text{ ft}(0.5) = 420\text{ ft}$$

Answer: C

REFERENCES

1. Occupational Safety and Health Administration (OSHA). Safety and Health Regulations for Construction (Standards – 29 CFR, Part 1926). https://www.osha.gov/laws-regs/regulations/standardnumber/1926.
2. Occupational Safety and Health Administration (OSHA). Occupational Safety and Health Standards (Standards – 29 CFR, Part 1910). https://www.osha.gov/laws-regs/regulations/standardnumber/1910.
3. Occupational Safety and Health Administration (OSHA). Occupational Safety and Health Standards Recording and Reporting Occupational Injuries and Illnesses (Standards – 29 CFR, Part 1904). https://www.osha.gov/laws-regs/regulations/standardnumber/1904.
4. US Department of Transportation, Federal Highway Administration. 2009. *Manual on Uniform Traffic Control Devices for Streets and Highways* (*MUTCD*). Washington, DC: Federal Highway Administration. https://mutcd.fhwa.dot.gov/pdfs/2009r1r2/mutcd2009r1r2edition.pdf.
5. American Association of State Highway and Transportation Officials (AASHTO). 2011. *Roadside Design Guide (4th ed.)*. Washington, DC: AASHTO.

Unit Conversions

A.1

FROM	TO	CONVERSION FACTOR (MULTIPLY)
ac	$chain^2$	10
ac	ft^2	43,560
ac	hectare	0.40469
ac	m^2	4,046.87
ac	mi^2	1/640
ac-ft	ft^3	43,560
ac-ft	mi^2-in	0.01875
ac-ft	m^3	1,233.5
ac-ft	gal	325,851
ac-ft/day	ft^3/sec	0.50416
ac-ft/mi^2	in (runoff)	0.01875
ac-in	L	102,790
ac-in/hr	ft^3/sec	1.0083
angstrom	m	1.0×10^{-10}
atm	bar	1.01325
atm	cm Hg	76
atm	ft water	33.90
atm	in Hg	29.921
atm	lbf/in^2	14.696
atm	kPa	101.33
atm	Pa	1.0133×10^5
bar	atm	0.9869
bar	Pa	1.0×10^5
Btu	ft-lbf	778.26
Btu	J	1,055.056
Btu	kW-hr	2.928×10^{-4}
Btu	therm	1.0×10^{-5}
Btu/hr	ft-lbf/sec	0.21611
Btu/hr	hp	3.929×10^{-4}
Btu/hr	W	0.29307
Btu/lbm	kJ/kg	2.326
Btu/lbm-°R	KJ/kg·K	4.1868
Btu/s	ft-lbf/sec	778.26
Btu/s	ft-lbf/min	46,680
Btu/s	hp	1.4148
Btu/s	kW	1.0545
cal	Btu	3.968×10^{-3}
cal	J	4.1868
chain	ft	66
chain	mi (statute)	1/80
chain	rod	4
chain	yd	22
$chain^2$	ac	1/10
cm	ft	0.03281
cm	in	0.3937

(continued)

FROM	TO	CONVERSION FACTOR (MULTIPLY)
cm/s	ft/min	1.9686
cm^2	in^2	0.155
cm^3	gal	2.6417×10^{-4}
cm^3	L	0.001
day (mean solar)	s	86,400
day (sidereal)	s	86,164.09
darcy	ft^2	1.0623×10^{-11}
degree (angular)	grad	1/0.9
degree (angular)	radian	2π/360
degree (angular)	mil	17.778
dyne	N	1.0×10^{-5}
eV	J	1.6022×10^{-19}
ft	chain	1/66
ft	m	0.3048
ft	mi	1/5,280
ft	rod	0.0606
ft of water	lbf/in^2	0.43328
ft-kips	kN·m	1.358
ft-lbf (energy)	Btu	1.2851×10^{-3}
ft-lbf (energy)	J	1.3558
ft-lbf (energy)	kW·h	3.766×10^{-7}
ft-lbf (torque)	N·m	1.3558
ft-lbf/min	Btu/s	1/46,680
ft-lbf/min	hp	1/33,000
ft-lbf/min	kW	2.2598×10^{-5}
ft-lbf/sec	Btu/s	1.2849×10^{-3}
ft-lbf/sec	hp	1/550
ft-lbf/sec	kW	1.3558×10^{-3}
ft/min	cm/s	0.50798
ft/sec	mph	0.68180
ft^2	ac	1/43,560
ft^2	darcy	9.4135×10^{10}
ft^2	m^2	0.0929
ft^3	ac-ft	1/43,560
ft^3	in^3	1,728
ft^3	gal	7.4805
ft^3	L	28.32
ft^3	yd^3	0.03704
ft^3/day	gal/min	0.005195
ft^3/mi^2	in runoff	4.3×10^{-7}
ft^3/min	gal/day	10,772
ft^3/sec	ac-ft/day	1.9835
ft^3/sec	ac-in/hr	0.99177
ft^3/sec	gal/day	646,317
ft^3/sec	gal/min	448.83
ft^3/sec	mi^2-in/day	0.03719
ft^3/sec	MGD	0.64632
g	oz	0.03527
g	lbm	0.002205
g/cm^3	kg/m^3	1,000
g/cm^3	lbm/ft^3	62.428
gal	ac-ft	1/325,851
gal	cm^3	3,785.4
gal	ft^3	0.13368
gal (Imperial)	gal (U.S.)	1.2

FROM	TO	CONVERSION FACTOR (MULTIPLY)
gal (U.S.)	gal (Imperial)	0.8327
gal	L	3.7854
gal	m^3	3.7854×10^{-3}
gal	MG	1.0×10^{-6}
gal	yd^3	0.00495
gal (of water)	lbm (of water)	8.34
gal/ac-day	MG/ft^2-day	0.04356
gal/day	ft^3/min	9.283×10^{-5}
gal/day	ft^3/sec	1.5472×10^{-6}
gal/day	gal/min	1/1,440
gal/day	MG/day	1.0×10^{-6}
gal/day-ft	m^3/m·day	0.01242
gal/day-ft^2	m^3/m^2·day	0.04075
gal/day-ft^2	Meinzer unit	1.0
gal/day-ft^2	MGD/ac (mgad)	0.04355
gal/min	ft^3/sec	0.002228
gal/min	ft^3/day	192.5
gal/min	gal/day	1,440
gal/min	L/s	0.06309
grad	degrees (angular)	0.9
grain	lbm	1.4286×10^{-4}
grain/gal	lbm/MG	142.86
grain/gal	ppm	17.118
grain/gal	mg/L	17.118
hectare	ac	2.4711
hectare	m^2	10,000
hp	ft-lbf/sec	550
hp	ft-lbf/min	33,000
hp	kW	0.7457
hp	Btu/hr	2,545
hp	Btu/s	0.70678
hp-hr	Btu	2,545.2
in	cm	2.54
in	m	0.0254
in	mm	25.4
in Hg	lbf/in^2	0.491
in Hg	lbf/ft^2	70.704
in Hg	in water	13.6
in runoff	ac-ft/mi^2	53.3
in runoff	ft^3/mi^2	2.3230×10^{6}
in water	lbf/ft^2	5.1990
in water	lbf/in^2	0.0361
in water	in Hg	0.07353
in-lbf	N·m	0.11298
in/ft	mm/m	1/0.012
in^2	cm^2	6.4516
in^3	ft^3	1/1,728
J	Btu	9.4778×10^{-4}
J	eV	6.2415×10^{18}
J	ft-lbf	0.73756
J	N·m	1.0
J/s	W	1.0
kg	lbm	2.2046
kg/m^3	lbm/ft^3	0.06243

(continued)

FROM	TO	CONVERSION FACTOR (MULTIPLY)
kip	kN	4.448
kip	lbf	1,000
kip	N	4,448
kip/ft	kN/m	14.594
kip/ft^2	kPa	47.88
kJ	Btu	0.94778
kJ	ft-lbf	737.56
kJ/kg	Btu/lbm	0.42992
kJ/kg·K	Btu/lbm-°R	0.23885
km	ft	3,280.8
km	mi	0.62138
km/hr	mph	0.62138
kN	kips	0.2248
kN·m	ft-kips	0.73757
kN/m	kips/ft	0.06852
kPa	atm	9.8692×10^{-3}
kPa	lbf/in^2	0.14504
kPa	Pa	1,000
kPa	kips/ft^2	0.02089
kPa	bar	0.01
ksi	Pa	6.8948×10^{6}
ksi	kPa	6,894.8
kW	ft-lbf/sec	737.56
kW	ft-lbf/min	44,250
kW	hp	1.341
kW	Btu/hr	3,413
kW	Btu/s	0.9483
kW-hr	Btu	3,413
kW-hr	J	3.6×10^{6}
L	ac-in	1/102,790
L	cm^3	1,000
L	ft^3	0.03531
L	gal	0.26417
L	in^3	61.024
L	m^3	0.001
L/s	ft^3/min	2.1189
L/s	gal/min	15.85
lbf	kips	0.001
lbf	N	4.4482
lbf/ft^2	in Hg	0.01414
lbf/ft^2	in water	0.19234
lbf/ft^2	lbf/in^2	0.00694
lbf/ft^2	Pa	47.88
lbf/ft^2	tons/ft^2	5×10^{-4}
lbf/in^2	atm	0.06805
lbf/in^2	lbf/ft^2	144
lbf/in^2	ft water	2.308
lbf/in^2	in water	27.7
lbf/in^2	in Hg	2.0370
lbf/in^2	Pa	6,894.8
lbf/in^2	tons/in^2	0.0005
lbf/in^2	tons/ft^2	0.072
lbm	grains	7,000
lbm	g	453.59

FROM	TO	CONVERSION FACTOR (MULTIPLY)
lbm	kg	0.45359
lbm	mg	4.5359×10^5
lbm	tons (mass)	5×10^{-4}
lbm (of water)	gal (of water)	0.12
lbm/ac-ft-day	lbm/1000 ft^3-day	0.02296
lbm/ft^3	g/cm^3	0.016018
lbm/ft^3	kg/m^3	16.018
lbm/1000 ft^3-day	lbm/ac-ft-day	43.56
lbm/1000 ft^3-day	lbm/MG-day	133.68
lbm/MG	grains/gal	0.007
lbm/MG	mg/L	0.11983
lbm/MG-day	lbm/1000 ft^3-day	0.00748
leagues	m	4,428
m	angstroms	1×10^{10}
m	ft	3.2808
m	in	39.37
m	leagues	2.2583×10^{-4}
m	yd	1.0936
m/s	ft/min	196.85
m^2	ac	2.4711×10^{-4}
m^2	ft^2	10.764
m^2	hectare	1/10,000
m^3	ac-ft	8.1071×10^{-4}
m^3/m·d	gal/day-ft	80.5196
m^3/m^2·d	gal/day-ft^2	24.542
Meinzer unit	gal/day-ft^2	1.0
mg	lbm	2.2046×10^{-6}
mg/L	ppm	1.0
mg/L	grains/gal	0.05842
mg/L	lbm/MG	8.3454
MG	gal	1×10^6
MG/ac-day	gal/ft^2-day	22.968
MGD	ft^3/sec	1.5472
MGD	gal/day	1×10^6
MGD/ac (mgad)	gal/day-ft^2	22.957
mi	ft	5,280
mi	chains	80
mi	km	1.6093
mi (statute)	miles (nautical)	0.86839
mi	rods	320
mi/hr	ft/sec	1.4667
mi^2	acres	640
mi^2-in	ac-ft	53.3
mi^2-in/day	ft^3/s	26.89
micron	m	1×10^{-6}
micron	mm	0.001
mil (angular)	degrees	0.05625
mil (angular)	min	3.375
min (angular)	mils	0.2963
min (angular)	radians	2.90888×10^{-4}
min (time, mean solar)	s	60
mm	in	1/25.4
mm	microns	1,000
mm/m	in/ft	0.012

(continued)

FROM	TO	CONVERSION FACTOR (MULTIPLY)
Mpa	Pa	1×10^6
N	lbf	0.22481
N	dynes	1×10^5
N·m	ft-lbf	0.73756
N·m	in-lbf	8.8511
N·m	J	1.0
N/m^2	Pa	1.0
oz	g	28.353
Pa	kPa	0.001
Pa	ksi	1.4504×10^{-7}
Pa	lbf/in^2	1.4504×10^{-4}
Pa	lbf/ft^2	0.02089
Pa	MPa	1×10^{-6}
Pa	N/m^2	1.0
ppm	grains/gal	0.05842
radian	degrees (angular)	180/π
radian	min (angular)	3,437.7
rod	chain	0.25
rod	ft	16.5
rod	mi	1/320
s (time)	day (mean solar)	1/86,400
s (time)	day (sidereal)	1.1605×10^{-5}
s (time)	min	1/60
therm	Btu	1×10^5
ton (force)	lbf	2,000
ton (mass)	lbm	2,000
ton/ft^2	lbf/ft^2	2,000
ton/ft^2	lbf/in^2	13.889
W	Btu/hr	3.413
W	ft-lbf/sec	0.73756
W	hp	1.341×10^{-3}
W	J/s	1.0
yd	chain	1/22
yd	m	0.9144
yd^3	ft^3	27
yd^3	gal	201.97

Engineering Economics

Anthony Oyatayo, PE, MBA

A.2

CONTENTS

COMMONLY USED SYMBOLS

A	annual amount
C	cost or present worth of costs
D	depreciation
D_j	depreciation year j
DR	present worth of after-tax depreciation recovery
EUAC	equivalent uniform annual cost
f	federal income tax rate
f	general inflation rate
F	future worth
G	uniform gradient amount
i	effective rate per period
i_e	annual effective interest rate
k	number of compounding periods per year
m	an integer
n	number of compounding periods or life of asset
P	present worth
r	nominal annual interest rate
s	state income tax rate
S_n	expected salvage value in year n
t	composite tax rate
t	time

INTRODUCTION

Engineering economics deals with the benefits and costs associated with projects so we can determine whether they're worthy of pursuing. Fundamental concepts, such as the worth of money at some point in time, interest rates, and depreciation, are included in this field. These fundamental concepts help engineers to effectively and systematically evaluate economic solutions to engineering problems. The successful exam taker should have a general understanding of the sections in this appendix. Practice the questions as many times as possible so you can become familiar with the subject and with all of the sources of information that may be brought to the exam.

A.2.1 CASH FLOW

Cash flow is the sum of money coming in (receipts) or money coming out (disbursements). A cash flow diagram represents the flow of cash as arrows on a timeline. Downward-pointing arrows represent expenses/disbursements, and upward-pointing arrows represent receipts.

The year-end convention in the example assumes that the expenses during the year occur at the end of the year. Unless explicitly stated, this is the assumption of all engineering economics problems.

Example A.1

A new machine costs $30,000, and its maintenance will cost $2,000 per year. The machine generates $5,000 in revenue per year for a period of five years. The salvage value is $4,000 at the end of the five-year period.

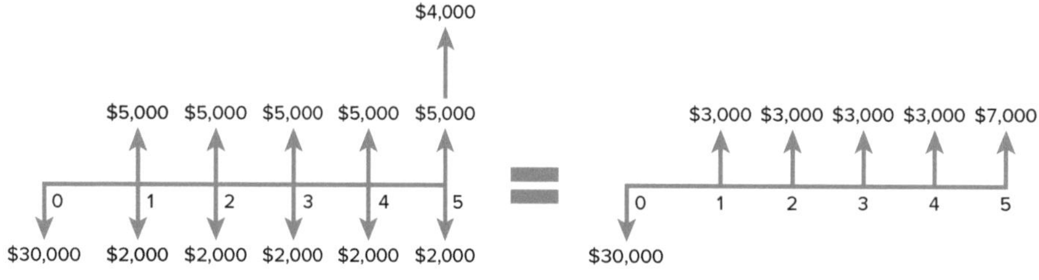

During the exam, you will be asked to perform various calculations. Some students prefer to use the formulas in Table A.1, and others prefer to use the specific factor tables, as shown in Table A.2.

A.2.2 SINGLE PAYMENT PRESENT WORTH

TABLE A.1

FACTOR NAME	CONVERTS	SYMBOL	FORMULA
Single Payment Compound Amount	to *F* given *P*	(*F/P*, *i*%, *n*)	$(1+i)^n$
Single Payment Present Worth	to *P* given *F*	(*P/F*, *i*%, *n*)	$(1+i)^{-n}$
Uniform Series Sinking Fund	to *A* given *F*	(*A/F*, *i*%, *n*)	$\frac{i}{(1+i)^n-1}$
Capital Recovery	to *A* given *P*	(*A/P*, *i*%, *n*)	$\frac{i(1+i)^n}{(1+i)^n-1}$
Uniform Series Compound Amount	to *F* given *A*	(*F/A*, *i*%, *n*)	$\frac{(1+i)^n-1}{i}$
Uniform Series Present Worth	to *P* given *A*	(*P/A*, *i*%, *n*)	$\frac{(1+i)^n-1}{i(1+i)^n}$
Uniform Gradient Present Worth	to *P* given *G*	(*P/G*, *i*%, *n*)	$\frac{(1+i)^n-1}{i^2(1+i)^n}-\frac{n}{i(1+i)^n}$
Uniform Gradient Future Worth[†]	to *F* given *G*	(*F/G*, *i*%, *n*)	$\frac{(1+i)^n-1}{i^2}-\frac{n}{i}$
Uniform Gradient Uniform Series	to *A* given *G*	(*A/G*, *i*%, *n*)	$\frac{1}{i}-\frac{n}{(1+i)^n-1}$

Source: *NCEES FE Reference Handbook, v. 9.5* (p. 136).

TABLE A.2

Factor Table for *i* = 6.00%									
N	*P/F*	*P/A*	*P/G*	*F/P*	*F/A*	*A/P*	*A/F*	*A/G*
1	0.9434	0.9434	0.0000	1.0600	1.0000	1.0600	1.0000	0.0000
2	0.8900	0.8334	0.8900	1.1236	2.0600	0.5454	0.4854	0.4854
3	0.8396	2.6730	2.5692	1.1910	3.1836	0.3741	0.3141	0.9612
4	0.7921	3.4651	4.9455	1.2625	4.3746	0.2886	0.2286	1.4272
5	**0.7473**	**4.2124**	**7.9345**	**1.3382**	**5.6371**	**0.2374**	**0.1774**	**1.8836**
6	0.7050	4.9173	11.4594	1.4185	6.9753	0.2034	0.1434	2.3304
7	0.6651	5.5824	15.4497	1.5036	8.3938	0.1791	0.1191	2.7676
8	0.6274	6.2098	19.8416	1.5938	9.8975	0.1610	0.1010	3.1952
9	0.5919	6.8017	24.5768	1.6895	11.4913	0.1470	0.0870	3.6133
10	**0.5584**	**7.3601**	**29.6023**	**1.7908**	**13.1808**	**0.1359**	**0.0759**	**4.0220**
11	0.5268	7.8869	34.8702	1.8983	14.9716	0.1268	0.0668	4.4213
12	0.4970	8.3838	40.3369	2.0122	16.8699	0.1193	0.0593	4.8113
13	0.4688	8.8527	45.9629	2.1329	18.8821	0.1130	0.0530	5.1920
14	0.4423	9.2950	51.7128	2.2609	21.0151	0.1076	0.0476	5.5635
15	**0.4173**	**9.7122**	**57.5546**	**2.3966**	**23.2760**	**0.1030**	**0.0430**	**5.5260**
16	0.3936	10.1059	63.4592	2.5404	25.6725	0.0990	0.0390	6.2794
17	0.3714	10.4773	69.4011	2.6928	28.2129	0.0954	0.0354	6.6240
18	0.3505	10.8276	75.3569	2.8543	30.9057	0.0924	0.0324	6.9597
19	0.3305	11.1581	81.3062	3.0256	33.7600	0.0896	0.0296	7.2867
20	**0.3118**	**11.4699**	**87.2304**	**3.2071**	**36.7856**	**0.0872**	**0.0272**	**7.6051**
21	0.2942	11.7641	93.1136	3.3996	39.9927	0.0850	0.0250	7.9151
22	0.2775	12.0416	98.9412	3.6035	43.3923	0.0830	0.0230	8.2166
23	0.2618	12.3034	104.7007	3.8197	46.9958	0.0813	0.0213	8.5099

TABLE A.2 *(continued)*

Factor Table for $i = 6.00\%$								
N	*P/F*	*P/A*	*P/G*	*F/P*	*F/A*	*A/P*	*A/F*	*A/G*
24	0.2470	12.5504	110.3812	4.0489	50.8156	0.0797	0.0197	8.7951
25	**0.2330**	**12.7834**	**115.9732**	**4.2919**	**54.8645**	**0.0782**	**0.0182**	**9.0722**
30	0.1741	13.7648	142.3588	5.7435	79.0582	0.0726	0.0126	10.3422
40	0.0972	15.0463	185.9568	10.2857	154.7620	0.0665	0.0065	12.3590
50	0.0543	15.7619	217.4574	18.4202	290.3359	0.0634	0.0034	13.7964
60	0.0303	16.1614	239.0428	32.9877	533.1282	0.0619	0.0019	14.7909
100	**0.0029**	**16.6175**	**272.0471**	**339.3021**	**5,638.681**	**0.0602**	**0.0002**	**16.3711**

Source: *NCEES FE Reference Handbook, v. 9.5* (p. 141).

The future value of a present amount of money is the amount that will be obtained at some point in the future using a specific rate of interest. The future value is the sum of the present value and the accrued interest.

Example A.2

If you would like to accumulate \$15,000 at the end of a five-year period, how much should you put into a high-yield savings account today, with a 6% annual interest rate?

Solution

$P = F/(1 + i)^n = \$15{,}000/(1 + 0.06)^5 = \$11{,}208.87$

The problem can also be solved using factor tables.

$(P/F, i\%, n) = 0.6209$

$\$15{,}000(P/F, 6\%, 5) = \$15{,}000(0.7473) = \$11{,}209.50$

Example A.3

A municipality is planning to build a new water treatment plant. It can be built at a reduced capacity now for \$10 million and could be expanded in 15 years (increasing to full capacity) for an additional \$20 million. An alternative is to build the plant at full capacity now for a total of \$35 million. At 6% interest, which alternative should be chosen based on the present worth (*P*) of cost?

Solution

For the two-stage construction:

P of cost = \$10 million + \$20 million $(P/F, 6\%, 15)$

= \$10 million + (\$20 × 0.4173 million)

= \$18.3 million

For the single-stage construction:

P of cost = \$25 million

The two-stage construction option should be chosen.

A.2.3 UNIFORM SERIES

Uniform series represents a series of payments that occur with the same amount and at the same interval (for example, a payment occurs every year). The uniform payment is noted as A and it can be used to find both the present and future worth that is equivalent to the series of uniform payments.

Example A.4

The cost of utilities, taxes, and maintenance on a home is \$5,000 per year. How much money would have to be invested today at an 8% interest rate to cover these expenses for the next five years? (Assume no inflation or tax increase.)

Solution

$\$5{,}000(P/A, 8\%, 5) = \$5{,}000 \times 3.9927 = \$19{,}964$

Example A.5

A company is considering purchasing a \$15,000 high-end 3D printer. There will be an estimated annual operating cost of \$2,000 per year. After four years, the operating cost is expected to rise to \$2,700 per year. The printer will be disposed of in 20 years, and there will be no salvage value. Using an interest rate of 6%, what is the present worth (P) cost of the printer?

Solution

$P = \$15{,}000 + \$2{,}000(P/A, 6\%, 4) + \$2{,}700(P/A, 6\%, 16)(P/F, 6\%, 4)$

$= \$15{,}000 + \$2{,}000(3.4651) + \$2{,}700(10.1059)(0.7921) = \$43{,}543.39$

Example A.6

A company must decide between the purchase of two machines. Machine 1 costs \$40,000 and has an estimated annual maintenance cost of \$4,000. Machine 1 can be sold for \$5,000 at the end of its useful life of three years. Machine 2 has an initial cost of \$45,000 and an estimated annual maintenance cost of \$3,500. It will be sold for \$7,000 at the end of its useful life of five years. According to the company maintenance program, there will be a two-year overhaul cost of \$7,500 and \$6,500 for machines 1 and 2, respectively. Using an interest rate of 6% per year, which machine will have the lowest annual cost?

Solution

Machine 1:

$\text{A} = \$40{,}000(A/P, 6\%, 3) - \$5{,}000(A/F, 6\%, 3) + \$4{,}000 + \$7{,}500(P/F, 6\%, 2)(A/P, 6\%, 3)$

$= \$40{,}000(0.3741) - \$5{,}000(0.3141) + \$4{,}000 + \$7{,}500(0.8900)(0.3741)$

$= \$19{,}890.62$

Machine 2:

$\text{A} = \$45{,}000(A/P, 6\%, 5) - \$7{,}000(A/F, 6\%, 5) + \$3{,}500 + \$6{,}500(P/F, 6\%, 2)$
$(A/P, 6\%, 5) + \$6{,}500(P/F, 6\%, 4)(A/P, 6\%, 5)$

$= \$45{,}000(0.2374) - \$7{,}000(0.1774) + \$3{,}500 + \$6{,}500(0.8900)(0.2374)$
$+ \$6{,}500(0.7921)(0.2374)$

$= \$15{,}536.85$

Machine 2 will have a lower annual cost.

A.2.4 UNIFORM GRADIENT

The uniform gradient factor applies to a uniformly increasing cash flow, which may be positive (revenue) or negative (costs). If the cash flow is in the proper form, one can determine its:

- Present worth (P) using the uniform gradient factor (P/G, $i\%$, n)
- Future worth (F) using the uniform gradient factor (F/G, $i\%$, n)
- Equivalent uniform series (A) using the uniform gradient factor (A/G, $i\%$, n)

Note that only one of the above three gradient factors is needed. Once one of the three cash flows—P, F, or A—is determined, the other two cash flows can easily be determined using other cash flow factors, such as P/F, F/P, A/P, and so forth. For the purpose of this discussion, only the P/G factor will be used. The uniform gradient factor can be used to find the present worth of a uniformly increasing or decreasing cash flow that starts in year 2.

Example A.7

The maintenance cost of a new car is $0 in year 1, but it is expected to increase by $250 per year for the next five years. What is the present worth of the next five years of maintenance costs of this car? Use an interest rate of 8%.

Solution

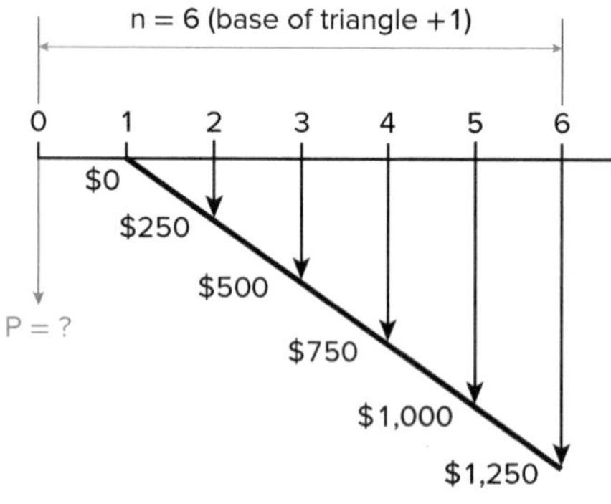

$$P = (P/G, i\%, n) = (\$250)(P/G, 8\%, 6) = (\$250)(10.523) = \$2,630.75$$

A.2.5 ANNUAL COST METHOD

The annual cost method can compare alternatives that accomplish the same purpose but have different useful lives. The calculated cost is known as the equivalent uniform annual cost (EUAC).

Example A.8

Which of the following alternatives is best if the interest rate is 7% over a 30-year period?

	A	B
Type	Brick	Wood
Life	30 years	10 years
Initial cost	$1,800	$450
Maintenance	$10/year	$20/year

Solution

$$\text{EUAC}(A) = (\$1,800)\ (A/P, 7\%, 30) + \$10 = (\$1,800)\ (0.0806) + 10 = \$155$$

$$\text{EUAC}(B) = (\$450)\ (A/P, 7\%, 10) + \$20 = (\$450)\ (0.1424) + 20 = \$84$$

Note that the maintenance cost is simply added since it is already an annual cost. The nonannual costs are annualized in the calculation.

A.2.6 NET PRESENT WORTH

The net present worth (NPW) can be used to compare alternatives based on the inclusion of annual benefits. It is defined as the present worth (P) of benefits minus the P of costs. Furthermore, the NPW must be greater than zero for the project to be considered acceptable.

Example A.9

Three options are being considered for streamlining company operations. The three options involve the purchase of one of three machines, each of which has an estimated useful life of 20 years. Using a 6% interest rate, estimate the NPW for each machine and determine which one should be selected.

MACHINE NUMBER	*P* COST ($)	ANNUAL BENEFIT ($)
1	$12,000	$ 950
2	$15,000	$2,000
3	$ 9,000	$1,200

Solution

$(P/A, i = 6\%, 20 \text{ years}) = 11.4699$

$\text{NPW}(1) = (\$950 \times 11.4699) - \$12,000 = -\$1,103.60$

$\text{NPW}(2) = (\$2,000 \times 11.4699) - \$15,000 = \$7,939.80$

$\text{NPW}(3) = (\$1,200 \times 11.4699) - \$9,000 = \$4,763.88$

The NPW for machine 1 is negative, so that automatically is not a viable option. Machine 2's NPW is the highest. Therefore, machine 2 should be purchased.

A.2.7 BENEFIT-COST RATIO METHOD

The benefit-cost (B/C) ratio method is often used on public works projects where benefits and costs are allocated to different parts of the community. Using this method, the present worth of all benefits is divided by the P of all costs. The project is considered acceptable if the B/C ratio equals or exceeds one.

$B/C = P$ of benefits/P of costs

Example A.10

A proposed project has an initial cost of $30 million. The present worth of the annual cost is $8 million, and the present worth of the benefit is $49 million. What is the benefit-cost ratio?

Solution

Benefit (B) = $49 million

Cost (C) = $30 million + $8 million

$$\frac{\text{Benefit}(B)}{\text{Cost}(C)} = \frac{\$49 \text{ million}}{\$38 \text{ million}} = 1.29$$

Example A.11

A municipality is trying to decide between two new traffic systems. System 1 costs $2,000, has a five-year useful life, and is expected to result in annual savings of $350. System 2 costs $3,000, has an eight-year useful life, and is expected to result in annual savings of $500. Both systems have no salvage value at the end of their useful lives. Using an interest rate of 8%, which system should the municipality purchase based on the B/C ratio?

Solution

System 1:

P of cost = $2,000

P of benefits = $350 ($P/A$, 8%, 5) = $350 × 3.9927 = $1,397

$B/C = \$1,397/\$2,000 = 0.70$

Example A.11 *(continued)*

System 2:

P of cost = \$3,000

P of benefits = \$500 $(P/A, 8\%, 8)$ = \$500 × 5.7466 = \$2,873

B/C = \$2,873/\$3,000 = 0.96

Although both systems have a B/C ratio of less than one, system 2 should be chosen over system 1.

A.2.8 RATE OF RETURN METHOD

The rate of return (ROR) is the interest paid on the unpaid balance of a loan. The payment schedule, when the final payment is made, makes the unpaid loan balance equal zero. To calculate the ROR of the investment, we must convert the various consequences of the investment into a cash flow (P of benefits – P of costs = 0; $B/C = 1$).

Example A.12

What is the ROR on a \$525 investment that pays back \$1,187 in 14 years?

Solution

$(F/P, i\%, 14) = F/P$

F/P = \$1,187/\$525 = 2.2609

Check the interest tables in Appendix A.3 to see which rate is close to a factor of 2.2609:

$i = 6\%$

A.2.9 DEPRECIATION

Depreciation is the allocation of the cost of an asset over its depreciable life. When depreciation is used in engineering economic analysis, there will be an increase in the after-tax present worth (profitability) of the asset being analyzed. The larger the depreciation amount, the greater the profitability will be. The depreciation basis of an asset is the part of its purchase price that is spread over its service life (depreciation period). In this appendix, the following three depreciation methods are discussed:

1. Straight-line depreciation
2. Sum-of-years digits (SOYD) depreciation
3. Modified accelerated cost recovery system (MACRS) depreciation

A.2.9.1 Straight-Line Depreciation

The straight-line method is used if the depreciation is the same each year. The depreciation basis is allocated to the years (n) during the period of depreciation. Depreciation in year j is calculated as:

$$D_j = \frac{C - S_n}{n}$$

Example A.13

The cost of an asset is \$900, and it has a useful life of five years. It is estimated that the end-of-life salvage value is \$70. Compute the annual depreciation using the straight-line method.

Solution

$C = \$900$

$S = \$70$

$n = 5$

$D = (\$900 - 70)/5 = \$166$ per year

A.2.9.2 Sum-of-Years Digits Depreciation

The sum-of-years digits (SOYD) depreciation method results in larger depreciation values in the early years of an asset's life and smaller values as the asset nears the end of its estimated useful life. In SOYD, the digits from one to n are summed. The total, T, can be calculated by $T = (\frac{1}{2})(n)(n+1)$. The depreciation in year j can be calculated using $D_j = (C - S_n)(n - j + 1)/T$.

Example A.14

The cost of an asset is \$900, and it has a useful life of five years. It is estimated that the end-of-life salvage value is \$70. Compute the depreciation schedule using the SOYD method.

Solution

$T = (\frac{1}{2})(5)(5 + 1) = 15$

Year 1: $(\$900 - \$70)(5 - 1 + 1)/15 = \$277$

Year 2: $(\$900 - \$70)(5 - 2 + 1)/15 = \$221$

Year 3: $(\$900 - \$70)(5 - 3 + 1)/15 = \$166$

Year 4: $(\$900 - \$70)(5 - 4 + 1)/15 = \$111$

Year 5: $(\$900 - \$70)(5 - 5 + 1)/15 = \$55$

A.2.9.3 MACRS Depreciation

Property placed into service must use MACRS, under which the cost recovery amount in year j of an asset's cost recovery period is calculated by multiplying the initial cost by a factor.

$D_j = C \times \text{factor}$

The table below shows the factor that is applied to a specific year, based on the recovery period.

TABLE A.3

MACRS FACTORS				
	RECOVERY PERIOD (YEARS)			
YEAR	3	5	7	10
	RECOVERY RATE (PERCENT)			
1	33.33	20.00	14.29	10.00
2	44.45	32.00	24.49	18.00
3	14.81	19.20	17.49	14.40
4	7.41	11.52	12.49	11.52
5		11.52	8.93	9.22
6		5.76	8.92	7.37
7			8.93	6.55
8			4.46	6.55
9				6.56
10				6.55
11				3.28

Source: *NCEES FE Reference Handbook, v. 9.5* (p. 137).

Example A.15

A company has purchased equipment that will be used on the production floor. The equipment cost the company \$50,000, and it falls under federal regulations of a seven-year recovery period. If the equipment was financed using a 6% interest rate, what is the present worth of the amount that will be depreciated in the third year of the recovery period?

Solution

Using a seven-year recovery period, the recovery rate in the third year is 17.5%.

$(P/F, 6\%, 3) = 0.8396$

$\$50{,}000(0.175)(0.8396) = \$7{,}346.50$

A.2.10 COMPOUND INTEREST

Compound interest refers to an interest rate that is compounded (accumulated) multiple times in a period of time. A compounded interest rate can be converted to an effective annual interest rate.

Example A.16

A car has been purchased with 100% financing. The interest rate is 10%, and it is compounded monthly per year. What is the effective annual interest rate of the loan?

Solution

Compute the effective interest rate per period.

$$i = \frac{r}{m}$$

Compute the effective annual interest rate.

$$i_e = \left(1 + \frac{r}{m}\right)^m - 1$$

r = annual interest rate

m = number of compounded periods per year

$i_e = [1 + (0.1/12)]^{12} - 1 = 10.5\%$

A.2.11 BREAKEVEN ANALYSIS

Breakeven analysis is a way to determine when the value of one option becomes equal to the value of another option. This type of analysis can be used to help determine the best of two or more alternatives. Breakeven analysis involves identifying the variable of interest and then finding the value of that variable that will lead to a breakeven condition.

Example A.17

Two methods of corrosion abatement are being considered. Method 1 involves lining at a cost of $20,000. OEM documentation says the lining is expected to last 20 years. Maintenance is expected to cost $1 per square foot per six months. Method 2 involves the use of a chemical spray, which costs $45 per gallon, with one gallon capable of treating 20 ft^2. The spraying equipment will cost about $1,500 and will have an estimated life of three years with a salvage value of $0. At an interest rate of 10% per year, (a) how many square feet must require treatment for both methods to break even? (b) If 500 ft^2 must be treated each year, which method should be selected?

Solution

a: $\$20{,}000(A/P, 10\%, 20) + 2x = \$1{,}500(A/P, 10\%, 3) + 45/20x$

$\$20{,}000(0.1175) + 2x = \$1{,}500(0.4021) + 2.25x$

$\$2{,}350 + 2x = \$603.15 + 2.25x$

$\$1{,}746.85 = 0.25x$

$x = 6{,}987.40\ \text{ft}^2/\text{year}$

b: Method 1 = $\$20{,}000(0.1175) + 2(500) = \$3{,}350$

Method 2 = $\$1{,}500(0.4021) + 2.25(500) = \$1{,}728$

At 500 ft^2 per year, the spray method has the lower cost.

Standard Cash Flow Equations/Diagrams

A.3

COMPUTE	GIVEN	EQUATION	GRAPHICS
P	F	$P = F(1+i)^{-n}$ $(P/F, i\%, n)$	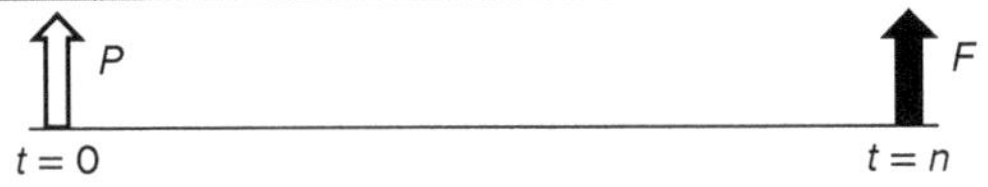 P at $t=0$; F at $t=n$
F	P	$F = P(1+i)^{n}$ $(F/P, i\%, n)$	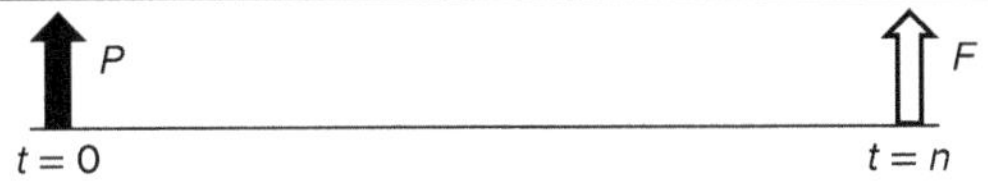 P at $t=0$; F at $t=n$
P	A	$P = A\left(\frac{(1+i)^n - 1}{i(1+i)^n}\right)$ $(P/A, i\%, n)$	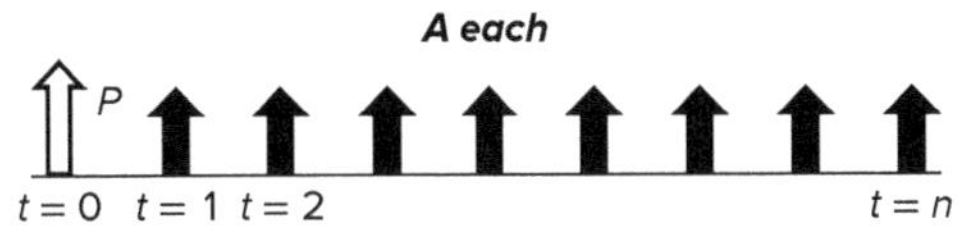 A each; P; $t=0$, $t=1$, $t=2$, $t=n$
A	P	$A = P\left(\frac{i(1+i)^n}{(1+i)^n - 1}\right)$ $(A/P, i\%, n)$	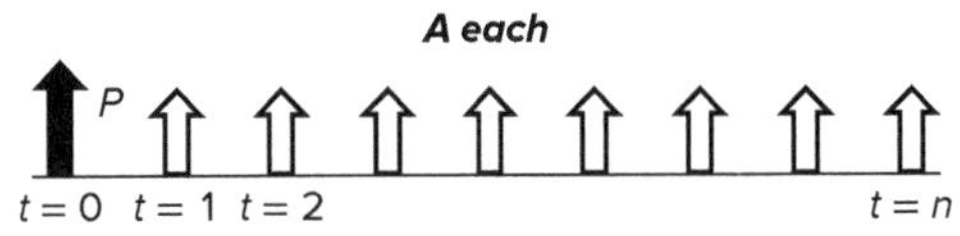 A each; P; $t=0$, $t=1$, $t=2$, $t=n$
F	A	$F = A\left(\frac{(1+i)^n - 1}{i}\right)$ $(F/A, i\%, n)$	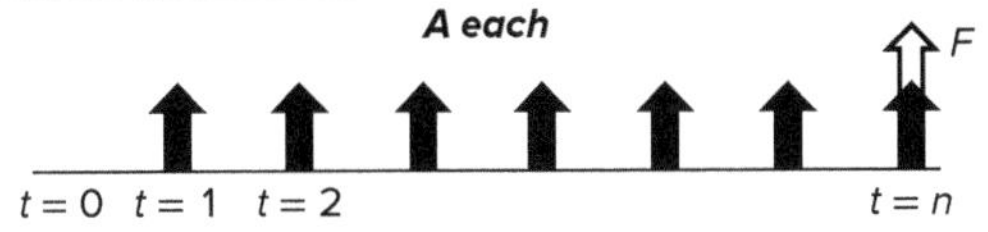 A each; F; $t=0$, $t=1$, $t=2$, $t=n$
A	F	$A = F\left(\frac{i}{(1+i)^n - 1}\right)$ $(A/F, i\%, n)$	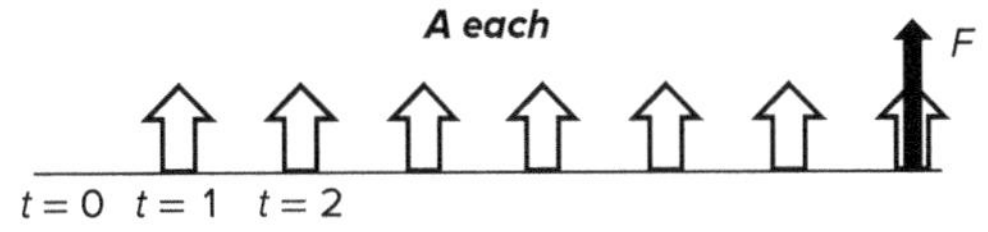 A each; F; $t=0$, $t=1$, $t=2$
P	G	$P = G\left(\frac{(1+i)^n - 1}{i^2(1+i)^n} - \frac{n}{i(1+i)^n}\right)$ $(P/G, i\%, n)$	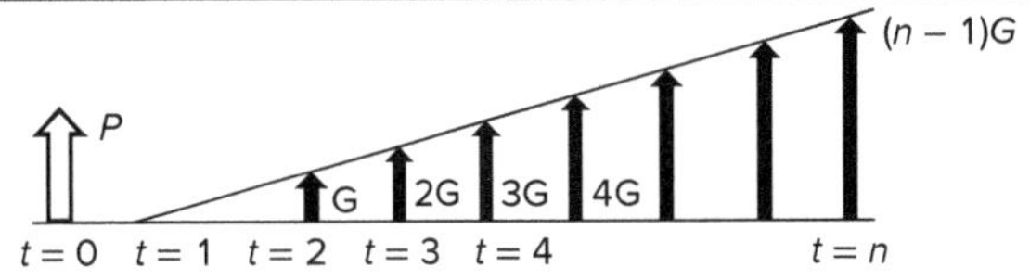 P; G, 2G, 3G, 4G, $(n-1)G$; $t=0$, $t=1$, $t=2$, $t=3$, $t=4$, $t=n$

Legend: Given Value (solid arrow) Computing Value (hollow arrow)

INTEREST RATE TABLES

Interest rate tables can be found in the *NCEES Reference Handbook, v. 9.5* (p.139–143).

Factor Table for $i = 0.50\%$

N	*P/F*	*P/A*	*P/G*	*F/P*	*F/A*	*A/P*	*A/F*	*A/G*
1	0.9950	0.9950	0.0000	1.0050	1.0000	1.0050	1.0000	0.0000
2	0.9901	1.9851	0.9901	1.0100	2.0050	0.5038	0.4988	0.4988
3	0.9851	2.9702	2.9604	1.0151	3.0150	0.3367	0.3317	0.9967
4	0.9802	3.9505	5.9011	1.0202	4.0301	0.2531	0.2481	1.4938
5	**0.9754**	**4.9259**	**9.8026**	**1.0253**	**5.0503**	**0.2030**	**0.1980**	**1.9900**
6	0.9705	5.8964	14.6552	1.0304	6.0755	0.1696	0.1646	2.4855
7	0.9657	6.8621	20.4493	1.0355	7.1059	0.1457	0.1407	2.9801
8	0.9609	7.8230	27.1755	1.0407	8.1414	0.1278	0.1228	3.4738
9	0.9561	8.7791	34.8244	1.0459	9.1821	0.1139	0.1089	3.9668
10	**0.9513**	**9.7304**	**43.3865**	**1.0511**	**10.2280**	**0.1028**	**0.0978**	**4.4589**
11	0.9466	10.6770	52.8526	1.0564	11.2792	0.0937	0.0887	4.9501
12	0.9419	11.6189	63.2136	1.0617	12.3356	0.0861	0.0811	5.4406
13	0.9372	12.5562	74.4602	1.0670	13.3972	0.0796	0.0746	5.9302
14	0.9326	13.4887	86.5835	1.0723	14.4642	0.0741	0.0691	6.4190
15	**0.9279**	**14.4166**	**99.5743**	**1.0777**	**15.5365**	**0.0694**	**0.0644**	**6.9069**
16	0.9233	15.3399	113.4238	1.0831	16.6142	0.0652	0.0602	7.3940
17	0.9187	16.2586	128.1231	1.0885	17.6973	0.0615	0.0565	7.8803
18	0.9141	17.1728	143.6634	1.0939	18.7858	0.0582	0.0532	8.3658
19	0.9096	18.0824	160.0360	1.0994	19.8797	0.0553	0.0503	8.8504
20	**0.9051**	**18.9874**	**177.2322**	**1.1049**	**20.9791**	**0.0527**	**0.0477**	**9.3342**
21	0.9006	19.8880	195.2434	1.1104	22.0840	0.0503	0.0453	9.8172
22	0.8961	20.7841	214.0611	1.1160	23.1944	0.0481	0.0431	10.2993
23	0.8916	21.6757	233.6768	1.1216	24.3104	0.0461	0.0411	10.7806
24	0.8872	22.5629	254.0820	1.1272	25.4320	0.0443	0.0393	11.2611
25	**0.8828**	**23.4456**	**275.2686**	**1.1328**	**26.5591**	**0.0427**	**0.0377**	**11.7407**
30	0.8610	27.7941	392.6324	1.1614	32.2800	0.0360	0.0310	14.1265
40	0.8191	36.1722	681.3347	1.2208	44.1588	0.0276	0.0226	18.8359
50	0.7793	44.1428	1,035.6966	1.2832	56.6452	0.0227	0.0177	23.4624
60	0.7414	51.7256	1,448.6458	1.3489	69.7700	0.0193	0.0143	28.0064
100	**0.6073**	**78.5426**	**3,562.7934**	**1.6467**	**129.3337**	**0.0127**	**0.0077**	**45.3613**

Factor Table for i = 1.00%

N	P/F	P/A	P/G	F/P	F/A	A/P	A/F	A/G
1	0.9901	0.9901	0.0000	1.0100	1.0000	1.0100	1.0000	0.0000
2	0.9803	1.9704	0.9803	1.0201	2.0100	0.5075	0.4975	0.4975
3	0.9706	2.9410	2.9215	1.0303	3.0301	0.3400	0.3300	0.9934
4	0.9610	3.9020	5.8044	1.0406	4.0604	0.2563	0.2463	1.4876
5	**0.9515**	**4.8534**	**9.6103**	**1.0510**	**5.1010**	**0.2060**	**0.1960**	**1.9801**
6	0.9420	5.7955	14.3205	1.0615	6.1520	0.1725	0.1625	2.4710
7	0.9327	6.7282	19.9168	1.0721	7.2135	0.1486	0.1386	2.9602
8	0.9235	7.6517	26.3812	1.0829	8.2857	0.1307	0.1207	3.4478
9	0.9143	8.5650	33.6959	1.0937	9.3685	0.1167	0.1067	3.9337
10	**0.9053**	**9.4713**	**41.8435**	**1.1046**	**10.4622**	**0.1056**	**0.0956**	**4.4179**
11	0.8963	10.3676	50.8067	1.1157	11.5668	0.0965	0.0865	4.9005
12	0.8874	11.2551	60.5687	1.1268	12.6825	0.0888	0.0788	5.3815
13	0.8787	12.1337	71.1126	1.1381	13.8093	0.0824	0.0724	5.8607
14	0.8700	13.0037	82.4221	1.1495	14.9474	0.0769	0.0669	6.3384
15	**0.8613**	**13.8651**	**94.4810**	**1.1610**	**16.0969**	**0.0721**	**0.0621**	**6.8143**
16	0.8528	14.7179	107.2734	1.1726	17.2579	0.0679	0.0579	7.2886
17	0.8444	15.5623	120.7834	1.1843	18.4304	0.0643	0.0543	7.7613
18	0.8360	16.3983	134.9957	1.1961	19.6147	0.0610	0.0510	8.2323
19	0.8277	17.2260	149.8950	1.2081	20.8109	0.0581	0.0481	8.7017
20	**0.8195**	**18.0456**	**165.4664**	**1.2202**	**22.0190**	**0.0554**	**0.0454**	**9.1694**
21	0.8114	18.8570	181.6950	1.2324	23.2392	0.0530	0.0430	9.6354
22	0.8034	19.6604	198.5663	1.2447	24.4716	0.0509	0.0409	10.0998
23	0.7954	20.4558	216.0660	1.2572	25.7163	0.0489	0.0389	10.5626
24	0.7876	21.2434	234.1800	1.2697	26.9735	0.0471	0.0371	11.0237
25	**0.7798**	**22.0232**	**252.8945**	**1.2824**	**28.2432**	**0.0454**	**0.0354**	**11.4831**
30	0.7419	25.8077	355.0021	1.3478	34.7849	0.0387	0.0277	13.7557
40	0.6717	32.8347	596.8561	1.4889	48.8864	0.0305	0.0205	18.1776
50	0.6080	39.1961	879.4176	1.6446	64.4632	0.0255	0.0155	22.4363
60	0.5504	44.9550	1,192.8061	1.8167	81.6697	0.0222	0.0122	26.5333
100	**0.3697**	**63.0289**	**2,605.7758**	**2.7048**	**170.4814**	**0.0159**	**0.0059**	**41.3426**

Factor Table for $i = 1.50\%$

N	P/F	P/A	P/G	F/P	F/A	A/P	A/F	A/G
1	0.9852	0.9852	0.0000	1.0150	1.0000	1.0150	1.0000	0.0000
2	0.9707	1.9559	0.9707	1.0302	2.0150	0.5113	0.4963	0.4963
3	0.9563	2.9122	2.8833	1.0457	3.0452	0.3434	0.3284	0.9901
4	0.9422	3.8544	5.7098	1.0614	4.0909	0.2594	0.2444	1.4814
5	**0.9283**	**4.7826**	**9.4229**	**1.0773**	**5.1523**	**0.2091**	**0.1941**	**1.9702**
6	0.9145	5.6972	13.9956	1.0934	6.2296	0.1755	0.1605	2.4566
7	0.9010	6.5982	19.4018	1.1098	7.3230	0.1516	0.1366	2.9405
8	0.8877	7.4859	26.6157	1.1265	8.4328	0.1336	0.1186	3.4219
9	0.8746	8.3605	32.6125	1.1434	9.5593	0.1196	0.1046	3.9008
10	**0.8617**	**9.2222**	**40.3675**	**1.1605**	**10.7027**	**0.1084**	**0.0934**	**4.3772**
11	0.8489	10.0711	48.8568	1.1779	11.8633	0.0993	0.0843	4.8512
12	0.8364	10.9075	58.0571	1.1956	13.0412	0.0917	0.0767	5.3227
13	0.8240	11.7315	67.9454	1.2136	14.2368	0.0852	0.0702	5.7917
14	0.8118	12.5434	78.4994	1.2318	15.4504	0.0797	0.0647	6.2582
15	**0.7999**	**13.3432**	**89.6974**	**1.2502**	**16.6821**	**0.0749**	**0.0599**	**6.7223**
16	0.7880	14.1313	101.5178	1.2690	17.9324	0.0708	0.0558	7.1839
17	0.7764	14.9076	113.9400	1.2880	19.2014	0.0671	0.0521	7.6431
18	0.7649	15.6726	126.9435	1.3073	20.4894	0.0638	0.0488	8.0997
19	0.7536	16.4262	140.5084	1.3270	21.7967	0.0609	0.0459	8.5539
20	**0.7425**	**17.1686**	**154.6154**	**1.3469**	**23.1237**	**0.0582**	**0.0432**	**9.0057**
21	0.7315	17.9001	169.2453	1.3671	24.4705	0.0559	0.0409	9.4550
22	0.7207	18.6208	184.3798	1.3876	25.8376	0.0537	0.0387	9.9018
23	0.7100	19.3309	200.0006	1.4084	27.2251	0.0517	0.0367	10.3462
24	0.6995	20.0304	216.0901	1.4295	28.6335	0.0499	0.0349	10.7881
25	**0.6892**	**20.7196**	**232.6310**	**1.4509**	**30.0630**	**0.0483**	**0.0333**	**11.2276**
30	0.6398	24.0158	321.5310	1.5631	37.5387	0.0416	0.0266	13.3883
40	0.5513	29.9158	524.3568	1.8140	54.2679	0.0334	0.0184	17.5277
50	0.4750	34.9997	749.9636	2.1052	73.6828	0.0286	0.0136	21.4277
60	0.4093	39.3803	988.1674	2.4432	96.2147	0.0254	0.0104	25.0930
100	**0.2256**	**51.6247**	**1,937.4506**	**4.4320**	**228.8030**	**0.0194**	**0.0044**	**37.5295**

Factor Table for $i = 2.00\%$

N	*P/F*	*P/A*	*P/G*	*F/P*	*F/A*	*A/P*	*A/F*	*A/G*
1	0.9804	0.9804	0.0000	1.0200	1.0000	1.0200	1.0000	0.0000
2	0.9612	1.9416	0.9612	1.0404	2.0200	0.5150	0.4950	0.4950
3	0.9423	2.8839	2.8458	1.0612	3.0604	0.3468	0.3268	0.9868
4	0.9238	3.8077	5.6173	1.0824	4.1216	0.2626	0.2426	1.4752
5	**0.9057**	**4.7135**	**9.2403**	**1.1041**	**5.2040**	**0.2122**	**0.1922**	**1.9604**
6	0.8880	5.6014	13.6801	1.1262	6.3081	0.1785	0.1585	2.4423
7	0.8706	6.4720	18.9035	1.1487	7.4343	0.1545	0.1345	2.9208
8	0.8535	7.3255	24.8779	1.1717	8.5830	0.1365	0.1165	3.3961
9	0.8368	8.1622	31.5720	1.1951	9.7546	0.1225	0.1025	3.8681
10	**0.8203**	**8.9826**	**38.9551**	**1.2190**	**10.9497**	**0.1113**	**0.0913**	**4.3367**
11	0.8043	9.7868	46.9977	1.2434	12.1687	0.1022	0.0822	4.8021
12	0.7885	10.5753	55.6712	1.2682	13.4121	0.0946	0.0746	5.2642
13	0.7730	11.3484	64.9475	1.2936	14.6803	0.0881	0.0681	5.7231
14	0.7579	12.1062	74.7999	1.3195	15.9739	0.0826	0.0626	6.1786
15	**0.7430**	**12.8493**	**85.2021**	**1.3459**	**17.2934**	**0.0778**	**0.0578**	**6.6309**
16	0.7284	13.5777	96.1288	1.3728	18.6393	0.0737	0.0537	7.0799
17	0.7142	14.2919	107.5554	1.4002	20.0121	0.0700	0.0500	7.5256
18	0.7002	14.9920	119.4581	1.4282	21.4123	0.0667	0.0467	7.9681
19	0.6864	15.6785	131.8139	1.4568	22.8406	0.0638	0.0438	8.4073
20	**0.6730**	**16.3514**	**144.6003**	**1.4859**	**24.2974**	**0.0612**	**0.0412**	**8.8433**
21	0.6598	17.0112	157.7959	1.5157	25.7833	0.0588	0.0388	9.2760
22	0.6468	17.6580	171.3795	1.5460	27.2990	0.0566	0.0366	9.7055
23	0.6342	18.2922	185.3309	1.5769	28.8450	0.0547	0.0347	10.1317
24	0.6217	18.9139	199.6305	1.6084	30.4219	0.0529	0.0329	10.5547
25	**0.6095**	**19.5235**	**214.2592**	**1.6406**	**32.0303**	**0.0512**	**0.0312**	**10.9745**
30	0.5521	22.3965	291.7164	1.8114	40.5681	0.0446	0.0246	13.0251
40	0.4529	27.3555	461.9931	2.2080	60.4020	0.0366	0.0166	16.8885
50	0.3715	31.4236	642.3606	2.6916	84.5794	0.0318	0.0118	20.4420
60	0.3048	34.7609	823.6975	3.2810	114.0515	0.0288	0.0088	23.6961
100	**0.1380**	**43.0984**	**1,464.7527**	**7.2446**	**312.2323**	**0.0232**	**0.0032**	**33.9863**

Factor Table for $i = 4.00\%$

N	*P/F*	*P/A*	*P/G*	*F/P*	*F/A*	*A/P*	*A/F*	*A/G*
1	0.9615	0.9615	0.0000	1.0400	1.0000	1.0400	1.0000	0.0000
2	0.9246	1.8861	0.9246	1.0816	2.0400	0.5302	0.4902	0.4902
3	0.8890	2.7751	2.7025	1.1249	3.1216	0.3603	0.3203	0.9739
4	0.8548	3.6299	5.2670	1.1699	4.2465	0.2755	0.2355	1.4510
5	**0.8219**	**4.4518**	**8.5547**	**1.2167**	**5.4163**	**0.2246**	**0.1846**	**1.9216**
6	0.7903	5.2421	12.5062	1.2653	6.6330	0.1908	0.1508	2.3857
7	0.7599	6.0021	17.0657	1.3159	7.8983	0.1666	0.1266	2.8433
8	0.7307	6.7327	22.1806	1.3686	9.2142	0.1485	0.1085	3.2944
9	0.7026	7.4353	27.8013	1.4233	10.5828	0.1345	0.0945	3.7391
10	**0.6756**	**8.1109**	**33.8814**	**1.4802**	**12.0061**	**0.1233**	**0.0833**	**4.1773**
11	0.6496	8.7605	40.3772	1.5395	13.4864	0.1141	0.0741	4.6090
12	0.6246	9.3851	47.2477	1.6010	15.0258	0.1066	0.0666	5.0343
13	0.6006	9.9856	54.4546	1.6651	16.6268	0.1001	0.0601	5.4533
14	0.5775	10.5631	61.9618	1.7317	18.2919	0.0947	0.0547	5.8659
15	**0.5553**	**11.1184**	**69.7355**	**1.8009**	**20.0236**	**0.0899**	**0.0499**	**6.2721**
16	0.5339	11.6523	77.7441	1.8730	21.8245	0.0858	0.0458	6.6720
17	0.5134	12.1657	85.9581	1.9479	23.6975	0.0822	0.0422	7.0656
18	0.4936	12.6593	94.3498	2.0258	25.6454	0.0790	0.0390	7.4530
19	0.4746	13.1339	102.8933	2.1068	27.6712	0.0761	0.0361	7.8342
20	**0.4564**	**13.5903**	**111.5647**	**2.1911**	**29.7781**	**0.0736**	**0.0336**	**8.2091**
21	0.4388	14.0292	120.3414	2.2788	31.9692	0.0713	0.0313	8.5779
22	0.4220	14.4511	129.2024	2.3699	34.2480	0.0692	0.0292	8.9407
23	0.4057	14.8568	138.1284	2.4647	36.6179	0.0673	0.0273	9.2973
24	0.3901	15.2470	147.1012	2.5633	39.0826	0.0656	0.0256	9.6479
25	**0.3751**	**15.6221**	**156.1040**	**2.6658**	**41.6459**	**0.0640**	**0.0240**	**9.9925**
30	0.3083	17.2920	201.0618	3.2434	56.0849	0.0578	0.0178	11.6274
40	0.2083	19.7928	286.5303	4.8010	95.0255	0.0505	0.0105	14.4765
50	0.1407	21.4822	361.1638	7.1067	152.6671	0.0466	0.0066	16.8122
60	0.0951	22.6235	422.9966	10.5196	237.9907	0.0442	0.0042	18.6972
100	**0.0198**	**24.5050**	**563.1249**	**50.5049**	**1,237.6237**	**0.0408**	**0.0008**	**22.9800**

Factor Table for $i = 5.00\%$

N	*P/F*	*P/A*	*P/G*	*F/P*	*F/A*	*A/P*	*A/F*	*A/G*
1	0.9524	0.9524	0.0000	1.0500	1.0000	1.0500	1.0000	0.0000
2	0.9070	1.8594	0.9070	1.1025	2.0500	0.5378	0.4878	0.4878
3	0.8638	2.7232	2.6347	1.1576	3.1525	0.3672	0.3172	0.9675
4	08.227	3.5460	5.1028	1.2155	4.3101	0.2820	0.2320	1.4391
5	**0.7835**	**4.3295**	**8.2369**	**1.2763**	**5.5256**	**0.2310**	**0.1810**	**1.9025**
6	0.7462	5.0757	11.9680	1.3401	6.8019	0.1970	0.1470	2.3579
7	0.7107	5.7864	16.2321	1.4071	8.1420	0.1728	0.1228	2.8052
8	0.6768	6.4632	20.9700	1.4775	9.5491	0.1547	0.1047	3.2445
9	0.6446	7.1078	26.1268	1.5513	11.0266	0.1407	0.0907	3.6758
10	**0.6139**	**7.7217**	**31.6520**	**1.6289**	**12.5779**	**0.1295**	**0.0795**	**4.0991**
11	0.5847	8.3064	37.4988	1.7103	14.2068	0.1204	0.0704	4.5144
12	0.5568	8.8633	43.6241	1.7959	15.9171	0.1128	0.0628	4.9219
13	0.5303	9.3936	49.9879	1.8856	17.7130	0.1065	0.0565	5.3215
14	0.5051	9.8986	56.5538	1.9799	19.5986	0.1010	0.0510	5.7133
15	**0.4810**	**10.3797**	**63.2880**	**2.0789**	**21.5786**	**0.0963**	**0.0463**	**6.0973**
16	0.4581	10.8378	70.1597	2.1829	23.6575	0.0923	0.0423	6.4736
17	0.4363	11.2741	77.1405	2.2920	25.8404	0.0887	0.03887	6.8423
18	0.4155	11.6896	84.2043	2.4066	28.1324	0.0855	0.0355	7.2034
19	0.3957	12.0853	91.3275	2.5270	30.5390	0.0827	0.0327	7.5569
20	**0.3769**	**12.4622**	**98.4884**	**2.6533**	**33.0660**	**0.0802**	**0.0302**	**7.9030**
21	0.3589	12.8212	105.6673	2.7860	35.7193	0.0780	0.0280	8.2416
22	0.3418	13.1630	112.8461	2.9253	38.5052	0.0760	0.0260	8.5730
23	0.3256	13.4886	120.0087	3.0715	41.4305	0.0741	0.0241	8.8971
24	0.3101	13.7986	127.1402	3.2251	44.5020	0.0725	0.0225	9.2140
25	**0.2953**	**14.0939**	**134.2275**	**3.3864**	**47.7251**	**0.0710**	**0.0210**	**9.5238**
30	0.2314	15.3725	168.6226	4.3219	66.4388	0.0651	0.0151	10.9691
40	0.1420	17.1591	229.5452	7.0400	120.7998	0.0583	0.0083	13.3775
50	0.0872	18.2559	277.9148	11.4674	209.3480	0.0548	0.0048	15.2233
60	0.0535	18.9293	314.3432	18.6792	353.5837	0.0528	0.0028	16.6062
100	**0.0076**	**19.8479**	**381.7492**	**131.5013**	**2,610.0252**	**0.0504**	**0.0004**	**19.2337**

Factor Table for $i = 6.00\%$

N	P/F	P/A	P/G	F/P	F/A	A/P	A/F	A/G
1	0.9434	0.9434	0.0000	1.0600	1.0000	1.0600	1.0000	0.0000
2	0.8900	1.8334	0.8900	1.1236	2.0600	0.5454	0.4854	0.4854
3	0.8396	2.6730	2.5692	1.1910	3.1836	0.3741	0.3141	0.9612
4	0.7921	3.4651	4.9455	1.2625	4.3746	0.2886	0.2286	1.4272
5	**0.7473**	**4.2124**	**7.9345**	**1.3382**	**5.6371**	**0.2374**	**0.1774**	**1.8836**
6	0.7050	4.9173	11.4594	1.4185	6.9753	0.2034	0.1434	2.3304
7	0.6651	5.5824	15.4497	1.5036	8.3938	0.1791	0.1191	2.7676
8	0.6274	6.2098	19.8416	1.5938	9.8975	0.1610	0.1010	3.1952
9	0.5919	6.8017	24.5768	1.6895	11.4913	0.1470	0.0870	3.6133
10	**0.5584**	**7.3601**	**29.6023**	**1.7908**	**13.1808**	**0.1359**	**0.0759**	**4.0220**
11	0.5268	7.8869	34.8702	1.8983	14.9716	0.1268	0.0668	4.4213
12	0.4970	8.3838	40.3369	2.0122	16.8699	0.1193	0.0593	4.8113
13	0.4688	8.8527	45.9629	2.1329	18.8821	0.1130	0.0530	5.1920
14	0.4423	9.2950	51.7128	2.2609	21.0151	0.1076	0.0476	5.5635
15	**0.4173**	**9.7122**	**57.5546**	**2.3966**	**23.2760**	**0.1030**	**0.0430**	**5.9260**
16	0.3936	10.1059	63.4592	2.5404	25.6725	0.0990	0.0390	6.2794
17	0.3714	10.4773	69.4011	2.6928	28.2129	0.0954	0.0354	6.6240
18	0.3505	10.8276	75.3569	2.8543	30.9057	0.0924	0.0324	6.9597
19	0.3305	11.1581	81.3062	3.0256	33.7600	0.0896	0.0296	7.2867
20	**0.3118**	**11.4699**	**87.2304**	**3.2071**	**36.7856**	**0.0872**	**0.0272**	**7.6051**
21	0.2942	11.7641	93.1136	3.3996	39.9927	0.0850	0.0250	7.9151
22	0.2775	12.0416	98.9412	3.6035	43.3923	0.0830	0.0230	8.2166
23	0.2618	12.3034	104.7007	3.8197	46.9958	0.0813	0.0213	8.5099
24	0.2470	12.5504	110.3812	4.0489	50.8156	0.0797	0.0197	8.7951
25	**0.2330**	**12.7834**	**115.9732**	**4.2919**	**54.8645**	**0.0782**	**0.0182**	**9.0722**
30	0.1741	13.7648	142.3588	5.7435	79.0582	0.0726	0.0126	10.3422
40	0.0972	15.0463	185.9568	10.2857	154.7620	0.0665	0.0065	12.3590
50	0.0543	15.7619	217.4574	18.4202	290.3359	0.0634	0.0034	13.7964
60	0.0303	16.1614	239.0428	32.9877	533.1282	0.0619	0.0019	14.7909
100	**0.0029**	**16.6175**	**272.0471**	**339.3021**	**5,638.3681**	**0.0602**	**0.0002**	**16.3711**

Factor Table for $i = 8.00\%$

N	P/F	P/A	P/G	F/P	F/A	A/P	A/F	A/G
1	0.9259	0.9259	0.0000	1.0800	1.0000	1.0800	1.0000	0.0000
2	0.8573	1.7833	0.8573	1.1664	2.0800	0.5608	0.4808	0.4808
3	0.7938	2.5771	2.4450	1.2597	3.2464	0.3880	0.3080	0.9487
4	0.7350	3.3121	4.6501	1.3605	4.5061	0.3019	0.2219	1.4040
5	**0.6806**	**3.9927**	**7.3724**	**1.4693**	**5.8666**	**0.2505**	**0.1705**	**1.8465**
6	0.6302	4.6229	10.5233	1.5869	7.3359	0.2163	0.1363	2.2763
7	0.5835	5.2064	14.0242	1.7138	8.9228	0.1921	0.1121	2.6937
8	0.5403	5.7466	17.8061	1.8509	10.6366	0.1740	0.0940	3.0985
9	0.5002	6.2469	21.8081	1.9990	12.4876	0.1601	0.0801	3.4910
10	**0.4632**	**6.7101**	**25.9768**	**2.1589**	**14.4866**	**0.1490**	**0.0690**	**3.8713**
11	0.4289	7.1390	30.2657	2.3316	16.6455	0.1401	0.0601	4.2395
12	0.3971	7.5361	34.6339	2.5182	18.9771	0.1327	0.0527	4.5957
13	0.3677	7.9038	39.0463	2.7196	21.4953	0.1265	0.0465	4.9402
14	0.3405	8.2442	43.4723	2.9372	24.2149	0.1213	0.0413	5.2731
15	**0.3152**	**8.5595**	**47.8857**	**3.1722**	**27.1521**	**0.1168**	**0.0368**	**5.5945**
16	0.2919	8.8514	52.2640	3.4259	30.3243	0.1130	0.0330	5.9046
17	0.2703	9.1216	56.5883	3.7000	33.7502	0.1096	0.0296	6.2037
18	0.2502	9.3719	60.8426	3.9960	37.4502	0.1067	0.0267	6.4920
19	0.2317	9.6036	65.0134	4.3157	41.4463	0.1041	0.0241	6.7697
20	**0.2145**	**9.8181**	**69.0898**	**4.6610**	**45.7620**	**0.1019**	**0.0219**	**7.0369**
21	0.1987	10.0168	73.0629	5.0338	50.4229	0.0998	0.0198	7.2940
22	0.1839	10.2007	76.9257	5.4365	55.4568	0.0980	0.0180	7.5412
23	0.1703	10.3711	80.6726	5.8715	60.8933	0.0964	0.0164	7.7786
24	0.1577	10.5288	84.2997	6.3412	66.7648	0.0950	0.0150	8.0066
25	**0.1460**	**10.6748**	**87.8041**	**6.8485**	**73.1059**	**0.0937**	**0.0137**	**8.2254**
30	0.0994	11.2578	103.4558	10.0627	113.2832	0.0888	0.0088	9.1897
40	0.0460	11.9246	126.0422	21.7245	259.0565	0.0839	0.0039	10.5699
50	0.0213	12.2335	139.5928	46.9016	573.7702	0.0817	0.0017	11.4107
60	0.0099	12.3766	147.3000	101.2571	1,253.2133	0.0808	0.0008	11.9015
100	**0.0005**	**12.4943**	**155.6107**	**2,199.7613**	**27,484.5157**	**0.0800**		**12.4545**

Appendix A.3

Factor Table for i = 10.00%

N	P/F	P/A	P/G	F/P	F/A	A/P	A/F	A/G
1	0.9091	0.9091	0.0000	1.1000	1.0000	1.1000	1.0000	0.0000
2	0.8264	1.7355	0.8264	1.2100	2.1000	0.5762	0.4762	0.4762
3	0.7513	2.4869	2.3291	1.3310	3.3100	0.4021	0.3021	0.9366
4	0.6830	3.1699	4.3781	1.4641	4.6410	0.3155	0.2155	1.3812
5	**0.6209**	**3.7908**	**6.8618**	**1.6105**	**6.1051**	**0.2638**	**0.1638**	**1.8101**
6	0.5645	4.3553	9.6842	1.7716	7.7156	0.2296	0.1296	2.2236
7	0.5132	4.8684	12.7631	1.9487	9.4872	0.2054	0.1054	2.6216
8	0.4665	5.3349	16.0287	2.1436	11.4359	0.1874	0.0874	3.0045
9	0.4241	5.7590	19.4215	2.3579	13.5735	0.1736	0.0736	3.3724
10	**0.3855**	**6.1446**	**22.8913**	**2.5937**	**15.9374**	**0.1627**	**0.0627**	**3.7255**
11	0.3505	6.4951	26.3962	2.8531	18.5312	0.1540	0.0540	4.0641
12	0.3186	6.8137	29.9012	3.1384	21.3843	0.1468	0.0468	4.3884
13	0.2897	7.1034	33.3772	3.4523	24.5227	0.1408	0.0408	4.6988
14	0.2633	7.3667	36.8005	3.7975	27.9750	0.1357	0.0357	4.9955
15	**0.2394**	**7.6061**	**40.1520**	**4.1772**	**31.7725**	**0.1315**	**0.0315**	**5.2789**
16	0.2176	7.8237	43.4164	4.5950	35.9497	0.1278	0.0278	5.5493
17	0.1978	8.0216	46.5819	5.0545	40.5447	0.1247	0.0247	5.8071
18	0.1799	8.2014	49.6395	5.5599	45.5992	0.1219	0.0219	6.0526
19	0.1635	8.3649	52.5827	6.1159	51.1591	0.1195	0.0195	6.2861
20	**0.1486**	**8.5136**	**55.4069**	**6.7275**	**57.2750**	**0.1175**	**0.0175**	**6.5081**
21	0.1351	8.6487	58.1095	7.4002	64.0025	0.1156	0.0156	6.7189
22	0.1228	8.7715	60.6893	8.1403	71.4027	0.1140	0.0140	6.9189
23	0.1117	8.8832	63.1462	8.9543	79.5430	0.1126	0.0126	7.1085
24	0.1015	8.9847	65.4813	9.8497	88.4973	0.1113	0.0113	7.2881
25	**0.0923**	**9.0770**	**67.6964**	**10.8347**	**98.3471**	**0.1102**	**0.0102**	**7.4580**
30	0.0573	9.4269	77.0766	17.4494	164.4940	0.1061	0.0061	8.1762
40	0.0221	9.7791	88.9525	45.2593	442.5926	0.1023	0.0023	9.0962
50	0.0085	9.9148	94.8889	117.3909	1,163.9085	0.1009	0.0009	9.5704
60	0.0033	9.9672	97.7010	304.4816	3,034.8164	0.1003	0.0003	9.8023
100	**0.0001**	**9.9993**	**99.9202**	**13,780.6123**	**137,796.1234**	**0.1000**		**9.9927**

Factor Table for $i = 12.00\%$

N	P/F	P/A	P/G	F/P	F/A	A/P	A/F	A/G
1	0.8929	0.8929	0.0000	1.1200	1.0000	1.1200	1.0000	0.0000
2	0.7972	1.6901	0.7972	1.2544	2.1200	0.5917	0.4717	0.4717
3	0.7118	2.4018	2.2208	1.4049	3.3744	0.4163	0.2963	0.9246
4	0.6355	3.0373	4.1273	1.5735	4.7793	0.3292	0.2092	1.3589
5	**0.5674**	**3.6048**	**6.3970**	**1.7623**	**6.3528**	**0.2774**	**0.1574**	**1.7746**
6	0.5066	4.1114	8.9302	1.9738	8.1152	0.2432	0.1232	2.1720
7	0.4523	4.5638	11.6443	2.2107	10.0890	0.2191	0.0991	2.5515
8	0.4039	4.9676	14.4714	2.4760	12.2997	0.2013	0.0813	2.9131
9	0.3606	5.3282	17.3563	2.7731	14.7757	0.1877	0.0677	3.2574
10	**0.3220**	**5.6502**	**20.2541**	**3.1058**	**17.5487**	**0.1770**	**0.0570**	**3.5847**
11	0.2875	5.9377	23.1288	3.4785	20.6546	0.1684	0.0484	3.8953
12	0.2567	6.1944	25.9523	3.8960	24.1331	0.1614	0.0414	4.1897
13	0.2292	6.4235	28.7024	4.3635	28.0291	0.1557	0.0357	4.4683
14	0.2046	6.6282	31.3624	4.8871	32.3926	0.1509	0.0309	4.7317
15	**0.1827**	**6.8109**	**33.9202**	**5.4736**	**37.2797**	**0.1468**	**0.0268**	**4.9803**
16	0.1631	6.9740	36.3670	6.1304	42.7533	0.1434	0.0234	5.2147
17	0.1456	7.1196	38.6973	6.8660	48.8837	0.1405	0.0205	5.4353
18	0.1300	7.2497	40.9080	7.6900	55.7497	0.1379	0.0179	5.6427
19	0.1161	7.3658	42.9979	8.6128	63.4397	0.1358	0.0158	5.8375
20	**0.1037**	**7.4694**	**44.9676**	**9.6463**	**72.0524**	**0.1339**	**0.0139**	**6.0202**
21	0.0926	7.5620	46.8188	10.8038	81.6987	0.1322	0.0122	6.1913
22	0.0826	7.6446	48.5543	12.1003	92.5026	0.1308	0.0108	6.3514
23	0.0738	7.7184	50.1776	13.5523	104.6029	0.1296	0.0096	6.5010
24	0.0659	7.7843	51.6929	15.1786	118.1552	0.1285	0.0085	6.6406
25	**0.0588**	**7.8431**	**53.1046**	**17.0001**	**133.3339**	**0.1275**	**0.0075**	**6.7708**
30	0.0334	8.0552	58.7821	29.9599	241.3327	0.1241	0.0041	7.2974
40	0.0107	8.2438	65.1159	93.0510	767.0914	0.1213	0.0013	7.8988
50	0.0035	8.3045	67.7624	289.0022	2,400.0182	0.1204	0.0004	8.1597
60	0.0011	8.3240	68.8100	897.5969	7,471.6411	0.1201	0.0001	8.2664
100		**8.3332**	**69.4336**	**83,522.2657**	**696,010.5477**	**0.1200**		**8.3321**

Factor Table for *i* = 15.00%

N	P/F	P/A	P/G	F/P	F/A	A/P	A/F	A/G
1	0.8696	0.8696	0.0000	1.1500	1.0000	1.1500	1.0000	0.0000
2	0.7561	1.6257	0.7561	1.3225	2.1500	0.6151	0.4651	0.4651
3	0.6575	2.2832	2.0712	1.5209	3.4725	0.4380	0.2880	0.9071
4	0.5718	2.8550	3.7864	1.7490	4.9934	0.3503	0.2003	1.3263
5	**0.4972**	**3.3522**	**5.7751**	**2.0114**	**6.7424**	**0.2983**	**0.1483**	**1.7228**
6	0.4323	3.7845	7.9368	2.3131	8.7537	0.2642	0.1142	2.0972
7	0.3759	4.1604	10.1924	2.6600	11.0668	0.2404	0.0904	2.4498
8	0.3269	4.4873	12.4807	3.0590	13.7268	0.2229	0.0729	2.7813
9	0.2843	4.7716	14.7548	3.5179	16.7858	0.2096	0.0596	3.0922
10	**0.2472**	**5.0188**	**16.9795**	**4.0456**	**20.3037**	**0.1993**	**0.0493**	**3.3832**
11	0.2149	5.2337	19.1289	4.6524	24.3493	0.1911	0.0411	3.6549
12	0.1869	5.4206	21.1849	5.3503	29.0017	0.1845	0.0345	3.9082
13	0.1625	5.5831	23.1352	6.1528	34.3519	0.1791	0.0291	4.1438
14	0.1413	5.7245	24.9725	7.0757	40.5047	0.1747	0.0247	4.3624
15	**0.1229**	**5.8474**	**26.9630**	**8.1371**	**47.5804**	**0.1710**	**0.0210**	**4.5650**
16	0.1069	5.9542	28.2960	9.3576	55.7175	0.1679	0.0179	4.7522
17	0.0929	6.0472	29.7828	10.7613	65.0751	0.1654	0.0154	4.9251
18	0.0808	6.1280	31.1565	12.3755	75.8364	0.1632	0.0132	5.0843
19	0.0703	6.1982	32.4213	14.2318	88.2118	0.1613	0.0113	5.2307
20	**0.0611**	**6.2593**	**33.5822**	**16.3665**	**102.4436**	**0.1598**	**0.0098**	**5.3651**
21	0.0531	6.3125	34.6448	18.8215	118.8101	0.1584	0.0084	5.4883
22	0.0462	6.3587	35.6150	21.6447	137.6316	0.1573	0.0073	5.6010
23	0.0402	6.3988	36.4988	24.8915	159.2764	0.1563	0.0063	5.7040
24	0.0349	6.4338	37.3023	28.6252	184.1678	0.1554	0.0054	5.7979
25	**0.0304**	**6.4641**	**38.0314**	**32.9190**	**212.7930**	**0.1547**	**0.0047**	**5.8834**
30	0.0151	6.5660	40.7526	66.2118	434.7451	0.1523	0.0023	6.2066
40	0.0037	6.6418	43.2830	267.8635	1,779.0903	0.1506	0.0006	6.5168
50	0.0009	6.6605	44.0958	1,083.6574	7,217.7163	0.1501	0.0001	6.6205
60	0.0002	6.6651	44.3431	4,383.9987	29,219.9916	0.1500	0.0000	6.6530
100	**0.0000**	**6.6667**	**44.4438**	**1,174,313.450**	**7,828,749.6713**	**0.1500**	**0.0000**	**6.6666**

Factor Table for $i = 18.00\%$

N	P/F	P/A	P/G	F/P	F/A	A/P	A/F	A/G
1	0.8475	0.8475	0.0000	1.1800	1.0000	1.1800	1.0000	0.0000
2	0.7182	1.5656	0.7182	1.3924	2.1800	0.6387	0.4587	0.4587
3	0.6086	2.1743	1.9354	1.6430	3.5724	0.4599	0.2799	0.8902
4	0.5158	2.6901	3.4828	1.9388	5.2154	0.3717	0.1917	1.2947
5	**0.4371**	**3.1272**	**5.2312**	**2.2878**	**7.1542**	**0.3198**	**0.1398**	**1.6728**
6	0.3704	3.4976	7.0834	2.6996	9.4423	0.2859	0.1059	2.0252
7	0.3139	3.8115	8.9670	3.1855	12.1415	0.2624	0.0824	2.3526
8	0.2660	4.0776	10.8292	3.7589	15.3270	0.2452	0.0652	2.6558
9	0.2255	4.3030	12.6329	4.4355	19.0859	0.2324	0.0524	2.9358
10	**0.1911**	**4.4941**	**14.3525**	**5.2338**	**23.5213**	**0.2225**	**0.0425**	**3.1936**
11	0.1619	4.6560	15.9716	6.1759	28.7551	0.2148	0.0348	3.4303
12	0.1372	4.7932	17.4811	7.2876	34.9311	0.2086	0.0286	3.6470
13	0.1163	4.9095	18.8765	8.5994	42.2187	0.2037	0.0237	3.8449
14	0.0985	5.0081	20.1576	10.1472	50.8180	0.1997	0.0197	4.0250
15	**0.0835**	**5.0916**	**21.3269**	**11.9737**	**60.9653**	**0.1964**	**0.0164**	**4.1887**
16	0.0708	5.1624	22.3885	14.1290	72.9390	0.1937	0.0137	4.3369
17	0.0600	5.2223	23.3482	16.6722	87.0680	0.1915	0.0115	4.4708
18	0.0508	5.2732	24.2123	19.6731	103.7403	0.1896	0.0096	4.5916
19	0.0431	5.3162	24.9877	23.2144	123.4135	0.1881	0.0081	4.7003
20	**0.0365**	**5.3527**	**25.6813**	**27.3930**	**146.6280**	**0.1868**	**0.0068**	**4.7978**
21	0.0309	5.3837	26.3000	32.3238	174.0210	0.1857	0.0057	4.8851
22	0.0262	5.4099	26.8506	38.1421	206.3448	0.1848	0.0048	4.9632
23	0.0222	5.4321	27.3394	45.0076	244.4868	0.1841	0.0041	5.0329
24	0.0188	5.4509	27.7725	53.1090	289.4944	0.1835	0.0035	5.0950
25	**0.0159**	**5.4669**	**28.1555**	**62.6686**	**342.6035**	**0.1829**	**0.0029**	**5.1502**
30	0.0070	5.5168	29.4864	143.3706	790.9480	0.1813	0.0013	5.3448
40	0.0013	5.5482	30.5269	750.3783	4,163.2130	0.1802	0.0002	5.5022
50	0.0003	5.5541	30.7856	3,927.3569	21,813.0937	0.1800		5.5428
60	0.0001	5.5553	30.8465	20,555.1400	114,189.6665	0.1800		5.5526
100		**5.5556**	**30.8642**	**15,424,131.91**	**85,689,616.17**	**0.1800**		**5.5555**

Diagrams and Equations for Beam Designs

A.4

A.4.1 SIMPLY SUPPORTED BEAM WITH POINT LOAD AT CENTER

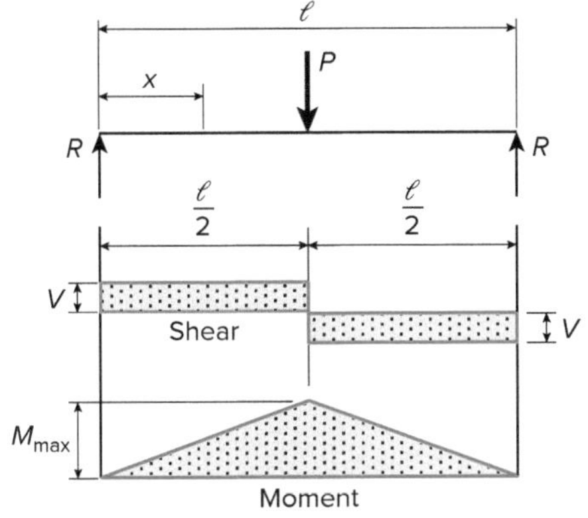

Reaction:

$$R = V = \frac{P}{2}$$

Moments:

$$M_{max}(\text{at point of load}) = \frac{P\ell}{4}$$

$$M_x\left(\text{when } x < \frac{\ell}{2}\right) = \frac{Px}{2}$$

Deflections:

$$\Delta_{max}(\text{at point of load}) = \frac{P\ell^3}{48EI}$$

$$\Delta_x\left(\text{when } x < \frac{\ell}{2}\right) = \frac{Px}{48EI}(3\ell^2 - 4x^2)$$

A.4.2 SIMPLY SUPPORTED BEAM WITH TWO EQUAL SYMMETRICALLY PLACED POINT LOADS

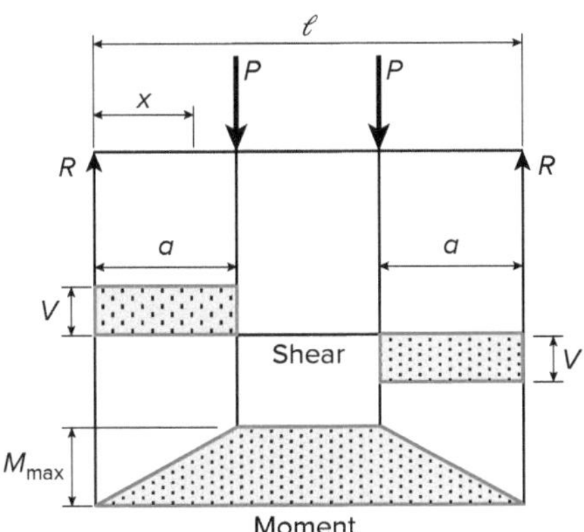

Reaction:

$$R = V = P$$

Moments:

$$M_{max}(\text{between loads}) = Pa$$
$$M_x(\text{when } x < a) = Px$$

Deflections:

$$\Delta_{max}(\text{at center}) = \frac{Pa}{24EI}(3\ell^2 - 4a^2)$$

$$\Delta_x(\text{when } x < a) = \frac{Px}{6EI}(3\ell a - 3a^2 - x^2)$$

$$\Delta_x[\text{when } x > a \text{ and } < (\ell - a)] = \frac{Pa}{6EI}(3\ell x - 3x^2 - a^2)$$

A.4.3 SIMPLY SUPPORTED BEAM WITH POINT LOAD PLACED AT AN ARBITRARY DISTANCE

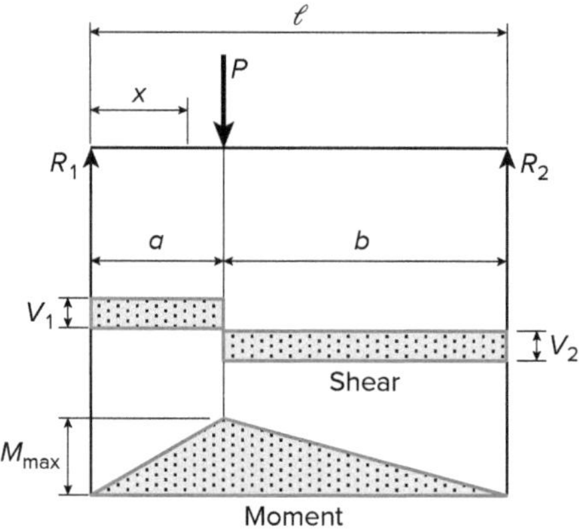

Reaction:

$$R_1 = V_1(\text{max when } a < b) = \frac{Pb}{\ell}$$

$$R_2 = V_2(\text{max when } a > b) = \frac{Pa}{\ell}$$

Moments:

$$M_{\max}(\text{at point of load}) = \frac{Pab}{\ell}$$

$$M_x(\text{when } x < a) = \frac{Pbx}{\ell}$$

Deflections:

$$\Delta_{\max}\left(\text{at } x = \sqrt{\frac{a(a+2b)}{3}} \text{ when } a > b\right) = \frac{Pab(a+2b)\sqrt{3a(a+2b)}}{27EI\ell}$$

$$\Delta_a(\text{at point of load}) = \frac{Pa^2b^2}{3EI\ell}$$

$$\Delta_x(\text{when } x < a) = \frac{Pbx}{6EI\ell}(\ell^2 - b^2 - x^2)$$

A.4.4 SIMPLY SUPPORTED BEAM WITH A UNIFORMLY DISTRIBUTED LOAD

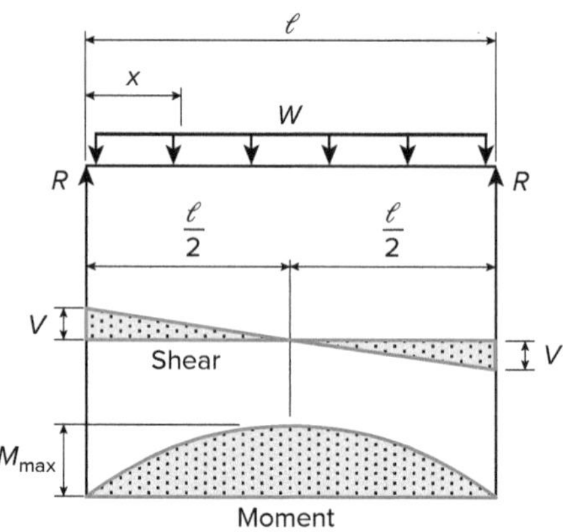

Reaction:

$$R = V = \frac{w\ell}{2}$$

Shear:

$$V_x = w\left(\frac{\ell}{2} - x\right)$$

Moments:

$$M_{\max}(\text{at center}) = \frac{w\ell^2}{8}$$

$$M_x = \frac{wx}{2}(\ell - x)$$

Deflections:

$$\Delta_{\max}(\text{at center}) = \frac{5w\ell^4}{384EI}$$

$$\Delta_x = \frac{wx}{24EI}(\ell^3 - 2\ell x + x^3)$$

A.4.5 SIMPLY SUPPORTED BEAM WITH PARTIAL UNIFORMLY DISTRIBUTED LOAD ARBITRARILY PLACED

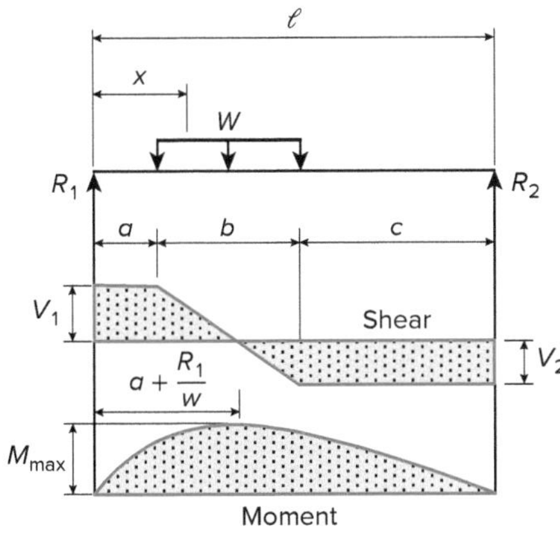

Reactions:

$$R_1 = V_1(\text{max when } a < c) = \frac{wb}{2\ell}(2c + b)$$

$$R_2 = V_2(\text{max when } a > c) = \frac{wb}{2\ell}(2a + b)$$

Shear:

$$V_x[\text{when } x > a \text{ and } < (a + b)] = R_1 - w(x - a)$$

Moments:

$$M_{max}\left(\text{at } x = a + \frac{R_1}{w}\right) = R_1\left(a + \frac{R_1}{2w}\right)$$

$$M_x(\text{when } x < a) = R_1 x$$

$$M_x[\text{when } x > a \text{ and } < (a + b)] = R_1 x - \frac{w}{2}(x - a)^2$$

$$M_x[\text{when } x > (a + b)] = R_2(\ell - x)$$

A.4.6 SIMPLY SUPPORTED BEAM WITH TWO DIFFERENT POINT LOADS ARBITRARILY PLACED

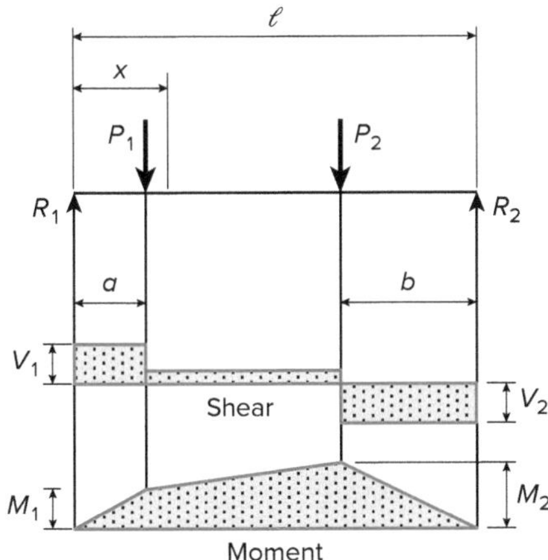

Reaction:

$$R_1 = V_1 = \frac{P_1(\ell - a) + P_2 b}{\ell}$$

$$R_2 = V_2 = \frac{P_1 a + P_2(\ell - b)}{\ell}$$

Shear:

$$V_x[\text{when } x > a \text{ and } < (\ell - b)] = R_1 - P_1$$

Moments:

$$M_1(\text{max when } R_1 < P_1) = R_1 a$$

$$M_2(\text{max when } R_2 < P_2) = R_2 b$$

$$M_x(\text{when } x < a) = R_1 x$$

$$M_x[\text{when } x > a \text{ and } < (\ell - b)] = R_1 x - P_1(x - a)$$

A.4.7 SIMPLY SUPPORTED BEAM WITH INCREASING LINEAR DISTRIBUTED LOAD

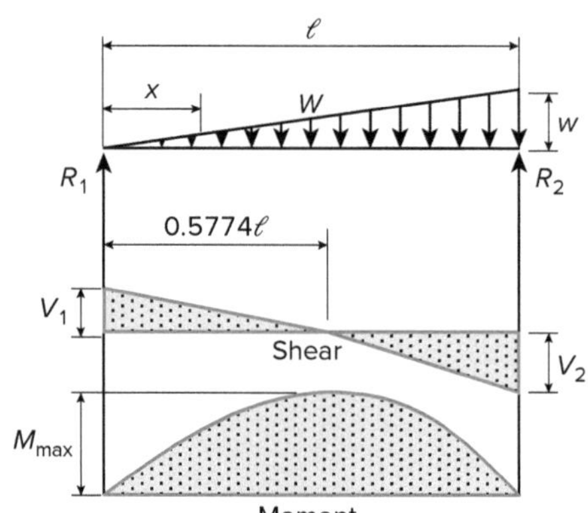

$$W = \frac{w\ell}{2}$$

Reactions:

$$R_1 = V_1 = \frac{W}{3}$$

$$R_2 = V_2(\text{max}) = \frac{2W}{3}$$

Shear:

$$V_x = \frac{W}{3} - \frac{Wx^2}{\ell^2}$$

Moments:

$$M_{\text{max}}\left(\text{at } x = \frac{\ell}{\sqrt{3}} = 0.5774\ell\right) = \frac{2W\ell}{9\sqrt{3}} = 0.1283W\ell$$

$$M_x = \frac{Wx}{3\ell^2}(\ell^2 - x^2)$$

Deflections:

$$\Delta_{\text{max}}\left(\text{at } x = \ell\sqrt{1 - \sqrt{\frac{8}{15}}} = 0.5193\ell\right) = 0.01304\frac{W\ell^3}{EI}$$

$$\Delta_x = \frac{Wx}{180EI\ell^2}(3x^4 - 10\ell^2x^2 + 7\ell^4)$$

A.4.8 SIMPLY SUPPORTED BEAM WITH TRIANGULAR DISTRIBUTED LOAD

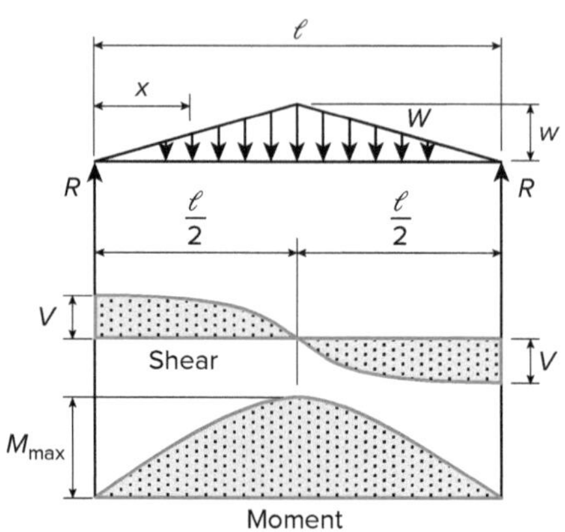

$$W = \frac{w\ell}{2}$$

Reactions:

$$R = V = \frac{W}{2}$$

Shear:

$$V_x\left(\text{when } x < \frac{\ell}{2}\right) = \frac{W}{2\ell^2}(\ell^2 - 4x^2)$$

Moments:

$$M_{\text{max}}(\text{at center}) = \frac{W\ell}{6}$$

$$M_x\left(\text{when } x < \frac{\ell}{2}\right) = Wx\left(\frac{1}{2} - \frac{2x^2}{3\ell^2}\right)$$

Deflections:

$$\Delta_{\text{max}}(\text{at center}) = \frac{W\ell^3}{60EI}$$

$$\Delta_x\left(\text{when } x < \frac{\ell}{2}\right) = \frac{Wx}{480EI\ell^2}(5\ell^2 - 4x^2)^2$$

A.4.9 SIMPLY SUPPORTED BEAM WITH MOMENT APPLIED AT ONE END ONLY

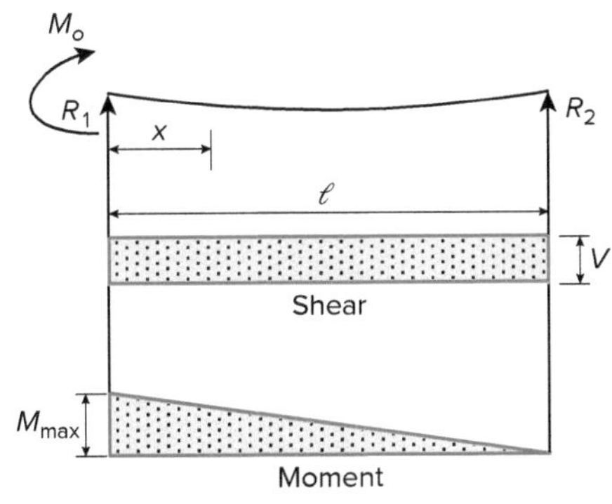

Reactions:

$$-R_1 = R_2 = V = \frac{M_o}{\ell}$$

Moments:

$$M_{max}(\text{at } R_1) = M_o$$

$$M_x = M_o - R_1 x = M_o\left(1 - \frac{x}{\ell}\right)$$

Deflections:

$$\Delta_{max}(\text{when } x = 0.423\ell) = 0.0642\frac{M_o\ell^2}{EI}$$

$$\Delta_x = \frac{M_o}{6EI}\left(3x^2 - \frac{x^3}{\ell} - 2\ell x\right)$$

Rotations:

$$\theta_1(\text{at } R_1) = \frac{M_o\ell}{3EI}$$

$$\theta_2(\text{at } R_2) = \frac{M_o\ell}{6EI}$$

A.4.10 SIMPLY SUPPORTED BEAM WITH MOMENT APPLIED AT AN ARBITRARY DISTANCE

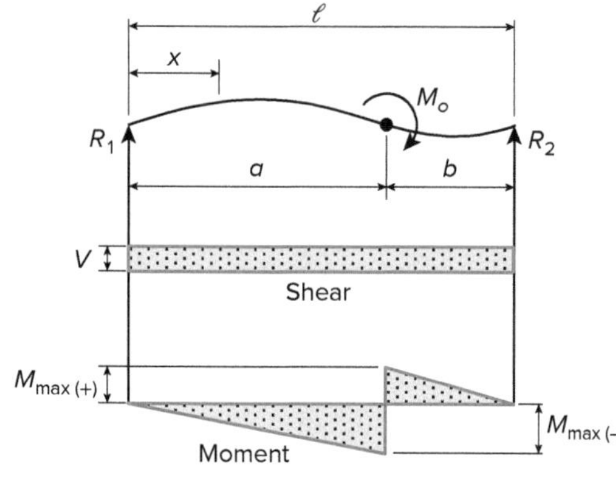

Reactions:

$$R_1 = V(\text{when } a > b) = \frac{M_o}{\ell}$$

$$R_2(\text{when } a > b) = \frac{M_o}{\ell}$$

Moments:

$$M_{\max(-)}(\text{at } x = a) = \frac{M_o a}{\ell}$$

$$M_{\max(+)}(\text{at } x = a) = M_o\left(1 - \frac{a}{\ell}\right)$$

$$M_x(\text{when } x < a) = \frac{M_o x}{\ell}$$

$$M_x(\text{when } x > a) = M_o\left(1 - \frac{x}{\ell}\right)$$

Deflections:

$$\Delta_x(\text{when } x < a) = \frac{M_o x}{6EI\ell}\left(\ell^2 - 3b^2 - x^2\right)$$

$$\Delta_x(\text{when } x > a) = \frac{M_o(\ell - x)}{6EI\ell}\left(3a^2 - 2\ell x + x^2\right)$$

$$\Delta_{\max}\left(\text{at } x = \sqrt{\frac{\ell^2 - 3b^2}{3}} \text{ if } a > 0.4226\ell\right) = \frac{M_o}{3EI\ell}\left(\frac{\ell^2 - 3b^2}{3}\right)^{3/2}$$

$$\Delta_{\max}\left(\text{at } x = \ell - \sqrt{\frac{\ell^2 - 3a^2}{3}} \text{ if } a > 0.5774\ell\right) = \frac{M_o}{3EI\ell}\left(\frac{\ell^2 - 3a^2}{3}\right)^{3/2}$$

$$M_Q \quad (\text{at center}) = -\frac{M_o}{2}$$

$$\Delta_Q \quad (\text{at center}) = \frac{M_o}{16EI}\left(\ell^2 - 4b^2\right)$$

$$\Delta_{\max}\left(\text{when } a = b = \frac{\ell}{2}, \text{ at } x = \frac{\sqrt{3}}{6}\ell = 0.28867\ell\right) = \frac{M_o \ell^2}{124.71EI}$$

Rotations:

$$\theta_Q(\text{at center}) = \frac{M_o \ell}{12EI}$$

Appendix A.4

School of PE

A.4.11 SIMPLY SUPPORTED BEAM WITH DIFFERENT END MOMENTS

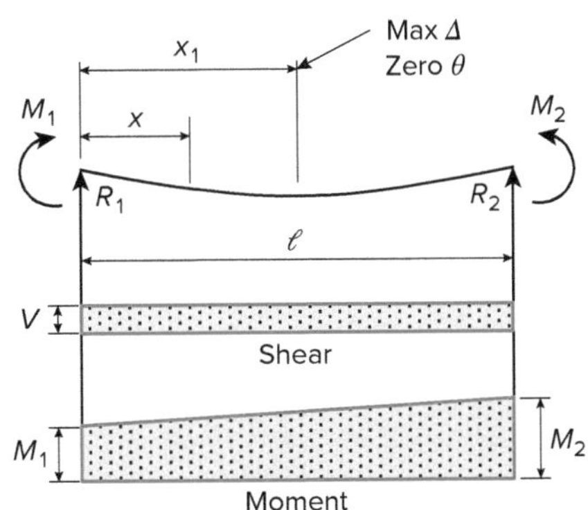

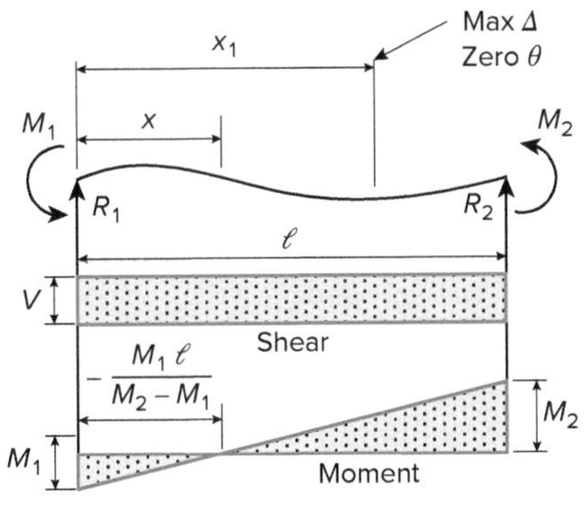

Reactions:

$$R_1 = -R_2 = V = \frac{M_2 - M_1}{\ell}$$

Moments:

$$M_x = (M_2 - M_1)\frac{x}{\ell} + M_1$$

Deflections:

$$\Delta_x = \frac{x(\ell - x)}{6EI\ell}[M_1(2\ell - x) + M_2(\ell + x)]$$

$$\text{at } x_1 = \frac{6M_1\ell \pm \sqrt{36M_1^2\ell^2 - 12(M_1 - M_2)\ell^2(2M_1 + M_2)}}{6(M_1 - M_2)},$$

$\Delta = \text{max}$ and $\theta = 0$

Rotations:

$$\theta_1(\text{at end}) = -\frac{\ell}{6EI}(2M_1 + M_2)$$

$$\theta_2(\text{at end}) = \frac{\ell}{6EI}(M_1 + 2M_2)$$

If M_1 and M_2 are opposite signs, then the above formulas hold. Use the actual sign of moment. M_x at point of contraflexure: X = insert equation = 0

$$\left(\text{where } x = \frac{M_1\ell}{M_2 - M_1}\right) = 0$$

A.4.12 CANTILEVER BEAM WITH UNIFORMLY DISTRIBUTED LOAD

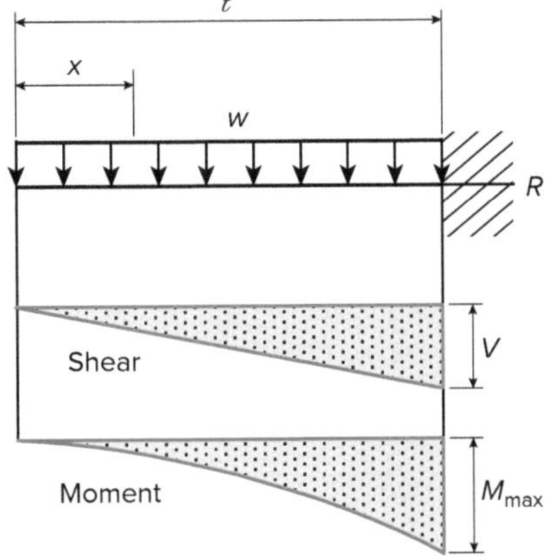

Reactions:

$$R = V = w\ell$$

Shear:

$$V_x = wx$$

Moments:

$$M_{max}(\text{at fixed end}) = \frac{w\ell^2}{2}$$

$$M_x = \frac{wx^2}{2}$$

Deflections:

$$\Delta_{max}(\text{at free end}) = \frac{w\ell^4}{8EI}$$

$$\Delta_x = \frac{w}{24EI}(x^4 - 4\ell^3 x + 3\ell^4)$$

A.4.13 CANTILEVER BEAM WITH LINEAR INCREASING DISTRIBUTED LOAD FROM FREE END

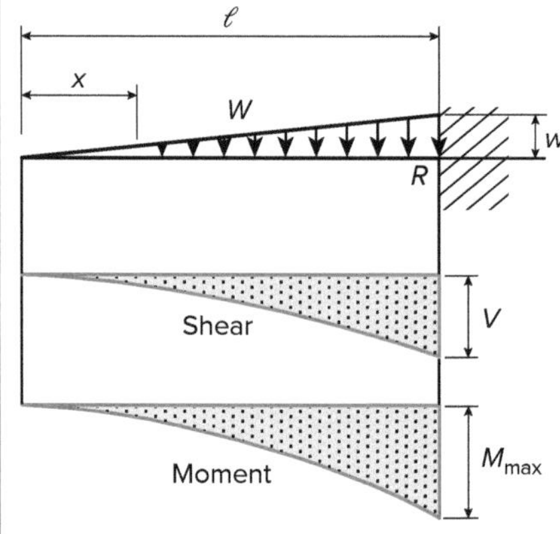

Reactions:

$$W = \frac{w\ell}{2}$$

$$R = V = W$$

Shear:

$$V_x = W\frac{x^2}{\ell^2}$$

Moments:

$$M_{max}(\text{at fixed end}) = \frac{W\ell}{3}$$

$$M_x = \frac{Wx^3}{3\ell^2}$$

Deflections:

$$\Delta_{max}(\text{at free end}) = \frac{W\ell^3}{15EI}$$

$$\Delta_x = \frac{W}{60EI\ell^2}\left(x^5 - 5\ell^4 x + 4\ell^5\right)$$

A.4.14 CANTILEVER BEAM WITH POINT LOAD AT FREE END

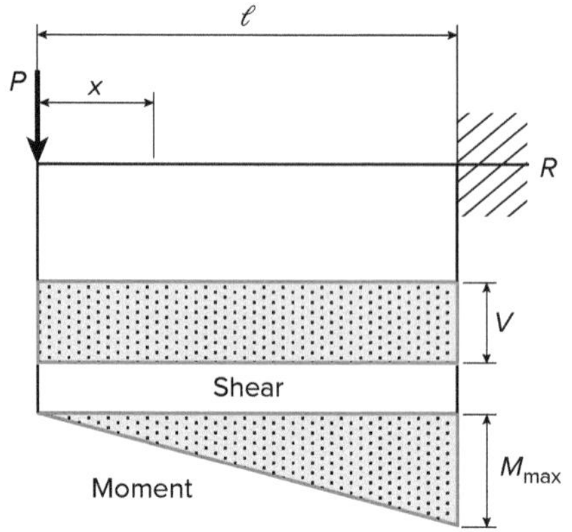

Reactions:

$$R = V = P$$

Moments:

$$M_{max}(\text{at fixed end}) = P\ell$$

$$M_x = Px$$

Deflections:

$$\Delta_{max}(\text{at free end}) = \frac{P\ell^3}{3EI}$$

$$\Delta_x = \frac{P}{6EI}\left(2\ell^3 - 3\ell^2 x + x^3\right)$$

A.4.15 CANTILEVER BEAM WITH POINT LOAD PLACED AT AN ARBITRARY DISTANCE

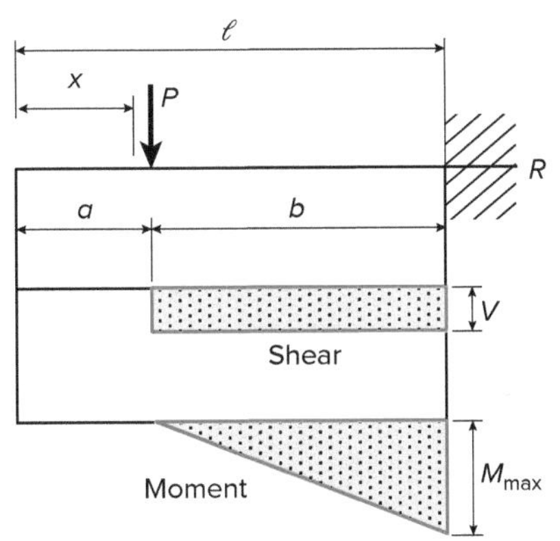

Reactions:

$$R = V = P$$

Moments:

$$M_{max}(\text{at fixed end}) = Pb$$

$$M_x(\text{when } x > a) = P(x - a)$$

Deflections:

$$\Delta_{max}(\text{at free end}) = \frac{Pb^2}{6El}(3\ell - b)$$

$$\Delta_a(\text{at point of load}) = \frac{Pb^3}{3El}$$

$$\Delta_x(\text{when } x < a) = \frac{Pb^2}{6El}(3\ell - 3x - b)$$

$$\Delta_x(\text{when } x > a) = \frac{P(\ell - x)^2}{6El}(3b - \ell + x)$$

A.4.16 SIMPLY SUPPORTED BEAM, FIXED AT ONE END - POINT LOAD AT CENTER

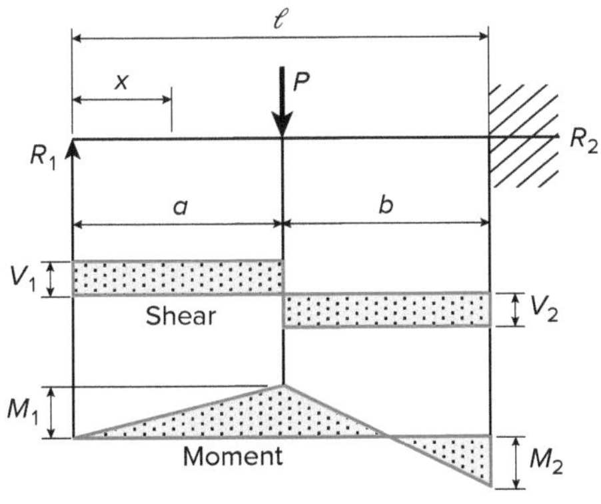

Reactions:

$$R_1 = V_1 = \frac{5P}{16}$$

$$R_2 = V_2(\max) = \frac{11P}{16}$$

Moments:

$$M_{max}(\text{at fixed end}) = \frac{3P\ell}{16}$$

$$M_1(\text{at point of load}) = \frac{5P\ell}{32}$$

$$M_x\left(\text{when } x < \frac{\ell}{2}\right) = \frac{5Px}{16}$$

$$M_x\left(\text{when } x > \frac{\ell}{2}\right) = P\left(\frac{\ell}{2} - \frac{11x}{16}\right)$$

Deflections:

$$\Delta_{max}\left(\text{at } x = \ell\sqrt{\frac{1}{5}} = 0.4472\ell\right) = \frac{P\ell^3}{48El\sqrt{5}} = 0.009317\frac{P\ell^3}{El}$$

$$\Delta_x(\text{at point of load}) = \frac{7P\ell^3}{768El}$$

$$\Delta_x\left(\text{when } x < \frac{\ell}{2}\right) = \frac{P_x}{96El}(3\ell^2 - 5x^2)$$

$$\Delta_x\left(\text{when } x > \frac{\ell}{2}\right) = \frac{P}{96El}(x - \ell)^2(11x - 2\ell)$$

A.4.17 SIMPLY SUPPORTED BEAM, FIXED AT ONE END - POINT LOAD PLACED AT AN ARBITRARY DISTANCE

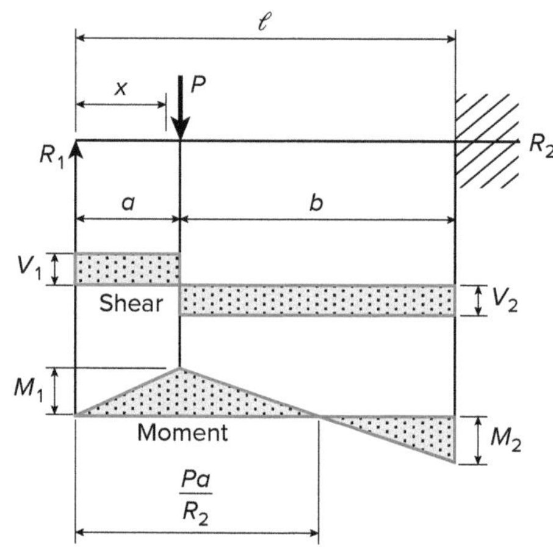

Reactions:

$$R_1 = V_1 = \frac{Pb^2}{2\ell^3}(a + 2\ell)$$

$$R_2 = V_2 = \frac{Pa}{2\ell^3}(3\ell^2 - a^2)$$

Moments:

$$M_1 \text{ (at point of load)} = R_1 a$$

$$M_2 \text{(at fixed end)} = \frac{Pab}{2\ell^2}(a + \ell)$$

$$M_x \text{ (when } x < a) = R_1 x$$

$$M_x \text{ (when } x > a) = R_1 x - P(x - a)$$

Deflections:

$$\Delta_{max}\left(\text{when } a < 0.414\ell, \text{ at } x = \ell\frac{\ell^2 + a^2}{3\ell^2 - a^2}\right) = \frac{Pa(\ell^2 - a^2)^3}{3EI(3\ell^2 - a^2)^2}$$

$$\Delta_{max}\left(\text{when } a > 0.414\ell, \text{ at } x = \ell\sqrt{\frac{a}{2\ell + a}}\right) = \frac{Pab^2}{6EI}\sqrt{\frac{a}{2\ell + a}}$$

$$\Delta_a \text{(at point of load)} = \frac{Pa^2b^3}{12EI\ell^3}(3\ell + a)$$

$$\Delta_x \text{(when } x < a) = \frac{Pb^2x}{12EI\ell^3}(3a\ell^2 - 2\ell x^2 - ax^2)$$

$$\Delta_x \text{(when } x > a) = \frac{Pa}{12EI\ell^2}(\ell - x)^2(3\ell^2 x - a^2 x - 2a^2\ell)$$

A.4.18 SIMPLY SUPPORTED BEAM, FIXED AT ONE END - UNIFORMLY DISTRIBUTED LOAD

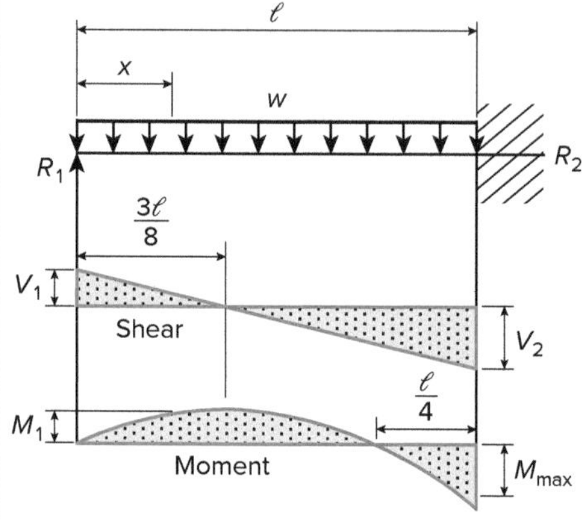

Reactions:

$$R_1 = V_1 = \frac{3w\ell}{8}$$

$$R_2 = V_2(\text{max}) = \frac{5w\ell}{8}$$

Shear:

$$V_x = R_1 - wx$$

Moments:

$$M_{max} = \frac{w\ell^2}{8}$$

$$M_1\left(\text{at } x = \frac{3}{8}\ell\right) = \frac{9}{128}w\ell^2$$

$$M_x = R_1 x - \frac{wx^2}{2}$$

Deflections:

$$\Delta_{max}\left(\text{at } x = \frac{\ell}{16}(1 + \sqrt{33}) = 0.4215\ell\right) = \frac{w\ell^4}{185EI}$$

$$\Delta_x = \frac{wx}{48EI}(\ell^3 - 3\ell x^2 + 2x^3)$$

A.4.19 SIMPLY SUPPORTED BEAM WITH OVERHANG - POINT LOAD PLACED AT OVERHANG

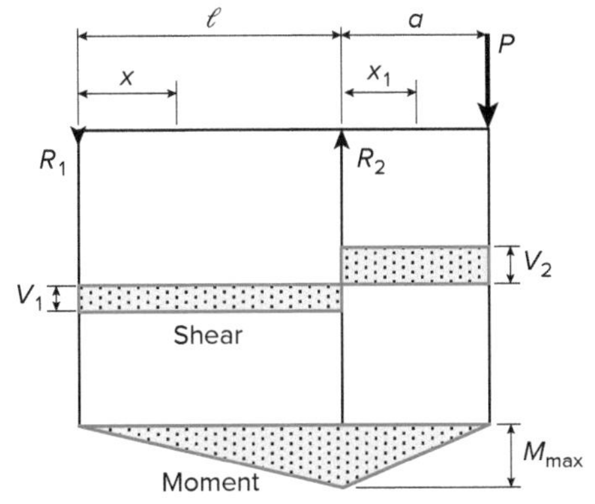

Reactions:

$$R_1 = V_1 = \frac{Pa}{\ell}$$

$$R_2 = V_1 + V_2 = \frac{P}{\ell}(\ell + a)$$

Shear:

$$V_2 = P$$

Moments:

$$M_{max}(\text{at } R_2) = Pa$$

$$M_x(\text{between supports}) = \frac{Pax}{\ell}$$

$$M_{x_1}(\text{for overhang}) = P(a - x_1)$$

Deflections:

$$\Delta_{max}\left(\text{between supports } x = \frac{\ell}{\sqrt{3}}\right) = \frac{Pa\ell^2}{9\sqrt{3}\,EI} = 0.06415\frac{Pa\ell^2}{EI}$$

$$\Delta_{max}(\text{for overhang at } x_1 = a) = \frac{Pa^2}{3EI}(\ell + a)$$

$$\Delta_x(\text{between supports}) = \frac{Pax}{6EI\ell}(\ell^2 - x^2)$$

$$\Delta_{x_1}(\text{for overhang}) = \frac{Px_1}{6EI}(2a\ell + 3ax_1 - x_1^2)$$

A.4.20 SIMPLY SUPPORTED BEAM WITH OVERHANG - POINT LOAD ARBITRARILY PLACED BETWEEN SUPPORTS

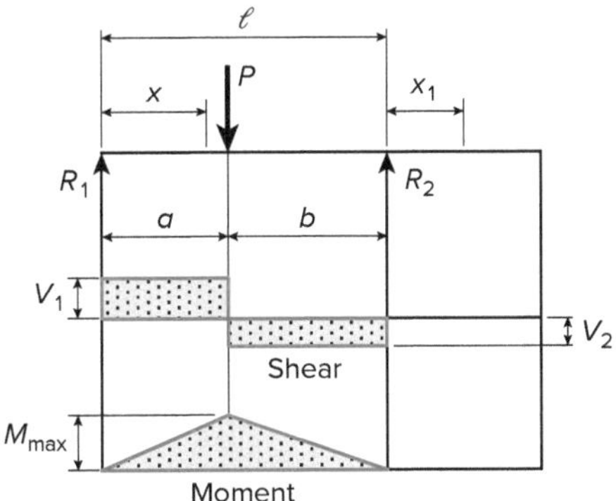

Reactions:

$$R_1 = V_1(\text{max when } a < b) = \frac{Pb}{\ell}$$

$$R_2 = V_2(\text{max when } a > b) = \frac{Pa}{\ell}$$

Moments:

$$M_{max}(\text{at point of load}) = \frac{Pab}{\ell}$$

$$M_x(\text{when } x < a) = \frac{Pbx}{\ell}$$

Deflections:

$$\Delta_{max}\left(\text{at } x = \sqrt{\frac{a(a+2b)}{3}} \text{ when } a > b\right) = \frac{Pab(a+2b)\sqrt{3a(a+2b)}}{27EI\ell}$$

$$\Delta_a(\text{at point of load}) = \frac{Pa^2b^2}{3EI\ell}$$

$$\Delta_x(\text{when } x < a) = \frac{Pbx}{6EI\ell}(\ell^2 - b^2 - x^2)$$

$$\Delta_x(\text{when } x > a) = \frac{Pa(\ell - x)}{6EI\ell}(2\ell x - x^2 - a^2)$$

$$\Delta_{x_1} = \frac{\text{Pabx}_1}{6EI\ell}(\ell + a)$$

A.4.21 SIMPLY SUPPORTED BEAM WITH OVERHANG - UNIFORMLY DISTRIBUTED LOAD

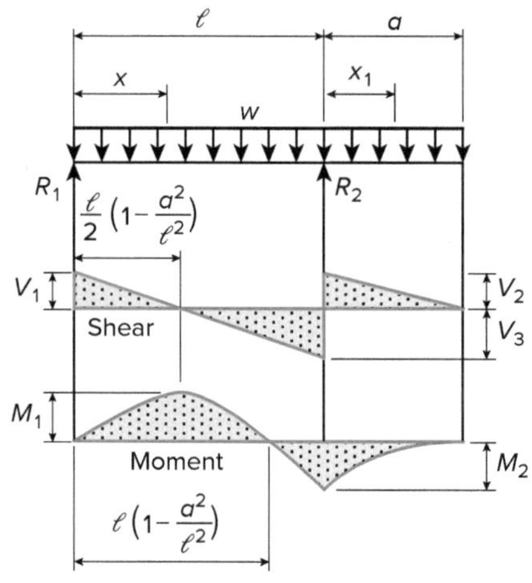

Reactions:

$$R_1 = V_1 = \frac{w}{2\ell}(\ell^2 - a^2)$$

$$R_2 = V_2 + V_3 = \frac{w}{2\ell}(\ell + a)^2$$

Shear:

$$V_2 = wa$$

$$V_3 = \frac{w}{2\ell}(\ell^2 + a^2)$$

$$V_x(\text{between supports}) = R_1 - wx$$

$$V_{x_1}(\text{for overhang}) = w(a - x_1)$$

Moments:

$$M_1\left(\text{at } x = \frac{\ell}{2}\left[1 - \frac{a^2}{\ell^2}\right]\right) = \frac{w}{8\ell^2}(\ell + a)^2(\ell - a)^2$$

$$M_2(\text{at } R_2) = \frac{wa^2}{2}$$

$$M_x(\text{between supports}) = \frac{wx}{2\ell}(\ell^2 - a^2 - x\ell)$$

$$M_{x_1}(\text{for overhang}) = \frac{w}{2}(a - x_1)^2$$

Deflections:

$$\Delta_x(\text{between supports}) = \frac{wx}{24El\ell}(\ell^4 - 2\ell^2x^2 + \ell x^3 - 2a^2\ell^2 + 2a^2x^2)$$

$$\Delta_{x_1}(\text{for overhang}) = \frac{wx_1}{24El}(4a^2\ell - \ell^3 + 6a^2x_1 - 4ax_1^2 + x_1^3)$$

A.4.22 SIMPLY SUPPORTED BEAM WITH OVERHANG - UNIFORMLY DISTRIBUTED LOAD APPLIED BETWEEN SUPPORTS

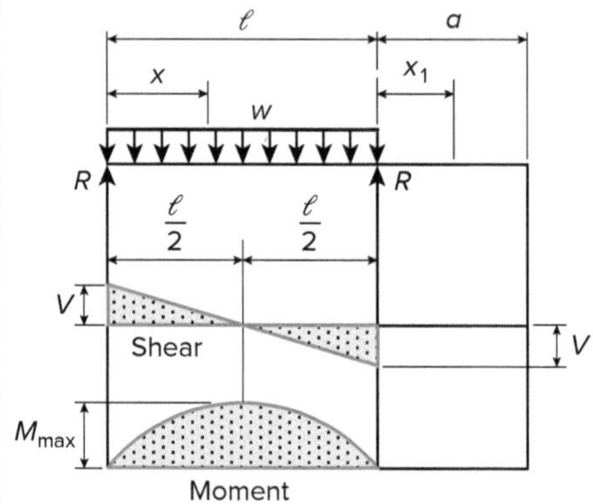

Reactions:

$$R = V = \frac{w\ell}{2}$$

Shear:

$$V_x = w\left(\frac{\ell}{2} - x\right)$$

Moments:

$$M_{max}(\text{at center}) = \frac{w\ell^2}{8}$$

$$M_x = \frac{wx}{2}(\ell - x)$$

Deflections:

$$\Delta_{max}(\text{at center}) = \frac{5w\ell^4}{384El}$$

$$\Delta_x = \frac{wx}{24El}(\ell^3 - 2\ell x^2 + x^3)$$

$$\Delta_{x_1} = \frac{w\ell^3 x_1}{24El}$$

A.4.23 SIMPLY SUPPORTED BEAM WITH OVERHANG - UNIFORMLY DISTRIBUTED LOAD APPLIED AT OVERHANG ONLY

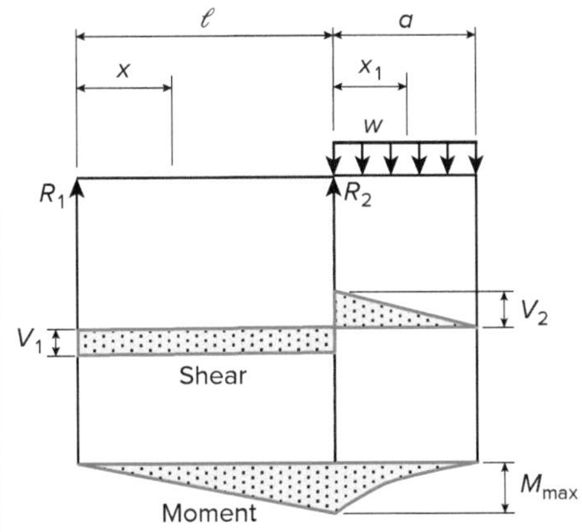

Reactions:

$$R_1 = V_1 = \frac{wa^2}{2\ell}$$

$$R_2 = V_1 + V_2 = \frac{wa}{2\ell}(2\ell + a)$$

Shear:

$$V_2 = wa$$

$$V_{x_1}(\text{for overhang}) = w(a - x_1)$$

Moments:

$$M_{max}(\text{at } R_2) = \frac{wa^2}{2}$$

$$M_x(\text{between supports}) = \frac{wa^2 x}{2\ell}$$

$$M_{x_1}(\text{for overhang}) = \frac{w}{2}(a - x_1)^2$$

Deflections:

$$\Delta_{max}\left(\text{between supports at } x = \frac{\ell}{\sqrt{3}}\right) = \frac{wa^2\ell^2}{18\sqrt{3}\,EI} = 0.03208\frac{wa^2\ell^2}{EI}$$

$$\Delta_{max}(\text{for overhang at } x_1 = a) = \frac{wa^3}{24EI}(4\ell + 3a)$$

$$\Delta_x(\text{between supports}) = \frac{wa^2 x}{12EI\ell}(\ell^2 - x^2)$$

$$\Delta_{x_1}(\text{for overhang}) = \frac{wx_1}{24EI}(4a^2\ell + 6a^2 x_1 - 4ax_1^2 + x_1^3)$$

A.4.24 SIMPLY SUPPORTED BEAM, FIXED AT BOTH ENDS - POINT LOAD APPLIED AT CENTER

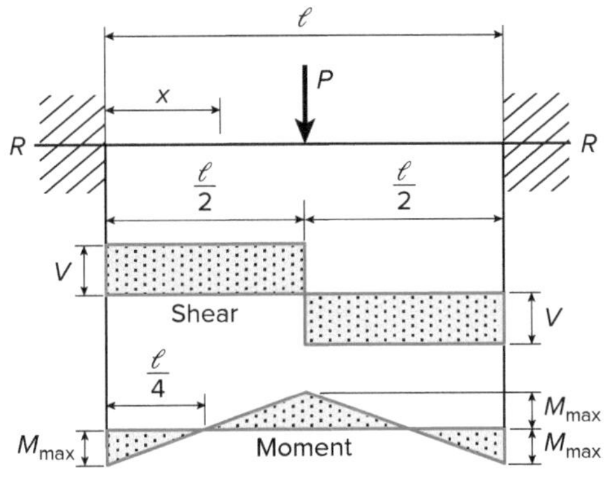

Reactions:

$$R = V = \frac{P}{2}$$

Moments:

$$M_{max}(\text{at center and ends}) = \frac{P\ell}{8}$$

$$M_x\left(\text{when } x < \frac{\ell}{2}\right) = \frac{P}{8}(4x - \ell)$$

Deflections:

$$\Delta_{max}(\text{at center}) = \frac{P\ell^3}{192EI}$$

$$\Delta_x\left(\text{when } x < \frac{\ell}{2}\right) = \frac{Px^2}{48EI}(3\ell - 4x)$$

A.4.25 SIMPLY SUPPORTED BEAM, FIXED AT BOTH ENDS - POINT LOAD APPLIED AT AN ARBITRARY DISTANCE

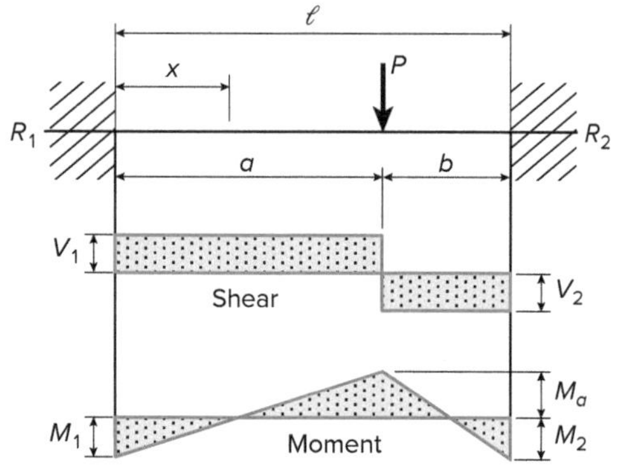

Reactions:

$$R_1 = V_1(\text{max when } a < b) = \frac{Pb^2}{\ell^3}(3a + b)$$

$$R_2 = V_2(\text{max when } a > b) = \frac{Pa^2}{\ell^3}(a + 3b)$$

Moments:

$$M_1(\text{max when } a < b) = \frac{Pab^2}{\ell^2}$$

$$M_2(\text{max when } a > b) = \frac{Pa^2b}{\ell^2}$$

$$M_a(\text{at point of load}) = \frac{2Pa^2b^2}{\ell^3}$$

$$M_x(\text{when } x < a) = R_1x - \frac{Pab^2}{\ell^2}$$

Deflections:

$$\Delta_{\max}\left(\text{when } a > b, \text{ at } x = \frac{2a\ell}{3a + b}\right) = \frac{2Pa^3b^2}{3EI(3a + b)^2}$$

$$\Delta_a(\text{at point of load}) = \frac{Pa^3b^3}{3EI\ell^3}$$

$$\Delta_x(\text{when } x < a) = \frac{Pb^2x^2}{6EI\ell^3}(3a\ell - 3ax - bx)$$

A.4.26 SIMPLY SUPPORTED BEAM, FIXED AT BOTH ENDS - UNIFORMLY DISTRIBUTED LOAD

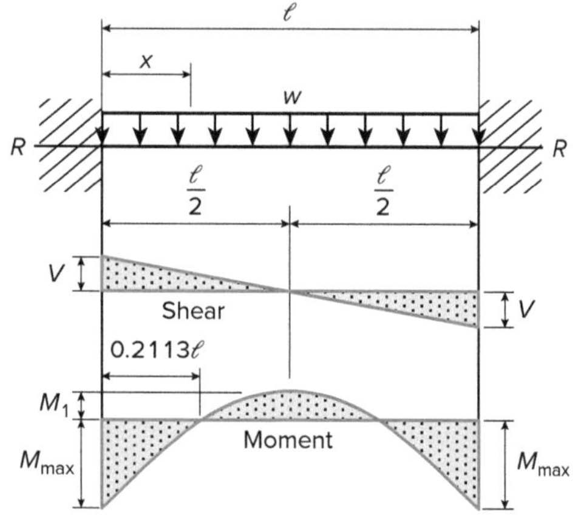

Reactions:

$$R = V = \frac{w\ell}{2}$$

Shear:

$$V_x = w\left(\frac{\ell}{2} - x\right)$$

Moments:

$$M_{\max}(\text{at ends}) = \frac{w\ell^2}{12}$$

$$M_1(\text{at center}) = \frac{w\ell^2}{24}$$

$$M_x = \frac{w}{12}(6\ell x - \ell^2 - 6x^2)$$

Deflections:

$$\Delta_{\max}(\text{at center}) = \frac{w\ell^4}{384EI}$$

$$\Delta_x = \frac{wx^2}{24EI}(\ell - x)^2$$

A.4.27 TWO SPAN CONTINUOUS BEAM WITH POINT LOAD APPLIED AT CENTER OF ONE SPAN

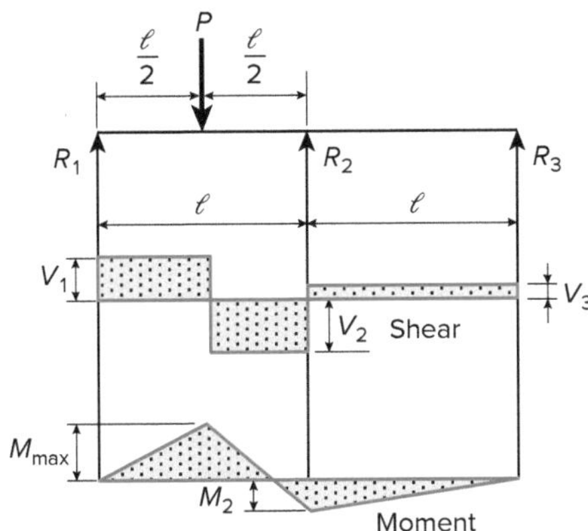

Reactions:

$$R_1 = V_1 = \frac{13}{32}P$$

$$R_2 = V_2 + V_3 = \frac{11}{16}P$$

$$R_3 = V_3 = -\frac{3}{32}P$$

Shear:

$$V_2 = \frac{19}{32}P$$

Moments:

$$M_{max}(\text{at point of load}) = \frac{13}{64}P\ell$$

$$M_2(\text{at } R_2) = \frac{3}{32}P\ell$$

A.4.28 TWO SPAN CONTINUOUS BEAM WITH POINT LOAD APPLIED AT AN ARBITRARY DISTANCE IN ONE SPAN

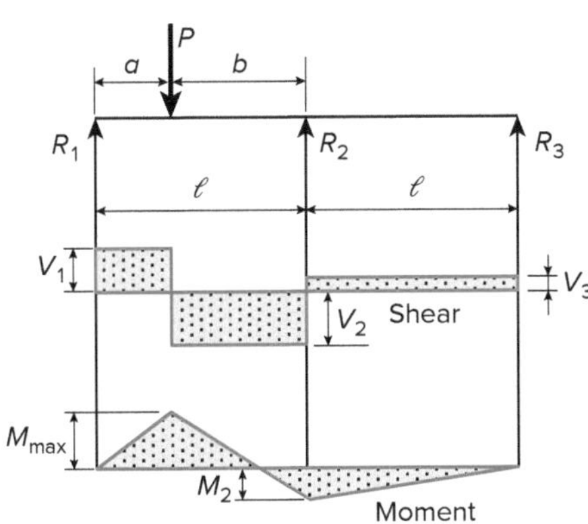

Reactions:

$$R_1 = V_1 = \frac{Pb}{4\ell^3}\left[4\ell^2 - a(\ell + a)\right]$$

$$R_2 = V_2 + V_3 = \frac{Pa}{2\ell^3}\left[2\ell^2 + b(\ell + a)\right]$$

$$R_3 = V_3 = \frac{Pab}{4\ell^3}(\ell + a)$$

Shear:

$$V_2 = \left(\frac{Pa}{4\ell^3}\right)\left[4\ell^2 + b(\ell + a)\right]$$

Moments:

$$M_{max}(\text{at point of load}) = \frac{Pab}{4\ell^3}\left[4\ell^2 - a(\ell + a)\right]$$

$$M_2(\text{at } R_2) = \frac{Pab}{4\ell^2}(\ell + a)$$

A.4.29 TWO SPAN CONTINUOUS BEAM WITH UNIFORMLY DISTRIBUTED LOAD PLACED IN ONE SPAN ONLY

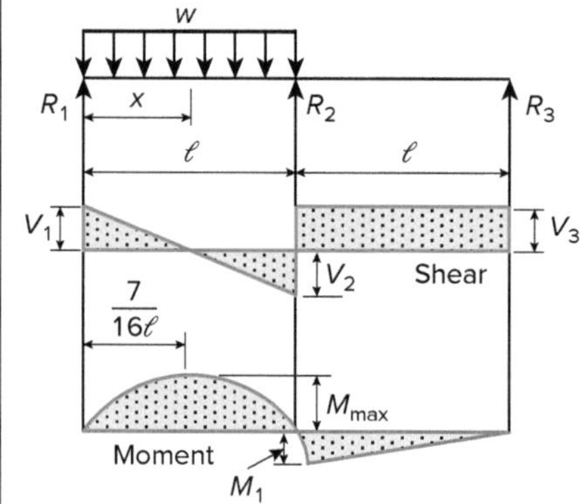

Reactions:

$$R_1 = V_1 = \frac{7}{16} w\ell$$

$$R_2 = V_2 + V_3 = \frac{5}{8} w\ell$$

$$R_3 = V_3 = -\frac{1}{16} w\ell$$

Shear:

$$V_2 = \frac{9}{16} w\ell$$

Moments:

$$M_{max}\left(\text{at } x = \frac{7}{16}\ell\right) = \frac{49}{512} w\ell^2$$

$$M_1(\text{at } R_2) = \frac{w\ell^2}{16}$$

$$M_x(\text{when } x < \ell) = \frac{wx}{16}(7\ell - 8x)$$

Fixed-End Beam Moments A.5

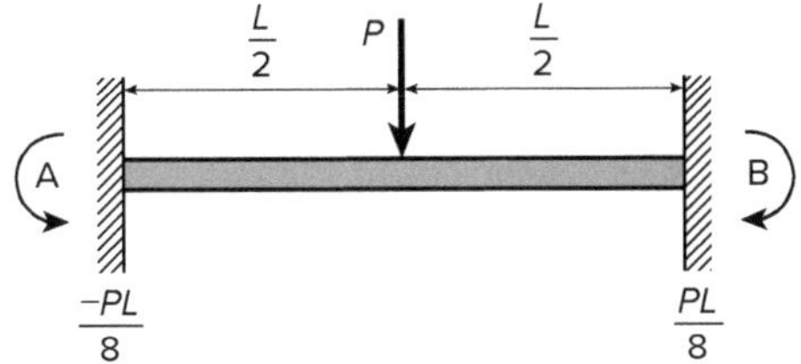

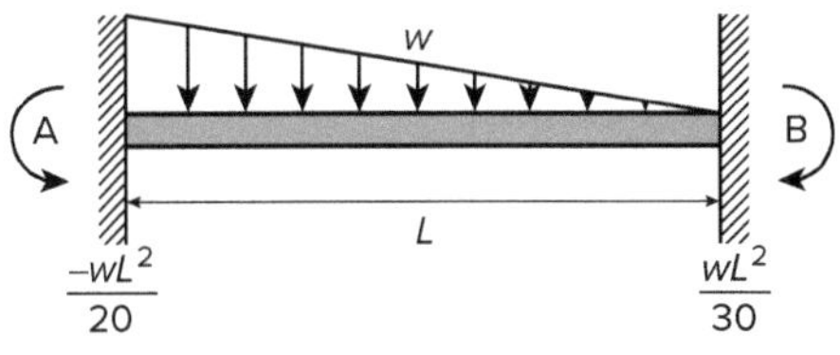

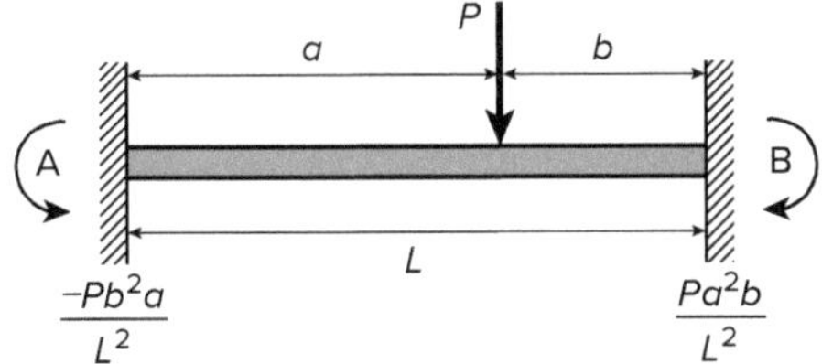

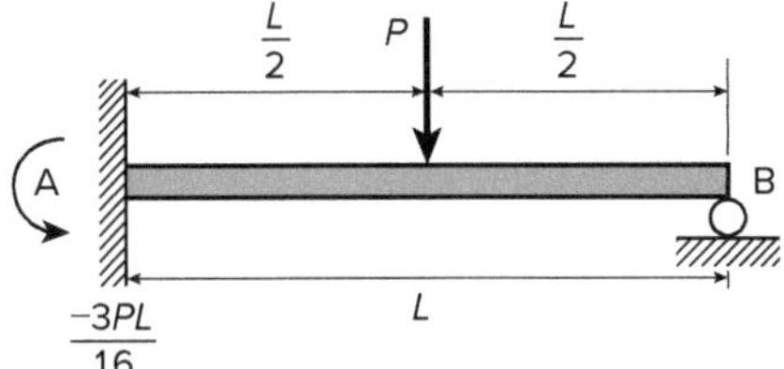

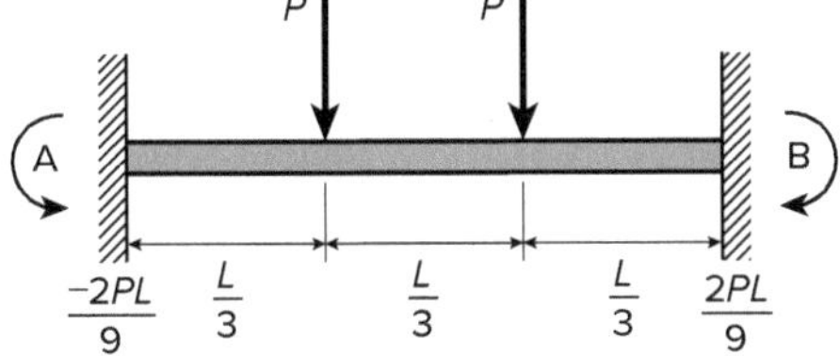

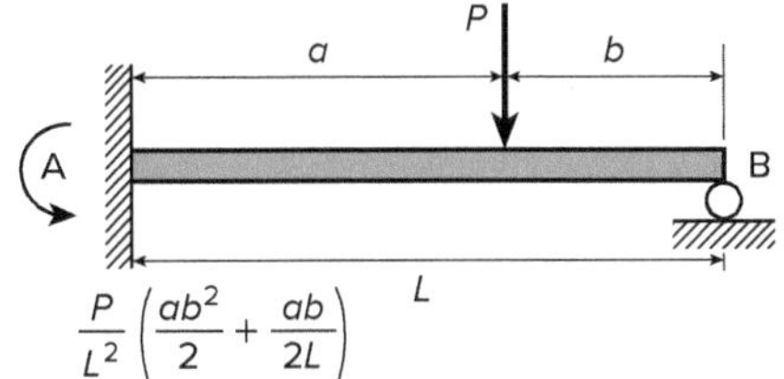

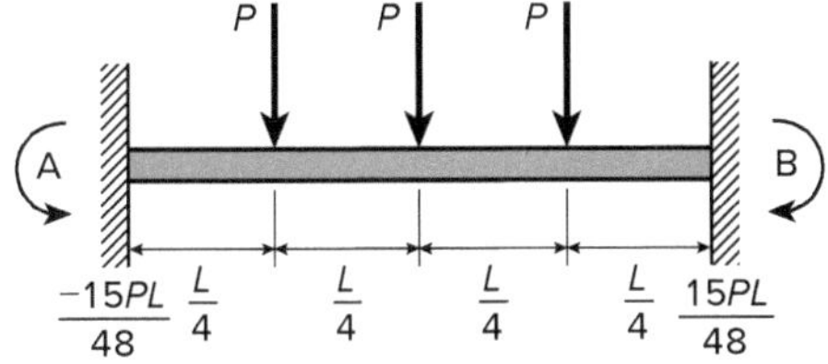

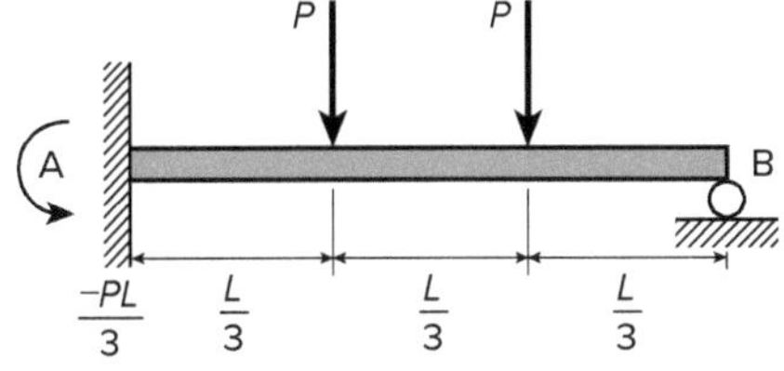

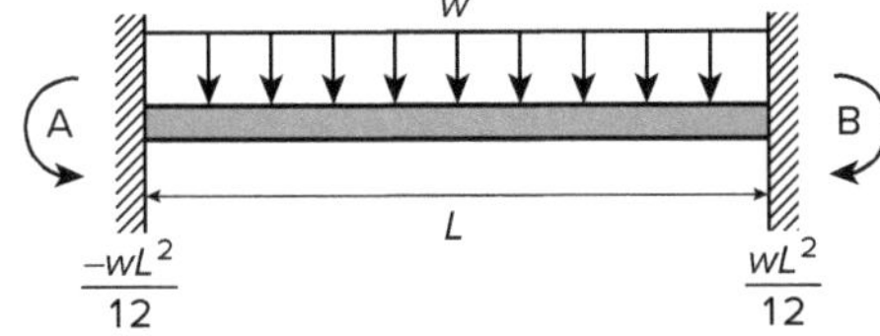

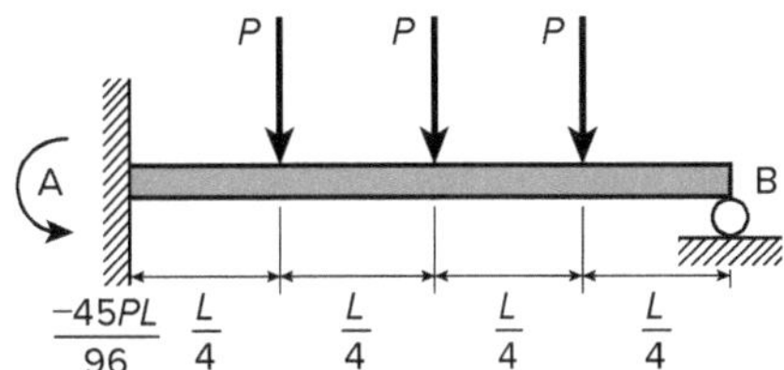

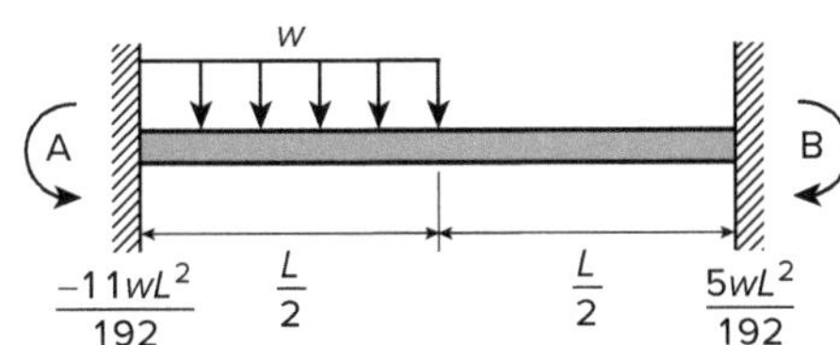

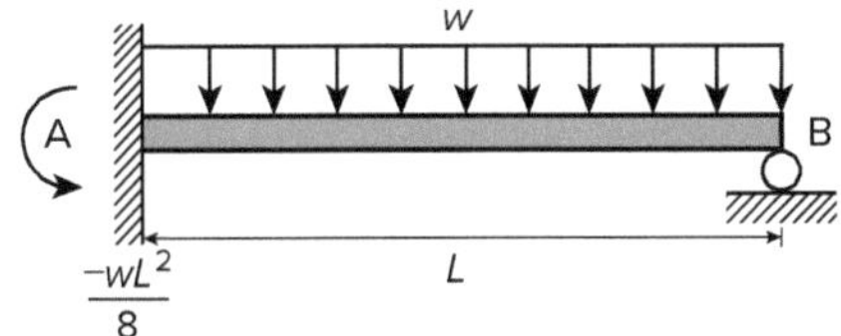

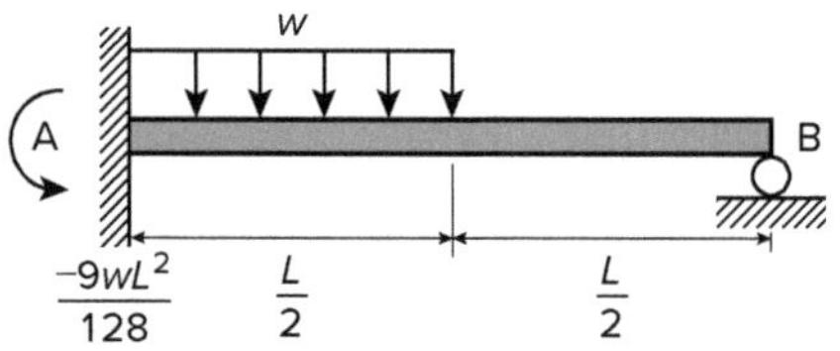
w
A
B
$\frac{-9wL^2}{128}$
$\frac{L}{2}$
$\frac{L}{2}$

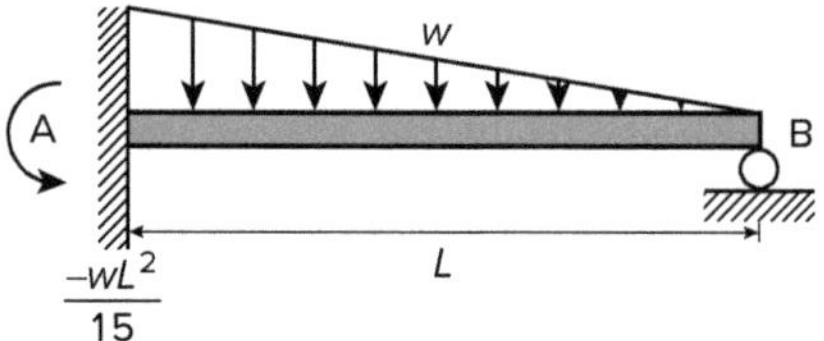
w
A
B
$\frac{-wL^2}{15}$
L

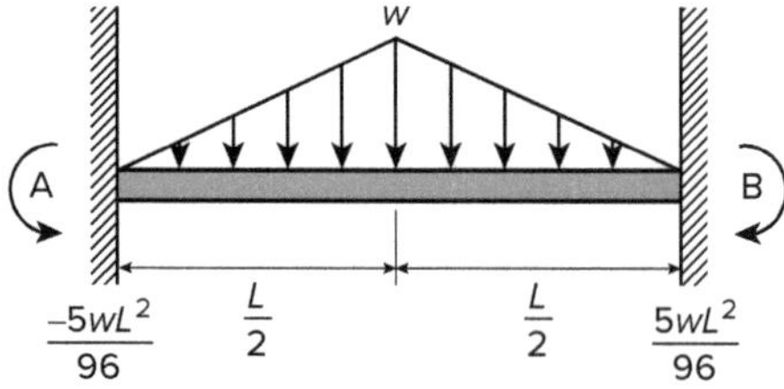
w
A
B
$\frac{-5wL^2}{96}$
$\frac{L}{2}$
$\frac{L}{2}$
$\frac{5wL^2}{96}$

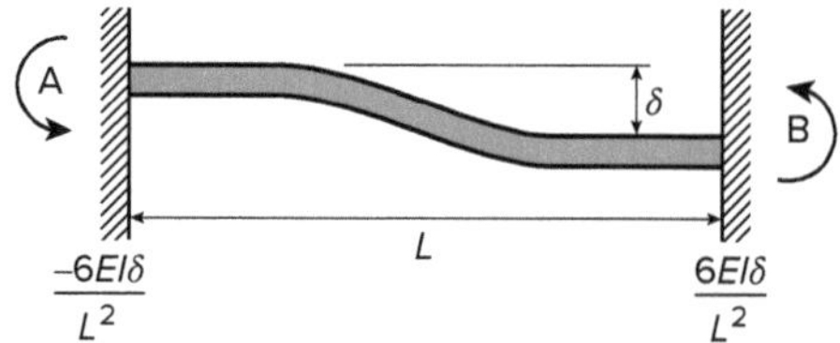
A
δ
B
$\frac{-6EI\delta}{L^2}$
L
$\frac{6EI\delta}{L^2}$

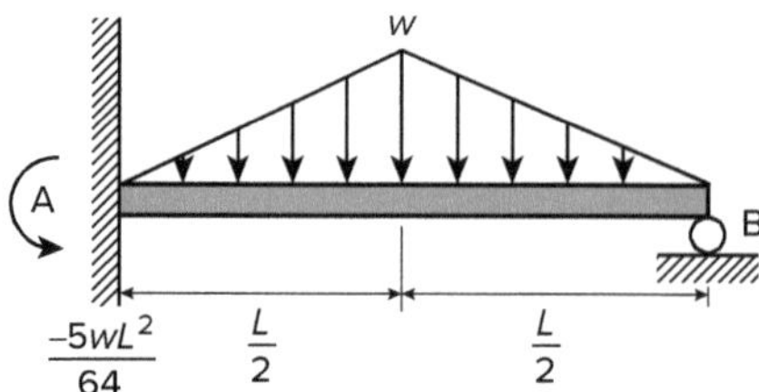
w
A
B
$\frac{-5wL^2}{64}$
$\frac{L}{2}$
$\frac{L}{2}$

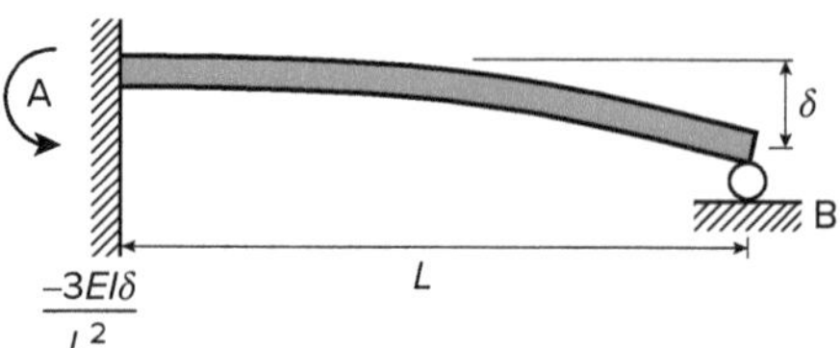
A
δ
B
$\frac{-3EI\delta}{L^2}$
L

Indeterminate Beams A.6

Uniformly distributed load:

w in load/unit length

Total load:

$W = wL$

Reactions:

$$R_1 = R_2 = \frac{W}{2}$$

Shear forces:

$$V_1 = +\frac{W}{2}$$

$$V_2 = -\frac{W}{2}$$

Maximum negative bending moment:

$$M_{\max} = -\frac{wL^2}{12} = -\frac{WL}{12} \text{ (at end)}$$

Maximum positive bending moment:

$$M_{\max} = \frac{wL^2}{24} = \frac{WL}{24} \text{ (at center)}$$

Maximum deflection:

$$\frac{wL^4}{384EI} = \frac{WL^3}{384EI} \text{ (at center)}$$

$$\delta = \left(\frac{wx^2}{24EI}\right)(L-x)^2,\ 0 \le x \le L$$

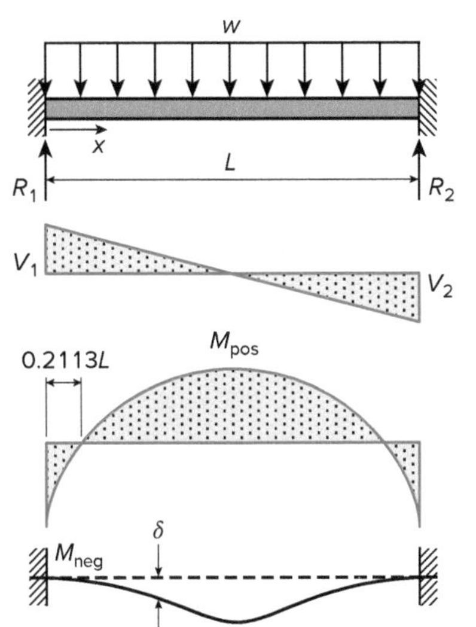

Concentrated load, P (at center)

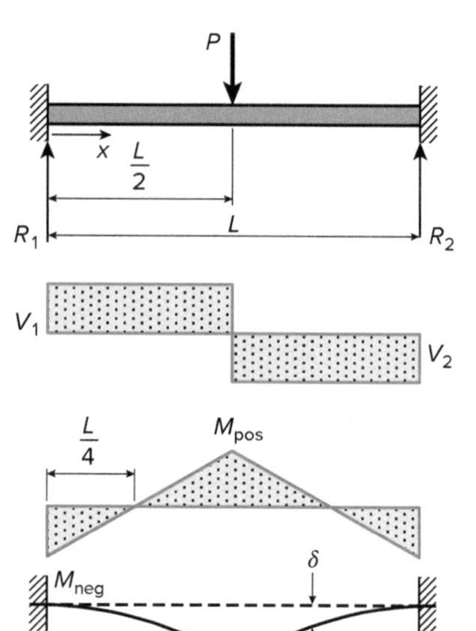

Reactions:

$$R_1 = R_2 = \frac{P}{2}$$

Shear forces:

$$V_1 = +\frac{P}{2};$$

$$V_2 = -\frac{P}{2}$$

Maximum positive bending moment:

$$M_{max} = \frac{PL}{8} \text{ (at center)}$$

Maximum negative bending moment:

$$M_{max} = -\frac{PL}{8} \text{ (at ends)}$$

Maximum deflection:

$$\frac{PL^3}{192EI} \text{ (at center)}$$

$$\delta = \left(\frac{Px^2}{48EI}\right)(3L - 4x),\ 0 \le x \le \frac{L}{2}$$

Concentrated load, P (at any point)

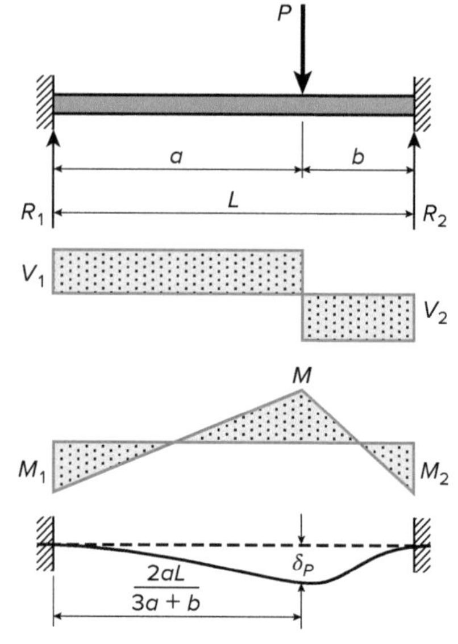

Reactions:

$$R_1 = \left(\frac{Pb^2}{L^3}\right)(3a + b)$$

$$R_2 = \left(\frac{Pa^2}{L^3}\right)(3b + a)$$

Shear forces:

$$V_1 = R_1;\ V_2 = -R_2$$

Bending moments:

$$M_1 = -\frac{Pab^2}{L^2}, \text{ maximum, when } a < b$$

$$M_2 = -\frac{Pa^2b}{L^2}, \text{ maximum, when } a > b$$

$$M_P = +\frac{2Pa^2b^2}{L^3} \text{ (at point of load)}$$

$$\delta_P: \frac{Pa^3b^3}{3EIL^3} \text{ (at point of load)}$$

$$\delta_{max} = \frac{2Pa^3b^2}{3EI(3a + b)^2}, \text{ at } x = \frac{2aL}{3a + b}, \text{ for } a > b$$

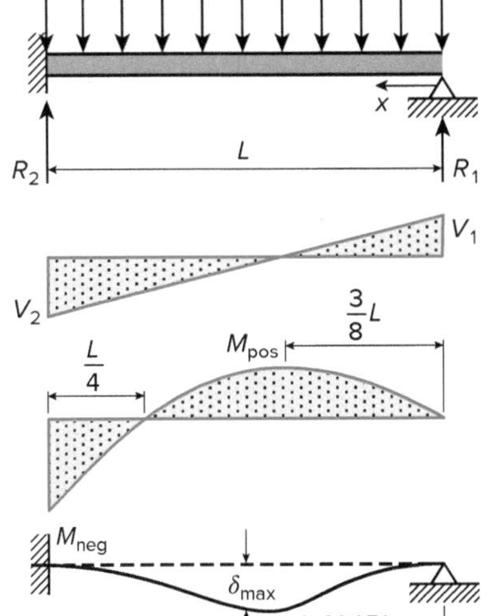

Uniformly distributed load:

w in load/unit length

Total load:

$W = wL$

Reactions:

$$R_1 = \frac{3wL}{8},$$

$$R_2 = \frac{5wL}{8}$$

Shear forces:

$V_1 = +R_1;$

$V_2 = -R_2$

Bending moments:

Maximum negative moment:

$-\frac{wL^2}{8}$ (at left end)

Maximum positive moment:

$$\frac{9}{128}wL^2,$$

$$x = \frac{3}{8}L$$

$$M = \frac{3wLx}{8} - \frac{wx^2}{2},\ 0 \le x \le L$$

Maximum deflection:

$$\frac{wL^4}{185EI},\ x = 0.4215L$$

$$\delta = \left(\frac{wx}{48EI}\right)(L^3 - 3Lx^2 + 2x^3),\ 0 \le x \le L$$

Appendix A.6

School of PE

Types of Admixtures A.7

TYPE OF ADMIXTURES	CHARACTERISTICS
Retarder admixtures	• Used in high-temperature applications where the mix sets faster than usual • Reduces initial setup of concrete and allows longer workability • Examples are calcium sulfate, gypsum, and sugar
Air-entrainment admixtures	• Provide additional air bubbles throughout the mix • Increase durability when freeze-thaw temperature cycles occur • Reduce segregation and bleeding of the mix
Accelerator admixtures	• Provide concrete strength in an increased time frame • Used for roadway work on heavily traveled corridors (rapid construction) • Increase rate of hydration, but may result in cracking when concrete begins to dry out • Examples are silica fume and calcium formate

Water-reducing admixtures	• Require less water and increase workability
	• Water demand reduction %
	  • Plasticizers 10%
	  • Midrange plasticizers 15%
	  • Superplasticizers 30%

Corrosion-inhibiting admixtures	• Used in locations exposed to water with high salinity or in areas with high chloride concentrations • Used where abrasive chemicals are present (industrial applications) • Examples are sodium nitrate, sodium nitrite, and sodium benzoate
Pozzolanic admixtures	• Composed of very fine microscopic particles that fill in smaller gaps • Less permeable and less susceptible to chemical infiltration • Much more water resistant • Reduce the heat of hydration during curing • Less shrinkage • Natural pozzolanic examples: clay, pumicite, and shale • Synthetic pozzolanic examples: silica fume and fly ash

Types of Portland Cement

TYPE OF PORTLAND CEMENT (PC)	CHARACTERISTICS
Type I	• Normal PC, general purpose • Used when heat hydration is not an issue
Type II	• Modified PC, moderate sulfate resistance • Used in warm weather for large structures near soil or water (such as drainage structures where sulfate in groundwater is higher than normal)
Type III	• High early strength • Achieves concrete strength in several days • Commonly used for roadways and bridges (rapid construction)
Type IV	• Low heat of hydration • Used for massive structures, like gravity dams, where heat generation must be kept to a minimum
Type V	• High sulfate resistance • Used when the water or soil is high in alkali

Equation Quick Reference

A.9

CHAPTER 1: PROJECT PLANNING

Geometric Properties of a Dimensional Shape		Page
$X_{\text{inside}} = X_{\text{outside}} - 2_{\text{thickness}}$	**Equation 1-1**	**7**
$Y_{\text{inside}} = Y_{\text{outside}} - 2_{\text{thickness}}$	**Equation 1-2**	**7**

Cross-sectional Area

$\text{Cross-sectional area} = 2_{(\text{thickness})}(X_{\text{outside}} + Y_{\text{inside}})$ $= 2_{(\text{thickness})}(X_{\text{inside}} + Y_{\text{outside}})$	**Equation 1-3**	**7**

Outside Perimeter of a Shape with Recesses Along a Face

Outside perimeter = 2 (length + width + recess)	**Equation 1-4**	**7**
Inside perimeter = outside perimeter – [4 (2 × thickness)]	**Equation 1-5**	**7**

Volume of a Shape

$\text{Mean perimeter} = \text{outside perimeter} - \left[4\left(2 \times \left(\frac{\text{thickness}}{2}\right)\right)\right]$	**Equation 1-6**	**8**

General Activity Duration

$\text{Duration}_{\text{activity}} = \frac{\text{Scope of Work}}{\text{Productivity} \times \text{Crew Size}}$	**Equation 1-7**	**18**

Network Analysis

EF = ES + Duration	**Equation 1-8**	**20**
LS = LF – Duration	**Equation 1-9**	**20**

Total Float

Float = LS – ES	**Equation 1-10**	**20**
Float = LF – EF	**Equation 1-11**	**20**
TF = LF – ES – Duration	**Equation 1-12**	**20**

Free Float

$\text{Free float}_{\text{Task A}} = \text{ES}_{\text{successor}} - \text{EF}_{\text{Task A}}$	**Equation 1-13**	**29**

CHAPTER 2: MEANS AND METHODS

Conditions for Static Equilibrium

$\Sigma F_x = 0, \quad \Sigma F_y = 0$ **Equation 2-1** **42**

Static Equilibrium

$\Sigma F = 0$ and $\Sigma M_O = 0$ **Equation 2-2** **42**

Stability

$\text{Stability} = (\text{lifting load} \times L_{\text{dis}}) + (W_{\text{boom}} \times Bcg) = (W_{\text{crane}} \times CG_{\text{tf}})$ **Equation 2-3** **49**

Darcy's Law

$Q = kiA$ **Equation 2-4** **51**

Factor of Safety

$\text{Factor of safety} = \dfrac{\text{resisting moments}}{\text{overturning moments}}$ **Equation 2-5** **56**

CHAPTER 3: SOIL MECHANICS

Lateral Earth Pressure Coefficient

$K = \dfrac{\sigma_h'}{\sigma_v'}$ **Equation 3-1** **62**

At-Rest Earth Pressure (Rankine Theory)

$K_O \approx 1 - \sin\phi'$[granular soils and normally consolidated (NC) clays] **Equation 3-2** **63**

Active Earth Pressure (Rankine Theory)

$p_a = K_a \gamma H - 2c\sqrt{K_a}$ **Equation 3-3** **63**

For saturated clay soils, $\phi = 0$, $K_a = 1$: $p_a = \gamma H - 2c$ **Equation 3-4** **64**

For granular soils, $c = 0$: $p_a = K_a \gamma H$ **Equation 3-5** **64**

Tension Crack Depth

$Z_{cr} = \dfrac{2c}{\gamma\sqrt{k_a}}$ **Equation 3-6** **64**

Active Earth Pressure, Level Backfill

$K_a = \dfrac{1}{K_p} = \dfrac{1 - \sin\phi}{1 + \sin\phi} = \tan^2\left(45° - \dfrac{\phi}{2}\right)$ **Equation 3-7** **65**

Active Earth Pressure, Sloping Backfill

$$K_a = \cos\beta\left(\frac{\cos\beta - \sqrt{\cos^2\beta - \cos^2\phi}}{\cos\beta + \sqrt{\cos^2\beta - \cos^2\phi}}\right)$$ **Equation 3-8** **65**

Passive Earth Pressure, Level Backfill (Rankine Theory)

$$p_p = K_p\gamma H + 2c\sqrt{K_p}$$ **Equation 3-9** **65**

For saturated clay soils, $\phi = 0$, $K_p = 1$: $p_p = \gamma H + 2c$ **Equation 3-10** **65**

For granular soils, $c = 0$: $p_p = K_p\gamma H$ **Equation 3-11** **65**

Passive Earth Pressure Coefficient, Level Backfill

$$K_p = \frac{1}{K_a} = \frac{1+\sin\phi}{1-\sin\phi} = \tan^2\left(45° + \frac{\phi}{2}\right)$$ **Equation 3-12** **65**

Passive Earth Pressure Coefficient, Sloping Backfill

$$K_p = \cos\beta\left(\frac{\cos\beta + \sqrt{\cos^2\beta - \cos^2\phi}}{\cos\beta - \sqrt{\cos^2\beta - \cos^2\phi}}\right)$$ **Equation 3-13** **65**

Active Earth Pressure Coefficient (Coulomb Theory)

$$K_a = \frac{\sin^2(\theta+\phi)}{\sin^2\theta\sin(\theta-\delta)\left(1+\sqrt{\frac{\sin(\phi+\delta)\sin(\phi-\beta)}{\sin(\theta-\delta)\sin(\theta+\beta)}}\right)^2}$$ **Equation 3-14** **66**

Passive Earth Pressure Coefficient (Coulomb Theory)

$$K_p = \frac{\sin^2(\theta-\phi)}{\sin^2\theta\sin(\theta+\delta)\left(1-\sqrt{\frac{\sin(\phi+\delta)\sin(\phi+\beta)}{\sin(\theta+\delta)\sin(\theta+\beta)}}\right)^2}$$ **Equation 3-15** **66**

Vertical Component of Active Earth Pressure, Sloping Backfill

$$(R_a)_v = R_a\sin(90 - \theta + \delta)$$ **Equation 3-16** **66**

Horizontal Component of Active Earth Pressure, Sloping Backfill

$$(R_a)_h = R_a\cos(90 - \theta + \delta)$$ **Equation 3-17** **66**

Retaining Walls Overturning Safety Factor

Overturning about toe: $FS_{OT} = \frac{\Sigma M_R}{\Sigma M_0}$ **Equation 3-18** **67**

Retaining Walls Sliding Safety Factor

Sliding: $FS_{SL} = \frac{\Sigma F_R}{\Sigma F_0}$ **Equation 3-19** **68**

Equivalent Fluid Density, Drained Backfill

$\gamma_{eq} = k\gamma_t$ | **Equation 3-20** | **70**

Equivalent Fluid Density, Saturated Backfill

$\gamma_{eq} = k\gamma_{sat} + (1 - k)\gamma_w$ | **Equation 3-21** | **70**

Overconsolidation Ratio (OCR)

$OCR = \frac{\sigma'_c}{\sigma'_o}$ | **Equation 3-22** | **73**

Primary Consolidation Settlement, Normally Consolidated (NC) Clays

$S_c = \sum\left(\frac{C_c}{1+e_o}\right)H\log\left(\frac{\sigma'_f}{\sigma'_o}\right)$ | **Equation 3-23** | **73**

Compression Index of NC Clays

$C_C = \frac{\Delta e}{\log\frac{\sigma'_f}{\sigma'_o}}$ | **Equation 3-24** | **74**

Primary Consolidation Settlement, Overconsolidated (OC) Clays (Final Effective Pressure Does Not Exceed Preconsolidation Pressure)

$S_c = \sum\left(\frac{C_r}{1+e_o}\right)H\log\left(\frac{\sigma'_f}{\sigma'_o}\right)$ | **Equation 3-25** | **75**

Primary Consolidation Settlement, Overconsolidated (OC) Clays (Final Effective Pressure Exceeds Preconsolidation Pressure)

$S_c = \sum\left[\left(\frac{C_r}{1+e_o}\right)H\log\left(\frac{\sigma'_c}{\sigma'_o}\right) + \left(\frac{C_c}{1+e_o}\right)H\log\left(\frac{\sigma'_f}{\sigma'_c}\right)\right]$ | **Equation 3-26** | **76**

$\Delta e_1 \Rightarrow$ $\Delta e_2 \Rightarrow$

Average Degree of Consolidation

$U = \frac{S_t}{S_c}\%$ | **Equation 3-27** | **76**

Nondimensional Time Factor

$T_v = \frac{C_v t}{H_d^2}$ | **Equation 3-28** | **76**

Total Vertical Stress

$\sigma_v = \Sigma\gamma_i Z_i$ | **Equation 3-29** | **78**

$\sigma_v = \gamma_1 Z_1 + \gamma_2 Z_2$ | **Equation 3-30** | **78**

Pore Water Pressure, Hydrostatic Condition

$u = \gamma_w Z_w$ | **Equation 3-31** | **79**

Pore Water Pressure, Seepage or Artesian Condition

$u = \gamma_w h_p$ Equation 3-32 79

Effective Vertical Stress

$\sigma'_v = \sigma_v - u$ Equation 3-33 80

Submerged (Effective) Unit Weight

$\gamma_{\text{submerged}} = \gamma' = (\gamma_{\text{saturated}} - \gamma_{\text{water}})$ Equation 3-34 80

Gross Bearing Pressure (Total Demand)

$Q_g = \dfrac{P_g}{A}$ Equation 3-35 83

Net Bearing Pressure (Net Demand)

$Q_{\text{net}} = \dfrac{P_{\text{net}}}{A}$ Equation 3-36 84

Soil Bearing Capacity – Design Equation

$(q_{\text{net}})_{\text{all}} = \left(\dfrac{q_{\text{net}}}{FS}\right)_{\text{capacity}} \geq (Q_{\text{net}})_{\text{applied}}$ Equation 3-37 84

$q_{\text{all}} = \left(\dfrac{q_{\text{ult}}}{FS}\right)_{\text{capacity}} \geq (Q_{\text{g}})_{\text{applied}}$ Equation 3-38 84

Terzaghi's Ultimate Bearing Capacity for Infinitely Long Foundations

$q_{\text{ult}} = cN_c + \gamma D_f N_q + 0.5\gamma B N_\gamma$ Equation 3-39 84

Net Ultimate Bearing Capacity

$q_{\text{net}} = q_{\text{ult}} - \gamma D_f$ Equation 3-40 84

Net Allowable Bearing Capacity

$(q_{\text{net}})_{\text{all}} = \dfrac{q_{\text{net}}}{FS}$ Equation 3-41 84

General Ultimate Bearing Capacity Equation

$q_{\text{ult}} = cN_c S_c + \gamma D_f N_q S_q + 0.5\gamma B N_\gamma S_\gamma$ Equation 3-42 85

Ultimate Bearing Capacity in Clay

$q_{\text{ult}} = cN_c S_c + \gamma D_f$ Equation 3-43 87

$q_{\text{net}} = cN_c S_c$ Equation 3-44 87

Undrained Shear Strength of Clay

$S_u = c = \dfrac{q_{\text{unc}}}{2}$ Equation 3-45 87

Ultimate Bearing Capacity in Sand

$q_{ult} = \gamma D_f + 0.5\gamma B N_\gamma S_\gamma$ **Equation 3-46** **87**

$q_{net} = \gamma D_f(N_q - 1) + 0.5\gamma B N_\gamma S_\gamma$ **Equation 3-47** **87**

Effective Vertical Stress at Foundation Level

$\sigma'_D = \gamma D_1 + \gamma' D_2$ **Equation 3-48** **88**

Submerged Unit Weight

$\gamma' = \gamma_{sat} - \gamma_w$ **Equation 3-49** **88**

Average Weighted Unit Weight

$\bar{\gamma} = \frac{d}{B}\gamma + \left(1 - \frac{d}{B}\right)\gamma'$ **Equation 3-50** **88**

Foundation Eccentricity

Eccentricity, $e = \frac{M}{P}$ **Equation 3-51** **90**

Contact Pressure with Eccentricity Inside Middle Third of Footing ($e < B/6$)

$q_{min} = \frac{P}{BL}\left(1 - \frac{6e}{B}\right),\ q_{max} = \frac{P}{BL}\left(1 + \frac{6e}{B}\right)$ **Equations 3-52 and 3-53** **90**

Contact Pressure with Eccentricity Outside Middle Third of Footing ($e > B/6$)

$q_{min} < 0,\ q_{max} = \frac{4P}{3L(B - 2e)}$ **Equation 3-54** **90**

Effective Footing Width for Bearing Capacity Calculations

$B' = B - 2e_B$ or $L' = L - 2e_L$ **Equation 3-55** **90**

Immediate Settlement (Modified Hough Method)

$\Delta H = \Sigma \frac{H}{C'} \log_{10} \frac{p_o + \Delta p}{p_o}$ **Equation 3-56** **92**

Increase in Vertical Stress – Point Load

$\Delta\sigma = \left(\frac{3P}{2\pi z^2}\right)\left(\frac{1}{1 + (r/z)^2}\right)^{2.5}$ **Equation 3-57** **95**

For $r = 0$: $\Delta\sigma = \left(\frac{3P}{2\pi z^2}\right)$ **Equation 3-58** **95**

Average Stress Increase Using 2:1 Method

For a rectangular area: $\overline{\Delta\sigma} = \frac{qBL}{(B + z)(L + z)} = \frac{P}{(B + z)(L + z)}$ **Equation 3-59** **97**

For a circular area: $\overline{\Delta\sigma} = \dfrac{qr^2}{\left(r+\frac{z}{2}\right)^2} = \dfrac{P}{\pi\left(r+\frac{z}{2}\right)^2}$ **Equation 3-60** **97**

For an infinitely long foundation wall: $\overline{\Delta\sigma} = \dfrac{P(\text{line load unit})}{(B+Z)(1)}$ **Equation 3-61** **97**

Influence Factor for Increase in Vertical Stress, Uniformly Loaded Area

$\Delta\sigma = qI$ **Equation 3-62** **97**

Safety Factor Against Slope Failure

$FS_{\text{slope stability}} = \dfrac{\text{available shear strength}}{\text{mobilized shear stresses}}$ **Equation 3-63** **99**

Safety Factor, Infinite Slopes in Dry Cohesionless Soil

$FS = \dfrac{S}{T} = \dfrac{N\tan\phi}{W\sin\beta} = \dfrac{(W\cos\beta)\tan\phi}{W\sin\beta} = \dfrac{\tan\phi}{\tan\beta}$ **Equation 3-64** **100**

Safety Factor, Infinite Slopes in c-ϕ Soil

$FS = \dfrac{c' + h(\gamma_{sat} - \gamma_w)(\cos^2\beta)\tan\phi'}{\gamma_{sat}\, h\sin\beta\cos\beta}$ **Equation 3-65** **101**

Safety Factor, Infinite Slopes in Clay Soil

$FS = \dfrac{\gamma'}{\gamma_{sat}}\dfrac{\tan\phi'}{\tan\beta}$ **Equation 3-66** **101**

Safety Factor, Slope Stability for Homogeneous Saturated Clay

$FS = \dfrac{N_o c}{\gamma_t H}$ **Equation 3-67** **101**

Depth Factor

$d = \dfrac{D}{H}$ **Equation 3-68** **102**

Safety Factor Against Overturning

$FS = \dfrac{\text{Sum of Resisting Forces} \times \text{Moment Arm (R)}}{\text{Sum of Driving Forces} \times \text{Moment Arm (R)}}$ **Equation 3-69** **103**

Safety Factor Against Sliding Along Failure Surface

$FS = \dfrac{\text{Sum of Resisting Forces}}{\text{Sum of Driving Forces}}$ **Equation 3-70** **103**

Safety Factor, Effective Stress Analysis

$F = \dfrac{\Sigma[c'l + (W\cos\alpha - ul)\tan\phi]}{\Sigma W\sin\alpha}$ **Equation 3-71** **104**

Safety Factor, Total Stress Analysis

$$F = \frac{\Sigma[c_T l + W \cos\alpha \tan\phi_T]}{\Sigma W \sin\alpha}$$ **Equation 3-72** **104**

CHAPTER 4: STRUCTURAL MECHANICS

Stress in Allowable Strength Design (ASD)

$f_{\text{Calculated}} \leq F_{\text{Allowable}}$ **Equation 4-1** **109**

LRFD Load Combinations

1.4D **Equation 4-2** **110**

$1.2D + 1.6L + 0.5(\text{Max}\ (L_r, S, R))$ **Equation 4-3** **110**

$1.2D + 1.6(\text{Max}(L_r, S, R)) + \text{Max}(1.0L, 0.5W)$ **Equation 4-4** **110**

$1.2D + 1.0W + 1.0L + 0.5(\text{Max}\ L_r, S, R)$ **Equation 4-5** **110**

$1.2D + 1.0E + 1.0L + 0.2S$ **Equation 4-6** **110**

$0.9D + 1.0W$ **Equation 4-7** **110**

$0.9D + 1.0E$ **Equation 4-8** **110**

Typical Moment Design LRFD Combinations

$M_u = 1.2 \times M_D + 1.6M_L$ **Equation 4-9** **110**

Resultant Trapezoidal Loading

$$\text{Resultant}_{\text{trapezoid}} = \frac{1}{2}(F_1 + F_2)h$$ **Equation 4-10** **115**

Equilibrium for 2D Beams and Trusses

$\Sigma F_v = 0$ (Summation of forces in the vertical direction = 0) **Equation 4-11** **115**

$\Sigma F_h = 0$ (Summation of forces in the horizontal direction = 0) **Equation 4-12** **116**

$\Sigma M_{\text{Point}\ x} = 0$ (Summation of moments about a point $x = 0$) **Equation 4-13** **116**

Yielding of the Gross Cross Section

$R_n = F_y \times A_g$ (*Steel Manual*, Sec. J4-1) **Equation 4-14** **132**

Fracture of the Net Cross Section

$R_n = F_u \times A_e$ (*Steel Manual*, Sec. J4-2) **Equation 4-15** **132**

Shear Distribution Over the Depth of a Cross Section

$$v = \frac{VQ}{Ib}$$ **Equation 4-16** **133**

First Moment of the Area Above the Neutral Axis

$$Q = \frac{bh^2}{8}$$ **Equation 4-17** **133**

Moment of Inertia

$$I = \frac{bh^3}{12}$$ **Equation 4-18** **133**

Flexural Stress

$$f_b = \frac{My}{I}$$ **Equation 4-19** **134**

Maximum Flexural Stress

$$f_{b_{max}} = \frac{Mc}{I}$$ **Equation 4-20** **134**

$$f_{b_{max}} = \frac{M}{S}$$ **Equation 4-21** **135**

$$S = \frac{bh^2}{6}$$ **Equation 4-22** **135**

Total Stress for Bending About One Axis

$$\sigma = \pm\frac{P}{A} \pm \frac{My}{I}$$ **Equation 4-23** **137**

$$\sigma = \pm\frac{P}{A} \pm \frac{Mc}{I}$$ **Equation 4-24** **137**

$$\sigma = \pm\frac{P}{A} \pm \frac{M}{S}$$ **Equation 4-25** **137**

Total Stress for Eccentric Loading About Two Axes

$$\sigma = \pm\frac{P}{A} \pm \frac{M_x y}{I_x} \pm \frac{M_y y}{I_y}$$ **Equation 4-26** **138**

Maximum Total Stress for Bending about Two Axes

$$\sigma = \pm\frac{P}{A} \pm \frac{M_x c}{I_x} \pm \frac{M_y c}{I_y}$$ **Equation 4-27** **138**

$$\sigma = \pm\frac{P}{A} \pm \frac{M_x}{S_x} \pm \frac{M_y}{S_y}$$ **Equation 4-28** **138**

Modulus of Rupture

$$f_t = 7.5\sqrt{f'_c}$$ **Equation 4-29** **139**

Concrete Design: Resultant Force in Steel and Concrete

$T_{\text{steel}} = A_s \times F_y$ and $C_{\text{concrete}} = 0.85 \times f'_c \times b \times a$ — **Equation 4-30** **140**

Concrete Depth of Compression Block for Rectangular Cross Sections

$$a = \frac{A_s \times f_y}{0.85 \times f'_c \times b}$$ **Equation 4-31** **140**

Concrete Beam Nominal Moment Capacity for Rectangular Cross Sections

$$M_n = A_s \times f_y \left(d - \frac{a}{2}\right)$$ **Equation 4-32** **140**

Concrete Shear Capacity

$$V_c = 2\sqrt{f'_c}\, b \times d$$ **Equation 4-33** **143**

General Design Equation for Shear

$$\phi V_n = \phi(V_c + V_s) \geq V_u$$ **Equation 4-34** **143**

General Design Equation for Shear, Steel Contribution

$$V_s \geq \frac{V_u}{\phi} - V_c$$ **Equation 4-35** **144**

Steel Contribution to Shear Strength

$$V_s = \frac{A_v \times f_y \times d}{s}$$ **Equation 4-36** **144**

Stirrup Spacing Based on Applied Loads

$$s_{\text{req}_{\text{strength}}} = \frac{A_v \times f_y \times d}{V_s}$$ **Equation 4-37** **144**

Stirrup Spacing Based on Beam Width and Concrete Strength

$$s_{\text{req}_{\text{width}}} = \text{smaller}\left(\frac{A_v \times f_y}{50 \times b_w}, \frac{A_v \times f_y}{0.75 \times b_w \sqrt{f'_c}}\right)$$ **Equation 4-38** **144**

Shear Force Limits for Use with Equation 4-37

$$\frac{\phi V_c}{2} < V_u \leq \phi V_c$$ **Equation 4-39** **144**

Isolated T Beam Effective Width

$$b_e \leq 4 \times b_w \leq 8 \times h_f$$ **Equation 4-40** **145**

Concrete Depth of Compression Block for T-Beam Cross Sections

$$a = \frac{A_s \times f_y}{0.85 \times f'_c \times b_e}$$ **Equation 4-41** **145**

Concrete Flange Compression Force for T-Beam Cross Sections

$$C_f = 0.85 \times f'_c \times h_f \times (b_e - b_w)$$ Equation 4-42 146

Concrete Web Compression Force for T-Beam Cross Sections

$$C_w = 0.85 \times f'_c \times a \times b_w$$ Equation 4-43 146

Concrete Compression Block within the Web Section b_w

$$a = \frac{(A_s \times f_y) - (0.85 \times f'_c \times h_f) \times (b_e - b_w)}{0.85 \times f'_c \times b_w}$$ Equation 4-44 146

Concrete Beam Nominal Moment Capacity for T-Beam Cross Sections with Compression Block 'a' in the Web (Used in Conjunction with Equation 4-44)

$$M_n = 0.85 \times f'_c \left[h_f \times (b_e - b_w) \times \left(d - \frac{h_f}{2} \right) + b_w \times a \times \left(d - \frac{a}{2} \right) \right]$$ Equation 4-45 146

Steel I-Beam Flange Compactness Criteria

$$\lambda_f = \frac{b_f}{2t_f} \le \lambda_{pf} = 0.38\sqrt{\frac{E}{F_y}}$$ Equation 4-46 147

Steel I-Beam Web Compactness Criteria

$$\lambda_w = \frac{h}{t_w} \le \lambda_{pw} = 3.76\sqrt{\frac{E}{F_y}}$$ Equation 4-47 147

Steel I-Beam Flange Compactness Criteria for F_y = 50 ksi

$$\frac{b_f}{2t_f} \le 9.15$$ Equation 4-48 148

Steel I-Beam Web Compactness Criteria for F_y = 50 ksi

$$\frac{h}{t_w} \le 90.6$$ Equation 4-49 148

ASD and LRFD Design Shear Stress

$$F_{v_{\text{allowable}}} = 0.6 \times \frac{F_y}{\Omega}$$ Equation 4-50 150

$$F_{v_{\text{ultimate}}} = \phi \times 0.6 \times F_y$$ Equation 4-51 150

$$F_{v_{\text{allowable}}} = 0.4 \times F_y$$ Equation 4-52 150

$$F_{v_{\text{ultimate}}} = 0.6 \times F_y$$ Equation 4-53 150

$$V_{\text{allowable}} = 0.4 \times F_y \times d \times t_w$$ Equation 4-54 150

$$V_n = 0.6 \times F_y \times d \times t_w$$ Equation 4-55 150

Euler Buckling Load for Columns

$$P_{cr} = \frac{\pi^2 \times E \times A}{\left(\frac{KL}{r}\right)^2}$$ Equation 4-56 151

Euler Buckling Stress for Columns

$$F_{cr} = \frac{\pi^2 \times E}{\left(\frac{KL}{r}\right)^2}$$ Equation 4-57 151

Radius of Gyration

$$r = \sqrt{\frac{I}{A}}$$ Equation 4-58 153

Slenderness Ratio Limit for Non-Sway/Braced Frame Concrete Columns

$$\frac{k_b \times L_u}{r} \leq 34 - 12\left(\frac{M_1}{M_2}\right) \leq 40$$ Equation 4-59 154

Slenderness Ratio Limit for Sway or Unbraced Frame Concrete Columns

$$\frac{k_b \times L_u}{r} \leq 22$$ Equation 4-60 154

Concrete Column Minimum Shear Reinforcement

$$0.75 \times \sqrt{f'_c} \times \frac{b_w s}{f_{yt}}$$ Equation 4-61 154

$$50 \times \frac{b_w s}{f_{yt}}$$ Equation 4-62 154

Steel Column Slenderness Limit for W-Shapes

$$\frac{0.5 \times b_f}{t_f} \leq 0.56 \times \sqrt{\frac{E}{F_y}}$$ Equation 4-63 155

Steel Column Critical Stress as a Function of the Slenderness Ratio

$$\text{If } \frac{K \cdot L}{r} \leq 4.71\sqrt{\frac{E}{F_y}}, \text{ then } F_{cr} = \left[0.658^{\frac{F_y}{F_e}}\right] \times F_y$$ Equation 4-64 155

$$\text{If } \frac{K \cdot L}{r} > 4.71\sqrt{\frac{E}{F_y}}, \text{ then } F_{cr} = 0.877 \times F_e$$ Equation 4-65 155

Steel Column Elastic Buckling Stress

$$F_e = \frac{\pi^2 E}{\left(\frac{KL}{r}\right)^2}$$ Equation 4-66 156

Concrete Slab Required Area of Temperature and Shrinkage Steel

$$\rho_g = \frac{A_s}{bh} = 0.0018$$ Equation 4-67 156

Spread Footing Maximum Area

$$A_{\max} = \frac{P_{\text{Total}}}{q_{\text{allowable}}}$$ **Equation 4-68** 159

Concrete Spread Footing One-Way Shear Capacity

$$\varphi V_c \geq V_u$$ **Equation 4-69** 159

$$(0.75)2bd\sqrt{f'_c} \geq q_u \times bx$$ **Equation 4-70** 159

Concrete Spread Footing Two-Way Shear Capacity

$$\varphi V_n = \varphi(V_c)$$ **Equation 4-71** 160

$$V_c = 4\lambda\sqrt{f'_c}\, b_0 d$$ **Equation 4-72** 160

$$V_c = \left(2 + \frac{4}{\beta}\right)\lambda\sqrt{f'_c}\, b_0 d$$ **Equation 4-73** 160

$$V_c = \left(2 + \frac{\alpha_s d}{b_0}\right)\lambda\sqrt{f'_c}\, b_0 d$$ **Equation 4-74** 160

Lateral Earth Pressure At-Rest Coefficient

$$K_0 = 1 - \sin(\phi)$$ **Equation 4-75** 161

Lateral Earth Pressure Active Coefficient

$$K_a = \tan^2\left(45 - \frac{\phi}{2}\right) = \frac{(1 - \sin(\phi))}{(1 + \sin(\phi))} = \frac{\sigma_a}{\sigma_v}$$ **Equation 4-76** 161

Lateral Earth Pressure Passive Coefficient

$$K_p = \tan^2\left(45 + \frac{\phi}{2}\right) = \frac{(1 + \sin(\phi))}{(1 - \sin(\phi))} = \frac{\sigma_p}{\sigma_v}$$ **Equation 4-77** 162

Horizontal Lateral Earth Pressure

$$\sigma_h = K \times \gamma_{\text{soil}} \times h$$ **Equation 4-78** 162

Force Due to Lateral Earth Pressure

$$P_{\text{Lateral}} = \frac{1}{2} \times \gamma_{\text{soil}} \times K \times h^2$$ **Equation 4-79** 162

CHAPTER 5: HYDRAULICS & HYDROLOGY

Hydraulics Radius

$R = A/P$ Equation 5-1 170

$R = D/4$ when a circular pipe is full or half-full

Critical Depth, Rectangular Channel

$$d_c = \left[\frac{C_1 Q}{b}\right]^{\frac{2}{3}}$$ Equation 5-2 171

$C_1 = 0.176$ (USCS), 0.319 (SI)

Critical Depth, Triangular Channel

$$d_c = C_2\left[\frac{Q}{z_1 + z_2}\right]^{\frac{2}{5}}$$ Equation 5-3 171

$C_2 = 0.757$ (USCS), 0.96 (SI)

Critical Depth, Other Shapes (Including Trapezoidal and Circular)

$$Q = \left[\frac{gA^3}{T}\right]^{\frac{1}{2}}$$ Equation 5-4 172

Critical Depth, Circular Channels (Approximate)

$$d_c = C_3\frac{Q^{\frac{1}{2}}}{D^{\frac{1}{4}}}$$ Equation 5-5 172

$C_3 = 0.42$ (USCS), 0.562 (SI)

Froude Number

$$Fr = \frac{v}{\sqrt{gD_h}}$$ Equation 5-6 172

Hydraulic Jump, Upstream Depth, Rectangular Channel

$$d_1 = -\frac{1}{2}d_2 + \left[\frac{2v_2^2 d_2}{g} + \frac{d_2^2}{4}\right]^{\frac{1}{2}}$$ Equation 5-7 173

Hydraulic Jump, Upstream Velocity, Rectangular Channel

$$v_1^2 = \left(\frac{gd_2}{2d_1}\right)(d_1 + d_2)$$ Equation 5-8 173

Manning's Equation

$$v = \frac{1.49}{n}R^{2/3}S^{1/2}$$ Equation 5-9 173

$$Q = \frac{1.49}{n}AR^{2/3}S^{1/2}$$ Equation 5-10 173

Friction Head Loss

$$h_f = \frac{Ln^2v^2}{2.208R^{4/3}} \text{ (USCS)}$$ Equation 5-11 175

Optimum Dimensions for Trapezoidal and Rectangular Channels

$$\frac{b}{d} = 2\left[\left(z^2+1\right)^{\frac{1}{2}} - z\right]$$ Equation 5-12 175

Contracted (Rectangular) and Suppressed Weirs, Volumetric Flow Rate

$$Q = \frac{2}{3}C_D B\left(H^{\frac{3}{2}}\right)\sqrt{2g}$$ Equation 5-13 177

Contracted (Rectangular) and Suppressed Weirs, Discharge Coefficient

$$C_D = 0.602 + 0.083(H/P)$$ Equation 5-14 177

Contracted (Rectangular) and Suppressed Weirs, Effective Width of Contracted Notch

$$B_{\text{effective}} = B_{\text{actual}} - 0.1NH$$ Equation 5-15 177

V-Notch (Triangular) Weir, Volumetric Flow Rate

$$Q = \frac{8}{15}C_e\left(\tan\frac{\theta}{2}\right)\left(h_e^{\frac{5}{2}}\right)\sqrt{2g}$$ Equation 5-16 178

V-Notch (Triangular) Weir, Volumetric Flow Rate, Notch Angle = 90°

$$Q = 2.49h_e^{2.48}$$ Equation 5-17 178

Cippoletti (Trapezoidal) Weir, Volumetric Flow Rate

$$Q = 3.36LH^{3/2}$$ Equation 5-18 179

Broad-Crested Weir, Volumetric Flow Rate, General

$$Q = \frac{2}{3}C_1L\left(H^{3/2}\right)\sqrt{2g}$$ Equation 5-19 179

Broad-Crested Weir, Volumetric Flow Rate, Horton Equation

$$Q = C_sLH^{3/2}$$ Equation 5-20 179

Broad-Crested Weir, Volumetric Flow Rate, Horton Equation with Approach Velocity

$$Q = C_sL\left(H + \frac{v^2}{2g}\right)^{3/2}$$ Equation 5-21 180

Parshall Flume, Volumetric Flow Rate, Free-Flow Conditions

$$Q = CH_a^n$$ Equation 5-22 180

Minimum Pipe Diameter, Full Flow

$$D = 1.335\left(\frac{nQ}{\sqrt{S}}\right)^{3/8}$$ Equation 5-23 181

Pipe Flow Rate, Full Flow

$$Q = \frac{0.463D^{\frac{8}{3}}\sqrt{S}}{n}$$ **Equation 5-24** **181**

Pipe Velocity, Full Flow

$$v = \frac{0.591D^{\frac{2}{3}}\sqrt{S}}{n}$$ **Equation 5-25** **181**

Flow Rate in a Gutter Section

$$Q = \frac{0.56}{n}S_x^{1.67}S^{0.5}T^{2.67}$$ **Equation 5-26** **183**

Minimum Gutter Width

$$T = \frac{(1.79Qn)^{\frac{3}{8}}}{S_x^{\frac{5}{8}}S^{\frac{3}{16}}}$$ **Equation 5-27** **183**

Flow Rate for *Handbook of Hydraulics* Table 7-10

$$Q = \frac{K}{n}D^{\frac{8}{3}}\sqrt{S} \Rightarrow K = \frac{Qn}{D^{\frac{8}{3}}\sqrt{S}}$$ **Equation 5-28** **184**

Flow Rate for *Handbook of Hydraulics* Table 7-11

$$Q = \frac{K'}{n}b^{\frac{8}{3}}\sqrt{S} \Rightarrow K' = \frac{Qn}{b^{\frac{8}{3}}\sqrt{S}}$$ **Equation 5-29** **185**

Culvert Head Loss, Full Flow

$$H_L = H_e + H_f + H_o$$ **Equation 5-30** **189**

Culvert Head Loss at Entrance, Full Flow

$$H_e = k_e(v^2/2g)$$ **Equation 5-31** **189**

Culvert Friction Loss Through Barrel, Full Flow

$$H_f = (K_u n^2 L R^{-1.33})(v^2/2g)$$ **Equation 5-32** **189**

Culvert Head Loss at Outlet, Full Flow

$$H_o = v^2/2g$$ **Equation 5-33** **189**

Culvert Energy Balance, Full Flow

$$HW_o + LS + (v_u^2/2g) = TW + (v_d^2/2g) + H_L$$ **Equation 5-34** **190**

Culvert Energy Balance, Full Flow, Negligible Velocity

$$HW_o = TW + H_L - LS$$ **Equation 5-35** **190**

Change in Stored Water

$\Delta S = \text{inflows} - \text{outflows}$ Equation 5-36 191

Arithmetic Mean Precipitation

$\overline{P} = \sum \frac{P_i}{n}$ Equation 5-37 194

Thiessen Polygon Method, Mean Precipitation

$\overline{P} = \frac{\sum_{i=1}^{n} P_i A_i}{\sum_{i=1}^{n} A_i} = \frac{P_1 A_1 + P_2 A_2 + P_3 A_3 + \cdots + P_n A_n}{A_1 + A_2 + A_3 + \cdots + A_n}$ Equation 5-38 195

Probability and Frequency Relationship

$p = \frac{1}{F}$ Equation 5-39 196

Probability of Event Occurring in *n* Years

$p = 1 - \left(1 - \frac{1}{F}\right)^n$ Equation 5-40 196

Probability of Event Occurring in *m* Consecutive Years

$p = p^m$ Equation 5-41 196

Probability of Event Not Occurring

$p_{\text{not}} = 1 - p$ Equation 5-42 196

Time of Concentration, General

$t = \frac{L}{v}$ Equation 5-43 198

Time of Concentration, Segmental Method

$t_c = t_{\text{sheet}} + t_{\text{shallow}} + t_{\text{channel}}$ Equation 5-44 199

Rational Method, Peak Flow Rate

$Q_{\text{peak}} = (C)(I)(A)$ Equation 5-45 202

Rational Method Runoff Coefficients, Weighted Value

$C_w = \Sigma C_i A_i / A_{\text{total}}$ Equation 5-46 203

NRCS Method, Runoff Depth

$Q = \frac{(P - I_a)^2}{(P - I_a) + S}$ Equation 5-47 206

Potential Maximum Retention

$S = \frac{1{,}000}{\text{CN}} - 10$ Equation 5-48 206

Initial Abstraction

$I_a = 0.2S$ **Equation 5-49** **206**

NRCS Method, Runoff Depth

$$Q = \frac{(P - 0.2S)^2}{(P + 0.8S)}$$ **Equation 5-50** **206**

Hydrostatic Pressure

$$P_{\text{hydrostatic}} = \frac{\text{weight}}{\text{area}} = \frac{\text{mg}}{A} = \frac{\rho Vg}{A} = \rho gh$$ **Equation 5-51** **215**

Fluid Height Equivalents for Pressure

US unit: P (psi) $= 0.43 \times h$ (ft) **Equation 5-52** **215**

SI metric units: P (kPa) $= 9.8 \times h$ (m) **Equation 5-53** **215**

Absolute and Total Pressure

$P_{\text{absolute}} = P_{\text{atmospheric}} + P_{\text{gauge}}$ **Equation 5-54** **215**

$P_{\text{total}} = P_{\text{static}} + P_{\text{dynamic}}$ **Equation 5-55** **215**

Continuity Equation

$Q = A_1 v_1 = A_2 v_2$ **Equation 5-56** **217**

Bernoulli Equation, Total Energy Head, Pressure Flow

$H(\text{ft}) = h_v + h_p + h_z$ **Equation 5-57** **218**

Velocity Head

$h_v = v^2/2g$ **Equation 5-58** **218**

Pressure Head

$h_p = p/\gamma$ **Equation 5-59** **218**

Elevation Head

$h_z = z$ **Equation 5-60** **218**

Reynolds Number

$$R_e = \frac{D_h v}{\nu}$$ **Equation 5-61** **219**

Darcy-Weisbach Equation, Friction Head

$$h_f = \frac{fLv^2}{2Dg}$$ **Equation 5-62** **220**

Laminar Flow

$f = 64/Re$ **Equation 5-63** **220**

$$v = \sqrt{\frac{2Dgh_f}{fL}}$$ Equation 5-64 221

$$h_f = \frac{\Delta p}{\gamma}$$ Equation 5-65 221

Hazen-Williams Equation, Friction Head

$$h_f = \frac{3.022v^{1.85}L}{C^{1.85}D^{1.17}}$$ Equation 5-66 221

$$h_f = \frac{10.44Q^{1.85}L}{C^{1.85}d^{4.87}}$$ Equation 5-67 221

$$v = \left(\frac{h_f C^{1.85}D^{1.17}}{3.022L}\right)^{0.54}$$ Equation 5-68 222

$$L = \frac{h_f C^{1.85}D^{1.17}}{3.022v^{1.85}}$$ Equation 5-69 222

$$D = \left(\frac{3.022v^{1.85}L}{C^{1.85}h_f}\right)^{0.855}$$ Equation 5-70 222

Minor Head Losses, Loss Coefficients

$$h_{\text{minor}} = K_L\frac{v^2}{2g}$$ Equation 5-71 223

Minor Head Loss, Equivalent Lengths

$$h_m = f\frac{L_e}{D}\frac{v^2}{2g}$$ Equation 5-72 223

$$L_e = \frac{kD}{f}$$ Equation 5-73 223

$$L_{\text{total}} = L_{\text{actual}} + \Sigma L_e$$ Equation 5-74 223

Total Head Loss Due to Friction

$$\Delta h_L = \Sigma h_{\text{major}} + \Sigma h_{\text{minor}} = \frac{v^2}{2g}\left(\frac{fL}{D} + \Sigma K_L\right)$$ Equation 5-75 224

Total Friction Head Loss for Pipes in a Series

$$h_{f\text{total}} = h_{fa} + h_{fb}$$ Equation 5-76 224

Flow in Parallel Branches

Flow divides so that friction loss is equal in the parallel pipes.

$$h_{f1} = h_{f2} = h_{f3}$$ Equation 5-77 225

Head loss between junctions can be calculated using any branch.

$h_{fA-B} = h_{f1} = h_{f2} = h_{f3}$ **Equation 5-78** **225**

Total flow between junctions is equal to the sum of flow in the branches.

$Q_A = Q_1 + Q_2 + Q_3 = Q_B$ **Equation 5-79** **225**

$Q_A = \frac{\pi}{4}\left(D_1^2 v_1 + D_2^2 v_2 + D_3^2 v_3\right) = Q_B$ **Equation 5-80** **225**

Energy Grade Line, Pressure Flow

$\text{EGL} = h_p + h_v + h_z$ **Equation 5-81** **226**

Hydraulic Grade Line, Pressure Flow

$\text{HGL} = h_p + h_z$ **Equation 5-82** **226**

Velocity of a Free Jet

$v_o = C_v\sqrt{2gh}$ **Equation 5-83** **230**

Flow Rate of a Free Jet

$\dot{V} = C_d A_o \sqrt{2gh}$ **Equation 5-84** **230**

Time to Empty a Tank

$t(s) = \dfrac{2A_t(\sqrt{z_1} - \sqrt{z_2})}{C_d A_o \sqrt{2g}}$ **Equation 5-85** **230**

Buoyant Force

$F_b = \gamma V$ **Equation 5-86** **233**

CHAPTER 6: GEOMETRICS

Latitude

$\text{Latitude } (\Delta y) = \text{Length } (L) \times \cos\theta$ **Equation 6-1** **245**

Departure

$\text{Departure } (\Delta x) = \text{Length } (L) \times \sin\theta$ **Equation 6-2** **245**

Inversing the Line

$\text{Latitude}_{GH} = N_H - N_G$ **Equation 6-3** **247**

$\text{Departure}_{GH} = E_H - E_G$ **Equation 6-4** **247**

$\text{Bearing Angle} = \tan^{-1}\left(\dfrac{\text{Departure}}{\text{Latitude}}\right)$ **Equation 6-5** **247**

Minimum Required Stopping Sight Distance

$$\text{SSD} = 1.47 \times V\,(\text{mph}) \times t_p + \frac{V^2(\text{mph})}{30\left(\frac{a}{32.2} \pm G\right)}$$ Equation 6-6 252

Circular Curve Calculations

$$R = \frac{5{,}729.58}{D}$$ Equation 6-7 255

$$R = \frac{LC}{2\sin\left(\frac{I}{2}\right)}$$ Equation 6-8 255

$$T = R\tan\left(\frac{I}{2}\right) = \frac{LC}{2\cos\left(\frac{I}{2}\right)}$$ Equation 6-9 255

$$L = RI\frac{\pi}{180} = \frac{I}{D}100$$ Equation 6-10 255

$$M = R\left[1 - \cos\left(\frac{I}{2}\right)\right]$$ Equation 6-11 255

$$\frac{R}{E+R} = \cos\left(\frac{I}{2}\right)$$ Equation 6-12 255

$$\frac{R-M}{R} = \cos\left(\frac{I}{2}\right)$$ Equation 6-13 255

$$c = 2R\sin\left(\frac{d}{2}\right)$$ Equation 6-14 255

$$l = Rd\left(\frac{\pi}{180}\right)$$ Equation 6-15 255

$$E = R\left[\frac{1}{\cos\left(\frac{I}{2}\right)} - 1\right]$$ Equation 6-16 255

$$\text{Deflection angle per 100 ft of arc length} = \left(\frac{D}{2}\right)$$

Degree of Curvature

$$D = 360° \frac{100\text{ ft}}{2\pi R} = \frac{5{,}729.578}{R}$$ Equation 6-17 256

Two-Centered Compound Curve

$$I = I_1 + I_2$$ Equation 6-18 257

$$X = R_2 \times \sin I + (R_1 - R_2) \times \sin I_1$$ Equation 6-19 257

Appendix A.9

School of PE

$Y = R_1 - R_2 \times \cos I - (R_1 - R_2) \times \cos I_1$	**Equation 6-20**	**258**
$T_L = \dfrac{R_2 - R_1 \times \cos I + (R_1 - R_2) \times \cos I_2}{\sin I}$	**Equation 6-21**	**258**
$T_S = \dfrac{R_1 - R_2 \times \cos I - (R_1 - R_2) \times \cos I_1}{\sin I}$	**Equation 6-22**	**258**

Centripetal Acceleration on a Horizontal Circular Curve

$a_n = \dfrac{V^2}{15R \text{ ft}}$	**Equation 6-23**	**262**

Minimum Radius of a Horizontal Curve

$0.01e + f = \dfrac{V^2}{15R} \Rightarrow R_{\min} = \dfrac{V^2}{15(0.01e_{\max} + f_{\max})}$	**Equation 6-24**	**263**

Length of a Spiral Curve

$L_s = \dfrac{3.15\ V^3}{RC}$	**Equation 6-25**	**265**

Horizontal Sightline Offset

$\text{HSO} = R \times \left(1 - \cos\dfrac{28.65 \times S}{R}\right)$	**Equation 6-26**	**267**

Actual Stopping Sight Distance

$S = \left(\dfrac{R}{28.65}\right)\left(\arccos\dfrac{R - \text{HSO}}{R}\right)$	**Equation 6-27**	**268**

Grades

$\text{Grade} = \dfrac{\Delta y}{\Delta x} \times 100$	**Equation 6-28**	**271**

Point of Vertical Curve Station

$\text{PVC sta} = \text{PVI sta} - \dfrac{L}{2}$	**Equation 6-29**	**274**

Point of Vertical Tangent Station

$\text{PVT sta} = \text{PVI sta} + \dfrac{L}{2}$	**Equation 6-30**	**274**

Support Equations for Elevations on a Vertical Curve

$Y_{\text{PVC}} = \text{PVI}_{\text{ELEV}} - G_1 \times \dfrac{L}{2}$	**Equation 6-31**	**274**
$Y_{\text{PVC}} + G_1 x = Y_{\text{PVI}} + G_2\left(x - \dfrac{L}{2}\right)$ (tangent elevation)	**Equation 6-32**	**274**

Equation	Reference	Page
$Y_{\text{PVC}} + G_1x + \left(\dfrac{G_2 - G_1}{2L}\right)x^2$ (curve elevation)	Equation 6-33	274
$y = ax^2$	Equation 6-34	274
$A = \lvert G_2 - G_1 \rvert$	Equation 6-35	274
$a = \dfrac{G_2 - G_1}{2L}$	Equation 6-36	274
$E = a(L/2)^2$	Equation 6-37	275
$r = \dfrac{G_2 - G_1}{L}$	Equation 6-38	275
$K = \dfrac{L}{A}$	Equation 6-39	275
x_m = horizontal distance to turning point $= \dfrac{-G_1}{2a} = \dfrac{G_1L}{G_1 - G_2}$	Equation 6-40	275

Minimum Required Length of a Vertical Curve

Equation	Reference	Page
$L = K \times \lvert A \rvert$	Equation 6-41	276

Crest Vertical Curve General Equation

Equation	Reference	Page
for $S \leq L$, $L = \dfrac{AS^2}{100\left(\sqrt{2h_1} + \sqrt{2h_2}\right)^2}$	Equation 6-42	278
for $S > L$, $L = 2S - \dfrac{200\left(\sqrt{h_1} + \sqrt{h_2}\right)^2}{A}$	Equation 6-43	278

Crest Vertical Curve Standard Criteria

Equation	Reference	Page
for $S \leq L$: $L = \dfrac{AS^2}{2{,}158} \Rightarrow S = \sqrt{\dfrac{2{,}158L}{A}}$	Equation 6-44	278
for $S > L$: $L = 2S - \dfrac{2{,}158}{A} \Rightarrow S = \dfrac{1}{2}\left(L + \dfrac{2{,}158}{A}\right)$	Equation 6-45	278

Actual Stopping Sight Distance on a Sag Vertical Curve, Difference of Grades

Equation	Reference	Page
for $S \leq L$, $L = \dfrac{AS^2}{400 + 3.5S}$	Equation 6-46	283
for $S > L$, $L = 2S - \left(\dfrac{400 + 3.5S}{A}\right)$	Equation 6-47	283

Actual Stopping Sight Distance on a Sag Vertical Curve at an Underpass

Equation	Reference	Page
$L = \dfrac{AS^2}{800\left(C - \dfrac{h_1 + h_2}{2}\right)}$	Equation 6-48	284

Equation	Reference	Page
$L = 2S - \frac{800}{A}\left(C - \frac{h_1 + h_2}{2}\right)$	Equation 6-49	284

Minimum Required Length of Sag Vertical Curve Based on Riding Comfort Criteria

Equation	Reference	Page
$L = \frac{AV^2}{46.5}$	Equation 6-50	284

Seasonal Factor

Equation	Reference	Page
$\text{SF} = \frac{\text{average daily traffic (vehicles)}}{\text{average annual daily traffic (vehicles)}}$	Equation 6-51	289

Design Hour Factor (*K* Factor)

Equation	Reference	Page
$K = \frac{\text{design hourly volume (vehicles)}}{\text{average annual daily traffic (vehicles)}}$	Equation 6-52	292

Peak Hour Factor

Equation	Reference	Page
$\text{PHF} = \frac{\text{actual hourly volume (vph)}}{\text{peak rate of flow (vph)}} = \frac{\text{peak hour volume}}{4 \times V_{15}}$	Equation 6-53	293

Directional Distribution

Equation	Reference	Page
$D = \frac{\text{peak direction volume}}{\text{total volume}}$	Equation 6-54	294

Directional Design Hourly Volume

Equation	Reference	Page
$\text{DDHV} = D \times K \times \text{AADT}$	Equation 6-55	295

Flow Rate, *v*

Equation	Reference	Page
$v = \frac{\text{vehicles}}{\text{unit time}} = D \times S$	Equation 6-56	298

Running Speed, *S*

Equation	Reference	Page
$\text{speed} = \frac{\text{distance}}{\text{time}}; S = \frac{d}{t}$	Equation 6-57	298

Density, *D*

Equation	Reference	Page
$\text{density} = \frac{\text{flow}}{\text{speed}}; D = \frac{v}{S}$	Equation 6-58	298

Time Mean Speed

Equation	Reference	Page
$S_t = \frac{\sum_{i=1}^{n} S_i}{n}$	Equation 6-59	300

Space Mean Speed

Equation	Reference	Page
$S_s = \frac{nL}{\sum_{i=1}^{n} t_i}$	Equation 6-60	301

Headway

$$h = \frac{3{,}600 \text{ s/hr}}{v \text{ vph}}$$ Equation 6-61 308

Capacity

$$c_{vph} = \frac{3{,}600_{\text{s/hr}}}{h_{\min}}$$ Equation 6-62 308

CHAPTER 7: MATERIALS

Group Index (GI):

$$\text{GI} = (\text{F}_{200} - 35)(0.2 + 0.005(\text{LL} - 40)) + 0.01(\text{F}_{200} - 15)(\text{PI} - 10)$$ Equation 7-1 316

$$\text{PGI} = 0.01(\text{F}_{200} - 15)(\text{PI} - 10)(\text{for soil groups A-2-6 and A-2-7})$$ Equation 7-2 316

Standardized *N* Value (Measured Field)

$$N_{60} = N_m \frac{E}{60}$$ Equation 7-3 324

Corrected *N* Value (Measured Field)

$$N_{\text{corr}} = N_{60}\sqrt{\frac{p_a}{\sigma'_v}}$$ Equation 7-4 324

Area Ratio

$$\text{Area ratio } (\%) = \frac{(OD)^2 - (ID)^2}{(ID)^2} \times 100\%$$ Equation 7-5 326

Rock Quality Designation

$$\text{RQD} = \frac{\Sigma \text{ lengths of intact pieces of core} > 100 \text{ mm}}{\text{length of core advance}} \times 100\%$$ Equation 7-6 328

Moisture Content

$$w = \frac{\text{weight of water}}{\text{weight of solids}} = \frac{W_w}{W_s} \times 100\%$$ Equation 7-7 330

Degree of Saturation

$$S = \frac{\text{volume of water}}{\text{volume of voids}} = \frac{V_w}{V_v} \times 100\%$$ Equation 7-8 330

Total Unit Weight

$$\gamma_t = \frac{\text{total weight}}{\text{total volume}} = \frac{W_t}{V_t}$$ Equation 7-9 331

Dry Unit Weight

$$\gamma_d = \frac{\text{weight of solids}}{\text{total volume}} = \frac{W_s}{V_t}$$ Equation 7-10 331

Appendix A.9

Solids Unit Weight

$$\gamma_s = \frac{\text{weight of solids}}{\text{volume of solids}} = \frac{W_S}{V_s}$$ **Equation 7-11** 331

Void Ratio

$$e = \frac{\text{volume of voids}}{\text{volume of solids}} = \frac{V_v}{V_s}$$ **Equation 7-12** 331

Porosity

$$n = \frac{\text{volume of voids}}{\text{total volume}} = \frac{V_v}{V_t}$$ **Equation 7-13** 331

Specific Gravity

$$G_s = \frac{\text{unit weight of solids}}{\text{unit weight of water}} = \frac{\gamma_{\text{solids}}}{\gamma_w}$$ **Equation 7-14** 331

$$G_s = \frac{W_s}{V_s \gamma_w} \quad \text{or} \quad G_s = \frac{M_s}{V_s \rho_w}$$ **Equation 7-15** 331

$$W = Mg, \quad \text{where } g = 32.2 \text{ ft/s}^2$$ **Equation 7-16** 331

Relative Density

$$D_r = \left(\frac{\gamma_d - (\gamma_d)_{\min}}{(\gamma_d)_{\max} - (\gamma_d)_{\min}}\right)\left(\frac{(\gamma_d)_{\max}}{\gamma_d}\right) \times 100\%$$ **Equation 7-17** 334

$$\gamma_d = \frac{(\gamma_d)_{\min}}{1 - \left(\frac{D_r}{(\gamma_d)_{\max}}\right)((\gamma_d)_{\max} - (\gamma_d)_{\min})}$$ **Equation 7-18** 334

$$D_r = \frac{e_{\max} - e}{e_{\max} - e_{\min}} \times 100\ \% \quad \rightarrow \quad e = e_{\max} - D_r(e_{\max} - e_{\min})$$ **Equation 7-19** 334

Coefficient of Uniformity

$$C_u = \frac{D_{60}}{D_{10}}$$ **Equation 7-20** 336

Coefficient of Curvature/Gradation

$$C_c \text{ (or } C_z) = \frac{(D_{30})^2}{D_{60} D_{10}}$$ **Equation 7-21** 336

Plasticity Index

$$\text{PI} = \text{LL} - \text{PL}$$ **Equation 7-22** 339

Soil Shear Strength is Expressed by Mohr-Coulomb Failure

$$\tau' = c' + \sigma' \tan' \phi'$$ **Equation 7-23** 339

Darcy's Law

$$Q = vA = kiA$$ Equation 7-24 339

Total Head

$$h = h_z + h_p$$ Equation 7-25 340

Seepage Velocity

$$v_s = \frac{ki}{n_e} = \frac{v}{n_e}$$ Equation 7-26 342

Coefficient of Permeability

$$k = \frac{VL}{\Delta hAt}$$ Equation 7-27 342

Coefficient of Permeability via Falling Head

$$k = \left(2.303\frac{aL}{At}\log_{10}\frac{h_0}{h_1}\right) = \left(\frac{aL}{At}\ln\frac{h_0}{h_1}\right)$$ Equation 7-28 343

$$k(\text{cm/s}) = 2.4622\left[(D_{10}\ \text{mm})^2\frac{e^3}{1+e}\right]^{0.7825}$$ Equation 7-29 343

Total Flow Rate

$$Q = k\Delta h\frac{N_f}{N_d}L$$ Equation 7-30 344

Overconsolidation Ratio

$$\text{OCR} = \frac{\sigma_c'}{\sigma_0'}$$ Equation 7-31 345

Coefficient of Consolidation

$$C_v = \frac{k}{\rho_w g}m_v$$ Equation 7-32 346

Compressive Strength of Concrete

$$f_c' = \frac{P}{A}$$ Equation 7-33 349

Tensile Strength of Concrete

$$f_{ct} = \frac{2P}{\pi DL}$$ Equation 7-34 350

Modulus of Rupture

$$f_r = 7.5 \times \sqrt{f_c'}$$ Equation 7-35 351

Water-to-Cement Ratio

$$w/c = \frac{\text{water content (lb)}}{\text{cement (lb)} + \text{pozzolan (lb)}}$$ Equation 7-36 351

Theoretical Energy (Proctor Laboratory Tests)

$$\text{Theoretical Energy} = \frac{(\text{Wt})(\text{Drop})(\text{Blows})(\text{Layers})}{\text{Volume}}$$ **Equation 7-37** **355**

Volume of Soil Excavated

$$V_{\text{hole}} = \frac{(W_o - W_f) - W_{\text{cone}}}{\gamma_{\text{sand}}} = \frac{(W_o - W_f)}{\gamma_{\text{sand}}} - V_{\text{cone}}$$ **Equation 7-38** **356**

Dry Unit Weight of Soil Excavated

$$\gamma_{\text{d}} = \frac{\gamma}{1 + w} = \frac{W_{\text{hole}}}{V_{\text{hole}}(1 + w)}$$ **Equation 7-39** **356**

Relative Compaction

$$RC = \frac{\text{field dry unit weight}}{\text{max dry unit weight}} = \frac{(\gamma_{\text{d}})_{\text{field}}}{(\gamma_{\text{d}})_{\text{max}}} \times 100\%$$ **Equation 7-40** **357**

California Bearing Ratio

$$\text{CBR} = \frac{\text{actual stress}}{\text{standard stress}} \times 100$$ **Equation 7-41** **358**

Resilient Modulus

$$M_R(\text{psi}) = (1{,}500)\text{CBR}$$ **Equation 7-42** **359**

$$M_R(\text{psi}) = (2{,}555)\text{CBR}^{0.64}$$ **Equation 7-43** **359**

CHAPTER 8: SITE DEVELOPMENT

Swell Factor

$$\text{SWF} = \frac{\text{BV}}{\text{LV}} - 1$$ **Equation 8-1** **365**

Shrink Factor

$$\text{SHF} = \frac{\text{CV}}{\text{BV}}$$ **Equation 8-2** **365**

Relative Compaction

$$\text{RC} = \left(\frac{\text{In Situ Dry Density}}{\text{Laboratory Maximum Dry Density}}\right) \times 100\%$$ **Equation 8-3** **370**

Dry Unit Density

$$\text{Dry Unit Density} = \frac{\text{Total Unit Weight In Situ}}{(1 + \text{Water Content})}$$ **Equation 8-4** **370**

Material Required to Complete Fill at Project-Specified Compaction Percentage

$$\text{Amt of Fill} = \frac{\text{(Fill Required)(Required Compaction \%)}}{\text{Relative Compaction}}$$ **Equation 8-5** **370**

Average End Area Method

$$V = \frac{L(A_1 + A_2)}{2}$$ **Equation 8-6** **371**

$$V = \frac{L(A_{base})}{3}$$ **Equation 8-7** **371**

Average Elevation Change

$$\text{Avg Elevation of Site} = \frac{\text{(Elevation of Points)}}{\text{(\# of Points)}}$$ **Equation 8-8** **373**

Conversion from DMS to Decimal Degree

$$\text{deg}^\circ\text{min}'\text{sec}'' = \text{deg}^\circ + \left(\frac{\text{min}' + \frac{\text{sec}''}{60}}{60}\right)$$ **Equation 8-9** **379**

Incident Rate

$$\text{Incident Rate (IR)} = \frac{\text{(number of recordable incidents} \times 200{,}000)}{\text{(total number of hours worked)}}$$ **Equation 8-10** **390**

Examples and Solutions Quick Reference A.10

Page

Chapter 3: Soil Mechanics

Chapter 4: Structural Mechanics

Chapter 5: Hydraulics and Hydrology

Conveyance Factors in Modified Manning's Equation

A.11

Values of K in Formula $Q = (K/n)\, D^{8/3}\, S^{1/2}$

Trapezoidal Channels (Table 7-10 from the *Handbook of Hydraulics*)

D = depth of water b = bottom width of channel

	Side slopes of channel, ratio of horizontal to vertical									
D/b	**Vertical**	**¼ - 1**	**½ - 1**	**¾ - 1**	**1 - 1**	**1 ½ - 1**	**2 - 1**	**2 ½ - 1**	**3 - 1**	**4 - 1**
0.05	27.9	28.4	28.9	29.2	29.5	30.0	30.5	30.9	31.2	32.0
0.10	13.2	13.7	14.1	14.4	14.8	15.3	15.7	16.2	16.6	17.4
0.15	8.32	8.80	9.22	9.57	9.88	10.4	10.9	11.4	11.8	12.7
0.20	5.94	6.40	6.81	7.16	7.47	8.01	8.50	8.97	9.43	10.3
0.25	4.54	4.98	5.38	5.73	6.04	6.58	7.08	7.56	8.03	8.95
0.30	3.62	4.05	4.44	4.78	5.09	5.64	6.15	6.63	7.10	8.04
0.35	2.98	3.40	3.78	4.12	4.43	4.98	5.49	5.97	6.45	7.39
0.40	2.51	2.92	3.29	3.62	3.93	4.48	4.99	5.48	5.96	6.91
0.45	2.15	2.55	2.91	3.24	3.55	4.10	4.61	5.11	5.59	6.54
0.50	1.87	2.26	2.61	2.94	3.25	3.80	4.31	4.81	5.29	6.24
0.55	1.65	2.02	2.37	2.70	3.00	3.55	4.07	4.56	5.05	6.00
0.60	1.46	1.83	2.17	2.50	2.80	3.35	3.86	4.36	4.84	5.80
0.65	1.31	1.67	2.01	2.33	2.63	3.18	3.69	4.19	4.68	5.63
0.70	1.18	1.53	1.87	2.19	2.48	3.03	3.55	4.04	4.53	5.49
0.75	1.08	1.41	1.75	2.06	2.36	2.91	3.42	3.92	4.41	5.36
0.80	.982	1.31	1.64	1.95	2.25	2.80	3.31	3.81	4.30	5.26
0.85	.902	1.23	1.55	1.86	2.15	2.70	3.22	3.71	4.20	5.16
0.90	.831	1.15	1.47	1.78	2.07	2.62	3.13	3.63	4.12	5.08
0.95	.769	1.08	1.40	1.70	1.99	2.54	3.05	3.55	4.04	5.00
1.00	.714	1.02	1.33	1.64	1.93	2.47	2.99	3.48	3.97	4.93
1.05	.666	.968	1.28	1.58	1.87	2.41	2.92	3.42	3.91	4.87
1.10	.622	.920	1.23	1.53	1.81	2.36	2.87	3.37	3.85	4.81
1.15	.583	.876	1.18	1.48	1.76	2.30	2.82	3.32	3.80	4.76
1.20	.548	.836	1.14	1.43	1.72	2.26	2.77	3.27	3.76	4.72
1.25	.516	.800	1.10	1.39	1.68	2.22	2.73	3.23	3.71	4.67
1.30	.487	.767	1.06	1.35	1.64	2.18	2.69	3.19	3.67	4.64
1.35	.460	.736	1.03	1.32	1.60	2.14	2.65	3.15	3.64	4.60
1.40	.436	.708	.998	1.29	1.57	2.11	2.62	3.12	3.60	4.57
1.45	.414	.682	.970	1.26	1.54	2.08	2.59	3.08	3.57	4.53
1.50	.393	.658	.944	1.23	1.51	2.05	2.56	3.06	3.54	4.50
1.55	.374	.636	.920	1.21	1.49	2.02	2.53	3.03	3.52	4.48
1.60	.357	.616	.897	1.18	1.46	2.00	2.51	3.00	3.49	4.45
1.65	.341	.596	.876	1.16	1.44	1.97	2.48	2.98	3.47	4.43
1.70	.326	.578	.856	1.14	1.42	1.95	2.46	2.96	3.44	4.41
1.75	.312	.561	.838	1.12	1.40	1.93	2.44	2.93	3.42	4.38
1.80	.298	.546	.820	1.10	1.38	1.91	2.42	2.91	3.40	4.36
1.85	.286	.531	.804	1.08	1.36	1.89	2.40	2.90	3.38	4.34
1.90	.275	.517	.788	1.07	1.34	1.87	2.38	2.88	3.37	4.33
1.95	.264	.504	.774	1.05	1.33	1.86	2.37	2.86	3.35	4.31
2.00	.254	.491	.760	1.04	1.31	1.84	2.35	2.84	3.33	4.29
2.05	.245	.480	.746	1.02	1.30	1.82	2.33	2.83	3.32	4.28
2.10	.236	.469	.734	1.01	1.28	1.81	2.32	2.81	3.30	4.26
2.15	.227	.458	.722	.996	1.27	1.80	2.31	2.80	3.29	4.25
2.20	.219	.448	.711	.984	1.26	1.78	2.29	2.79	3.28	4.24
2.25	.212	.439	.700	.973	1.24	1.77	2.28	2.77	3.26	4.22
∞	.000	.091	.274	.500	.743	1.24	1.74	2.23	2.71	3.67

Values of K' in Formula $Q = (K'/n)b^{8/3}S^{1/2}$

Trapezoidal Channels (Table 7-11 from the *Handbook of Hydraulics*)

D = depth of water b = bottom width of channel

D/b	Side slopes of channel, ratio of horizontal to vertical									
	Vertical	¼ - 1	½ - 1	¾ - 1	1 - 1	1 ½ - 1	2 - 1	2 ½ - 1	3 - 1	4 - 1
0.05	.00946	.00964	.00979	.00991	.01002	.01019	.01033	.01047	.01060	.01086
0.10	.0284	.0294	.0304	.0311	.0318	.0329	.0339	.0348	.0358	.0376
0.15	.0528	.0559	.0585	.0608	.0627	.0662	.0692	.0721	.0749	.0805
0.20	.0812	.0876	.0931	.0979	.1021	.1096	.1163	.1227	.1290	.1414
0.25	.1125	.1236	.133	.142	.150	.163	.176	.188	.199	.222
0.30	.146	.163	.179	.193	.205	.228	.248	.267	.287	.324
0.35	.181	.207	.230	.251	.269	.303	.334	.363	.392	.450
0.40	.218	.253	.286	.315	.341	.389	.434	.476	.518	.600
0.45	.256	.303	.346	.386	.422	.488	.549	.607	.664	.777
0.50	.295	.355	.412	.463	.511	.598	.679	.757	.833	.983
0.55	.335	.410	.482	.548	.609	.722	.826	.926	1.025	1.22
0.60	.375	.468	.557	.640	.717	.858	.990	1.117	1.24	1.49
0.65	.416	.529	.637	.738	.833	1.008	1.17	1.33	1.48	1.79
0.70	.457	.592	.722	.844	.959	1.17	1.37	1.56	1.75	2.12
0.75	.499	.657	.811	.957	1.095	1.35	1.59	1.82	2.05	2.49
0.80	.542	.725	.906	1.078	1.24	1.54	1.83	2.10	2.37	2.90
0.85	.585	0.796	1.006	1.21	1.40	1.75	2.08	2.41	2.72	3.35
0.90	.628	0.869	1.11	1.34	1.56	1.98	2.36	2.74	3.11	3.83
0.95	.671	0.944	1.22	1.49	1.74	2.22	2.66	3.10	3.52	4.36
1.00	.714	1.022	1.33	1.64	1.93	2.47	2.99	3.48	3.97	4.93
1.05	.758	1.10	1.46	1.80	2.13	2.75	3.33	3.90	4.45	5.55
1.10	.802	1.19	1.58	1.97	2.34	3.04	3.70	4.34	4.97	6.21
1.15	.846	1.27	1.71	2.14	2.56	3.35	4.09	4.81	5.52	6.92
1.20	.891	1.36	1.85	2.33	2.79	3.67	4.51	5.32	6.11	7.67
1.25	.935	1.45	1.99	2.52	3.04	4.02	4.95	5.85	6.73	8.48
1.30	.980	1.54	2.14	2.73	3.30	4.38	5.41	6.41	7.40	9.33
1.35	1.024	1.64	2.29	2.94	3.57	4.77	5.91	7.01	8.10	10.24
1.40	1.07	1.74	2.45	3.16	3.85	5.17	6.42	7.64	8.84	11.2
1.45	1.11	1.84	2.61	3.39	4.15	5.59	6.97	8.31	9.62	12.2
1.50	1.16	1.94	2.78	3.63	4.46	6.04	7.54	9.01	10.45	13.3
1.55	1.20	2.05	2.96	3.88	4.78	6.50	8.15	9.74	11.3	14.4
1.60	1.25	2.16	3.14	4.14	5.12	6.99	8.78	10.52	12.2	15.6
1.65	1.29	2.27	3.33	4.41	5.47	7.50	9.44	11.3	13.2	16.8
1.70	1.34	2.38	3.52	4.69	5.83	8.02	10.13	12.2	14.2	18.1
1.75	1.39	2.50	3.73	4.98	6.21	8.58	10.8	13.1	15.2	19.5
1.80	1.43	2.62	3.93	5.28	6.60	9.15	11.6	14.0	16.3	20.9
1.85	1.48	2.74	4.15	5.59	7.01	9.75	12.4	14.9	17.5	22.4
1.90	1.52	2.86	4.36	5.91	7.43	10.37	13.2	15.9	18.6	24.0
1.95	1.57	2.99	4.59	6.24	7.87	11.0	14.0	17.0	19.9	25.6
2.00	1.61	3.12	4.82	6.58	8.32	11.7	14.9	18.1	21.2	27.3
2.05	1.66	3.25	5.06	6.93	8.79	12.4	15.8	19.2	22.5	29.0
2.10	1.70	3.39	5.31	7.30	9.27	13.1	16.8	20.4	23.9	30.8
2.15	1.75	3.53	5.56	7.67	9.77	13.8	17.8	21.6	25.3	32.7
2.20	1.80	3.67	5.82	8.06	10.28	14.6	18.8	22.8	26.8	34.7
2.25	1.84	3.81	6.09	8.46	10.8	15.4	19.8	24.1	28.4	36.7

Adapted from King, H. W. *Handbook of Hydraulics (5th ed.).* McGraw-Hill, 1963. Used with permission.

Index

A

B

C

D

E

F

S